Annals of Mathematics Studies

Number 156

Radon Transforms and the Rigidity
of the Grassmannians

JACQUES GASQUI
AND
HUBERT GOLDSCHMIDT

PRINCETON UNIVERSITY PRESS

Princeton and Oxford

2004

Published by Princeton University Press, 41 William Street,
Princeton, New Jersey 08540
In the United Kingdom: Princeton University Press, 3 Market Place,
Woodstock, Oxfordshire OX20 1SY

All Rights Reserved
Library of Congress Control Number 2003114656
ISBN: 0-691-11898-1
0-691-11899-X (paper)
British Library Cataloging-in-Publication Data is available
The publisher would like to acknowledge the authors of this
volume for providing the camera-ready copy from which this
book was printed

Printed on acid-free paper

pup.princeton.edu

Printed in the United States of America

10 9 8 7 6 5 4 3 2 1

TABLE OF CONTENTS

CHAPTER V
The complex quadric

CHAPTER VI
The rigidity of the complex quadric

CHAPTER VII
The rigidity of the real Grassmannians

CHAPTER VIII
The complex Grassmannians

CHAPTER IX

The rigidity of the complex Grassmannians

CHAPTER X

Products of symmetric spaces

INTRODUCTION

This monograph is motivated by a fundamental rigidity problem in Riemannian geometry: determine whether the metric of a given Riemannian symmetric space of compact type can be characterized by means of the spectrum of its Laplacian. An infinitesimal isospectral deformation of the metric of such a symmetric space belongs to the kernel of a certain Radon transform defined in terms of integration over the flat totally geodesic tori of dimension equal to the rank of the space. Here we study an infinitesimal version of this spectral rigidity problem: determine all the symmetric spaces of compact type for which this Radon transform is injective in an appropriate sense. We shall both give examples of spaces which are not infinitesimally rigid in this sense and prove that this Radon transform is injective in the case of most Grassmannians.

At present, it is only in the case of spaces of rank one that infinitesimal rigidity in this sense gives rise to a characterization of the metric by means of its spectrum. In the case of spaces of higher rank, there are no analogues of this phenomenon and the relationship between the two rigidity problems is not yet elucidated. However, the existence of infinitesimal deformations belonging to the kernel of the Radon transform might lead to non-trivial isospectral deformations of the metric.

Here we also study another closely related rigidity question which arises from the Blaschke problem: determine all the symmetric spaces for which the X-ray transform for symmetric 2-forms, which consists in integrating over all closed geodesics, is injective in an appropriate sense. In the case of spaces of rank one, this problem coincides with the previous Radon transform question. The methods used here for the study of these two problems are similar in nature.

Let (X, g) be a Riemannian symmetric space of compact type. Consider a family of Riemannian metrics $\{g_t\}$ on X, for $|t| < \varepsilon$, with $g_0 = g$. The family $\{g_t\}$ is said to be an isospectral deformation of g if the spectrum of the Laplacian of the metric g_t is independent of t. We say that the space (X, g) is infinitesimally spectrally rigid (i.e., spectrally rigid to first-order) if, for every such isospectral deformation $\{g_t\}$ of g, there is a one-parameter family of diffeomorphisms $\{\varphi_t\}$ of X such that $g_t = \varphi_t^* g$ to first-order in t at $t = 0$, or equivalently if the symmetric 2-form, which is equal to the infinitesimal deformation $\frac{d}{dt} g_t|_{t=0}$ of $\{g_t\}$, is a Lie derivative of the metric g.

In [35], Guillemin introduced a criterion for infinitesimal spectral rigidity which may be expressed as follows. We say that a symmetric p-form u on X satisfies the Guillemin condition if, for every maximal flat totally geodesic torus Z contained in X and for all parallel vector fields ζ on Z, the integral

$$\int_Z u(\zeta, \zeta, \ldots, \zeta) \, dZ$$

vanishes, where dZ is the Riemannian measure of Z. A symmetric 2-form, which is a Lie derivative of the metric, always satisfies the Guillemin condition. Guillemin proved that a symmetric 2-form, which is equal to the infinitesimal deformation of an isospectral deformation of g, satisfies the Guillemin condition. We say that the space (X, g) is rigid in the sense of Guillemin if the following property holds: the only symmetric 2-forms on X satisfying the Guillemin condition are the Lie derivatives of the metric g. Thus if the symmetric space X is rigid in the sense of Guillemin, it is infinitesimally spectrally rigid.

We are interested in determining which symmetric spaces of compact type are infinitesimally spectrally rigid; in particular, we wish to find those spaces which are rigid in the sense of Guillemin. We show that an arbitrary non-trivial product of irreducible symmetric spaces of compact type, which are not equal to Lie groups, is not rigid in the sense of Guillemin. Consequently, we shall restrict our attention to irreducible spaces. We shall also see below that, in order for an irreducible space to be rigid in the sense of Guillemin, it must be equal to its adjoint space.

Although much work has been done on the problem of isospectrality, there are still very few results for positively curved spaces. All the previously known spectral rigidity results for symmetric spaces with positive curvature concern spaces of rank one. In fact, we will see below that the real projective space $\mathbb{RP}^n$, with $n \geq 2$, is spectrally rigid; on the other hand, for $2 \leq n \leq 6$, the spectral rigidity of the sphere S^n was established by Berger and Tanno (see [4] and [51]). The Guillemin rigidity of the spaces of rank one (i.e., the projective spaces) which are not spheres was first proved by Michel [45] for the real projective spaces $\mathbb{RP}^n$, with $n \geq 2$, and by Michel [45] and Tsukamoto [53] for the other projective spaces. As we shall explain below, spectral rigidity results for these other projective spaces can be derived from their Guillemin rigidity by means of Kiyohara's work [38].

In contrast to the case of negatively curved spaces, at present the problem of isospectrality for positively curved spaces does not admit any truly effective general approach. While the study of the symmetric spaces considered here requires a case by case analysis, we have nevertheless been able to develop criteria for rigidity which can be applied to numerous situations. Several fundamental aspects of differential geometry – the theory of overdetermined partial differential equations, deformation theory of Einstein manifolds, harmonic analysis on symmetric spaces of compact type, the geometry of the Grassmannians and their totally geodesic submanifolds – enter into the elaboration of these criteria and their application to the various spaces. Many of the results, which we present in the process, are of considerable interest in their own right outside the context of deformation theory and spectral rigidity.

In this monograph, we introduce new methods for studying the Guillemin rigidity of irreducible symmetric spaces of compact type. The theory of linear overdetermined partial differential equations of [28] provides us with a fundamental ingredient of these methods, namely a certain differential operator which allows us to encode properties of the space under consideration. Quite remarkably, these methods lead us to a criterion for the Guillemin rigidity of a space in which neither this operator nor the theory of overdetermined partial differential equations appear. We apply this criterion to the Grassmannians of rank ≥ 2 and we determine all those which are rigid in the sense of Guillemin. In fact, we extend and complete our previous work on the real Grassmannians of rank 2 undertaken in [23]. Harmonic analysis on homogeneous spaces and results concerning the infinitesimal deformations of Einstein metrics also play an important role here.

Let $\mathbb{K}$ be a division algebra over $\mathbb{R}$ (i.e., $\mathbb{K}$ is equal to $\mathbb{R}$, $\mathbb{C}$ or the quaternions $\mathbb{H}$). For $m, n \geq 1$, the Grassmannian $G^{\mathbb{K}}_{m,n}$ of all $\mathbb{K}$-planes of dimension m in $\mathbb{K}^{m+n}$ is a symmetric space of rank $\min(m, n)$. The Grassmannians are irreducible and of compact type, with the exception of $G^{\mathbb{R}}_{1,1} = S^1$ and of $G^{\mathbb{R}}_{2,2}$ whose universal covering space is $S^2 \times S^2$. The Grassmannian $G^{\mathbb{K}}_{1,n}$ is the projective space $\mathbb{K}P^n$. The main result presented in this monograph may be stated as follows:

THEOREM 1. *Let $\mathbb{K}$ be a division algebra over $\mathbb{R}$ and $m, n \geq 1$ be given integers. The Grassmannian $G^{\mathbb{K}}_{m,n}$ is rigid in the sense of Guillemin if and only if $m \neq n$.*

All the known examples of spaces which are rigid in the sense of Guillemin are described in this theorem. When m or n is equal to 1, Theorem 1 gives us the results of Michel and Tsukamoto mentioned above concerning the projective spaces. This theorem implies that the Grassmannians $G^{\mathbb{K}}_{m,n}$, with $m, n \geq 2$ and $m \neq n$, are infinitesimally spectrally rigid, and provides us with the first examples of symmetric spaces of compact type of arbitrary rank > 1 having this property.

Let (X, g) be a Riemannian symmetric space of compact type. The symmetric space X is a homogeneous space of a compact semi-simple Lie group G, which acts on X by isometries. The space Ξ of all maximal flat totally geodesic tori of X is a homogeneous space of G. The maximal flat Radon transform of X considered by Grinberg in [33] and [34] is a G-equivariant linear mapping from the space of functions on X to the space of functions on Ξ; it assigns to a function f on X the function $\hat{f}$ on Ξ whose value at a torus Z of Ξ is the integral of f over Z. In view of Guillemin's criterion, we define a maximal flat Radon transform for symmetric p-forms, which is a G-equivariant linear mapping I from the space of symmetric p-forms on X to the space of sections of a certain homogeneous vector bundle over Ξ. Its kernel consists of those symmetric p-forms on X satisfying the Guillemin condition. On functions, it coincides with the one considered by Grinberg. Determining whether the space X is rigid in the sense of Guillemin may be viewed as a problem concerning this Radon transform for symmetric 2-forms.

We recall that the adjoint space of X is the symmetric space which admits X as a Riemannian cover and which is itself not a Riemannian cover of another symmetric space. For example, the adjoint space of the n-sphere S^n is the real projective space $\mathbb{RP}^n$. For these spaces of rank one, the maximal flat tori are the closed geodesics. The kernel of the maximal flat Radon transform for functions on S^n is the space of all odd functions on S^n. In fact, this Radon transform is injective when restricted to the even functions on S^n; this is equivalent to the classic fact that the Radon transform for functions on $\mathbb{RP}^n$ is injective. In [33] and [34], Grinberg generalized these results and proved that the maximal flat Radon transform for functions on X is injective if and only if the space X is equal to its adjoint space.

Suppose that X possesses an involutive isometry σ which has no fixed points, and that the quotient of X by the group of isometries of order 2 generated by σ is also a symmetric space. Then X is not equal to its adjoint space. In this case, it is easily seen that an arbitrary odd symmetric p-form u on X (i.e., which satisfies the relation $\sigma^* u = -u$) satisfies the Guillemin condition, and we can construct odd symmetric 2-forms which are not Lie derivatives of the metric. It follows directly that the maximal flat Radon transform for functions on X is not injective and that X is not rigid in the sense of Guillemin. In particular, this situation applies to the sphere S^n together with the anti-podal involution.

We now suppose that the space X is irreducible. If X is not isometric to a sphere and is rigid in the sense of Guillemin, we show that the maximal flat Radon transform for functions on X is injective. Since the sphere S^n is not rigid in this sense, from Grinberg's result we infer that, if the irreducible symmetric space X is rigid in the sense of Guillemin, it must necessarily be equal to its adjoint space. The Grassmannian $G_{m,n}^{\mathbb{K}}$, with $m, n \geq 1$, is equal to its adjoint space if and only if $m \neq n$. Therefore by Theorem 1, we see that a Grassmannian, which is not flat, is rigid in the sense of Guillemin if and only if it is equal to its adjoint space.

We now consider the Grassmannian $G_{n,n}^{\mathbb{K}}$, with $n \geq 2$. This space possesses an involutive isometry Ψ which sends an n-plane of $\mathbb{K}^{2n}$ into its orthogonal complement. The quotient space $\bar{G}_{n,n}^{\mathbb{K}}$ of $G_{n,n}^{\mathbb{K}}$ by the group of isometries generated by Ψ is a symmetric space of rank n, which is equal to the adjoint space of $G_{n,n}^{\mathbb{K}}$. According to the discussion which appears above, the space $G_{n,n}^{\mathbb{K}}$ is not rigid in the sense of Guillemin.

We observe that $\bar{G}_{2,2}^{\mathbb{R}}$ is isometric to the product $\mathbb{RP}^2 \times \mathbb{RP}^2$; hence this space is not rigid in the sense of Guillemin. All the other spaces $\bar{G}_{n,n}^{\mathbb{K}}$ are irreducible. On the other hand, the space $\bar{G}_{2,2}^{\mathbb{C}}$, which is isometric to the Grassmannian $G_{2,4}^{\mathbb{R}}$, and the space $\bar{G}_{2,2}^{\mathbb{H}}$ are rigid.

The following theorem describes our results concerning the Guillemin rigidity of these spaces:

THEOREM 2. *Let $n_0 \geq 3$ be a given integer.*

(i) *If the symmetric space $\bar{G}_{n_0,n_0}^{\mathbb{R}}$ is rigid in the sense of Guillemin, so are all the spaces $\bar{G}_{n,n}^{\mathbb{R}}$, with $n \geq n_0$.*

(ii) *If the symmetric space $\bar{G}_{n_0,n_0}^{\mathbb{C}}$ is rigid in the sense of Guillemin, so are all the spaces $\bar{G}_{n,n}^{\mathbb{K}}$, with $n \geq n_0$ and $\mathbb{K} = \mathbb{C}$ or $\mathbb{H}$.*

In conjunction with the Blaschke conjecture, Michel had previously introduced another notion of rigidity for symmetric spaces; it coincides with Guillemin rigidity for spaces of rank one. We say that a symmetric p-form on an arbitrary symmetric space X satisfies the zero-energy condition if all its integrals over the closed geodesics of X vanish. The space X is said to be infinitesimally rigid if the only symmetric 2-forms on X satisfying the zero-energy condition are the Lie derivatives of the metric. The infinitesimal rigidity of a flat torus of dimension ≥ 2 was established by Michel in [46].

The canonical metric g of a projective space X equal to $\mathbb{KP}^n$, with $n \geq 2$, or to the Cayley plane is a C_π-metric, i.e., a metric all of whose geodesics are closed and of the same length π. An important question which arises from the Blaschke conjecture consists in determining whether the metric g is the only C_π-metric of X, up to an isometry. Green and Berger have answered this question in the affirmative in the case of the real projective spaces (see [5]). The infinitesimal deformation of g by C_π-metrics satisfies the zero-energy condition. Thus the infinitesimal rigidity of X may be interpreted as the rigidity to first-order for the deformation problem of g by C_π-metrics. In [11], Duistermaat and Guillemin proved that a metric g' on X, whose spectrum is equal to the spectrum of the metric g, is a C_π-metric. In the case of the real projective space $\mathbb{RP}^n$, with $n \geq 2$, the positive resolution of the Blaschke conjecture then implies that the metric g' is isometric to the metric g and, therefore, that this space is spectrally rigid. For the other projective spaces, which are not spheres, in [38] Kiyohara gave a partial answer to our question; in fact, he used the infinitesimal rigidity of X in order to show that a C_π-metric g' on X, which is sufficiently close to g, is isometric to g. In all cases, Kiyohara's work can be combined with the above-mentioned result of Duistermaat and Guillemin to give us the following spectral rigidity result: a metric g' on X, whose spectrum is equal to that of g and which is sufficiently close to g, is isometric to g.

We now return to the study of the symmetric spaces of compact type of arbitrary rank. We show that a space which is rigid in the sense of Guillemin is also infinitesimally rigid. Thus Theorem 1 implies that the Grassmannian $G_{m,n}^{\mathbb{K}}$, with $m, n \geq 1$ and $m \neq n$, is infinitesimally rigid.

The real Grassmannian $\widetilde{G}_{m,n}^{\mathbb{R}}$ of oriented m-planes in $\mathbb{R}^{m+n}$ is the simply-connected double cover of the Grassmannian $G_{m,n}^{\mathbb{R}}$. In fact, when $m \neq n$, its adjoint space is the Grassmannian $G_{m,n}^{\mathbb{R}}$. We may identify the Grassmannian $\widetilde{G}_{1,n}^{\mathbb{R}}$ with the sphere S^n. On the other hand, the Grassmannian $\widetilde{G}_{2,n}^{\mathbb{R}}$, with $n \geq 2$, is isometric to the complex quadric Q_n, which is a hypersurface of $\mathbb{CP}^{n+1}$.

All the known results concerning the infinitesimal rigidity of irreducible symmetric spaces are given by the following:

THEOREM 3. *Let $m, n \geq 1$ be given integers.*

(i) *Suppose that $m + n > 2$. Then the real Grassmannian $\widetilde{G}_{m,n}^{\mathbb{R}}$ is infinitesimally rigid if and only if $m, n \geq 2$ and $m + n \geq 5$.*

(ii) *If $\mathbb{K}$ is equal to $\mathbb{R}$ or $\mathbb{C}$, the Grassmannian $G_{m,n}^{\mathbb{K}}$ is infinitesimally rigid if and only if $m + n > 2$.*

(iii) *If $(m, n) \neq (1, 1)$ and $(2, 2)$, the Grassmannian $G_{n,n}^{\mathbb{H}}$ is infinitesimally rigid.*

Theorem 3 tells us that any Grassmannian, which is not isometric to a sphere, or to a product of spheres, or to $G_{2,2}^{\mathbb{H}}$, is infinitesimally rigid. The infinitesimal rigidity of such a Grassmannian, as long as it is not isometric to a projective space or to the Grassmannian $G_{2,2}^{\mathbb{R}}$, is proved by means of the methods used to demonstrate Theorem 1.

We now present the various methods for proving the Guillemin rigidity or the infinitesimal rigidity of an irreducible symmetric space of compact type. The first one requires techniques based on the harmonic analysis on homogeneous spaces of compact Lie groups. We used it in [14] to establish the infinitesimal rigidity of the complex projective space $\mathbb{CP}^n$, with $n \geq 2$, and in [23] to prove the Guillemin rigidity of the Grassmannian $G_{2,3}^{\mathbb{R}}$. The proofs of the infinitesimal rigidity of the complex quadric Q_3 of dimension three, given in Chapter VI, and of the Grassmannian $G_{2,2}^{\mathbb{R}}$, given in Chapter X, are similar in nature.

In [13], the theory of linear overdetermined partial differential equations of [28] is used to construct the resolution of the sheaf of Killing vector fields on a symmetric space. This resolution plays a fundamental role in our study of rigidity; in particular, one of its differential operators can be used to encode properties of families of totally geodesic submanifolds of our space and the prior knowledge of the rigidity of these submanifolds.

One approach to infinitesimal rigidity, which appears in [18], relies on a resolution of the sheaf of Killing vector fields and leads to a new proof of the infinitesimal rigidity of the complex projective space $\mathbb{CP}^n$, with $n \geq 2$. For this space, this approach requires a minimal use of harmonic analysis; it also allows us to deduce the infinitesimal rigidity of the complex quadric Q_n, with $n \geq 5$, from that of its totally geodesic submanifolds isometric to the complex projective plane $\mathbb{CP}^2$ or to a flat 2-torus.

In Section 8 of Chapter II, we introduce a new approach to our rigidity problems which is partially inspired by the one developed in [22] for the study of the complex quadric Q_4 of dimension four. We give criteria both for the Guillemin rigidity and for the infinitesimal rigidity of an irreducible space X of compact type which exploit the fact that X is an Einstein manifold. The relationship between the resolution of the sheaf of Killing vector fields and the finite-dimensional space $E(X)$ of infinitesimal Einstein deformations of X introduced by Berger and Ebin [3] provides us with one of the main ingredients of the proofs of these criteria. We still require some results from harmonic analysis in the proofs of these criteria, but only in a limited way. Also the fact that the Lichnerowicz Laplacian acting on the space of symmetric forms is equal to a Casimir operator, which was proved by Koiso in [41], plays an important role. Although the theory of overdetermined partial differential equations enters in an essential way into the proofs of our criteria, it should again be emphasized that it does not appear in any form in their final statements.

We apply our criteria to the Grassmannians of rank ≥ 2 in order to prove Theorems 1 and 3 for these spaces. On such a Grassmannian which is equal to its adjoint space, the injectivity of the Radon transform for functions on the real projective plane is used to prove that the Guillemin condition is hereditary with respect to certain totally geodesic submanifolds. For our proofs, we also require the Guillemin rigidity of complex projective plane $\mathbb{CP}^2$ and the real Grassmannian $G_{2,3}^{\mathbb{R}}$. Moreover, in the case of the complex Grassmannians, we must show that an infinitesimal Einstein deformation satisfying the Guillemin condition vanishes. This last fact, which is always a necessary condition for Guillemin rigidity, is proved in Chapter VIII for the Grassmannians $G_{m,n}^{\mathbb{C}}$, with $m \neq n$, by computing the integrals of specific symmetric 2-forms over certain closed geodesics.

In Chapters VII and VIII, we introduce an averaging process which assigns to a symmetric p-form on the space $\bar{G}_{n+1,n+1}^{\mathbb{K}}$, with $n \geq 2$ and $\mathbb{K} = \mathbb{R}$ or $\mathbb{C}$, a class of symmetric p-forms on the space $\bar{G}_{n,n}^{\mathbb{K}}$. This process has the following property, which enables it to play an essential role in the proof of Theorem 2: if the p-form u on $\bar{G}^{\mathbb{K}}_{n+1,n+1}$ satisfies the Guillemin condition, so do all the p-forms on $\bar{G}^{\mathbb{K}}_{n,n}$ associated to u. In fact, this process is used to show that an infinitesimal Einstein deformation of the complex Grassmannian $G^{\mathbb{C}}_{n,n}$ satisfying the Guillemin condition vanishes. This last assertion and several others concerning the spaces $G^{\mathbb{K}}_{n,n}$ are proved by induction on n. To demonstrate Theorem 2, we also exploit to a considerable extent the methods which enter into the proofs of our rigidity criteria of Chapter II.

The study of 1-forms on the Grassmannians satisfying the Guillemin condition is of independent interest. Clearly, an exact 1-form always satisfies this condition. In fact, we have the following converse:

THEOREM 4. *Let $\mathbb{K}$ be a division algebra over $\mathbb{R}$. Let X be a symmetric space equal to one of the following spaces:*
  (i) $G^{\mathbb{K}}_{m,n}$, *with $m + n > 1$ and $m \neq n$.*
  (ii) $\bar{G}^{\mathbb{K}}_{n,n}$, *with $n \geq 2$ and $\mathbb{K} = \mathbb{C}$ or $\mathbb{H}$.*
Then a 1-form on X satisfying the Guillemin condition is exact.

By means of our methods, we are able to give elementary and direct proofs of Grinberg's result concerning the maximal flat Radon transform for functions on all the irreducible symmetric spaces of compact type considered above that are equal to their adjoint spaces.

In this monograph, all the known results concerning our rigidity problems for symmetric spaces, which are either of compact type, or flat tori, or products of such manifolds, are presented in a unified way. For the irreducible spaces, we give proofs which either are complete or omit only certain technical details.

We wish to point out that, in Chapters I and III, several results and formulas of Riemannian geometry are presented or derived in a particularly simple way. Moreover, the result concerning conformal Killing vector fields on Einstein manifolds given by Proposition 1.6 is new; it is required for the proof of Proposition 2.16. The latter proposition leads us to the necessary condition for Guillemin rigidity of an irreducible symmetric space described above.

We now proceed to give a brief description of the contents of the chapters of this monograph. In Chapter I, we introduce various differential operators on a Riemannian manifold (X, g) arising from the curvature and a complex of differential operators related to the Killing vector fields, which includes the differential operator mentioned above. When X is a compact Einstein manifold, the space $E(X)$ of infinitesimal Einstein deformations of the metric g contains the cohomology of this complex. In [41] and [42], Koiso determined the irreducible symmetric spaces X of compact type for which the space $E(X)$ vanishes; this result gives us the exactness of this complex for these spaces. The study of the Radon transforms, the tools derived from harmonic analysis on symmetric spaces and our criteria for rigidity are to be found in Chapter II. In the following chapter, we present the infinitesimal rigidity results for flat tori and the projective spaces, which are not equal to spheres. In particular, in the case of the real projective spaces, we give the proof of their infinitesimal rigidity due to Bourguignon (see [5]) and a variant of the one due to Michel [45].

In Chapter IV, we study the differential geometry of real Grassmannians $\widetilde{G}_{m,n}^{\mathbb{R}}$ and $G_{m,n}^{\mathbb{R}}$ and view them as symmetric spaces and homogeneous spaces of the orthogonal group $SO(m + n)$. We show that the Guillemin condition for forms on the Grassmannian $G_{m,n}^{\mathbb{R}}$, with $m \neq n$, is hereditary with respect to certain totally geodesic submanifolds.

Chapter V is devoted to the geometry of the complex quadric. We view this quadric Q_n of dimension n as a hypersurface of $\mathbb{CP}^{n+1}$, develop the local formalism of Kähler geometry on this space and describe its totally geodesic submanifolds. We also identify the quadric Q_n with the Grassmannian $\widetilde{G}_{2,n}^{\mathbb{R}}$ of oriented 2-planes in $\mathbb{R}^{n+2}$ and use the harmonic analysis on Q_n viewed as a homogeneous space of $SO(n + 2)$ to examine the space of complex symmetric 2-forms on Q_n. The various proofs of the infinitesimal rigidity of the complex quadric Q_n, with $n \geq 3$, and the proof of the Guillemin rigidity of the Grassmannian $G_{2,n}^{\mathbb{R}}$, with $n \geq 3$, are presented in Chapter VI.

In Chapter VII, we give the proofs of the rigidity of the real Grassmannians of rank ≥ 3 and introduce the averaging process for symmetric forms on $G_{n,n}^{\mathbb{R}}$, which leads to the result given by Theorem 2 for the adjoint spaces $\bar{G}_{n,n}^{\mathbb{R}}$. In Chapter VIII, we study the differential geometry of the complex Grassmannians $G_{m,n}^{\mathbb{C}}$ and we view them as symmetric spaces and homogeneous spaces. We introduce certain explicit functions and symmetric 2-forms on these spaces, which enter into our analysis of the space of infinitesimal Einstein deformations of these Grassmannians. We also define the averaging process for symmetric forms on the Grassmannian $G_{n,n}^{\mathbb{C}}$, which is used here to prove properties of its space of infinitesimal Einstein deformations and in the next chapter to study the rigidity of its adjoint space $\bar{G}_{n,n}^{\mathbb{C}}$. Chapter IX is mainly devoted to the proofs of the rigidity of the complex and the quaternionic Grassmannians of rank ≥ 2.

In Chapter X, we prove the non-rigidity of the product of irreducible symmetric spaces, which we mentioned above. We also present results from [19] concerning the geometry of products of symmetric spaces and their infinitesimal rigidity. The study of the real Grassmannian $G_{2,2}^{\mathbb{R}}$ is to be found here.

CHAPTER I

SYMMETRIC SPACES AND EINSTEIN MANIFOLDS

§1. Riemannian manifolds

Let X be a differentiable manifold of dimension n, whose tangent and cotangent bundles we denote by $T = T_X$ and $T^* = T_X^*$, respectively. Let $C^\infty(X)$ be the space of complex-valued functions on X. By $\bigotimes^k E$, $S^l E$, $\bigwedge^j E$, we shall mean the k-th tensor product, the l-th symmetric product and the j-th exterior product of a vector bundle E over X, respectively. We shall identify $S^k T^*$ and $\bigwedge^k T^*$ with sub-bundles of $\bigotimes^k T^*$ by means of the injective mappings

$$S^k T^* \to \bigotimes^k T^*, \qquad \bigwedge^k T^* \to \bigotimes^k T^*,$$

sending the symmetric product $\beta_1 \cdot \ldots \cdot \beta_k$ into

$$\sum_{\sigma \in \mathfrak{S}_k} \beta_{\sigma(1)} \otimes \cdots \otimes \beta_{\sigma(k)}$$

and the exterior product $\beta_1 \wedge \cdots \wedge \beta_k$ into

$$\sum_{\sigma \in \mathfrak{S}_k} \operatorname{sgn} \sigma \cdot \beta_{\sigma(1)} \otimes \cdots \otimes \beta_{\sigma(k)},$$

where $\beta_1, \ldots, \beta_k \in T^*$ and $\mathfrak{S}_k$ is the group of permutations of $\{1, \ldots, k\}$ and $\operatorname{sgn} \sigma$ is the signature of the element σ of $\mathfrak{S}_k$. If $\alpha, \beta \in T^*$, the symmetric product $\alpha \cdot \beta$ is identified with the element $\alpha \otimes \beta + \beta \otimes \alpha$ of $\bigotimes^2 T^*$. If $\xi \in T$, $h \in S^2 T^*$, let $\xi \lrcorner h$ be the element of T^* defined by

$$(\xi \lrcorner h)(\eta) = h(\xi, \eta),$$

for $\eta \in T$. If $h \in S^2 T^*$, we denote by

$$h^\flat : T \to T^*$$

the mapping sending $\xi \in T$ into $\xi \lrcorner h$. If h is non-degenerate, then $h^\flat$ is an isomorphism, whose inverse will be denoted by $h^\sharp$.

Let E be a vector bundle over X; we denote by $E_{\mathbb{C}}$ its complexification, by $\mathcal{E}$ the sheaf of sections of E over X and by $C^\infty(E)$ the space of global sections of E over X. We write $S^l E_{\mathbb{C}}$ and $\bigwedge^j E_{\mathbb{C}}$ for the complexifications of $S^l E$ and $\bigwedge^j E$. We consider the vector bundle $J_k(E)$ of k-jets of sections of E, whose fiber at $x \in X$ is the quotient of the space $C^\infty(E)$ by its subspace consisting of the sections of E which vanish to order $k+1$ at x. If s is a section of E over X, the k-jet $j_k(s)(x)$ of s at $x \in X$ is the equivalence class of s in $J_k(E)_x$. The mapping $x \mapsto j_k(s)(x)$ is a section $j_k(s)$ of $J_k(E)$ over X. We denote by $\pi_k : J_{k+l}(E) \to J_k(E)$ the natural projection sending $j_{k+l}(s)(x)$ into $j_k(s)(x)$, for $x \in X$. We shall identify $J_0(E)$ with E and we set $J_k(E) = 0$, for $k < 0$. The morphism of vector bundles

$$\varepsilon : S^k T^* \otimes E \to J_k(E)$$

determined by

$$\varepsilon(((df_1 \cdot \ldots \cdot df_k) \otimes s)(x)) = j_k\left(\left(\prod_{i=1}^k f_i\right)\right)(x),$$

where $f_1, \ldots, f_k$ are real-valued functions on X vanishing at $x \in X$ and s is a section of E over X, is well-defined since the function $\prod_{i=1}^k f_i$ vanishes to order $k-1$ at x. One easily verifies that the sequence

$$0 \to S^k T^* \otimes E \xrightarrow{\varepsilon} J_k(E) \xrightarrow{\pi_{k-1}} J_{k-1}(E) \to 0$$

is exact, for $k \geq 0$.

Let E and F be vector bundles over X. If $D : \mathcal{E} \to \mathcal{F}$ is a differential operator of order k, there exists a unique morphism of vector bundles

$$p(D) : J_k(E) \to F$$

such that

$$Ds = p(D)j_k(s),$$

for all $s \in \mathcal{E}$. The symbol of D is the morphism of vector bundles

$$\sigma(D) : S^k T^* \otimes E \to F$$

equal to $p(D) \circ \varepsilon$. If $x \in X$ and $\alpha \in T_x^*$, let

$$\sigma_\alpha(D) : E_x \to F_x$$

be the linear mapping defined by

$$\sigma_\alpha(D)u = \frac{1}{k!}\,\sigma(D)(\alpha^k \otimes u),$$

where $u \in E_x$ and α^k denotes the k-th symmetric product of α. We say that D is elliptic if, for all $x \in X$ and $\alpha \in T_x^*$, the mapping $\sigma_\alpha(D)$ is injective. If D is elliptic and X is compact, then it is well-known that the kernel of $D : C^\infty(E) \to C^\infty(F)$ is finite-dimensional.

If ξ is a vector field on X and β is a section of $\bigotimes^k T^*$ over X, we denote by $\mathcal{L}_\xi\beta$ the Lie derivative of β along ξ. For $x \in X$, let ρ denote the representation of the Lie algebra $(T^*\otimes T)_x$ on T_x and also the representation induced by ρ on $\bigotimes^k T_x^*$. If ξ is an element of T_x satisfying $\xi(x) = 0$, and if u is the unique element of $(T^* \otimes T)_x$ determined by the relation $\varepsilon(u) = j_1(\xi)(x)$, then we have

$$\rho(u)\eta(x) = -[\xi, \eta](x),$$

for all $\eta \in T_x$, and

(1.1) $$\rho(u)\beta(x) = -(\mathcal{L}_\xi\beta)(x),$$

for $\beta \in \bigotimes^k T_x^*$.

We endow X with a Riemannian metric g and we associate various objects to the Riemannian manifold (X, g). The mappings $g^\flat : T \to T^*$, $g^\sharp : T^* \to T$ are the isomorphisms determined by the metric g; we shall sometimes write $\xi^\flat = g^\flat(\xi)$ and $\alpha^\sharp = g^\sharp(\alpha)$, for $\xi \in T$ and $\alpha \in T^*$. The metric g induces scalar products on the vector bundle $\bigotimes^p T^* \otimes \bigotimes^q T$ and its sub-bundles. We denote by dX the Riemannian measure of the Riemannian manifold (X, g). If X is compact, the volume $\mathrm{Vol}\,(X, g)$ of (X, g) is equal to the integral $\int_X dX$.

Let E and F be vector bundles over X endowed with scalar products and $D : \mathcal{E} \to \mathcal{F}$ be a differential operator of order k. We consider the scalar products on $C^\infty(E)$ and $C^\infty(F)$, defined in terms of these scalar products on E and F and the Riemannian measure of X, and the formal adjoint $D^* : \mathcal{F} \to \mathcal{E}$ of D, which is a differential operator of order k. If D is elliptic and X is compact, then $DC^\infty(E)$ is a closed subspace of $C^\infty(F)$ and we have the orthogonal decomposition

(1.2) $$C^\infty(F) = DC^\infty(E) \oplus \{\, u \in C^\infty(F) \mid D^*u = 0\,\}.$$

Let $B = B_X$ be the sub-bundle of $\bigwedge^2 T^* \otimes \bigwedge^2 T^*$ consisting of those tensors $u \in \bigwedge^2 T^* \otimes \bigwedge^2 T^*$ which satisfy the first Bianchi identity

$$u(\xi_1, \xi_2, \xi_3, \xi_4) + u(\xi_2, \xi_3, \xi_1, \xi_4) + u(\xi_3, \xi_1, \xi_2, \xi_4) = 0,$$

for all $\xi_1, \xi_2, \xi_3, \xi_4 \in T$. It is easily seen that, if an element u of B satisfies the relation

$$u(\xi_1, \xi_2, \xi_1, \xi_2) = 0,$$

for all $\xi_1, \xi_2 \in T$, then u vanishes. We consider the morphism of vector bundles

$$\tau_B : S^2 T^* \otimes S^2 T^* \to B$$

defined by

$$(\tau_B u)(\xi_1, \xi_2, \xi_3, \xi_4) = \tfrac{1}{2}\{u(\xi_1, \xi_3, \xi_2, \xi_4) + u(\xi_2, \xi_4, \xi_1, \xi_3)$$
$$- u(\xi_1, \xi_4, \xi_2, \xi_3) - u(\xi_2, \xi_3, \xi_1, \xi_4)\},$$

for all $u \in S^2 T^* \otimes S^2 T^*$ and $\xi_1, \xi_2, \xi_3, \xi_4 \in T$; it is well-known that this morphism is an epimorphism (see Lemma 3.1 of [13]). Let

$$\sigma : T^* \otimes B \to \textstyle\bigwedge^3 T^* \otimes \bigwedge^2 T^*$$

be the restriction of the morphism of vector bundles

$$T^* \otimes \textstyle\bigwedge^2 T^* \otimes \bigwedge^2 T^* \to \bigwedge^3 T^* \otimes \bigwedge^2 T^*,$$

which sends $\alpha \otimes \theta_1 \otimes \theta_2$ into $(\alpha \wedge \theta_1) \otimes \theta_2$, for $\alpha \in T^*$, $\theta_1, \theta_2 \in \bigwedge^2 T^*$. The kernel H of this morphism σ is equal to the sub-bundle of $T^* \otimes B$ consisting of those tensors $v \in T^* \otimes B$ which satisfy the relation

$$v(\xi_1, \xi_2, \xi_3, \xi_4, \xi_5) + v(\xi_2, \xi_3, \xi_1, \xi_4, \xi_5) + v(\xi_3, \xi_1, \xi_2, \xi_4, \xi_5) = 0,$$

for all $\xi_1, \xi_2, \xi_3, \xi_4, \xi_5 \in T$.

Let

$$\mathrm{Tr} = \mathrm{Tr}_g = \mathrm{Tr}_X : S^2 T^* \to \mathbb{R}, \qquad \mathrm{Tr} = \mathrm{Tr}_g : \textstyle\bigwedge^2 T^* \otimes \bigwedge^2 T^* \to \bigotimes^2 T^*$$

be the trace mappings defined by

$$\mathrm{Tr}\, h = \sum_{j=1}^n h(t_j, t_j), \qquad (\mathrm{Tr}\, u)(\xi, \eta) = \sum_{j=1}^n u(t_j, \xi, t_j, \eta),$$

for $h \in S^2 T_x^*$, $u \in (\bigwedge^2 T^* \otimes \bigwedge^2 T^*)_x$ and $\xi, \eta \in T_x$, where $x \in X$ and $\{t_1, \ldots, t_n\}$ is an orthonormal basis of T_x. It is easily seen that

$$\mathrm{Tr}\, B \subset S^2 T^*.$$

We denote by $S_0^2 T^*$ the sub-bundle of $S^2 T^*$ equal to the kernel of the trace mapping $\mathrm{Tr} : S^2 T^* \to \mathbb{R}$ and by B^0 the sub-bundle of B equal to the kernel of the trace mapping $\mathrm{Tr} : B \to S^2 T^*$.

We consider the morphism of vector bundles

$$\hat{\tau}_B : S^2 T^* \to B$$

defined by

$$\hat{\tau}_B(h) = \tau_B(h \otimes g),$$

for $h \in S^2T^*$. If $h \in S^2T^*$, we easily verify that

(1.3)
$$\operatorname{Tr} \hat{\tau}_B(h) = \tfrac{1}{2}((\operatorname{Tr} h) \cdot g + (n-2)h),$$

and so we have

$$\operatorname{Tr} \cdot \operatorname{Tr} \hat{\tau}_B(h) = (n-1)\operatorname{Tr} h.$$

When $n \geq 3$, from the preceding formulas we infer that the morphism $\hat{\tau}_B$ is injective, that the morphism $\operatorname{Tr} : B \to S^2T^*$ is surjective and that B^0 is the orthogonal complement of $\hat{\tau}_B(S^2T^*)$ in B.

We introduce various differential operators and objects associated with the Riemannian manifold (X, g). First, let $\nabla = \nabla^g$ be the Levi-Civita connection of (X, g). If f is a real-valued function on X, we denote by $\operatorname{Hess} f = \nabla df$ the Hessian of f. If $d^* : \bigwedge^j T^* \to \bigwedge^{j-1} T^*$ is the formal adjoint of the exterior differential operator $d : \bigwedge^{j-1} T^* \to \bigwedge^j T^*$, we consider the de Rham Laplacian $\Delta = dd^* + d^*d$ acting on $\bigwedge^j T^*$. The Laplacian $\Delta = \Delta_g = \Delta_X$ acting on $C^\infty(X)$ is also determined by the relation $\Delta f = -\operatorname{Tr} \operatorname{Hess} f$, for $f \in C^\infty(X)$. The spectrum $\operatorname{Spec}(X, g)$ of the metric g on X is the sequence of eigenvalues (counted with multiplicities)

$$0 = \lambda_0 < \lambda_1 \leq \lambda_2 \leq \cdots \leq \lambda_m \leq \cdots$$

of the Laplacian Δ_g acting on the space $C^\infty(X)$.

The Killing operator

$$D_0 = D_{0,X} : T \to S^2T^*$$

of (X, g), which sends $\xi \in T$ into $\mathcal{L}_\xi g$, and the symmetrized covariant derivative

$$D^1 : T^* \to S^2T^*,$$

defined by

$$(D^1\theta)(\xi, \eta) = \tfrac{1}{2}((\nabla\theta)(\xi, \eta) + (\nabla\theta)(\eta, \xi)),$$

for $\theta \in T^*$, $\xi, \eta \in T$, are related by the formula

(1.4)
$$\tfrac{1}{2} D_0 \xi = D^1 g^\flat(\xi),$$

for $\xi \in T$. By (1.4), the conformal Killing operator

$$D_0^c : T \to S_0^2T^*$$

of (X, g) is determined by

$$D_0^c \xi = D_0 \xi - \frac{1}{n} \left(\operatorname{Tr} D_0 \xi \right) \cdot g = D_0 \xi + \frac{2}{n} (d^* \xi^b) \cdot g,$$

for $\xi \in \mathcal{T}$. The Killing (resp. conformal Killing) vector fields of (X, g) are the solutions $\xi \in C^\infty(T)$ of the equation $D_0 \xi = 0$ (resp. $D_0^c \xi = 0$). A vector field ξ on X is a conformal Killing vector field if and only if there is a real-valued function f such that $D_0 \xi = fg$. According to (1.4), a Killing vector field ξ on X satisfies the relation $d^* g^b(\xi) = 0$, and a real-valued function f on X satisfies the relation

(1.5) $$D_0 (df)^\sharp = 2 \operatorname{Hess} f.$$

If ϕ is a local isometry of X defined on an open subset U of X and ξ is a vector field on U, according to (1.4) we see that

(1.6) $$\phi^* D_0 \xi = D_0 (\phi_*^{-1} \xi).$$

LEMMA 1.1. *Let Y be a totally geodesic submanifold of the Riemannian manifold (X, g). Let $i : Y \to X$ be the natural imbedding and $g_Y = i^* g$ be the Riemannian metric on Y induced by g. Let ξ be a vector field on X. If η is the vector field on Y whose value at $x \in Y$ is equal to the orthogonal projection of $\xi(x)$ onto the subspace $T_{Y,x}$ of T_x, then we have*

$$i^* g^b(\xi) = g_Y^b(\eta), \qquad i^* \mathcal{L}_\xi g = \mathcal{L}_\eta g_Y.$$

PROOF: The lemma is a consequence of the relation (1.4) and the vanishing of the second fundamental form of the imbedding i.

By (1.1), the symbol

$$\sigma(D_0) : T^* \otimes T \to S^2 T^*$$

of D_0 sends $u \in T^* \otimes T$ into $-\rho(u)g$. It follows that

$$\sigma(D_0)(\alpha \otimes \xi) = \alpha \cdot g^b(\xi),$$

for $\alpha \in T^*$ and $\xi \in \mathcal{T}$. This last equality implies that $\sigma(D_0)$ is an epimorphism and that the differential operator D_0 is elliptic. We verify that

$$\sigma(D_0^c)(\alpha \otimes \xi) = \alpha \cdot g^b(\xi) - \frac{1}{n} \langle \xi, \alpha \rangle \, g,$$

for $\alpha \in T^*$ and $\xi \in \mathcal{T}$, and we easily see that the differential operator D_0^c is elliptic when $n \geq 2$. The kernel of $\sigma(D_0)$ is equal to the sub-bundle

$$g_1 = \{ u \in T^* \otimes T \mid \rho(u)g = 0 \}$$

of $T^* \otimes T$. We consider the isomorphism of vector bundles

$$\iota : T^* \otimes T^* \to T^* \otimes T$$

equal to $\mathrm{id} \otimes g^\sharp$, which is also determined by the relation

$$g(\iota(u)\xi, \eta) = u(\xi, \eta),$$

for $u \in T^* \otimes T^*$ and $\xi, \eta \in T$, and its restrictions

$$\iota : S^2 T^* \to T^* \otimes T, \qquad \iota : \bigwedge^2 T^* \to T^* \otimes T.$$

Then it is easily verified that

$$(1.7) \qquad\qquad \sigma(D_0)\iota(h) = 2\iota(h),$$

for $h \in S^2 T^*$, and that the image of the morphism $\iota : \bigwedge^2 T^* \to T^* \otimes T$ is equal to g_1; thus

$$\iota : \bigwedge^2 T^* \to g_1$$

is an isomorphism of vector bundles (see [13, §3]).

We consider the divergence operator

$$\mathrm{div} : S^2 T^* \to T^*,$$

which is the first-order differential operator defined by

$$(\mathrm{div}\, h)(\xi) = -\sum_{j=1}^{n} (\nabla h)(t_j, t_j, \xi),$$

for $h \in C^\infty(S^2 T^*)$, $\xi \in T_x$, where $x \in X$ and $\{t_1, \ldots, t_n\}$ is an orthonormal basis of T_x. Let f be a real-valued function on X and let h be an element of $C^\infty(S^2 T^*)$. Then we have

$$(1.8) \qquad\qquad \mathrm{div}\,(fh) = f\,\mathrm{div}\, h - (df)^\sharp \,\lrcorner\, h;$$

thus we see that

$$(1.9) \qquad D^1 \mathrm{div}\,(fh) = f D^1 \mathrm{div}\, h + \tfrac{1}{2} df \cdot \mathrm{div}\, h - D^1((df)^\sharp \,\lrcorner\, h),$$

and the section $D^1((df)^\sharp \,\lrcorner\, h)$ of $S^2 T^*$ is determined by

$$(1.10) \quad \begin{aligned} D^1((df)^\sharp \,\lrcorner\, h)(\xi, \eta) = \tfrac{1}{2}\{ &h((\xi \,\lrcorner\, \mathrm{Hess}\, f)^\sharp, \eta) + h((\eta \,\lrcorner\, \mathrm{Hess}\, f)^\sharp, \xi) \\ &+ (\nabla h)(\xi, (df)^\sharp, \eta) + (\nabla h)(\eta, (df)^\sharp, \xi) \}, \end{aligned}$$

for $\xi, \eta \in T$.

The formal adjoint of D_0 is equal to $2g^\sharp \cdot \mathrm{div} : S^2 T^* \to T$. It follows that the formal adjoint of $D_0^c : T \to S_0^2 T^*$ is equal to $2g^\sharp \cdot \mathrm{div} : S_0^2 T^* \to T$. When X is compact, since the operator D_0 is elliptic, by (1.2) we therefore have the orthogonal decomposition

$$(1.11) \qquad C^\infty(S^2 T^*) = D_0 C^\infty(T) \oplus \{\, h \in C^\infty(S^2 T^*) \mid \mathrm{div}\, h = 0 \,\}$$

(see [3]). When X is a compact manifold of dimension ≥ 2, since the differential operator D_0^c is elliptic, by (1.2) we have the decomposition

$$(1.12) \qquad C^\infty(S_0^2 T^*) = D_0^c C^\infty(T) \oplus \{\, h \in C^\infty(S_0^2 T^*) \mid \mathrm{div}\, h = 0 \,\}.$$

The curvature tensor $\widetilde{\mathcal{R}}(g)$ of g is the section of $\bigwedge^2 T^* \otimes T^* \otimes T$ determined by

$$\widetilde{\mathcal{R}}(g)(\xi, \eta, \zeta) = (\nabla_\xi^g \nabla_\eta^g - \nabla_\eta^g \nabla_\xi^g)\zeta,$$

for $\xi, \eta, \zeta \in T$. The Riemann curvature tensor of g is the section $\mathcal{R}(g)$ of $\bigwedge^2 T^* \otimes \bigwedge^2 T^*$ determined by

$$\mathcal{R}(g)(\xi_1, \xi_2, \xi_3, \xi_4) = g(\xi_4, \widetilde{\mathcal{R}}(g)(\xi_1, \xi_2, \xi_3)),$$

for $\xi_1, \xi_2, \xi_3, \xi_4 \in T$; according to the first Bianchi identity, $\mathcal{R}(g)$ is a section of B. The second Bianchi identity tells us that $(\mathcal{D}\mathcal{R})(g) = \nabla^g \mathcal{R}(g)$ is a section of H. The Ricci tensor $\mathrm{Ric}(g)$ of the metric g is the section of $S^2 T^*$ equal to $-\mathrm{Tr}_g \mathcal{R}(g)$, while the scalar curvature $r(g)$ of the metric g is the function $\mathrm{Tr}_g \mathrm{Ric}(g)$ on X. We shall write $\widetilde{R} = \widetilde{R}_X = \widetilde{\mathcal{R}}(g)$, $R = R_X = \mathcal{R}(g)$ and $\mathrm{Ric} = \mathrm{Ric}(g)$.

Let ξ be a vector field on X. According to the second Bianchi identity and the relation (1.4), we see that

$$(1.13) \qquad d^* \mathrm{Ric}^\flat(\xi) = -\tfrac{1}{2}\,(\xi \cdot r(g) + \langle \mathrm{Ric}, D_0\xi \rangle),$$

where $\langle\,,\,\rangle$ is the scalar product on $S^2 T^*$ induced by the metric g. We easily verify that

$$(1.14) \qquad \mathrm{div}\, D_0 \xi = (\Delta + dd^*)\xi^\flat - 2\,\mathrm{Ric}^\flat(\xi);$$

thus by (1.8), we have

$$(1.15) \qquad \mathrm{div}\, D_0^c \xi = \Delta \xi^\flat - 2\,\mathrm{Ric}^\flat(\xi) + \frac{n-2}{n}\, dd^* \xi^\flat.$$

From the preceding equation and (1.13), we obtain

$$(1.16) \qquad d^* \mathrm{div}\, D_0^c \xi = \frac{2(n-1)}{n}\,\Delta d^* \xi^\flat + \xi \cdot r(g) + \langle \mathrm{Ric}, D_0\xi \rangle.$$

If the scalar curvature $r(g)$ of (X, g) is constant and equal to $\mu \in \mathbb{R}$ and if ξ is a conformal Killing vector field, then by (1.16) we see that the real-valued function $f = -(2/n) \cdot d^* \xi^\flat$ satisfying $D_0 \xi = fg$ is a solution of the equation

$$(1.17) \qquad\qquad (n-1)\Delta f = \mu f.$$

If X is compact, we know that the eigenvalues of the Laplacian Δ acting on $C^\infty(X)$ are ≥ 0; hence from the previous observation, we obtain the following result due to Lichnerowicz [43, §83]:

LEMMA 1.2. *Assume that (X, g) is a Riemannian manifold of dimension $n \geq 2$ and that its scalar curvature is constant and equal to $\mu \in \mathbb{R}$. If ξ is a conformal vector field on X and if the real-valued function f on X determined by the relation $\mathcal{L}_\xi g = fg$ is non-zero, then f is an eigenfunction of Δ with eigenvalue $\mu/(n-1)$. If X is compact and if $\mu < 0$, then a conformal Killing vector field on X is a Killing vector field.*

Let

$$\mathcal{R}'_g : S^2 T^* \to \mathcal{B}, \qquad (\mathcal{D}\mathcal{R})'_g : S^2 T^* \to \mathcal{H}, \qquad \mathrm{Ric}'_g : S^2 T^* \to S^2 T^*$$

be the linear differential operators, which are the linearizations along g of the non-linear operators $h \mapsto \mathcal{R}(h)$, $h \mapsto (\mathcal{D}\mathcal{R})(h)$ and $h \mapsto \mathrm{Ric}(h)$, respectively, where h is a Riemannian metric on X. Let h be a section of $S^2 T^*$ over X. For $|t| < \epsilon$, we know that $g_t = g + th$ is a Riemannian metric on a neighborhood of x; by definition, we have

$$\mathcal{R}'_g h = \frac{d}{dt} \mathcal{R}(g + th)_{|t=0}, \qquad (\mathcal{D}\mathcal{R})'_g h = \frac{d}{dt} (\mathcal{D}\mathcal{R})(g + th)_{|t=0},$$

$$\mathrm{Ric}'_g h = \frac{d}{dt} \mathrm{Ric}(g + th)_{|t=0}.$$

The differential operators $\mathcal{R}'_g$ and Ric'_g are of order 2, while the operator $(\mathcal{D}\mathcal{R})'_g$ is of order 3. The invariance of the three operators $h \mapsto \mathcal{R}(h)$, $h \mapsto (\mathcal{D}\mathcal{R})(h)$ and $h \mapsto \mathrm{Ric}(h)$ leads us to the formulas

$$(1.18) \quad \mathcal{R}'_g(\mathcal{L}_\xi g) = \mathcal{L}_\xi R, \quad (\mathcal{D}\mathcal{R})'_g(\mathcal{L}_\xi g) = \mathcal{L}_\xi \nabla R, \quad \mathrm{Ric}'_g(\mathcal{L}_\xi g) = \mathcal{L}_\xi \mathrm{Ric},$$

for all $\xi \in \mathcal{T}$. If $h \in C^\infty(S^2 T^*)$ and $h(x) = 0$, with $x \in X$, then we have

$$(1.19) \qquad\qquad \mathrm{Tr}\, (\mathcal{R}'_g h)(x) = -(\mathrm{Ric}'_g h)(x).$$

It is easily verified that

$$
\begin{aligned}
(1.20) \quad (\mathcal{R}'_g h)&(\xi_1, \xi_2, \xi_3, \xi_4) \\
&= \tfrac{1}{2}\{(\nabla^2 h)(\xi_1, \xi_3, \xi_2, \xi_4) + (\nabla^2 h)(\xi_2, \xi_4, \xi_1, \xi_3) \\
&\quad - (\nabla^2 h)(\xi_1, \xi_4, \xi_2, \xi_3) - (\nabla^2 h)(\xi_2, \xi_3, \xi_1, \xi_4) \\
&\quad + h(\widetilde{R}(\xi_1, \xi_2, \xi_3), \xi_4) - h(\widetilde{R}(\xi_1, \xi_2, \xi_4), \xi_3)\},
\end{aligned}
$$

for $h \in S^2 T^*$ and $\xi_1, \xi_2, \xi_3, \xi_4 \in T$ (see Lemma 4.3 of [13]). By formula (1.20), we see that the morphism of vector bundles

$$
\sigma(\mathcal{R}'_g) : S^2 T^* \otimes S^2 T^* \to B
$$

is equal to τ_B.

The following result is given by Proposition 4.1 of [13].

PROPOSITION 1.3. *Let Y be a totally geodesic submanifold of X. Let $i : Y \to X$ be the natural imbedding and $g_Y = i^* g$ be the Riemannian metric on Y induced by g. If h is a section of $S^2 T^*$ over X, then the equality*

$$
i^* \mathcal{R}'_g h = \mathcal{R}'_{g_Y}(i^* h)
$$

holds on Y.

PROOF: If $t \in \mathbb{R}$, we write $g_t = g + th$ and $\tilde{g}_t = g_Y + ti^* h$. Let y be a point of Y. Then there exist $\varepsilon > 0$ and neighborhoods $\tilde{U}$ of y in Y and U of $i(y)$ in X such that $i(\tilde{U}) \subset U$ and such that $\tilde{g}_t$ and g_t are Riemannian metrics on $\tilde{U}$ and U, respectively, for $|t| < \varepsilon$. Since $i^* g_t = \tilde{g}_t$, for $|t| < \varepsilon$ we denote by B_t the second fundamental form of the imbedding i, viewed as a mapping from the Riemannian manifold $(\tilde{U}, \tilde{g}_t)$ to the Riemannian manifold (U, g_t). By the Gauss equation, we have

$$
\begin{aligned}
(i^* \mathcal{R}(g_t) - \mathcal{R}(\tilde{g}_t))(\xi_1, \xi_2, \xi_3, \xi_4) &= g_t(B_t(\xi_1, \xi_3), B_t(\xi_2, \xi_4)) \\
&\quad - g_t(B_t(\xi_1, \xi_4), B_t(\xi_2, \xi_3)),
\end{aligned}
$$

for all $\xi_1, \xi_2, \xi_3, \xi_4 \in T_{\tilde{U}}$. According to our hypothesis, we have $B_0 = 0$; thus if we differentiate both sides of the above equation with respect to t, at $t = 0$, we obtain

$$
\frac{d}{dt}(i^* \mathcal{R}(g_t) - \mathcal{R}(\tilde{g}_t))_{|t=0} = i^* \mathcal{R}'_g h - \mathcal{R}'_{g_Y}(i^* h) = 0
$$

on $\tilde{U}$ and hence at the point y.

Let

$$D_g : S^2T^* \to \bigwedge^2 T^* \otimes \bigwedge^2 T^*$$

be the linear differential operator defined by

(1.21)
$$(D_g h)(\xi_1, \xi_2, \xi_3, \xi_4) = (\mathcal{R}'_g h)(\xi_1, \xi_2, \xi_3, \xi_4) - h(\tilde{R}(\xi_1, \xi_2, \xi_3), \xi_4)$$
$$+ h(\tilde{R}(\xi_1, \xi_2, \xi_4), \xi_3),$$

for $h \in S^2T^*$ and $\xi_1, \xi_2, \xi_3, \xi_4 \in T$. Thus if h is a section of S^2T^* over X vanishing at $x \in X$, we see that $(D_g h)(x) = (\mathcal{R}'_g h)(x)$ belongs to B_x. If f is a real-valued function on X, according to (1.20) and (1.21) we see that

(1.22) $\quad \mathcal{R}'_g(fg) = \tau_B(\text{Hess } f \otimes g) + fR, \quad D_g(fg) = \tau_B(\text{Hess } f \otimes g) - fR.$

We consider the sub-bundle $\tilde{B} = \tilde{B}_X$ of B, with variable fiber, whose fiber at $x \in X$ is

$$\tilde{B}_x = \{ (\mathcal{L}_\xi R)(x) \mid \xi \in T_x \text{ with } (\mathcal{L}_\xi g)(x) = 0 \},$$

and we denote by $\alpha : B \to B/\tilde{B}$ the canonical projection. By (1.1), the infinitesimal orbit of the curvature

$$\{ \rho(u)R \mid u \in \mathfrak{g}_1 \}$$

is a sub-bundle of $\tilde{B}$ with variable fiber.

The following result is given by Lemma 5.3 of [13].

LEMMA 1.4. *For all $x \in X$, we have*

$$(\nabla R)(x) \in (T^* \otimes \tilde{B})_x.$$

We now suppose that $\tilde{B}$ is a vector bundle. We consider the second-order differential operator

$$D_1 = D_{1,X} : S^2T^* \to B/\tilde{B}$$

introduced in [13] and determined by

$$(D_1 h)(x) = \alpha(\mathcal{R}'_g(h - \mathcal{L}_\xi g))(x),$$

for $x \in X$ and $h \in S^2T_x^*$, where ξ is an element of T_x satisfying $h(x) = (\mathcal{L}_\xi g)(x)$ whose existence is guaranteed by the surjectivity of $\sigma(D_0)$. By the first relation of (1.18), we see that this operator is well-defined. Clearly, the sequence

$$(1.23) \qquad T \xrightarrow{\ D_0\ } S^2T^* \xrightarrow{\ D_1\ } B/\tilde{B}$$

is a complex, and so we may consider the complex

$$(1.24) \qquad C^\infty(T) \xrightarrow{\ D_0\ } C^\infty(S^2T^*) \xrightarrow{\ D_1\ } C^\infty(B/\tilde{B}).$$

If h is a section of S^2T^* vanishing at $x \in X$, we have

$$(1.25) \qquad (D_1 h)(x) = \alpha(\mathcal{R}'_g h)(x) = \alpha(D_g h)(x).$$

We infer that the morphism of vector bundles

$$\sigma(D_1) : J_1(S^2T^*) \to B/\tilde{B}$$

is equal to $\alpha \circ \sigma(\mathcal{R}'_g)$ and is therefore an epimorphism.
     Let

$$\kappa : S^2T^* \to B/\tilde{B}$$

be the morphism of vector bundles determined by

$$\kappa(h) = \tfrac{1}{2}\alpha(\rho(\iota(h)))R,$$

for $h \in S^2T^*$, $\xi, \eta \in T$. If ξ is vector field on X and h is the section $\mathcal{L}_\xi g$ of S^2T^*, from (1.7) we infer that

$$p(D_0)(j_1(\xi) - \varepsilon\lambda(h)) = 0;$$

hence by (1.1) and the definition of $\tilde{B}$, we see that

$$\kappa(h) = -\alpha(\mathcal{L}_\xi R).$$

Thus by the first relation of (1.18), we obtain the equality

$$(1.26) \qquad D_1 h = \alpha(\mathcal{R}'_g(h)) + \kappa(h),$$

for $h \in S^2T^*$.
     Let f be a real-valued function on X. Since (1.23) is a complex, by (1.5) we see that

$$(1.27) \qquad D_1 \operatorname{Hess} f = 0.$$

We easily verify that

$$\rho(\iota(g)) = -4R;$$

from this last equality, (1.26) and (1.22), we obtain the relations

(1.28) $$D_1(fg) = \alpha(\tau_B(\mathrm{Hess}\, f \otimes g) - fR) = \alpha D_g(fg).$$

When X is compact, according to the decomposition (1.11) we see that the natural mapping from the space

$$H(X) = \{\, h \in C^\infty(S^2 T^*) \mid \mathrm{div}\, h = 0,\; D_1 h = 0 \,\}$$

to the cohomology of the complex (1.24) is an isomorphism.

We no longer make any assumption on $B/\tilde{B}$. For $h, h' \in S^2 T^*$, we define an element $h \lrcorner h'$ of $S^2 T^*$ by

$$(h \lrcorner h')(\xi, \eta) = h'(g^\sharp \cdot h^\flat(\xi), \eta) + h'(g^\sharp \cdot h^\flat(\eta), \xi),$$

for $\xi, \eta \in T$. We consider the morphism of vector bundles $L : S^2 T^* \to S^2 T^*$ determined by

$$L(\alpha \cdot \beta)(\xi, \eta) = R(\xi, g^\sharp\alpha, \eta, g^\sharp\beta) + R(\xi, g^\sharp\beta, \eta, g^\sharp\alpha)$$
$$+ R(\eta, g^\sharp\alpha, \xi, g^\sharp\beta) + R(\eta, g^\sharp\beta, \xi, g^\sharp\alpha),$$

for $\alpha, \beta \in T^*$ and $\xi, \eta \in T$. We have

(1.29) $$g \lrcorner h = 2h, \qquad Lg = -2\,\mathrm{Ric},$$

for $h \in S^2 T^*$.

The Laplacian

$$\overline{\Delta} = \nabla^* \nabla : \bigotimes^p T^* \to \bigotimes^p T^*$$

is determined by

$$(\overline{\Delta} u)(\xi_1, \ldots, \xi_p) = -\sum_{j=1}^{n} (\nabla^2 u)(t_j, t_j, \xi_1, \ldots, \xi_p),$$

for $u \in C^\infty(\bigotimes^p T^*)$, $\xi_1, \ldots, \xi_p \in T_x$, where $x \in X$ and $\{t_1, \ldots, t_n\}$ is an orthonormal basis of T_x. The Lichnerowicz Laplacian

$$\Delta = \Delta_g : \bigotimes^p T^* \to \bigotimes^p T^*$$

of [44] can be written as $\Delta = \overline{\Delta} + \kappa_p$, where $\kappa_p : \bigotimes^p T^* \to \bigotimes^p T^*$ is a morphism of vector bundles (for an intrinsic definition of Δ, see [29, §4]). For $\alpha \in T^*$ and $u \in \bigotimes^p T^*$, we easily see that

$$\sigma_\alpha(\Delta)u = \sigma_\alpha(\overline{\Delta})u = -||\alpha||^2 \cdot u,$$

where $||\alpha||$ is the norm of α (with respect to the metric g); it follows that the Lichnerowicz Laplacian $\Delta : \bigotimes^p T^* \to \bigotimes^p T^*$ is elliptic. We recall that the restriction of the Lichnerowicz Laplacian to $\bigwedge^p T^*$ is equal to the de Rham Laplacian and that $\Delta(S^p T^*) \subset S^p T^*$. The Lichnerowicz Laplacian

$$\Delta : S^2 T^* \to S^2 T^*$$

is determined by

$$\Delta h = \overline{\Delta} h + \mathrm{Ric} \lrcorner\, h + Lh,$$

for $h \in S^2 T^*$, and satisfies

(1.30) $$\mathrm{Tr}\, \Delta h = \Delta \mathrm{Tr}\, h,$$

for all $h \in S^2 T^*$. If λ is a positive real number, then the Lichnerowicz Laplacian $\Delta_{g'}$ corresponding to the Riemannian metric $g' = \lambda g$ is related to Δ_g by

$$\Delta_g = \lambda \Delta_{g'}.$$

The operator Ric'_g can be expressed in terms of the Lichnerowicz Laplacian; in fact, we have

(1.31) $$\mathrm{Ric}'_g h = \tfrac{1}{2}(\Delta h - \mathrm{Hess}\, \mathrm{Tr}\, h) - D^1 \mathrm{div}\, h,$$

for $h \in S^2 T^*$ (see, for example, Theorem 1.174 of [6]). Hence by (1.30), we obtain the equality

(1.32) $$\mathrm{Tr}\, \mathrm{Ric}'_g h = \Delta \mathrm{Tr}\, h - d^* \mathrm{div}\, h,$$

for $h \in S^2 T^*$. Thus if $h \in S^2 T^*$ satisfies $\mathrm{div}\, h = 0$, then we have

(1.33) $$\mathrm{Ric}'_g h = \tfrac{1}{2}(\Delta h - \mathrm{Hess}\, \mathrm{Tr}\, h), \qquad \mathrm{Tr}\, \mathrm{Ric}'_g h = \Delta \mathrm{Tr}\, h$$

(see [22, §1]). If f is a real-valued function on X, by (1.8) and (1.32) we see that the relation

(1.34) $$\mathrm{Tr}\, \mathrm{Ric}'_g(fg) = (n-1)\Delta f$$

holds.

§2. Einstein manifolds

We say that the Riemannian manifold (X, g) is an Einstein manifold if there is a real number λ such that $\mathrm{Ric} = \lambda g$. In this section, we suppose that g is an Einstein metric, i.e., that there is a real number λ such that $\mathrm{Ric} = \lambda g$; the scalar curvature $r(g)$ of (X, g) is constant and equal to $n\lambda$. By (1.29), we have

$$\mathrm{Ric} \lrcorner\, h = 2\lambda h,$$

for $h \in S^2 T^*$. In this case, we see that

$$\Delta h - 2\lambda h = \overline{\Delta} h + Lh,$$

for $h \in S^2 T^*$. If f is a real-valued function on X, it is easily seen that

$$(1.35) \qquad\qquad \mathrm{div}\,\mathrm{Hess}\, f = d\Delta f - \lambda df.$$

Let ξ be a vector field on X. According to (1.14) and (1.15), we have

$$(1.36) \qquad\qquad \mathrm{div}\, D_0 \xi = (\Delta + dd^*)\xi^\flat - 2\lambda \xi^\flat,$$

$$(1.37) \qquad\qquad \mathrm{div}\, D_0^c \xi = \Delta \xi^\flat - 2\lambda \xi^\flat + \frac{n-2}{n}\, dd^* \xi^\flat.$$

By the last relation of (1.18), we see that

$$(1.38) \qquad\qquad (\mathrm{Ric}'_g - \lambda \,\mathrm{id})\mathcal{L}_\xi g = \mathcal{L}_\xi \mathrm{Ric} - \lambda \mathcal{L}_\xi g = 0.$$

Hence if the section $h = D_0 \xi$ of $S^2 T^*$ satisfies $\mathrm{Tr}\, h = 0$, by (1.31) we have

$$(1.39) \qquad\qquad \tfrac{1}{2}\Delta h - D^1 \mathrm{div}\, h - \lambda h = 0.$$

Now suppose that ξ is a conformal Killing vector field. If the dimension of X is ≥ 2 and if the real-valued function f on X determined by the relation $\mathcal{L}_\xi g = fg$ is non-zero, then f is an eigenfunction of Δ with eigenvalue $n\lambda/(n-1)$; by (1.38), we see that

$$\mathrm{Ric}'_g(fg) = \lambda fg,$$

and so the fact that f is an eigenfunction of Δ may also be viewed as a consequence of formula (1.34).

Assertion (ii) of the following lemma is proved in [42, p. 649].

LEMMA 1.5. *Assume that (X, g) is a compact Einstein manifold and that $\mathrm{Ric} = \lambda g$, where $\lambda \neq 0$. Let ξ be a vector field on X.*

(i) *If ξ is a Killing vector field, it verifies the relation $\Delta \xi^\flat = 2\lambda \xi^\flat$.*

(ii) *If the dimension of X is ≥ 3 and if ξ is a conformal Killing vector field satisfying the relation $\Delta \xi^\flat = 2\lambda \xi^\flat$, then ξ is a Killing vector field.*

PROOF: If ξ is a Killing vector field, by (1.4) we know that $d^*\xi^\flat = 0$; then the relation $\Delta\xi^\flat = 2\lambda\xi^\flat$ is a consequence of (1.36). If $n \geq 3$ and ξ is a conformal Killing vector field satisfying the relation $\Delta\xi^\flat = 2\lambda\xi^\flat$, from formula (1.37) we deduce that $dd^*\xi^\flat = 0$; this last relation clearly implies that $d^*\xi^\flat = 0$, and so $D_0\xi = D_0^c\xi = 0$.

We shall consider the n-sphere S^n, with $n \geq 2$, viewed as the unit sphere in $\mathbb{R}^{n+1}$ endowed with its metric g_0 of constant curvature 1 induced by the Euclidean metric of $\mathbb{R}^{n+1}$. By an n-sphere of dimension n, we shall mean the sphere S^n endowed with the metric μg_0 of constant curvature μ, where μ is a positive real number.

In §4, Chapter II, we shall require the following result:

PROPOSITION 1.6. *Assume that (X, g) is a compact, connected Einstein manifold and that* $\mathrm{Ric} = \lambda g$, *where $\lambda > 0$. If X is not isometric to a sphere, then a conformal Killing vector field on X is a Killing vector field.*

PROOF: Let ξ be a conformal Killing vector field on X and let f be the real-valued function on X satisfying $\mathcal{L}_\xi g = fg$. Assume that ξ is not a Killing vector field; in other words, we suppose that the function f does not vanish. By Lemma 1.2, we know that f is an eigenfunction of the Laplacian Δ with eigenvalue $n\lambda/(n-1)$. Then Obata's theorem (see [48] or Theorem D.I.6 in Chapter III of [4]) tells us that X is isometric to a sphere.

LEMMA 1.7. *If (X, g) is an Einstein manifold, then we have*

$$\mathrm{Tr}\,\tilde{B} = \{0\}.$$

PROOF: If $x \in X$ and $\xi \in T_x^*$ satisfy $(\mathcal{L}_\xi g)(x) = 0$, then we have

$$\mathrm{Tr}\,(\mathcal{L}_\xi R)(x) = (\mathcal{L}_\xi \mathrm{Tr}\,R)(x) = -(\mathcal{L}_\xi \lambda g)(x) = 0.$$

According to Lemma 1.7, the trace mapping $\mathrm{Tr} : B \to S^2T^*$ induces, by passage to the quotient, a morphism of vector bundles (with variable fibers)

$$\mathrm{Tr} : B/\tilde{B} \to S^2T^*.$$

PROPOSITION 1.8. *Assume that (X, g) is an Einstein manifold and that* $\mathrm{Ric} = \lambda g$, *with $\lambda \in \mathbb{R}$. If $\tilde{B}$ is a vector bundle, then the diagram*

(1.40)
$$
\begin{array}{ccc}
S^2T^* & \xrightarrow{\ D_1\ } & B/\tilde{B} \\
\downarrow{\scriptstyle \mathrm{id}} & & \downarrow{\scriptstyle -\mathrm{Tr}} \\
S^2T^* & \xrightarrow{\mathrm{Ric}_g' - \lambda\,\mathrm{id}} & S^2T^*
\end{array}
$$

is commutative.

PROOF: Let h be a section of S^2T^* over X. If $x \in X$ and $h(x) = 0$, then by (1.20) and (1.21) we have

$$-\text{Tr}\,(D_1 h)(x) = -\text{Tr}\,\alpha(\mathcal{R}'_g h)(x) = -\text{Tr}\,(\mathcal{R}'_g h)(x) = (\text{Ric}'_g h)(x).$$

If ξ is a vector field on X, then we know that $D_1 \mathcal{L}_\xi g = 0$; on the other hand, by the last relation of (1.18), we know that (1.38) holds. If $x \in X$, by the surjectivity of $\sigma(D_0)$ we know that there exists a vector field ξ on X such that $(\mathcal{L}_\xi g)(x) = h(x)$. The commutativity of the diagram (1.40) is now a consequence of the previous observations.

Lemma 1.7 and Proposition 1.8 are proved in [25, §1]. If $\tilde{B}$ is a vector bundle, from Proposition 1.8 and relation (1.32) we obtain the equality

(1.41) $$-\text{Tr}\cdot\text{Tr}\,D_1 h = \Delta\text{Tr}\,h - \lambda\text{Tr}\,h + d^*\text{div}\,h,$$

for $h \in S^2T^*$.

LEMMA 1.9. Assume that (X, g) is a compact, connected Einstein manifold and that $\text{Ric} = \lambda g$, where $\lambda \neq 0$. Let h be an element of $C^\infty(S^2T^*)$ satisfying $\text{div}\,h = 0$ and

(1.42) $$\text{Tr}\,(\text{Ric}'_g h - \lambda h) = 0.$$

Then we have $\text{Tr}\,h = 0$.

PROOF: Since $\text{div}\,h = 0$, from (1.33) and (1.42) we obtain the equality

(1.43) $$\Delta\text{Tr}\,h = \lambda\text{Tr}\,h.$$

When λ is positive, Lichnerowicz's theorem (see [43, p. 135] or Theorem D.I.1 in Chapter III of [4]) tells us that the first non-zero eigenvalue of the Laplacian Δ acting on $C^\infty(X)$ is $\geq n\lambda/(n-1)$. Therefore from our hypothesis that $\lambda \neq 0$ and (1.43), we deduce that $\text{Tr}\,h = 0$.

The proof of Lemma 1.9 can be found in [3, §7]. The following lemma is a generalization of Proposition 3.2 of [22].

LEMMA 1.10. Assume that (X, g) is a compact, connected Einstein manifold and that $\text{Ric} = \lambda g$, where $\lambda \neq 0$. Suppose that $\tilde{B}$ is a vector bundle. Let N be a sub-bundle of B containing $\tilde{B}$ and E be a sub-bundle of S^2T^* satisfying $\text{Tr}\,N \subset E$ and $\text{Tr}\,E = \{0\}$. Let h be an element of $C^\infty(S^2T^*)$ satisfying

$$\text{div}\,h = 0, \qquad D_1 h \in C^\infty(N/\tilde{B}).$$

Then we have $\text{Tr}\,h = 0$ and

$$\Delta h - 2\lambda h \in C^\infty(E).$$

PROOF: Since $D_1 h$ is a section of $N/\tilde{B}$, by Proposition 1.8 our hypotheses on N and E imply that

$$(1.44) \qquad \mathrm{Ric}'_g\, h - \lambda h = -\mathrm{Tr}\, D_1 h \in C^\infty(E)$$

and hence that (1.42) holds. By Lemma 1.9, we have $\mathrm{Tr}\, h = 0$. Thus by (1.31) and (1.44), we see that

$$\Delta h - 2\lambda h = 2(\mathrm{Ric}'_g\, h - \lambda h) \in C^\infty(E).$$

The following lemma is due to Berger and Ebin [3] and Koiso [41].

LEMMA 1.11. *Assume that* (X, g) *is a compact, connected Einstein manifold and that* $\mathrm{Ric} = \lambda g$, *with* $\lambda \in \mathbb{R}$. *Let* $\varepsilon > 0$ *and* g_t *be a one-parameter family of Einstein metrics on* X *defined for* $|t| < \varepsilon$, *with* $g_0 = g$. *If the symmetric 2-form* $h = \frac{d}{dt} g_t{}_{|t=0}$ *satisfies the conditions*

$$(1.45) \qquad \mathrm{div}\, h = 0, \qquad \int_X \mathrm{Tr}\, h \cdot dX = 0,$$

then we have

$$(1.46) \qquad \Delta h = 2\lambda h, \qquad \mathrm{Tr}\, h = 0.$$

PROOF: For $|t| < \varepsilon$, we write $\mathrm{Ric}(g_t) = \lambda_t g_t$, where $\lambda_t \in \mathbb{R}$ and $\lambda_0 = \lambda$. This relation implies that

$$(1.47) \qquad \mathrm{Ric}'_g(h) = \lambda h + \lambda' g,$$

where $\lambda' = \frac{d}{dt} \lambda_t{}_{|t=0}$. Since $\mathrm{div}\, h = 0$, by the second formula of (1.33) we obtain

$$\Delta \mathrm{Tr}\, h = \lambda \mathrm{Tr}\, h + n\lambda'.$$

Since the integral of the function $\Delta \mathrm{Tr}\, h$ over X vanishes, from the second relation of (1.45) we deduce that $\lambda' = 0$. If $\lambda \neq 0$, by Lemma 1.9 we see that $\mathrm{Tr}\, h = 0$. If $\lambda = 0$, we have $\Delta \mathrm{Tr}\, h = 0$; therefore the function $\mathrm{Tr}\, h$ is constant and the second relation of (1.45) implies that it vanishes. By (1.33) and (1.47), with $\lambda' = 0$, we obtain the first equation of (1.46).

We now suppose that X is compact and connected, and we consider the space $\mathcal{M}$ of all Riemannian metrics on X. The space of elements h of $C^\infty(S^2 T^*)$ satisfying (1.45) can be identified with the "tangent space" to the subset of $\mathcal{M}$ consisting of all Riemannian metrics $\tilde{g}$ on X, which satisfy $\mathrm{Vol}(X, \tilde{g}) = \mathrm{Vol}(X, g)$ and belong to a subset of $\mathcal{M}$ transversal to the orbit of g under the group of diffeomorphisms of X. In fact, if g_t is a one-parameter family of Einstein metrics on X defined for $|t| < \varepsilon$, with $\varepsilon > 0$ and $g_0 = g$, satisfying $\mathrm{Vol}\,(X, g_t) = \mathrm{Vol}\,(X, g)$, then the infinitesimal deformation $h = \frac{d}{dt} g_t|_{t=0}$ of the family g_t satisfies the second relation of (1.45). In view of these remarks, the decomposition (1.11) and the above lemma, the space $E(X)$ of infinitesimal Einstein deformations of the metric g, introduced by Berger and Ebin in [3], is defined by

$$E(X) = \{\, h \in C^\infty(S^2 T^*) \mid \mathrm{div}\, h = 0, \ \mathrm{Tr}\, h = 0, \ \Delta h = 2\lambda h \,\}$$

(see also Koiso [41]). By definition, the space $E(X)$ is contained in an eigenspace of the Lichnerowicz Laplacian Δ, which is an elliptic operator, and is therefore finite-dimensional.

According to Lemma 1.7, when $\tilde{B}$ is a vector bundle, we may take $N = \tilde{B}$ and $E = \{0\}$ in Lemma 1.10; from the latter lemma and the relation between the space $H(X)$ and the cohomology of the sequence (1.24), we obtain the following result:

LEMMA 1.12. *Assume that (X, g) is a compact, connected Einstein manifold and that $\mathrm{Ric} = \lambda g$, where $\lambda \neq 0$. Suppose that $\tilde{B}$ is a vector bundle. Then the space $H(X)$ is finite-dimensional and is a subspace of $E(X)$. Moreover if $E(X) = \{0\}$, then the sequence (1.24) is exact.*

§3. Symmetric spaces

We say that the Riemannian manifold (X, g) is a locally symmetric space if $\nabla R = 0$. According to Lemma 1.4, if the equality

$$(1.48) \qquad\qquad H \cap (T^* \otimes \tilde{B}) = \{0\}$$

holds, then the manifold (X, g) is locally symmetric. Throughout this section, we shall suppose that the manifold (X, g) is a connected locally symmetric space. Since the set of local isometries of X acts transitively on X, we see that $\tilde{B}$ is a vector bundle. According to [13], the infinitesimal orbit of the curvature is equal to $\tilde{B}$, and so we have:

LEMMA 1.13. *Suppose that (X, g) is a connected locally symmetric space. Then $\tilde{B}$ is a vector bundle equal to the infinitesimal orbit of the curvature*

$$\{\, \rho(u)R \mid u \in g_1 \,\}.$$

We now suppose that (X, g) has constant curvature K; then we have

$$R(\xi_1, \xi_2, \xi_3, \xi_4) = K(g(\xi_1, \xi_4)g(\xi_2, \xi_3) - g(\xi_1, \xi_3)g(\xi_2, \xi_4)),$$

for $\xi_1, \xi_2, \xi_3, \xi_4 \in T$, and $\mathrm{Ric} = (n-1)Kg$. It follows directly from the definition of the vector bundle $\tilde{B}$ that it vanishes in this case. Thus if h

is a section of $S^2 T^*$ vanishing at $x \in X$, then by (1.25) we see that the equality $D_g h = D_1 h$ holds at $x \in X$. According to [13, §6], we know that

$$(1.49) \qquad\qquad D_g \circ D_0 = 0.$$

Since the mapping $\sigma(D_0)$ is an epimorphism, it follows that the operator D_g takes its values in B and that D_1 is equal to D_g. In this case, the complex (1.23) becomes the sequence

$$(1.50) \qquad\qquad T \xrightarrow{D_0} S^2 T^* \xrightarrow{D_g} B$$

introduced by Calabi [8], while the complex (1.24) becomes

$$(1.51) \qquad\qquad C^\infty(T) \xrightarrow{D_0} C^\infty(S^2 T^*) \xrightarrow{D_g} C^\infty(B).$$

The Lichnerowicz Laplacian Δ acting on symmetric 2-forms is given by

$$(1.52) \qquad\qquad \Delta h = \overline{\Delta} h + 2nKh - 2K(\operatorname{Tr} h) \cdot g,$$

for $h \in S^2 T^*$.

We assume moreover that X is a surface. Then we have $\operatorname{Ric} = \lambda g$, with $\lambda = K$, and B is a line bundle. Therefore the mapping

$$\operatorname{Tr} \cdot \operatorname{Tr} : B_x \to \mathbb{R}$$

is an isomorphism, for all $x \in X$. Thus according to the relation (1.41), if $h \in S^2 T^*$, the relation $D_1 h = 0$ holds if and only if the right-hand side of (1.41) vanishes; in particular, if $h \in S^2 T^*$ satisfies $\operatorname{div} h = 0$, then the relation $D_1 h = 0$ is equivalent to the equality $\Delta \operatorname{Tr} h = \lambda \operatorname{Tr} h$. If X is compact and $K \neq 0$, by Lemma 1.12 we obtain the equality

$$(1.53) \qquad H(X) = \{\, h \in C^\infty(S_0^2 T^*) \mid \operatorname{div} h = 0 \,\};$$

thus we see that the equality $H(X) = \{0\}$ implies that $E(X) = \{0\}$. On the other hand, if X is compact and $K = 0$, we obtain the equality

$$(1.54) \qquad H(X) = \{\, h \in C^\infty(S_0^2 T^*) \mid \operatorname{div} h = 0 \,\} \oplus \mathbb{R} \cdot g.$$

We now again suppose that (X, g) is an arbitrary connected locally symmetric space. If X is a surface, we have just seen that the vector bundle B is a line bundle and that the vector bundle $\tilde{B}$ vanishes. According to Lemma 1.13, we know that the vector bundle $\tilde{B}$ always satisfies

$$\operatorname{rank} \tilde{B} \leq \operatorname{rank} g_1 = \frac{n(n-1)}{2}.$$

If the dimension of X is ≥ 3, in §1 we saw that $\operatorname{Tr} B = S^2 T^*$, and so we obtain the inequalities

$$\operatorname{rank} \tilde{B} \leq \frac{n(n-1)}{2} < \frac{n(n+1)}{2} \leq \operatorname{rank} B.$$

Thus we see that, if the dimension of X is ≥ 2, the vector bundle $B/\tilde{B}$ is always non-zero.

Let Y be a totally geodesic submanifold of X. Let $i : Y \to X$ be the natural imbedding and $g_Y = i^* g$ be the Riemannian metric on Y induced by g. Then (Y, g_Y) is a locally symmetric space; its Riemann curvature tensor R_Y is equal to the section $i^* R$ of B_Y and the infinitesimal orbit of Y is equal to $\tilde{B}_Y$. For $x \in Y$, the diagram

$$
\begin{array}{ccc}
\bigwedge^2 T_x^* & \longrightarrow & B_x \\
\downarrow{\scriptstyle i^*} & & \downarrow{\scriptstyle i^*} \\
\bigwedge^2 T_{Y,x}^* & \longrightarrow & B_{Y,x}
\end{array}
$$

whose horizontal arrows send $\beta_1 \in \bigwedge^2 T_x^*$ and $\beta_2 \in \bigwedge^2 T_{Y,x}^*$ onto $\rho(\iota(\beta_1))R$ and $\rho(\iota(\beta_2))R_Y$, respectively, is commutative. Therefore we have the relation

$$(1.55) \qquad\qquad i^* \tilde{B}_x \subset \tilde{B}_{Y,x},$$

for $x \in Y$. Thus if Y is connected, the imbedding i induces a morphism of vector bundles

$$i^* : (B/\tilde{B})_{|Y} \to B_Y/\tilde{B}_Y.$$

If Y has constant curvature, from (1.55) and the fact that $\tilde{B}_Y = \{0\}$ we infer that

$$(1.56) \qquad\qquad i^* \tilde{B} = \{0\}.$$

PROPOSITION 1.14. *Suppose that X is a connected locally symmetric space. Let Y be a totally geodesic submanifold of X; let $i : Y \to X$ be the natural imbedding and $g_Y = i^* g$ be the Riemannian metric on Y induced by g. Let h be a section of $S^2 T^*$ over X.*

(i) *The equality*

$$(1.57) \qquad\qquad i^* D_g h = D_{g_Y}(i^* h)$$

holds on Y, and, when Y is connected, we have

$$(1.58) \qquad\qquad i^*(D_1 h)_{|Y} = D_{1,Y} i^* h.$$

(ii) *Assume that the manifold Y has constant curvature. If $x \in Y$ and $\xi_1, \xi_2, \xi_3, \xi_4 \in T_x$ are tangent to Y and if u is an element of B_x satisfying $\alpha u = (D_1 h)(x)$, then the equality*

$$(1.59) \qquad\qquad i^* u = i^*(D_g h)(x)$$

holds and we have

$$(1.60) \qquad\qquad u(\xi_1, \xi_2, \xi_3, \xi_4) = (D_g h)(\xi_1, \xi_2, \xi_3, \xi_4).$$

PROOF: Let ξ_1, ξ_2, ξ_3 be vectors of T_x, with $x \in Y$, which are tangent to Y; according to Theorems 7.2 and 4.2 in Chapter IV of [36], we know that the vector $\tilde{R}(\xi_1, \xi_2)\xi_3$ of T_x is tangent to Y. Since $R_Y = i^* R$, we see that

$$\tilde{R}_Y(\xi_1, \xi_2)\xi_3 = \tilde{R}(\xi_1, \xi_2)\xi_3.$$

The equality (1.57) now follows from Proposition 1.3 and formula (1.21). Now suppose that Y is connected and let h be a section of $S^2 T^*$. Let x be a point of Y and let u be an element of B_x satisfying $\alpha u = (D_1 h)(x)$. First, suppose that h vanishes at the point x of Y. Then $(D_g h)(x)$ is an element of B_x; by (1.25) and (1.58), we see that the vector $u - (D_g h)(x)$ of B_x belongs to $\tilde{B}_x$ and that

$$i^*(D_1 h)(x) = i^*(\alpha D_g h)(x) = \alpha(D_{g_Y} i^* h)(x) = (D_{1,Y} i^* h)(x).$$

If Y has constant curvature, by (1.56) we obtain the relation (1.59). Next, suppose there is a vector field ξ on X satisfying $\mathcal{L}_\xi g = h$. Then we have $D_1 h = 0$ and, if η is the vector field on Y determined by ξ, according to Lemma 1.1 we have $i^* h = \mathcal{L}_\eta g_Y$; thus we see that $D_{1,Y} i^* h = 0$. If Y has constant curvature, by (1.49) we know that $D_{g_Y} \mathcal{L}_\eta g_Y = 0$; thus by (1.57) we have $i^*(D_g h) = 0$. Therefore, under either one of the two assumptions imposed on h, we know that the equality (1.58) holds at x and that, if Y has constant curvature, the relation (1.59) also holds. As the mapping $\sigma(D_0)$ is an epimorphism, the preceding observation implies that the equality (1.58) always holds and that, if Y has constant curvature, the relation (1.59) also holds. The relation (1.60) is a consequence of (1.58).

LEMMA 1.15. *Assume that (X, g) is a connected locally symmetric space. Let Y be a totally geodesic submanifold of X of constant curvature. Let h be a section of $S^2 T^*$ over X. Let $x \in Y$ and u be an element of B_x such that $(D_1 h)(x) = \alpha u$. If the restriction of h to the submanifold Y is a Lie derivative of the metric on Y induced by g, then the restrictions of $D_g h$ and u to the submanifold Y vanish.*

PROOF: Let $i : Y \to X$ be the natural imbedding and $g_Y = i^* g$ be the Riemannian metric on Y induced by g. Assume that the restriction $i^* h$ of

h to the submanifold Y is equal to the Lie derivative $\mathcal{L}_\eta g_Y$ of g_Y along a vector field η on Y. Since Y has constant curvature, by (1.49) and (1.57) we have

$$i^* D_g h = D_{g_Y}(i^* h) = D_{g_Y} \mathcal{L}_\xi g_Y = 0.$$

From Proposition 1.14,(ii) and (1.59), we infer that $i^* u = 0$.

From Lemmas 1.1 and 1.15, we deduce the following result:

LEMMA 1.16. *Assume that (X, g) is a connected locally symmetric space. Let Y, Z be totally geodesic submanifolds of X; suppose that Z is a submanifold of Y of constant curvature. Let h be a section of $S^2 T^*$ over X. Let $x \in Z$ and u be an element of B_x such that $(D_1 h)(x) = \alpha u$. If the restriction of h to the submanifold Y is a Lie derivative of the metric on Y induced by g, then the restriction of u to the submanifold Z vanishes.*

We consider the third-order differential operator

$$D_2 = (\mathcal{D}\mathcal{R})'_g : S^2 T^* \to \mathcal{H}$$

and the differential operator

$$(D_2, D_1) : S^2 T^* \to \mathcal{H} \oplus \mathcal{B}/\tilde{\mathcal{B}}.$$

According to (1.18), we see that

$$(\mathcal{D}\mathcal{R})'_g(\mathcal{L}_\xi g) = 0,$$

for all vector fields ξ on X; thus we may consider the complex

(1.61) $$T \xrightarrow{D_0} S^2 T^* \xrightarrow{(D_2, D_1)} \mathcal{H} \oplus \mathcal{B}/\tilde{\mathcal{B}}.$$

We consider the connection

$$\nabla : \mathcal{B} \to T^* \otimes \mathcal{B}$$

and the first-order differential operator

$$\sigma \nabla : \mathcal{B} \to \wedge^3 T^* \otimes \wedge^2 T^*,$$

where σ is the morphism of vector bundles defined in §1. Since $\nabla R = 0$, by Lemma 1.13 we easily see that

$$\nabla \tilde{\mathcal{B}} \subset T^* \otimes \tilde{\mathcal{B}}$$

(see Lemma 7.3,(i) of [13]). Therefore the connection ∇ induces by passage to the quotient a connection

$$\nabla : \mathcal{B}/\tilde{\mathcal{B}} \to T^* \otimes \mathcal{B}/\tilde{\mathcal{B}}$$

in the vector bundle $B/\tilde{B}$. According to [13, §7], we have

$$(1.62) \qquad \nabla D_1 h = (\mathrm{id} \otimes \alpha) D_2 h,$$

for $h \in S^2 T^*$. Since X is locally homogeneous, the restriction

$$\sigma : T^* \otimes \tilde{B} \to {\textstyle\bigwedge}^3 T^* \otimes {\textstyle\bigwedge}^2 T^*$$

of the morphism σ has constant rank; we denote by B_1 its cokernel and by $\alpha : \bigwedge^3 T^* \otimes \bigwedge^2 T^* \to B_1$ the canonical projection. We easily see that there exists a unique first-order differential operator $D_1' : B/\tilde{B} \to B_1$ such that the diagram

$$
\begin{array}{ccc}
\mathcal{B} & \xrightarrow{\ \sigma\nabla\ } & \bigwedge^3 T^* \otimes \bigwedge^2 T^* \\
\downarrow{\scriptstyle \alpha} & & \downarrow{\scriptstyle \alpha} \\
\mathcal{B}/\tilde{\mathcal{B}} & \xrightarrow{\ D_1'\ } & \mathcal{B}_1
\end{array}
$$

commutes. The following result is given by Lemma 1.3 of [18].

LEMMA 1.17. *Suppose that* (X, g) *is a connected locally symmetric space. Then the sequence*

$$\mathcal{T} \xrightarrow{\ D_0\ } S^2 T^* \xrightarrow{\ D_1\ } \mathcal{B}/\tilde{\mathcal{B}} \xrightarrow{\ D_1'\ } \mathcal{B}_1$$

is a complex.

PROOF: Since the sequence (1.23) is a complex, it suffices to show that $D_1' \cdot D_1 = 0$. Let h be an element of $S^2 T^*$ and u be an element of $\tilde{B}$ satisfying $D_1 h = \alpha u$. Then we have $\nabla D_1 h = (\mathrm{id} \otimes \alpha) \nabla u$, and so by (1.62) we see that $\nabla u - D_2 h$ is an element of $T^* \otimes \tilde{B}$. Since $\sigma D_2 h = 0$, we have

$$D_1' D_1 h = \alpha \sigma \nabla u = \alpha \sigma (\nabla u - D_2 h) = 0.$$

According to Theorem 7.2 of [13], the complex (1.61) is exact and formally exact in the sense of [7, Chapter X] (see also [28] and [13]); hence the differential operator (D_2, D_1) is the compatibility condition of the Killing operator D_0. Therefore if Θ is the sheaf of Killing vector fields on X, that is, the kernel of the operator $D_0 : \mathcal{T} \to S^2 T^*$, the cohomology of the complex

$$(1.63) \qquad C^\infty(T) \xrightarrow{\ D_0\ } C^\infty(S^2 T^*) \xrightarrow{(D_2, D_1)} C^\infty(H \oplus B/\tilde{B})$$

is isomorphic to the cohomology group $H^1(X, \Theta)$. According to [13, §7], the sheaf Θ is locally constant and, when X is simply-connected, the sequence (1.63) is exact. The natural injective mapping from the cohomology of the complex (1.63) to the cohomology of the sequence (1.24) induces an injective mapping from the cohomology group $H^1(X, \Theta)$ to the cohomology of the complex (1.24).

According to Theorem 7.3 of [13], if the equality (1.48) holds, a section h of $S^2 T^*$ satisfying $D_1 h = 0$ also satisfies the equation $D_2 h = 0$. Therefore if (1.48) is true, the complex (1.23) is exact and the two mappings considered above involving cohomology groups are isomorphisms. Moreover, if the equality (1.48) holds and X is simply-connected, the sequence (1.24) is exact; on the other hand, if the equality (1.48) holds and X is compact, the cohomology group $H^1(X, \Theta)$ is isomorphic to $H(X)$.

The universal covering manifold $\tilde{X}$ of X endowed with the Riemannian metric induced by g is a locally symmetric space. If $\tilde{X}$ is a finite covering of X, then X is equal to the quotient $\tilde{X}/\Gamma$, where Γ is a finite subgroup of the group of isometries of $\tilde{X}$ which acts without fixed points; since the sequence (1.63) for the manifold $\tilde{X}$ is exact, the sequence (1.63) for the manifold X is also exact. If X is a symmetric space of compact type, the manifold $\tilde{X}$ is also a symmetric space of compact type and the covering mapping $\tilde{X} \to X$ is finite (see Chapters IV and V of [36]). These observations, together with the discussion which follows Lemma 1.17, give us the following:

THEOREM 1.18. *Suppose that (X, g) is a connected locally symmetric space satisfying one of the following conditions:*
   (i) *the covering mapping $\tilde{X} \to X$ is finite;*
   (ii) *(X, g) is a symmetric space of compact type.*
Then the sequence (1.63) is exact. If the equality (1.48) holds, the sequence (1.24) is also exact.

We now suppose that (X, g) has constant curvature K; then the vector bundle $\tilde{B}$ vanishes and so the equality (1.48) holds. Thus the complex (1.50) is exact. If X is simply-connected, the sequence (1.51) is also exact. These two results were first proved by Calabi [8] (see also [2]); other direct proofs are given in [13, §6]. Furthermore if (X, g) is a compact surface of constant curvature K, from the equalities (1.53) and (1.54) we obtain the following proposition, whose first assertion is given by [2, p. 24].

PROPOSITION 1.19. *Let (X, g) be a compact surface of constant curvature K.*
   (i) *The cohomology group $H^1(X, \Theta)$ is isomorphic to the space*
$$\{ h \in C^\infty(S_0^2 T^*) \mid \operatorname{div} h = 0 \}$$
when $K \neq 0$, and to the space
$$\{ h \in C^\infty(S_0^2 T^*) \mid \operatorname{div} h = 0 \} \oplus \mathbb{R} \cdot g$$
when $K = 0$.

(ii) If $K \neq 0$ and the cohomology group $H^1(X, \Theta)$ vanishes, then we have $E(X) = \{0\}$.

We now suppose that X is equal to the sphere (S^n, g_0) of constant curvature 1, which is a symmetric space of compact type. We know that the sequences (1.24) and (1.51) are exact and that the cohomology groups $H^1(X, \Theta)$ and $H(X)$ vanish. When X is equal to the sphere (S^2, g_0), according to Proposition 1.19,(i) and the decomposition (1.12) we see that

$$(1.64) \qquad\qquad C^\infty(S_0^2 T^*) = D_0^c C^\infty(T);$$

therefore, for $h \in C^\infty(S^2 T^*)$, we may write

$$h = \mathcal{L}_\xi g + f g,$$

where ξ is a vector field and f is a real-valued function on $X = S^2$. Moreover, by Proposition 1.19,(ii), we have $E(X) = \{0\}$ when $X = S^2$.

If (X, g) is a compact manifold with positive constant curvature, then its universal covering manifold is isometric to $(S^n, \mu g_0)$, where μ is a positive real number. Thus from Theorem 1.18 and the above results, we obtain the following:

PROPOSITION 1.20. *Let (X, g) is a compact manifold with positive constant curvature. Then cohomology group $H^1(X, \Theta)$ and the space $H(X)$ vanish, and the sequence (1.51) is exact.*

We now suppose that (X, g) is a symmetric space of compact type. Then there is a Riemannian symmetric pair (G, K) of compact type, where G is a compact, connected semi-simple Lie group and K is a closed subgroup of G, such that the space X is isometric to the homogeneous space G/K endowed with a G-invariant metric. We identify X with G/K, and let x_0 be the point of X corresponding to the coset of the identity element of G in G/K. If $\mathfrak{g}_0$ and $\mathfrak{k}_0$ are the Lie algebras of G and K, respectively, we consider the Cartan decomposition $\mathfrak{g}_0 = \mathfrak{k}_0 \oplus \mathfrak{p}_0$ corresponding to the Riemannian symmetric pair (G, K), where $\mathfrak{p}_0$ is a subspace of $\mathfrak{g}_0$. We identify $\mathfrak{p}_0$ with the tangent space to X at the point x_0. If B is the Killing form of the Lie algebra $\mathfrak{g}_0$, the restriction of $-B$ to $\mathfrak{p}_0$ induces a G-invariant Riemannian metric g_0 on X. According to Theorem 7.73 of [6], we know that

$$(1.65) \qquad\qquad \mathrm{Ric}(g_0) = \tfrac{1}{2} g_0.$$

If X is an irreducible symmetric space, the metric g is a positive multiple of g_0 and is therefore an Einstein metric. The Ricci tensor of g is equal to $\mathrm{Ric} = \lambda g$, where λ is a positive real number; by (1.65), we see that $g_0 = 2\lambda g$. Thus the Lichnerowicz Laplacians Δ and Δ_{g_0} corresponding to the metrics g and g_0, respectively, are related by $\Delta = 2\lambda \Delta_{g_0}$.

LEMMA 1.21. *Let (X, g) be an irreducible symmetric space of compact type. Then g is an Einstein metric and there is a positive real number λ such that* $\mathrm{Ric} = \lambda g$; *moreover, the metric g_0 induced by the Killing form of $\mathfrak{g}_0$ is equal to $2\lambda g$.*

In [41] and [42], Koiso proved the following:

THEOREM 1.22. *Let X be an irreducible symmetric space of compact type whose universal covering manifold is not equal to one of the following:*
- (i) $SU(n+1)$, *with* $n \geq 2$;
- (ii) $SU(n)/SO(n)$, *with* $n \geq 3$;
- (iii) $SU(2n)/Sp(n)$, *with* $n \geq 3$;
- (iv) $SU(p+q)/S(U(p) \times U(q))$, *with* $p, q \geq 2$;
- (v) E_6/F_4.

Then we have $E(X) = \{0\}$.

Some of the methods used by Koiso to prove Theorem 1.22 will be described in §7, Chapter II; in fact, we shall give an outline of the proof of this theorem for an irreducible symmetric space X which is not equal to a simple Lie group. According to the remarks preceding Theorem 1.18, from Lemma 1.12 and Theorem 1.22 we deduce:

THEOREM 1.23. *Let X be an irreducible symmetric space of compact type whose universal covering manifold is not equal to one of the spaces* (i)–(v) *of Theorem 1.22. Then the sequence* (1.24) *is exact.*

In [13] and [10], the equality (1.48) is proved for the complex projective space $\mathbb{CP}^n$, with $n \geq 2$, and the complex quadric of dimension ≥ 3 (see Propositions 3.32 and 5.14). Thus from Theorem 1.18, without the use of the space of infinitesimal Einstein deformations, we obtain the exactness of the sequence (1.24) when X is an irreducible symmetric space of compact type equal either to a sphere, to a real or complex projective space, or to the complex quadric of dimension ≥ 3. Theorem 1.23 also gives us the exactness of the sequence (1.24) for these irreducible symmetric spaces other than the complex quadric of dimension 4. In fact, *we conjecture that the equality* (1.48) *holds for any irreducible symmetric space.*

§4. Complex manifolds

In this section, we suppose that X is a complex manifold endowed with a Hermitian metric g. We consider the sub-bundles T' and T'' of $T_{\mathbb{C}}$ of complex vector fields of type $(1, 0)$ and $(0, 1)$, respectively; then we have the decomposition

$$T_{\mathbb{C}} = T' \oplus T''.$$

The complex structure J of X induces involutions

$$J : \bigwedge^2 T^* \to \bigwedge^2 T^*, \qquad J : S^2 T^* \to S^2 T^*, \qquad J : T^* \otimes T \to T^* \otimes T$$

defined by

$$\beta^J(\xi,\eta) = \beta(J\xi, J\eta), \qquad h^J(\xi,\eta) = h(J\xi, J\eta), \qquad u^J(\xi) = -Ju(J\xi),$$

for $\beta \in \bigwedge^2 T^*$, $h \in S^2 T^*$, $u \in T^* \otimes T$ and $\xi, \eta \in T$. Then g_1 is stable under the involution J of $T^* \otimes T$ and the sub-bundle B of $\bigwedge^2 T^* \otimes \bigwedge^2 T^*$ is stable under the involution

$$J = J \otimes J : \bigwedge^2 T^* \otimes \bigwedge^2 T^* \to \bigwedge^2 T^* \otimes \bigwedge^2 T^*.$$

We then obtain the orthogonal decompositions

$$(1.66) \qquad
\begin{aligned}
\bigwedge^2 T^* &= T_{\mathbb{R}}^{1,1} \oplus (\bigwedge^2 T^*)^-, & S^2 T^* &= (S^2 T^*)^+ \oplus (S^2 T^*)^-, \\
B &= B^+ \oplus B^-, & g_1 &= g_1^+ \oplus g_1^-
\end{aligned}$$

into direct sums of the eigenbundles $T_{\mathbb{R}}^{1,1}$, $(\bigwedge^2 T^*)^-$, $(S^2 T^*)^+$, $(S^2 T^*)^-$, B^+, B^-, g_1^+ and g_1^- corresponding to the eigenvalues $+1$ and -1, respectively, of the involutions J. In fact, $T_{\mathbb{R}}^{1,1}$ is the bundle of real forms of type $(1,1)$, while $(S^2 T^*)^+$ and $(S^2 T^*)^-$ are the bundles of (real) Hermitian and skew-Hermitian symmetric 2-forms on X, respectively. It is easily verified that

$$\operatorname{Tr}(B^+) \subset (S^2 T^*)^+, \qquad \operatorname{Tr}(B^-) \subset (S^2 T^*)^-.$$

Clearly we have

$$g_1^+ = \{\, u \in g_1 \mid u \circ J = J \circ u \,\};$$

hence the fiber of this vector bundle g_1^+ at $x \in X$ is equal to the Lie algebra of the unitary group of the Hermitian vector space $(T_x, J(x), g(x))$.

We consider the morphism of vector bundles

$$(1.67) \qquad\qquad \bigotimes^2 T^* \to \bigotimes^2 T^*,$$

sending $u \in \bigotimes^2 T^*$ into the element $\breve{u}$ of $\bigotimes^2 T^*$ defined by

$$\breve{u}(\xi,\eta) = u(J\xi, \eta),$$

for all $\xi, \eta \in T$. Clearly, the square of this morphism is equal to $-\mathrm{id}$ and so (1.67) is an isomorphism. We easily verify that the isomorphism (1.67) induces isomorphisms of vector bundles

$$
\begin{aligned}
(S^2 T^*)^+ &\to T_{\mathbb{R}}^{1,1}, & T_{\mathbb{R}}^{1,1} &\to (S^2 T^*)^+, \\
(1.68) \qquad (S^2 T^*)^- &\to (S^2 T^*)^-, & (\bigwedge^2 T^*)^- &\to (\bigwedge^2 T^*)^-.
\end{aligned}$$

The squares of the isomorphisms (1.68) are equal to $-\mathrm{id}$. The metric g is a section of $(S^2T^*)^+$ and its image under the isomorphism (1.67) is a section of $T_{\mathbb{R}}^{1,1}$, which is the Kähler form ω of X.

Let $S^{p,q}T^*$ (resp. $\bigwedge^{p,q}T^*$) be the bundle of complex differential symmetric (resp. exterior differential) forms of degree $p + q$ and of type (p, q) on X. Then $S^{1,1}T^*$ and $\bigwedge^{1,1}T^*$ are the complexifications of the bundles $(S^2T^*)^+$ and $T_{\mathbb{R}}^{1,1}$, respectively. The eigenbundles corresponding to the eigenvalues $+i$ and $-i$ of the endomorphism of $(S^2T^*)_{\mathbb{C}}^-$ (resp. of $(\bigwedge^2T^*)_{\mathbb{C}}^-$) induced by the mapping (1.68) are the bundles $S^{2,0}T^*$ and $S^{0,2}T^*$ (resp. the bundles $\bigwedge^{2,0}T^*$ and $\bigwedge^{0,2}T^*$), respectively. We write $T^{1,0} = \bigwedge^{1,0}T^*$ and $T^{0,1} = \bigwedge^{0,1}T^*$. Then we have the orthogonal decompositions

$$(S^2T^*)_{\mathbb{C}}^- = S^{2,0}T^* \oplus S^{0,2}T^*, \qquad (\textstyle\bigwedge^2T^*)_{\mathbb{C}}^- = \bigwedge^{2,0}T^* \oplus \bigwedge^{0,2}T^*,$$

$$(1.69) \quad T_{\mathbb{C}}^* = T^{1,0} \oplus T^{0,1}, \qquad S^2T_{\mathbb{C}}^* = S^{2,0}T^* \oplus S^{1,1}T^* \oplus S^{0,2}T^*.$$

The isomorphism of vector bundles $g^\flat : T \to T^*$ determined by the Hermitian metric g induces isomorphisms of vector bundles

$$(1.70) \qquad g^\flat : T' \to T^{0,1}, \qquad g^\flat : T'' \to T^{1,0}.$$

We consider the natural projections

$$\pi_+ : S^2T^* \to (S^2T^*)^+, \quad \pi' : S^2T_{\mathbb{C}}^* \to S^{2,0}T^*, \quad \pi'' : S^2T_{\mathbb{C}}^* \to S^{0,2}T^*$$

determined by the decompositions (1.66) and (1.69). Since the metric g is a section of $(S^2T^*)^+$, if f is a real-valued function on X, we see that

$$(1.71) \qquad \operatorname{Tr} \pi_+ \operatorname{Hess} f = \operatorname{Tr} \operatorname{Hess} f = -\Delta f.$$

The isomorphism of vector bundles $\iota : \bigwedge^2T^* \to \mathfrak{g}_1$ induces isomorphisms

$$\iota : T_{\mathbb{R}}^{1,1} \to \mathfrak{g}_1^+, \qquad \iota : (\textstyle\bigwedge^2T^*)^- \to \mathfrak{g}_1^-.$$

If β is an element of $(\bigwedge^2T^*)^-$, we easily verify that

$$(1.72) \qquad \rho(\iota(\beta))\omega = 2\check{\beta}.$$

Let π be the endomorphism of the vector bundle $\bigwedge^2T^* \otimes \bigwedge^2T^*$ defined by

$$\pi(\theta_1 \otimes \theta_2)(\xi_1, \xi_2, \xi_3, \xi_4) = 2\theta_1(\xi_1, \xi_2)\theta_2(\xi_3, \xi_4) + 2\theta_1(\xi_3, \xi_4)\theta_2(\xi_1, \xi_2)$$
$$+ \theta_1(\xi_1, \xi_3)\theta_2(\xi_2, \xi_4) + \theta_1(\xi_2, \xi_4)\theta_2(\xi_1, \xi_3)$$
$$- \theta_1(\xi_2, \xi_3)\theta_2(\xi_1, \xi_4) - \theta_1(\xi_1, \xi_4)\theta_2(\xi_2, \xi_3),$$

for $\theta_1, \theta_2 \in \bigwedge^2 T^*$ and $\xi_1, \xi_2, \xi_3, \xi_4 \in T$. We easily verify that the image of the morphism π is contained in B. We consider the morphism of vector bundles

$$\psi : \bigwedge^2 T^* \to B$$

defined by

$$\psi(\theta) = \pi(\omega \otimes \theta),$$

for $\theta \in \bigwedge^2 T^*$. Clearly, we have

$$\psi(T_{\mathbb{R}}^{1,1}) \subset B^+, \qquad \psi((\bigwedge^2 T^*)^-) \subset B^-.$$

LEMMA 1.24. *The morphism ψ is injective.*

PROOF: Let θ be an element of $\bigwedge^2 T^*$ and let ξ, η be vectors of T. Then we have

$$\psi(\theta)(\xi, \eta, \xi, J\xi) = 3(\omega(\xi, \eta)\theta(\xi, J\xi) + \omega(\xi, J\xi)\theta(\xi, \eta)).$$

This formula implies that

$$\psi(\theta)(\xi, J\xi, \xi, J\xi) = 6\omega(\xi, J\xi)\theta(\xi, J\xi)$$

and, when ξ is orthogonal to $J\eta$, that

$$\psi(\theta)(\xi, \eta, \xi, J\xi) = 3\omega(\xi, J\xi)\theta(\xi, \eta).$$

We now suppose that θ belongs to the kernel of ψ. From these two formulas, we infer that $\theta(\xi, J\xi) = 0$, for all $\xi \in T$, and that $\theta(\xi, \eta) = 0$, for all $\xi, \eta \in T$, whenever η is orthogonal to $J\xi$. Thus we see that $\theta = 0$.

We shall consider the morphism of vector bundles

$$\check{\psi} : (\bigwedge^2 T^*)^- \to B$$

defined by

$$\check{\psi}(\beta) = \psi(\check{\beta}),$$

for $\beta \in (\bigwedge^2 T^*)^-$.

We now suppose that (X, g) is a Kähler manifold. If h is a section of $(S^2 T^*)^+$, we easily verify that

(1.73) $$\operatorname{div} h = i(\bar{\partial}^* - \partial^*)\check{h}.$$

The following result is given by Lemma 1.1 of [21].

LEMMA 1.25. *Suppose that (X, g) is a Kähler manifold and let f be an element of $C^\infty(X)$. If h is the section $\pi_+ \text{Hess} f$ of $(S^2 T^*)_{\mathbb{C}}^+$, then we have*

$$\check{h} = i \partial \bar{\partial} f.$$

Let f be an element of $C^\infty(X)$. We easily verify that

(1.74)
$$D^1 \partial f = \pi' \text{Hess} f + \tfrac{1}{2} \pi_+ \text{Hess} f,$$
$$D^1 \bar{\partial} f = \pi'' \text{Hess} f + \tfrac{1}{2} \pi_+ \text{Hess} f.$$

Therefore, we have

(1.75)
$$\pi_+ D^1 \partial f = \pi_+ D^1 \bar{\partial} f = \tfrac{1}{2} \pi_+ \text{Hess} f,$$
$$\pi' \text{Hess} f - \pi'' \text{Hess} f = D^1 (\partial f - \bar{\partial} f).$$

By (1.4), from this last equality we infer that the section $\pi' \text{Hess} f - \pi'' \text{Hess} f$ belongs to the space $D_0 C^\infty(T_{\mathbb{C}})$. According to (1.73) and Lemma 1.25, we easily see that

(1.76)
$$\text{div} \, \pi_+ \text{Hess} f = \tfrac{1}{2} d\Delta f.$$

Thus if (X, g) is a Kähler-Einstein manifold and $\text{Ric} = \lambda g$, where $\lambda \in \mathbb{R}$, by (1.35) we have

(1.77) $\quad \text{div} \, \pi' \text{Hess} f = \tfrac{1}{2} \partial \Delta f - \lambda \partial f, \qquad \text{div} \, \pi'' \text{Hess} f = \tfrac{1}{2} \bar{\partial} \Delta f - \lambda \bar{\partial} f.$

Suppose that (X, g) is a Hermitian symmetric space. Since (X, g) is a Kähler manifold, its curvature $\tilde{R}$ verifies the relation

$$\tilde{R}(\xi, \eta) = R(\xi, \eta),$$

for all $\xi, \eta \in T$, and we have

(1.78)
$$\rho(J)R = 0,$$
$$(\rho(u)R)^J = \rho(u^J)R,$$

for $u \in T^* \otimes T$. From Lemma 1.13 and the decompositions (1.66), we obtain the equality

(1.79)
$$\tilde{B} = \tilde{B}^+ \oplus \tilde{B}^-,$$

where

$$\tilde{B}^+ = \tilde{B} \cap B^+ = \rho(g_1^+)R, \qquad \tilde{B}^- = \tilde{B} \cap B^- = \rho(g_1^-)R.$$

CHAPTER II

RADON TRANSFORMS ON SYMMETRIC SPACES

§1. Outline

In this chapter, we introduce the Radon transforms for functions and symmetric forms on a symmetric space (X, g) of compact type, namely the X-ray transform and the maximal flat transform. In §2, we present results concerning harmonic analysis on homogeneous spaces and use them to study these Radon transforms in §5 and to describe properties of certain spaces of symmetric forms in §7. The notions of rigidity in the sense of Guillemin and of infinitesimal rigidity of the space X are introduced in §3; in this section, we also state the fundamental result of Guillemin [35] concerning isospectral deformations of the metric g of X (Theorem 2.14). In §4, we present Grinberg's theorem concerning the injectivity of the maximal flat Radon transform for functions on X; when the space (X, g) is irreducible, from this result we infer that, if the space X is rigid in the sense of Guillemin, it is necessarily equal to its adjoint space. In §5, criteria for the rigidity of the space X are given in terms of harmonic analysis. Some lemmas concerning irreducible G-modules, where G is a compact semi-simple Lie group, proved in §6 are used in our study of symmetric forms on an irreducible symmetric space presented in §7. Results concerning the space of infinitesimal Einstein deformations of an irreducible symmetric space can be found in §7. Our criteria for the infinitesimal rigidity or the rigidity in the sense of Guillemin of an irreducible symmetric space are given in §8.

§2. Homogeneous vector bundles and harmonic analysis

Let (X, g) be a Riemannian manifold which may be written as a homogeneous space G/K, where G is a compact Lie group and K is a closed subgroup of G. We assume that the group G acts by isometries on the Riemannian manifold X. If F is a homogeneous vector bundle over X, then the space $C^\infty(F)$ is a G-module.

Let F be a complex homogeneous vector bundle over X endowed with a Hermitian scalar product. We endow the space $C^\infty(F)$ of sections of F over X with the Hermitian scalar product obtained from the scalar product on F and the Riemannian measure dX of X. If the vector bundle F is unitary in the sense of [56, §2.4], then the space $C^\infty(F)$ is a unitary G-module. Let x_0 be the point of X corresponding to the coset of the identity element of G. The action of G on the fiber F_0 of F at the point x_0 of X induces a representation τ of K on F_0. Then F is isomorphic to the homogeneous vector bundle $G \times_\tau F_0$ and we shall identify these two homogeneous vector bundles. The representation τ is unitary if and only if the vector bundle F is unitary.

We henceforth suppose that F is a unitary homogeneous vector bundle. If $C^\infty(G; F_0)$ is the space of functions on G with values in F_0, we consider its subspace

$$C^\infty(G; \tau) = \{\, f \in C^\infty(G; F_0) \mid f(ak) = \tau(k)^{-1} f(a), \text{ for } a \in G,\ k \in K \,\}$$

and we write

$$(\pi(a_1) f)(a) = f(a_1^{-1} a),$$

for $a, a_1 \in G$ and $f \in C^\infty(G; \tau)$. Then π is representation of G on the space $C^\infty(G; \tau)$ and the mapping

$$A : C^\infty(F) \to C^\infty(G; \tau),$$

defined by

$$(Au)(a) = a^{-1} u(aK),$$

for $u \in C^\infty(F)$ and $a \in G$, is an isomorphism of G-modules. In particular, if K is the subgroup $\{e\}$ of G, where e is the identity element of G, and $F_0 = \mathbb{C}$, then π is a representation of G on the space $C^\infty(G)$ which gives us a structure of G-module on $C^\infty(G)$.

Let $\hat{G}$ be the dual of the group G, that is, the set of equivalence classes of irreducible G-modules over $\mathbb{C}$. For $\gamma \in \hat{G}$, let V_γ be a representative of γ; the mapping

$$\iota_\gamma : V_\gamma \otimes \operatorname{Hom}_K(V_\gamma, F_0) \to C^\infty(G; \tau),$$

defined by

$$\iota_\gamma(v \otimes \varphi)(a) = \varphi(a^{-1} v),$$

for all $v \in V_\gamma$, $\varphi \in \operatorname{Hom}_K(V_\gamma, F_0)$ and $a \in G$, is injective. According to the Frobenius reciprocity theorem, the image $C^\infty_\gamma(F)$ of the mapping $A^{-1} \circ \iota_\gamma$ is a finite-dimensional G-submodule of $C^\infty(F)$, which depends only on γ and is isomorphic to the direct sum of m copies of V_γ, where m is the integer $\dim \operatorname{Hom}_K(V_\gamma, F_0)$. If W is a G-submodule of $C^\infty(F)$, the image of the mapping

$$V_\gamma \otimes \operatorname{Hom}_G(V_\gamma, W) \to C^\infty(F),$$

which sends $v \otimes \varphi$ into $\varphi(v)$, for $v \in V_\gamma$ and $\varphi \in \operatorname{Hom}_G(V_\gamma, W)$, is a G-submodule of W called the isotypic component of W corresponding to γ. In fact, the isotypic component of $C^\infty(F)$ corresponding to γ is equal to $C^\infty_\gamma(F)$. A G-submodule W of $C^\infty_\gamma(F)$ is therefore isomorphic to the direct sum of k copies of V_γ, with $k \leq \dim \operatorname{Hom}_K(V_\gamma, F_0)$; this integer k is called the multiplicity of the G-module W and denoted by $\operatorname{Mult} W$. According to Schur's lemma, if the representation τ of K is irreducible, the multiplicity $\dim \operatorname{Hom}_K(V_\gamma, F_0)$ of $C_\gamma^\infty(F)$ is equal to the multiplicity of the representation τ in the decomposition of V_γ into irreducible K-modules. For $\gamma, \gamma' \in \hat{G}$, with $\gamma \neq \gamma'$, the submodules $C_\gamma^\infty(F)$ and $C_{\gamma'}^\infty(F)$ of $C^\infty(F)$ are orthogonal (see [56, §5.3]). For $\gamma \in \hat{G}$, we denote by P_γ the orthogonal projection of $C^\infty(F)$ onto its G-submodule $C_\gamma^\infty(F)$. The following proposition is a direct consequence of Theorem 5.3.6 of [56].

PROPOSITION 2.1. *The direct sum* $\bigoplus_{\gamma \in \hat{G}} C_\gamma^\infty(F)$ *is a dense submodule of* $C^\infty(F)$.

The vector bundle $\bigotimes^p T^* \otimes \bigotimes^q T$ is a homogeneous vector bundle and its complexification is a unitary homogeneous vector bundle. Thus the vector bundles $T_\mathbb{C}$, $\bigotimes^k T_\mathbb{C}^*$, $\bigwedge^j T_\mathbb{C}^*$ and $S^p T_\mathbb{C}^*$ are unitary homogeneous G-vector bundles over X, and we consider the unitary G-modules $C^\infty(T_\mathbb{C})$ and $C^\infty(\bigotimes^k T_\mathbb{C}^*)$. Moreover, we know that $C^\infty(T)$ is a G-submodule of $C^\infty(T_\mathbb{C})$, while

$$C^\infty(\textstyle\bigwedge^k T_\mathbb{C}^*), \quad C^\infty(S^k T_\mathbb{C}^*), \quad C^\infty(\bigotimes^k T^*), \quad C^\infty(\bigwedge^k T^*), \quad C^\infty(S^k T^*)$$

are G-submodules of $C^\infty(\bigotimes^k T_\mathbb{C}^*)$. For all $\gamma \in \hat{G}$, the isomorphism of vector bundles $g^\flat : T \to T^*$ induces isomorphisms of G-modules

$$g^\flat : C_\gamma^\infty(T_\mathbb{C}) \to C_\gamma^\infty(T_\mathbb{C}^*).$$

Let $\mathfrak{g}$ and $\mathfrak{k}$ be the Lie algebras of G and K, respectively. In this section, we henceforth suppose that G/K is a reductive homogeneous space; this means that there is an $\operatorname{Ad}(K)$-invariant complement of $\mathfrak{k}$ in $\mathfrak{g}$. This assumption always holds when the compact group G is connected and semisimple and (G, K) is a Riemannian symmetric pair of compact type. Let F_1, F_2, F_3 be complex homogeneous vector bundles over X endowed with Hermitian scalar products. Assume that these vector bundles are unitary. Let

$$D : \mathcal{F}_1 \to \mathcal{F}_2$$

be a homogeneous differential operator. Then we have

(2.1) $$DP_\gamma = P_\gamma D,$$

for $\gamma \in \hat{G}$; therefore the morphism of G-modules

$$D : C^\infty(F_1) \to C^\infty(F_2)$$

induces by restriction a morphism of G-modules

$$D : C_\gamma^\infty(F_1) \to C_\gamma^\infty(F_2),$$

for $\gamma \in \hat{G}$. We consider the subspace

$$\operatorname{Ker} D = \{ u \in C^\infty(F_1) \mid Du = 0 \}$$

of $C^\infty(F_1)$; according to the relation (2.1), if u is an element of $\operatorname{Ker} D$, then $P_\gamma u$ also belongs to $\operatorname{Ker} D$, for all $\gamma \in \hat{G}$. If D is elliptic, we recall that $DC^\infty(F_1)$ is a closed subspace of $C^\infty(F_2)$; the following proposition is a consequence of the results of [56, §5.7] and in particular its Lemma 5.7.7.

PROPOSITION 2.2. *Let $D : C^\infty(F_1) \to C^\infty(F_2)$ be a homogeneous differential operator. Then the following assertions hold:*

(i) *The direct sum $\bigoplus_{\gamma \in \hat{G}} (C^\infty(F_1) \cap \operatorname{Ker} D)$ is a dense subspace of $\operatorname{Ker} D$; in fact, an element u of $C^\infty(F_1)$ belongs to $\operatorname{Ker} D$ if and only if $P_\gamma u$ belongs to $\operatorname{Ker} D$, for all $\gamma \in \hat{G}$.*

(ii) *The direct sum $\bigoplus_{\gamma \in \hat{G}} DC_\gamma^\infty(F_1)$ is a dense subspace of $DC^\infty(F_1)$.*

(iii) *If D is elliptic, then the closure of the direct sum $\bigoplus_{\gamma \in \hat{G}} DC_\gamma^\infty(F_1)$ in $C^\infty(F_2)$ is equal to $DC^\infty(F_1)$; in fact, if u is an element of $C^\infty(F_2)$ and $P_\gamma u$ belongs to $DC_\gamma^\infty(F_1)$, for all $\gamma \in \hat{G}$, then u belongs to $DC^\infty(F_1)$.*

The following proposition is given by Proposition 2.3 of [14].

PROPOSITION 2.3. *Let*

$$Q_1 : C^\infty(F_1) \to C^\infty(F_2), \qquad Q_2 : C^\infty(F_2) \to C^\infty(F_3)$$

be homogeneous differential operators satisfying $Q_2 \circ Q_1 = 0$. Suppose that the operator Q_1 is either elliptic or equal to 0. Then the following assertions are equivalent:

(i) *The complex*

$$C^\infty(F_1) \xrightarrow{Q_1} C^\infty(F_2) \xrightarrow{Q_2} C^\infty(F_3)$$

is exact.

(ii) *For all $\gamma \in \hat{G}$, the complex*

$$C_\gamma^\infty(F_1) \xrightarrow{Q_1} C_\gamma^\infty(F_2) \xrightarrow{Q_2} C_\gamma^\infty(F_3)$$

is exact.

(iii) *For all $\gamma \in \hat{G}$, we have*

$$\operatorname{Mult} Q_2(C_\gamma^\infty(F_2)) \geq \operatorname{Mult} C_\gamma^\infty(F_2) - \operatorname{Mult} C_\gamma^\infty(F_1)$$
$$+ \operatorname{Mult}(C_\gamma^\infty(F_1) \cap \operatorname{Ker} Q_1).$$

PROOF: First, suppose that assertion (ii) holds. Then according to Proposition 2.2,(i) the subspace $\operatorname{Ker} Q_2$ is equal to the closure of

$$\bigoplus_{\gamma \in \hat{G}} (C_\gamma^\infty(F_2) \cap \operatorname{Ker} Q_2),$$

and hence to the closure of $\bigoplus_{\gamma \in \hat{G}} Q_1(C_\gamma^\infty(F_1))$. Since Q_1 either vanishes or is elliptic, by Proposition 2.2,(iii) this last space is equal to $Q_1(C^\infty(F_1)$, and so (i) holds. The equivalence of assertions (ii) and (iii) is immediate.

Now suppose that X is a complex manifold, that g is a Hermitian metric and that the group G acts by holomorphic isometries on X. Then the vector bundles T' and T'' are homogeneous sub-bundles of $T_{\mathbb{C}}$, while the vector bundles $T^{1,0}$ and $T^{0,1}$ are homogeneous sub-bundles of $T_{\mathbb{C}}^*$. The isomorphisms of vector bundles (1.70) are G-equivariant. Therefore for all $\gamma \in \hat{G}$, the isomorphism of vector bundles $g^\flat : T_{\mathbb{C}} \to T_{\mathbb{C}}^*$ induces isomorphisms of G-modules

$$(2.2) \qquad g^\flat : C_\gamma^\infty(T') \to C_\gamma^\infty(T^{0,1}), \qquad C_\gamma^\infty(T'') \to C_\gamma^\infty(T^{1,0}).$$

§3. The Guillemin and zero-energy conditions

Let (X, g) be a Riemannian manifold. For $p \geq 0$, we consider the symmetrized covariant derivative

$$D^p = D_X^p : S^p T^* \to S^{p+1} T^*,$$

which is the first-order differential operator defined by

$$(D^p u)(\xi_1, \dots, \xi_{p+1}) = \frac{1}{p+1} \sum_{j=1}^{p+1} (\nabla u)(\xi_j, \xi_1, \dots, \hat{\xi}_j, \dots, \xi_{p+1}),$$

for $u \in S^p T^*$ and $\xi_1, \dots, \xi_{p+1} \in T$. The operator D^0 is equal to the exterior differential operator d on functions, and the operator D^1 was already introduced in §1, Chapter I.

If (X, g) is a flat torus, then we easily see that

$$(2.3) \qquad \int_X \zeta \cdot f \, dX = 0,$$

for all $f \in C^\infty(X)$ and all parallel vector fields ζ on X; therefore, if u is a symmetric p-form on X, we have

$$\int_X (D^p u)(\zeta_1, \dots, \zeta_{p+1}) \, dX = 0,$$

for all parallel vector fields $\zeta_1, \ldots, \zeta_{p+1}$ on X. By formula (1.4), we thus see that

$$\int_X (\mathcal{L}_\xi g)(\zeta_1, \zeta_2)\, dX = 0,$$

for all vector fields ξ on X and all parallel vector fields ζ_1, ζ_2 on X.

The following lemma is a consequence of the preceding remarks and formula (1.4).

LEMMA 2.4. *Let (X, g) be a Riemannian manifold and let Y be a totally geodesic flat torus contained in X.*

(i) *Let u be a symmetric p-form on X. Then for all parallel vector fields $\zeta_1, \ldots, \zeta_{p+1}$ on Y, the integral*

$$\int_X (D^p u)(\zeta_1, \ldots, \zeta_{p+1})\, dY$$

vanishes.

(ii) *Let ξ be a vector field on X. Then for all parallel vector fields ζ_1, ζ_2 on Y, the integral*

$$\int_Y (\mathcal{L}_\xi g)(\zeta_1, \zeta_2)\, dY$$

vanishes.

Let $\gamma : [0, L] \to X$ be a closed geodesic of X of length L parametrized by its arc-length s; we denote by $\dot\gamma(s)$ the tangent vector to the geodesic γ at the point $\gamma(s)$. If u is a symmetric p-form on X, we consider the integral

$$\int_\gamma u = \int_0^L u(\dot\gamma(s), \dot\gamma(s), \ldots, \dot\gamma(s))\, ds$$

of u over γ.

DEFINITION 2.5. *We say that a symmetric p-form u on X satisfies the zero-energy condition if, for every closed geodesic γ of X, the integral of u over γ vanishes.*

Let Y be a totally geodesic submanifold of X; clearly, if u is a symmetric p-form on X satisfying the zero-energy condition, then the restriction of u to Y also satisfies the zero-energy condition. From Lemma 2.4, we obtain the following result:

LEMMA 2.6. *If u is a symmetric p-form on X, then the symmetric $(p+1)$-form $D^p u$ satisfies the zero-energy condition. A symmetric 2-form on X, which is equal to a Lie derivative of the metric g, satisfies the zero-energy condition.*

DEFINITION 2.7. *We say that the Riemannian metric g on X is a C_L-metric if all its geodesics are periodic and have the same length L.*

If g is a C_L-metric, we say that X is a C_L-manifold; then the geodesic flow on the unit tangent bundle of (X, g) is periodic with least period L.

The following proposition is due to Michel (Proposition 2.2.4 of [45]; see also Proposition 5.86 of [5]).

PROPOSITION 2.8. *Let g_t be a one-parameter family of C_L-metrics on X, for $|t| < \varepsilon$, with $g_0 = g$. Then the infinitesimal deformation $h = \frac{d}{dt} g_t|_{t=0}$ of $\{g_t\}$ satisfies the zero-energy condition.*

PROOF: Let $\gamma : [0, L] \to X$ be a closed geodesic of the Riemannian manifold (X, g). Then there exists a real number $0 < \delta \le \varepsilon$ and a differentiable family of closed curves $\gamma_t : [0, L] \to X$, for $|t| < \delta$, which possesses the following properties: for each t, with $|t| < \delta$, the curve γ_t is a geodesic of the metric g_t parametrized by arc-length, and the curve γ_0 is equal to γ. We consider the variation of the family $\{\gamma_t\}$ which is the vector field ξ along the curve γ defined by $\xi(s) = \frac{d}{dt} \gamma_t(s)|_{t=0}$, for $0 \le s \le L$. Since g_t is a C_L-metric, we know that

$$\int_0^L g_t(\dot\gamma_t(s), \dot\gamma_t(s))\, ds = L,$$

for $|t| < \delta$. We differentiate the left-hand side of the above equality with respect to t, evaluate it at $t = 0$, and then obtain the relation

$$\int_0^L \left(h(\dot\gamma_0(s), \dot\gamma_0(s)) + 2g(\dot\gamma_0(s), \xi(s)) \right) ds = 0.$$

On the other hand, according to the first variation formula, since γ is geodesic of the metric g, the derivative

$$\frac{d}{dt} \int_0^L g(\dot\gamma_t(s), \dot\gamma_t(s))\, ds$$

vanishes at $t = 0$; this gives us the relation

$$\int_0^L g(\dot\gamma_0(s), \xi(s))\, ds = 0.$$

From the previous equalities, we infer that

$$\int_0^L h(\dot\gamma_0(s), \dot\gamma_0(s))\, ds = 0,$$

and so h satisfies the zero-energy condition.

DEFINITION 2.9. *We say that a symmetric p-form u on a compact locally symmetric space X satisfies the Guillemin condition if, for every maximal flat totally geodesic torus Z contained in X and for all unitary parallel vector fields ζ on Z, the integral*

$$(2.4) \qquad\qquad \int_Z u(\zeta, \zeta, \ldots, \zeta)\, dZ$$

vanishes.

From Lemma 2.4, we obtain:

LEMMA 2.10. *Let X be a compact locally symmetric space. If u is a symmetric p-form on X, then the symmetric $(p+1)$-form $D^p u$ satisfies the Guillemin condition. If ξ is a vector field on X, the symmetric 2-form $\mathcal{L}_\xi g$ on X satisfies the Guillemin condition.*

Thus every exact one-form on a compact locally symmetric space satisfies the Guillemin and zero-energy conditions.

LEMMA 2.11. *Let X be a flat torus. A symmetric p-form on X satisfying the zero-energy condition also satisfies the Guillemin condition.*

PROOF: Let u be a symmetric p-form on X. It is easily seen that the form u satisfies the Guillemin condition if and only if the integral (2.4) vanishes for all unitary parallel vector fields ζ on X, all of whose orbits are closed geodesics. Let ζ be such a vector field and let $\varphi_t = \exp t\zeta$, for $t \in \mathbb{R}$, be the family of diffeomorphisms of X determined by ζ; then there exists a real number $L > 0$ such that $\varphi_{t+L} = \varphi_t$ and such that, for all $x \in X$, the mapping $\gamma_x : [0, L] \to X$ defined by $\gamma_x(t) = \varphi_t(x)$, with $0 \le t \le L$, is a closed geodesic of X of length L parametrized by its arc-length. We suppose that L is the smallest such number. There is a flat torus X' of dimension $m - 1$ and a Riemannian fibration $\pi : X \to X'$ whose fibers are equal to the family of all these closed geodesics; in fact, if $x_1, x_2 \in X$, we have $\pi(x_1) = \pi(x_2)$ if and only if we can write $x_2 = \varphi_t(x_1)$, for some $0 \le t \le L$. If f is the function defined on X by

$$f(x) = \int_{\gamma_x} u,$$

for $x \in X$, clearly there is a function f' on X' satisfying $\pi^* f' = f$. Then we have

$$\int_Z u(\zeta, \zeta, \ldots, \zeta) \, dZ = \int_{X'} f' \, dX'.$$

If u satisfies the zero-energy condition, then the functions f and f' vanish, and so, by the preceding formula, u satisfies the Guillemin condition.

If (X, g) is a compact locally symmetric space, according to Lemma 2.11 a symmetric p-form on X satisfying the zero-energy condition also satisfies the Guillemin condition.

DEFINITION 2.12. *We say that a compact locally symmetric space X is rigid in the sense of Guillemin (resp. infinitesimally rigid) if the only symmetric 2-forms on X satisfying the Guillemin (resp. the zero-energy) condition are the Lie derivatives of the metric g.*

If X is a compact locally symmetric space X and $p \geq 0$ is an integer, we consider the space $\mathcal{Z}_p$ of all sections of $C^\infty(S^p T^*)$ satisfying the zero-energy condition. According to Lemma 2.6, we have the inclusion

$$D^p C^\infty(S^p T^*) \subset \mathcal{Z}_{p+1}.$$

By formula (1.4), we see that the infinitesimal rigidity of the compact locally symmetric space X is equivalent to the equality $D^1 C^\infty(T^*) = \mathcal{Z}_2$. On the other hand, the equality $D^0 C^\infty(T^*) = \mathcal{Z}_1$ means that every differential form of degree 1 on X satisfying the zero-energy condition is exact.

PROPOSITION 2.13. *Let X be a compact locally symmetric space. Then the following assertions are equivalent:*

(i) Every symmetric 2-form h on X, which satisfies the Guillemin (resp. the zero-energy) condition and the relation div $h = 0$, vanishes.

(ii) The space X is rigid in the sense of Guillemin (resp. is infinitesimally rigid).

PROOF: First assume that assertion (i) holds. Let h be a symmetric 2-form on X satisfying the Guillemin (resp. the zero-energy) condition. According to the decomposition (1.11), we may write

$$h = h_0 + D_0 \xi,$$

where h_0 is an element of $C^\infty(S^2 T^*)$ satisfying div $h_0 = 0$ and $\xi \in C^\infty(T)$. Clearly, by Lemma 2.10 (resp. Lemma 2.6), the symmetric 2-form h_0 also satisfies the Guillemin (resp. the zero-energy) condition; our assumption implies that h_0 vanishes, and so h is a Lie derivative of the metric. Therefore (ii) is true. According to the decomposition (1.11), we see that assertion (i) is a direct consequence of (ii).

We now assume that (X, g) is a symmetric space of compact type. If the space X is rigid in the sense of Guillemin, it is also infinitesimally rigid. If X is a space of rank one, the closed geodesics of X are the maximal flat totally geodesic tori of X, and so the notions of Guillemin rigidity and infinitesimal rigidity for X are equivalent.

Consider a family of Riemannian metrics $\{g_t\}$ on X, for $|t| < \varepsilon$, with $g_0 = g$. We say that $\{g_t\}$ is an isospectral deformation of g if the spectrum $\mathrm{Spec}(X, g_t)$ of the metric g_t is equal to $\mathrm{Spec}(X, g)$, for all $|t| < \varepsilon$. We say that the space (X, g) is infinitesimally spectrally rigid (i.e., spectrally rigid to first-order) if, for every such isospectral deformation $\{g_t\}$ of g, there is a one-parameter family of diffeomorphisms φ_t of X such that $g_t = \varphi_t^* g$ to first-order in t at $t = 0$, or equivalently if the infinitesimal deformation $\frac{d}{dt} g_t|_{t=0}$ of $\{g_t\}$ is a Lie derivative of the metric g.

In [35], Guillemin proved the following result:

THEOREM 2.14. *A symmetric 2-form on a symmetric space (X, g) of compact type, which is equal to the infinitesimal deformation of an isospectral deformation of g, satisfies the Guillemin condition.*

This theorem leads us to Guillemin's criterion for the infinitesimal spectral rigidity of a symmetric space of compact type which may be expressed as follows:

THEOREM 2.15. *If a symmetric space of compact type is rigid in the sense of Guillemin, it is infinitesimally spectrally rigid.*

§4. Radon transforms

Let (X, g) be a symmetric space of compact type. Then there is a Riemannian symmetric pair (G, K) of compact type, where G is a compact, connected semi-simple Lie group and K is a closed subgroup of G such that the space X is isometric to the homogeneous space G/K endowed with a G-invariant metric. We identify X with G/K, and let x_0 be the point of X corresponding to the coset of the identity element of G in G/K. Since the maximal flat totally geodesic tori of X are conjugate under the action of G on X, the space Ξ of all such tori is a homogeneous space of G. We also consider the set Ξ' of all closed geodesics of X; when the rank of X is equal to one, then Ξ' is equal to Ξ.

A Radon transform for functions on X assigns to a function on X its integrals over a class of totally geodesic submanifolds of X of a fixed dimension. Here we shall consider two such Radon transforms, the maximal flat Radon transform and the X-ray transform.

The maximal flat Radon transform for functions on X assigns to a real-valued function f on X the function $\hat{f}$ on Ξ, whose value at a torus

$Z \in \Xi$ is the integral

$$\hat{f}(Z) = \int_Z f \, dZ$$

of f over Z. Clearly this transform is injective if every function on X satisfying the Guillemin condition vanishes. The X-ray transform for functions on X assigns to a real-valued function f on X the function $\check{f}$ on Ξ', whose value at a closed geodesic $\gamma \in \Xi'$ is the integral

$$\check{f}(\gamma) = \int_\gamma f.$$

Clearly this transform is injective if every function on X satisfying the zero-energy condition vanishes. If the rank of X is equal to one, the maximal flat Radon transform for functions on X coincides with the X-ray transform for functions on X.

Let f be a real-valued function on X. If Z is a torus belonging to Ξ and if ζ is a unitary parallel vector field on Z, then we see that

$$\int_Z (fg)(\zeta, \zeta) \, dZ = \hat{f}(Z).$$

On the other hand, if γ is a closed geodesic of X, we have

$$\int_\gamma fg = \check{f}(\gamma).$$

Thus the maximal flat Radon (resp. the X-ray) transform of f vanishes if and only if the symmetric 2-form fg satisfies the Guillemin (resp. the zero-energy) condition.

If X is an irreducible symmetric space of compact type, we recall that g is an Einstein metric and that $\mathrm{Ric} = \lambda g$, where λ is a positive real number; moreover, the space $E(X)$ of infinitesimal Einstein deformations of the metric g is a G-submodule of $C^\infty(S^2 T^*)$.

PROPOSITION 2.16. *Let X be an irreducible symmetric space of compact type, which is not isometric to a sphere. If X is rigid in the sense of Guillemin (resp. is infinitesimally rigid), then the maximal flat Radon (resp. the X-ray) transform for functions on X is injective.*

PROOF: Assume that X is rigid in the sense of Guillemin (resp. is infinitesimally rigid). Let f be real-valued function on the Einstein manifold X; suppose that the function $\hat{f}$ (resp. the function $\check{f}$) vanishes. Then the symmetric 2-form fg on X satisfies the Guillemin (resp. the zero-energy) condition. Therefore we may write $fg = \mathcal{L}_\xi g$, where ξ is a vector field on X. According to Proposition 1.6, the function f vanishes, and so the corresponding Radon transform for functions is injective.

Let Λ be a finite group of isometries of X of order q. If F is a vector bundle equal either to a sub-bundle of $T_{\mathbb{C}}$ or to a sub-bundle of $S^pT_{\mathbb{C}}^*$ invariant under the group Λ, we denote by $C^\infty(F)^\Lambda$ the space consisting of all Λ-invariant sections of F; if the vector bundle F is also invariant under the group G and if the isometries of Λ commute with the action of G, then $C^\infty(F)^\Lambda$ is a G-submodule of $C^\infty(F)$. If F is the trivial complex line bundle, we consider the G-submodule $C^\infty(X)^\Lambda = C^\infty(F)^\Lambda$ of Λ-invariant functions on X.

We suppose that the group Λ acts without fixed points. Then the quotient $Y = X/\Lambda$ is a manifold and the natural projection $\varpi : X \to Y$ is a covering projection. Thus the metric g induces a Riemannian metric g_Y on Y such that $\varpi^* g_Y = g$. Clearly the space Y is locally symmetric.

A symmetric p-form u on X is invariant under the group Λ if and only if there is a symmetric p-form $\hat{u}$ on Y such that $u = \varpi^*\hat{u}$. The projection ϖ induces an isomorphism

$$(2.5) \qquad \varpi^* : C^\infty(Y, S^pT_{Y,\mathbb{C}}^*) \to C^\infty(S^pT_{\mathbb{C}}^*)^\Lambda,$$

sending $u \in C^\infty(Y, S^pT_{Y,\mathbb{C}}^*)$ into ϖ^*u. A vector field ξ on X is invariant under Λ if and only if it is ϖ-projectable, i.e., if there exists a vector field $\hat{\xi}$ on Y such that $\varpi_*\xi(x) = \hat{\xi}(\varpi(x))$, for all $x \in X$. If ξ is a Λ-invariant vector field on X, then the Lie derivative $D_0\xi$ is a Λ-invariant symmetric 2-form on X and, if $\hat{\xi}$ denotes the vector field on Y induced by ξ, we see that

$$(2.6) \qquad D_0\xi = \varpi^*(D_{0,Y}\hat{\xi}).$$

If X is an irreducible symmetric space, then X and Y are Einstein manifolds; according to the definition of the spaces $E(X)$ and $E(Y)$ of infinitesimal Einstein deformations, we see that the projection ϖ and the isomorphism (2.5) induce an isomorphism

$$(2.7) \qquad \varpi^* : E(Y) \to E(X)^\Lambda,$$

where

$$E(X)^\Lambda = E(X) \cap C^\infty(S^2T^*)^\Lambda.$$

Throughout the remainder of this section, we also suppose that the isometries of Λ commute with the action of G on X; then Y is a homogeneous space of G. Assume furthermore that there is a subgroup K' of G containing K and a G-equivariant diffeomorphism $\varphi : Y \to G/K'$ which have the following properties:

(i) (G, K') is a Riemannian symmetric pair;

(ii) when we identify X with G/K, the projection $\varphi \circ \varpi$ is equal to the natural projection $G/K \to G/K'$.

Under these conditions, the space (Y, g_Y) is isometric to the symmetric space G/K' of compact type endowed with a G-invariant metric.

Let Z be a maximal flat totally geodesic torus of X. Then $\varpi(Z)$ is a flat torus of Y. On the other hand, if Z' is a maximal flat totally geodesic torus of Y, then $\varpi^{-1}Z'$ is a totally geodesic flat torus of X. From these observations, it follows that $Z = \varpi^{-1}(Z')$, where $Z' = \varpi(Z)$; we also see that the rank of Y is equal to the rank of X and that the induced mapping $\varpi : Z \to Z'$ is q-fold covering. Moreover, the torus Z is invariant under the group Λ. A parallel vector field ξ on Z is ϖ-projectable, i.e., if there exists a parallel vector field $\hat{\xi}$ on $Z' = \varpi(Z)$ such that $\varpi_* \xi(x) = \hat{\xi}(\varpi(x))$, for all $x \in Z$. Conversely, any parallel vector field on Z' is of the form $\varpi_* \xi$, for some parallel vector field ξ on Z.

Let u be a symmetric p-form on X invariant under Λ and let $\hat{u}$ be the symmetric p-form on Y such that $u = \varpi^* \hat{u}$. Let ξ be a parallel vector field on Z and $\hat{\xi}$ be the parallel vector field on Z' such that $\varpi_* \xi = \hat{\xi}$. Then the function $u(\xi, \xi, \ldots, \xi)$ on Z is invariant under Λ and satisfies

$$u(\xi, \xi, \ldots, \xi) = \varpi^* \hat{u}(\hat{\xi}, \hat{\xi}, \ldots, \hat{\xi});$$

thus we obtain the equality

$$\int_Z u(\xi, \xi, \ldots, \xi) \, dZ = q \int_{Z'} \hat{u}(\hat{\xi}, \hat{\xi}, \ldots, \hat{\xi}) \, dZ'.$$

If γ is a closed geodesic of Y, it is easily seen that there is a closed geodesic γ' of X and an integer $1 \le q_1 \le q$ such that the image of the mapping $\varpi \circ \gamma'$ is equal to the image of γ and such that the equality

$$\int_{\gamma'} \varpi^* u = q_1 \int_{\gamma} u$$

holds for all symmetric p-forms u on Y.

From the above observations, we deduce the following:

LEMMA 2.17. *Suppose that the quotient* $Y = X/\Lambda$ *is a symmetric space. Then a symmetric p-form u on Y satisfies the Guillemin (resp. the zero-energy) condition if and only if the symmetric p-form $\varpi^* u$ on X, which is invariant under the group Λ, satisfies the Guillemin (resp. the zero-energy) condition.*

By Lemma 2.17, we see that the maximal flat Radon (resp. the X-ray) transform for functions on Y is injective if and only if the restriction of the

the maximal flat Radon (resp. the X-ray) transform for functions on X to the space $C^\infty(X)^\Lambda$ is injective. From Lemma 2.17 and the equality (2.6), we deduce the following three results:

PROPOSITION 2.18. *Suppose that the quotient $Y = X/\Lambda$ is a symmetric space. Then the following assertions are equivalent:*

(i) *Every symmetric 2-form on the space X, which is invariant under the group Λ and satisfies the Guillemin condition, is a Lie derivative of the metric.*

(ii) *The space Y is rigid in the sense of Guillemin.*

PROPOSITION 2.19. *Suppose that the quotient $Y = X/\Lambda$ is a symmetric space. Then the following assertions are equivalent:*

(i) *Every symmetric 2-form on the space X, which is invariant under the group Λ and satisfies the zero-energy condition, is a Lie derivative of the metric.*

(ii) *The space Y is infinitesimally rigid.*

PROPOSITION 2.20. *Suppose that the quotient $Y = X/\Lambda$ is a symmetric space. Then the following assertions are equivalent:*

(i) *Every differential form of degree 1 on the space X, which is invariant under the group Λ and satisfies the Guillemin (resp. the zero-energy) condition, is exact.*

(ii) *Every differential form of degree 1 on the space Y, which satisfies the Guillemin (resp. the zero-energy) condition, is exact.*

Let F be a G-invariant sub-bundle of $S^p T^*_{\mathbb{C}}$, which is also invariant under the group Λ. Then there exists a unique G-invariant sub-bundle F_Y of $S^p T^*_{Y,\mathbb{C}}$ such that, for all $x \in X$, the isomorphism $\varpi^* : S^p T^*_{Y,y} \to S^p T^*_x$, where $y = \varpi(x)$, induces an isomorphism $\varpi^* : F_{Y,y} \to F_x$. A symmetric p-form u on Y is a section of F_Y if and only if the Λ-invariant symmetric p-form $\varpi^* u$ on X is a section of F. Then the mapping

$$\varpi^* : C^\infty(Y, F_Y) \to C^\infty(F)^\Lambda,$$

induced by (2.5), is an isomorphism of G-modules.

The following proposition is a consequence of Lemma 2.17.

PROPOSITION 2.21. *Suppose that the quotient $Y = X/\Lambda$ is a symmetric space. Let F be a sub-bundle of $S^p T^*$ invariant under the groups G and Λ, and let F_Y be the G-invariant sub-bundle of $S^p T^*_Y$ induced by F. Then the following assertions are equivalent:*

(i) *Any section of the vector bundle F over the space X, which is invariant under the group Λ and satisfies the Guillemin (resp. the zero-energy) condition, vanishes.*

(ii) *Any section of the vector bundle F_Y over the space Y, which satisfies the Guillemin (resp. the zero-energy) condition, vanishes.*

We now suppose that the group Λ is equal to the group $\{\mathrm{id}, \tau\}$ of order 2, where τ is an involutive isometry of X. We say that a symmetric p-form u on X is even (resp. odd) with respect to τ if $\tau^* u = \varepsilon u$, where $\varepsilon = 1$ (resp. $\varepsilon = -1$). A vector field ξ on X is even (resp. odd) with respect to τ if $\tau_* \xi = \xi$ (resp. $\tau_* \xi = -\xi$). Any of these tensors on X is even if and only if it is invariant under the group Λ. If F is a G-invariant sub-bundle of $T_{\mathbb{C}}$ or of $S^p T_{\mathbb{C}}^*$, which is also invariant under τ, the space $C^\infty(F)^{\mathrm{ev}}$ (resp. $C^\infty(F)^{\mathrm{odd}}$) consisting of all even (resp. odd) sections of F over X is a G-submodule of $C^\infty(F)$. Clearly, we have the equality $C^\infty(F)^{\mathrm{ev}} = C^\infty(F)^\Lambda$ and the decomposition of G-modules

$$(2.8) \qquad C^\infty(F) = C^\infty(F)^{\mathrm{ev}} \oplus C^\infty(F)^{\mathrm{odd}}.$$

In particular, if F is the trivial complex line bundle, we obtain the decomposition

$$C^\infty(X) = C^\infty(X)^{\mathrm{ev}} \oplus C^\infty(X)^{\mathrm{odd}},$$

where $C^\infty(X)^{\mathrm{ev}}$ (resp. $C^\infty(X)^{\mathrm{odd}}$) is the G-submodule of $C^\infty(X)$ consisting of all even (resp. odd) functions on X; in fact, the space $C^\infty(X)^{\mathrm{ev}}$ is isomorphic to the space of all complex-valued functions on Y. Moreover, we have the inclusion

$$dC^\infty(X)^{\mathrm{ev}} \subset C^\infty(T^*)^{\mathrm{ev}}, \qquad dC^\infty(X)^{\mathrm{odd}} \subset C^\infty(T^*)^{\mathrm{odd}}.$$

By (1.6), we have the inclusions

$$(2.9) \quad D_0 C^\infty(T)^{\mathrm{ev}} \subset C^\infty(S^2 T^*)^{\mathrm{ev}}, \qquad D_0 C^\infty(T)^{\mathrm{odd}} \subset C^\infty(S^2 T^*)^{\mathrm{odd}}.$$

If X is an irreducible symmetric space, we consider the G-submodules

$$E(X)^{\mathrm{ev}} = E(X) \cap C^\infty(S^2 T^*)^{\mathrm{ev}}, \qquad E(X)^{\mathrm{odd}} = E(X) \cap C^\infty(S^2 T^*)^{\mathrm{odd}}$$

of $E(X)$; then we have the equality $E(X)^{\mathrm{ev}} = E(X)^\Lambda$ and the decomposition of G-modules

$$(2.10) \qquad E(X) = E(X)^{\mathrm{ev}} \oplus E(X)^{\mathrm{odd}}.$$

Let Z be a maximal flat totally geodesic torus of X. Since τ preserves Z, if f is an odd function on X, we see that the integral of f over Z vanishes. Therefore the odd functions on X satisfy the Guillemin condition, and so belong to the kernel of the maximal flat Radon transform for functions.

PROPOSITION 2.22. *We suppose that the group Λ is equal to the group $\{id, \tau\}$ of order 2, where τ is an involutive isometry of X, and that the quotient $Y = X/\Lambda$ is a symmetric space. Then an odd symmetric p-form on X satisfies the Guillemin condition, and the maximal flat Radon transform for functions on X is not injective. Moreover, the space X is not rigid in the sense of Guillemin.*

PROOF: Let u be an odd symmetric p-form on X and let Z be a maximal flat totally geodesic torus of X. If ξ is a parallel vector field on Z, then the function $u(\xi, \xi, \dots, \xi)$ on Z is odd, that is,

$$u(\xi, \xi, \dots, \xi)(\tau(x)) = -u(\xi, \xi, \dots, \xi)(x),$$

for all $x \in Z$; hence its integral over Z vanishes. We now construct an odd symmetric 2-form h' on X which is not a Lie derivative of the metric. Let x be a point of X and U be a open neighborhood of x for which $U \cap \tau(U) = \emptyset$. By Lemma 1.13 and remarks made in §3, Chapter I, we know that the infinitesimal orbit of the curvature $\tilde{B}$ is a vector bundle and that the quotient bundle $B/\tilde{B}$ is non-zero. According to §1, Chapter I, the morphism $\sigma(D_1) : S^2 T^* \otimes S^2 T^* \to B/\tilde{B}$ is surjective; hence we may choose a symmetric 2-form h on X whose support is contained in U and which satisfies $(D_1 h)(x) \neq 0$. We know that h is not a Lie derivative of the metric on any neighborhood of x. The symmetric 2-form $h' = h - \tau^* h$ on X is odd and its restriction to U is equal to h. Hence the form h' satisfies the Guillemin condition, and so the space X is not rigid in the sense of Guillemin.

We now consider an example of the above situation. The n-sphere (S^n, g_0), with $n \geq 2$, is an irreducible symmetric space of rank one; in fact, the group $SO(n+1)$ acts transitively on S^n, and the sphere S^n is isometric to the homogeneous space $SO(n+1)/SO(n)$ (see §2, Chapter III and also §10 in Chapter XI of [40]). The anti-podal involution τ of S^n is an isometry which commutes with the action of $SO(n+1)$ on $X = S^n$. If Λ is the group $\{id, \tau\}$ of isometries of X, the quotient $Y = X/\Lambda$ is equal to the real projective space $\mathbb{RP}^n$ endowed with the metric g_0 of constant curvature 1 induced by the metric g_0 of S^n. The natural projection $\varpi : S^n \to \mathbb{RP}^n$ is a two-fold covering. The Riemannian manifold $\mathbb{RP}^n$ is also an irreducible symmetric space of compact type of rank one. The closed geodesics of the sphere S^n are the great circles, and the maximal flat totally geodesic tori of S^n and $\mathbb{RP}^n$ are the closed geodesics. In fact, the metric g_0 on S^n (resp. on $\mathbb{RP}^n$) is a C_L-metric, where $L = 2\pi$ (resp. $L = \pi$). We easily see directly, or by Proposition 2.22, that the odd symmetric p-forms (with respect to τ) on S^n satisfy the zero-energy condition. Moreover, according to Proposition 2.22 the sphere S^n is not infinitesimally rigid.

In §3, Chapter III, we shall prove that the X-ray transform for functions on the sphere S^n, with $n \geq 2$, is injective on the space of all even functions (see Proposition 3.17). Clearly, this result is equivalent to assertion (i) of the following theorem. By Lemma 2.17, we know that assertions (i) and (ii) of this theorem are equivalent. We point out that assertion (i) of this theorem in the case of the 2-sphere S^2 is a classic result due to Funk.

THEOREM 2.23. (i) *The kernel of the X-ray transform for functions on the sphere (S^n, g_0), with $n \geq 2$, is equal to the space of all odd functions on S^n.*

(ii) *The X-ray transform for functions on the real projective space $(\mathbb{RP}^n, g_0)$, with $n \geq 2$, is injective.*

The adjoint space of the symmetric space X is the symmetric space which admits X as a Riemannian covering and is itself not a Riemannian covering of another symmetric space. For example, the adjoint space of the n-sphere S^n, with $n \geq 2$, is the real projective space $\mathbb{RP}^n$.

In [34], Grinberg generalized Theorem 2.23 and proved the following:

THEOREM 2.24. *The maximal flat Radon transform for functions on a symmetric space X of compact type is injective if and only if X is equal to its adjoint space.*

By Proposition 2.22, the sphere S^n is not infinitesimally rigid. Hence from Proposition 2.16 and Theorem 2.24, we obtain the following necessary condition for Guillemin rigidity:

THEOREM 2.25. *Let X be an irreducible symmetric space of compact type. If X is rigid in the sense of Guillemin, then X is equal to its adjoint space.*

In Chapter III, we shall show that the X-ray transform for functions on a flat torus of dimension > 1 is injective; this result is due to Michel [46] (see Proposition 3.5). If the symmetric space X is of rank q, each point of X is contained in a totally geodesic flat torus of dimension q of X (see Theorem 6.2 in Chapter V of [36]). Thus from the injectivity of the X-ray transform for functions on a flat torus, we deduce the following:

PROPOSITION 2.26. *The X-ray transform for functions on a symmetric space X of compact type of rank > 1 is injective.*

We now extend the definitions of the maximal flat Radon transform and the X-ray transform to symmetric p-forms. Let L be the vector bundle over Ξ whose fiber at a point $Z \in \Xi$ is the space of all parallel vector fields on the flat torus Z. This vector bundle is a homogeneous G-bundle over Ξ and its rank is equal to the rank of the symmetric space X. We consider the p-th symmetric product $S^p L^*$ of the dual L^* of L. The space $C^\infty(S^p T^*)$ of all symmetric p-forms on X and the space $C^\infty(\Xi, S^p L^*)$ of all sections of $S^p L^*$ over Ξ are G-modules. The maximal flat Radon transform for symmetric p-forms on X is the morphism of G-modules

$$I_p : C^\infty(S^p T^*) \to C^\infty(\Xi, S^p L^*),$$

which assigns to a symmetric p-form u on X the section $I_p(u)$ of $S^p L^*$ whose value at the point $Z \in \Xi$ is determined by

$$I_p(u)(\zeta_1, \zeta_2, \ldots, \zeta_p) = \int_Z u(\zeta_1, \zeta_2, \ldots, \zeta_p)\, dZ,$$

where $\zeta_1, \zeta_2, \ldots, \zeta_p$ are elements of L_Z. The kernel $\mathcal{N}_p$ of this mapping I_p is the G-submodule of $C^\infty(S^p T^*)$ equal to the space consisting of all symmetric p-forms on X which satisfy the Guillemin condition. The complexification $\mathcal{N}_{p,\mathbb{C}}$ of the space $\mathcal{N}_p$ shall be viewed as the G-submodule of $C^\infty(S^p T^*_\mathbb{C})$ equal to the kernel of the morphism of G-modules

$$(2.11) \qquad I_p : C^\infty(S^p T^*_\mathbb{C}) \to C^\infty(\Xi, S^p L^*_\mathbb{C})$$

induced by the mapping I_p. The mapping I_0 coincides with the maximal flat Radon transform for functions defined above, while the mapping I_2 was introduced in [23].

The X-ray transform for symmetric p-forms on X is the linear mapping I'_p sending an element $u \in C^\infty(S^p T^*)$ into the real-valued function $\check{u}$ on Ξ' whose value at the closed geodesic γ is the integral

$$\int_\gamma u.$$

The kernel of this mapping I'_p is equal to the space $\mathcal{Z}_p$ of all symmetric p-forms on X satisfying the zero-energy condition. Then according to Lemma 2.11, we have

$$\mathcal{Z}_p \subset \mathcal{N}_p.$$

Let γ be a closed geodesic of X; if ϕ is an element of G, we consider the closed geodesic $\gamma^\phi = \phi \circ \gamma$. The mapping

$$\Phi_\gamma : C^\infty(S^p T^*_\mathbb{C}) \to C^\infty(G),$$

which sends the complex symmetric p-form u into the complex-valued function $\Phi_\gamma(u)$ on G defined by

$$\Phi_\gamma(u)(\phi) = \int_{\gamma^\phi} u = \int_\gamma \phi^* u,$$

for $\phi \in G$, is a morphism of G-modules. The complexification $\mathcal{Z}_{p,\mathbb{C}}$ of the space $\mathcal{Z}_p$ shall be viewed as the G-submodule

$$\bigcap_{\gamma \in \Xi'} \mathrm{Ker}\, \Phi_\gamma$$

of $C^\infty(S^p T^*_{\mathbb{C}})$ consisting of all complex symmetric p-forms on X which satisfy the zero-energy condition.

When the rank of X is equal to one, the vector bundle L is a line bundle; in this case, the X-ray transform for symmetric p-forms, which may be viewed as a morphism of G-modules

$$I'_p : C^\infty(S^p T^*) \to C^\infty(\Xi),$$

determines the maximal flat Radon transform for symmetric p-forms.

§5. Radon transforms and harmonic analysis

We consider the symmetric space (X, g) of compact type of §4 and the compact Lie groups G and K introduced there. We denote by Γ the dual $\hat{G}$ of the group G. As we mentioned in §2, the vector bundles $T_{\mathbb{C}}$ and $\bigotimes^p T^*_{\mathbb{C}}$ endowed with the Hermitian scalar products induced by the metric g are homogeneous and unitary. Let F be a homogeneous complex vector bundle over X; assume that F either is a G-invariant complex sub-bundle of $T_{\mathbb{C}}$ or can be written in the form E/E', where E and E' are G-invariant complex sub-bundles of $\bigotimes^p T^*_{\mathbb{C}}$ satisfying $E' \subset E$. We endow F with the Hermitian scalar product obtained from the Hermitian scalar product on $T_{\mathbb{C}}$ or on $\bigotimes^p T^*_{\mathbb{C}}$. Clearly, the vector bundle F is unitary, and the space $C^\infty(F)$ endowed with the Hermitian scalar product obtained from the Hermitian scalar product on F and the G-invariant Riemannian measure dX of X is a unitary G-module. As in §2, we denote by $C^\infty_\gamma(F)$ the isotypic component of the G-module $C^\infty(F)$ corresponding to $\gamma \in \Gamma$, and by P_γ the orthogonal projection of $C^\infty(F)$ onto its submodule $C^\infty_\gamma(F)$.

Throughout the remainder of this section, we shall assume that F is a G-invariant complex sub-bundle of $T_{\mathbb{C}}$ or of $S^p T^*_{\mathbb{C}}$. For $\gamma \in \Gamma$, we consider the orthogonal projections P_γ of $C^\infty(T_{\mathbb{C}})$ onto its submodule $C^\infty_\gamma(T_{\mathbb{C}})$, and of $C^\infty(S^p T^*_{\mathbb{C}})$ onto its submodule $C^\infty_\gamma(S^p T^*_{\mathbb{C}})$; then the restriction of P_γ to $C^\infty(F)$ is the orthogonal projection of $C^\infty(F)$ onto $C^\infty_\gamma(F)$. Thus if F is a sub-bundle of $T_{\mathbb{C}}$ (resp. $S^p T^*_{\mathbb{C}}$), an element u of $C^\infty(T_{\mathbb{C}})$ (resp. $C^\infty(S^p T^*_{\mathbb{C}})$) is a section of F if and only if $P_\gamma u$ is an element of $C^\infty_\gamma(F)$, for all $\gamma \in \Gamma$ (see Chapter 5 of [56]).

The vector bundle $S^0 T^*_{\mathbb{C}}$ is the trivial complex line bundle and we shall identify the G-modules $C^\infty(S^0 T^*_{\mathbb{C}})$ and $C^\infty(X)$; we shall denote by $C^\infty_\gamma(X)$ the isotypic component of the G-module $C^\infty(X)$ corresponding to $\gamma \in \Gamma$.

Let Γ_0 be the subset of Γ consisting of those elements γ of Γ for which the G-module $C^\infty_\gamma(X)$ is non-zero. It is well-known that, for $\gamma \in \Gamma_0$, the G-module $C^\infty_\gamma(X)$ is irreducible (see Theorem 4.3 in Chapter V of [37]).

We endow the homogeneous space Ξ with a G-invariant Riemannian metric and the homogeneous vector bundle $S^p L^*_\mathbb{C}$ with a G-invariant Hermitian scalar product. The space $C^\infty(\Xi, S^p L^*_\mathbb{C})$, endowed with the Hermitian scalar product obtained from the Hermitian scalar product on $S^p L^*_\mathbb{C}$ and the Riemannian measure of the Riemannian manifold Ξ, is a unitary G-module. As in §2, for $\gamma \in \Gamma$, we denote by $C^\infty_\gamma(\Xi, S^p L^*_\mathbb{C})$ the isotypic component of the G-module $C^\infty(\Xi, S^p L^*_\mathbb{C})$ corresponding to γ and by P_γ the orthogonal projection of $C^\infty(\Xi, S^p L^*_\mathbb{C})$ onto its G-submodule $C^\infty_\gamma(\Xi, S^p L^*_\mathbb{C})$. Since the mapping (2.11) is a morphism of G-modules, its kernel $\mathcal{N}_{p,\mathbb{C}}$ is a closed G-submodule of $C^\infty(S^p T^*_\mathbb{C})$ and we have the inclusion

$$I_p(C^\infty_\gamma(S^p T^*_\mathbb{C})) \subset C^\infty_\gamma(\Xi, S^p L^*_\mathbb{C});$$

moreover the relation

$$I_p P_\gamma = P_\gamma I_p$$

holds for all $\gamma \in \Gamma$. Thus we have

(2.12) $$P_\gamma \mathcal{N}_{p,\mathbb{C}} \subset \mathcal{N}_{p,\mathbb{C}},$$

for all $\gamma \in \Gamma$.

We choose a left-invariant volume form on G and endow the G-module $C^\infty(G)$ of all complex-valued functions on the group G with the corresponding G-invariant Hermitian scalar product. For $\gamma \in \Gamma$, we denote by P_γ the orthogonal projection of $C^\infty(G)$ onto the isotypic component $C^\infty_\gamma(G)$ of $C^\infty(G)$ corresponding to γ.

If δ is a closed geodesic of X, the kernel of the morphism Φ_δ is a closed subspace of $C^\infty(S^p T^*_\mathbb{C})$, and so $\mathcal{Z}_{p,\mathbb{C}}$ is a closed G-submodule of $C^\infty(S^p T^*_\mathbb{C})$. Since Φ_δ is a morphism of G-modules, we also have the inclusion

$$\Phi_\delta(C^\infty_\gamma(S^p T^*_\mathbb{C})) \subset C^\infty_\gamma(G)$$

and the relation

(2.13) $$\Phi_\delta P_\gamma = P_\gamma \Phi_\delta$$

holds, for all $\gamma \in \Gamma$. By (2.13), we see that

(2.14) $$P_\gamma \mathcal{Z}_{p,\mathbb{C}} \subset \mathcal{Z}_{p,\mathbb{C}},$$

for all $\gamma \in \Gamma$.

Now let Σ be a finite set of isometries of X which commute with the action of G on X and let ε be a real number equal to ± 1. The space $C^\infty(T_{\mathbb{C}})^{\Sigma,\varepsilon}$, which consists of all elements ξ of $C^\infty(T_{\mathbb{C}})$ satisfying

$$\sigma_*\xi = \varepsilon\xi,$$

for all $\sigma \in \Sigma$, is a G-submodule of $C^\infty(T_{\mathbb{C}})$, while the space $C^\infty(S^pT_{\mathbb{C}}^*)^{\Sigma,\varepsilon}$ which consists of all sections u of $S^pT_{\mathbb{C}}^*$ over X satisfying

$$\sigma^*u = \varepsilon u,$$

for all $\sigma \in \Sigma$, is a G-submodule of $C^\infty(S^pT_{\mathbb{C}}^*)$. For all $\sigma \in \Sigma$, the actions of σ_* on $C^\infty(T_{\mathbb{C}})$ and of σ^* on $C^\infty(S^pT_{\mathbb{C}}^*)$ are unitary and we have

$$P_\gamma\sigma_*\xi = \sigma_*P_\gamma\xi, \qquad P_\gamma\sigma^*u = \sigma^*P_\gamma u,$$

for all $\xi \in C^\infty(T_{\mathbb{C}})$, $u \in C^\infty(S^pT_{\mathbb{C}}^*)$ and $\gamma \in \Gamma$.

Suppose that the vector bundle F is invariant under the isometries of Σ. Then we consider the G-submodule $C^\infty(F)^{\Sigma,\varepsilon}$ of $C^\infty(F)$, which is defined by

$$C^\infty(F)^{\Sigma,\varepsilon} = C^\infty(F) \cap C^\infty(T_{\mathbb{C}})^{\Sigma,\varepsilon}$$

whenever F is a sub-bundle of $T_{\mathbb{C}}$, and by

$$C^\infty(F)^{\Sigma,\varepsilon} = C^\infty(F) \cap C^\infty(S^pT_{\mathbb{C}}^*)^{\Sigma,\varepsilon}$$

whenever F is a sub-bundle of $S^pT_{\mathbb{C}}^*$. For $\gamma \in \Gamma$, we set

$$C_\gamma^\infty(F)^{\Sigma,\varepsilon} = C_\gamma^\infty(F) \cap C^\infty(F)^{\Sigma,\varepsilon}.$$

If $\gamma \in \Gamma$ and $\sigma \in \Sigma$, we know that the action of σ preserves $C_\gamma^\infty(F)$. From the above equalities, we obtain the inclusions

$$(2.15) \quad P_\gamma C^\infty(T_{\mathbb{C}})^{\Sigma,\varepsilon} \subset C_\gamma^\infty(T_{\mathbb{C}})^{\Sigma,\varepsilon}, \quad P_\gamma C^\infty(S^pT_{\mathbb{C}}^*)^{\Sigma,\varepsilon} \subset C_\gamma^\infty(S^pT_{\mathbb{C}}^*)^{\Sigma,\varepsilon},$$

for all $\gamma \in \Gamma$, and we see that $C_\gamma^\infty(F)^{\Sigma,\varepsilon}$ is the isotypic component of the G-module $C^\infty(F)^{\Sigma,\varepsilon}$ corresponding to γ. If F is the trivial complex line bundle $S^0T_{\mathbb{C}}^*$, we set

$$C^\infty(X)^{\Sigma,\varepsilon} = C^\infty(F)^{\Sigma,\varepsilon}, \qquad C_\gamma^\infty(X)^{\Sigma,\varepsilon} = C_\gamma^\infty(F)^{\Sigma,\varepsilon},$$

for $\gamma \in \Gamma$.

If τ is an involutive isometry of X which commutes with the action of G and if Σ is the set $\{\tau\}$, we have the relations

$$C^\infty(F)^{\Sigma,+1} = C^\infty(F)^{\mathrm{ev}}, \qquad C^\infty(F)^{\Sigma,-1} = C^\infty(F)^{\mathrm{odd}}$$

involving the G-modules defined in §4; moreover, for $\gamma \in \Gamma$, we have the orthogonal decomposition of G-modules

$$(2.16) \qquad C_\gamma^\infty(F) = C_\gamma^\infty(F)^{\mathrm{ev}} \oplus C_\gamma^\infty(F)^{\mathrm{odd}}.$$

According to (1.6) and observations made in §1, Chapter I, the differential operator $D_0 : \mathcal{T}_\mathbb{C} \to S^2 T_\mathbb{C}^*$ is homogeneous and elliptic. Hence according to §2, we have

$$D_0 C_\gamma^\infty(\mathcal{T}_\mathbb{C}) \subset C_\gamma^\infty(S^2 T_\mathbb{C}^*),$$

for all $\gamma \in \Gamma$. Since D_0 is an elliptic operator and $\varepsilon = \pm 1$, we easily see that

$$(2.17) \qquad D_0 C^\infty(\mathcal{T}_\mathbb{C})^{\Sigma,\varepsilon} = D_0 C^\infty(\mathcal{T}_\mathbb{C}) \cap C^\infty(S^2 T_\mathbb{C}^*)^{\Sigma,\varepsilon}.$$

The differential operators

$$d : S^0 T_\mathbb{C}^* \to T_\mathbb{C}^*, \qquad \mathrm{div} : S^2 T_\mathbb{C}^* \to T_\mathbb{C}^*$$

and the Lichnerowicz Laplacian $\Delta : S^p T_\mathbb{C}^* \to S^p T_\mathbb{C}^*$ are homogeneous, and so we have

$$d C_\gamma^\infty(X) \subset C_\gamma^\infty(T_\mathbb{C}^*), \qquad \mathrm{div}\, C_\gamma^\infty(S^2 T_\mathbb{C}^*) \subset C_\gamma^\infty(T_\mathbb{C}^*),$$

$$\Delta C_\gamma^\infty(S^p T_\mathbb{C}^*) \subset C_\gamma^\infty(S^p T_\mathbb{C}^*),$$

for all $\gamma \in \Gamma$. We also see that

$$d C^\infty(X)^{\Sigma,\varepsilon} = d C^\infty(X) \cap C^\infty(T_\mathbb{C}^*)^{\Sigma,\varepsilon}.$$

We obtain the following two propositions from the inclusions (2.12), (2.14) and (2.15), and from Theorem 5.3.6 of [56].

PROPOSITION 2.27. *Let F be a complex sub-bundle of $S^p T_\mathbb{C}^*$ on the symmetric space (X, g) which is invariant under the group G and the isometries of Σ. Then if ε a real number equal to ± 1, the submodule*

$$\bigoplus_{\gamma \in \Gamma} \left(\mathcal{N}_{p,\mathbb{C}} \cap C_\gamma^\infty(F)^{\Sigma,\varepsilon} \right)$$

is a dense subspace of $\mathcal{N}_{p,\mathbb{C}} \cap C^\infty(F)^{\Sigma,\varepsilon}$.

PROPOSITION 2.28. *Let F be a complex sub-bundle of $S^p T_\mathbb{C}^*$ on the symmetric space (X, g) which is invariant under the group G and the isometries of Σ. Then if ε a real number equal to ± 1, the submodule*

$$\bigoplus_{\gamma \in \Gamma} \left(\mathcal{Z}_{p,\mathbb{C}} \cap C_\gamma^\infty(F)^{\Sigma,\varepsilon} \right)$$

is a dense subspace of $\mathcal{Z}_{p,\mathbb{C}} \cap C^\infty(F)^{\Sigma,\varepsilon}$.

The following proposition is a direct consequence of Propositions 2.27 and 2.28.

PROPOSITION 2.29. *Let (X, g) be a symmetric space of compact type. Let Σ be a finite set of isometries of X which commute with the action of G on X and let ε be a real number equal to ± 1.*

(i) *The restriction of the maximal flat Radon transform for functions on X to the space $C^\infty(X)^{\Sigma,\varepsilon}$ is injective if and only if the equality*

$$\mathcal{N}_{0,\mathbb{C}} \cap C_\gamma^\infty(X)^{\Sigma,\varepsilon} = \{0\}$$

holds for all $\gamma \in \Gamma$.

(ii) *The restriction of the X-ray transform for functions on X to the space $C^\infty(X)^{\Sigma,\varepsilon}$ is injective if and only if the equality*

$$\mathcal{Z}_{0,\mathbb{C}} \cap C_\gamma^\infty(X)^{\Sigma,\varepsilon} = \{0\}$$

holds for all $\gamma \in \Gamma$.

Proposition 2.29 implies the following assertions:

(i) The maximal flat Radon transform for functions on X is injective if and only if the equality

$$\mathcal{N}_{0,\mathbb{C}} \cap C_\gamma^\infty(X) = \{0\}$$

holds for all $\gamma \in \Gamma$.

(ii) The X-ray transform for functions on X is injective if and only if the equality

$$\mathcal{Z}_{0,\mathbb{C}} \cap C_\gamma^\infty(X) = \{0\}$$

holds for all $\gamma \in \Gamma$.

As we mentioned above, for $\gamma \in \Gamma_0$, the G-module $C_\gamma^\infty(X)$ is irreducible. Thus in order to prove that the maximal flat Radon (resp. the X-ray) transform for functions on X is injective, it suffices to carry out the following steps:

(i) For all $\gamma \in \Gamma_0$, describe an explicit non-zero vector f_γ of the G-module $C_\gamma^\infty(X)$.

(ii) For $\gamma \in \Gamma_0$, prove that the function $\hat{f}_\gamma$ (resp. the function $\check{f}_\gamma$) is non-zero.

By Lemmas 2.6 and 2.11, we see that

$$(2.18) \qquad D_0 C^\infty(T_{\mathbb{C}}) \subset \mathcal{Z}_{2,\mathbb{C}} \subset \mathcal{N}_{2,\mathbb{C}}, \qquad dC^\infty(X) \subset \mathcal{Z}_{1,\mathbb{C}} \subset \mathcal{N}_{1,\mathbb{C}}.$$

Clearly, the space X is rigid in the sense of Guillemin (resp. is infinitesimally rigid) if and only if the equality

$$\mathcal{N}_{2,\mathbb{C}} = D_0 C^\infty(T_{\mathbb{C}})$$

holds (resp. the equality

$$\mathcal{Z}_{2,\mathbb{C}} = D_0 C^\infty(T_\mathbb{C})$$

holds). Also, the equality

$$\mathcal{N}_{1,\mathbb{C}} = d C^\infty(X)$$

implies (resp. the equality

$$\mathcal{Z}_{1,\mathbb{C}} = d C^\infty(X)$$

implies) that any differential form of degree 1 on X satisfying the Guillemin (resp. the zero-energy) condition is exact.

PROPOSITION 2.30. *Let (X, g) be a symmetric space of compact type. Let Σ be a finite set of isometries of X which commute with the action of G on X and let ε be a real number equal to ± 1.*
  (i) *The equality*

$$\mathcal{N}_{2,\mathbb{C}} \cap C^\infty(S^2 T_\mathbb{C}^*)^{\Sigma,\varepsilon} = D_0 C^\infty(T_\mathbb{C})^{\Sigma,\varepsilon}$$

holds if and only if

$$\mathcal{N}_{2,\mathbb{C}} \cap C_\gamma^\infty(S^2 T_\mathbb{C}^*)^{\Sigma,\varepsilon} = D_0 C_\gamma^\infty(T_\mathbb{C})^{\Sigma,\varepsilon},$$

for all $\gamma \in \Gamma$.
  (ii) *The equality*

$$\mathcal{Z}_{2,\mathbb{C}} \cap C^\infty(S^2 T_\mathbb{C}^*)^{\Sigma,\varepsilon} = D_0 C^\infty(T_\mathbb{C})^{\Sigma,\varepsilon}$$

holds if and only if

$$\mathcal{Z}_{2,\mathbb{C}} \cap C_\gamma^\infty(S^2 T_\mathbb{C}^*)^{\Sigma,\varepsilon} = D_0 C_\gamma^\infty(T_\mathbb{C})^{\Sigma,\varepsilon},$$

for all $\gamma \in \Gamma$.

PROOF: Since D_0 is an elliptic homogeneous differential operator, the assertions of the proposition follow from the first inclusion of (2.15), the equality (2.17) and from Propositions 2.2,(iii), 2.27 and 2.28.

Proposition 2.30 gives us the following criteria for the Guillemin rigidity and the infinitesimal rigidity of X, which are analogous to the criteria for the injectivity of the Radon transforms for functions on X obtained from Proposition 2.29:

PROPOSITION 2.31. *Let (X,g) be a symmetric space of compact type.*
(i) *The space X is rigid in the sense of Guillemin if and only if*

$$\mathcal{N}_{2,\mathbb{C}} \cap C_\gamma^\infty(S^2 T_\mathbb{C}^*) = D_0 C_\gamma^\infty(T_\mathbb{C}),$$

for all $\gamma \in \Gamma$.
(ii) *The space X is infinitesimally rigid if and only if*

$$\mathcal{Z}_{2,\mathbb{C}} \cap C_\gamma^\infty(S^2 T_\mathbb{C}^*) = D_0 C_\gamma^\infty(T_\mathbb{C}),$$

for all $\gamma \in \Gamma$.

PROPOSITION 2.32. *Let (X,g) be a symmetric space of compact type. Let Σ be a finite set of isometries of X which commute with the action of G on X and let ε be a real number equal to ± 1.*
(i) *The equality*

$$\mathcal{N}_{1,\mathbb{C}} \cap C^\infty(T_\mathbb{C}^*)^{\Sigma,\varepsilon} = dC^\infty(X)^{\Sigma,\varepsilon}$$

holds if and only if

$$\mathcal{N}_{1,\mathbb{C}} \cap C_\gamma^\infty(T_\mathbb{C}^*)^{\Sigma,\varepsilon} = dC_\gamma^\infty(X)^{\Sigma,\varepsilon},$$

for all $\gamma \in \Gamma$.
(ii) *The equality*

$$\mathcal{Z}_{1,\mathbb{C}} \cap C^\infty(T_\mathbb{C}^*)^{\Sigma,\varepsilon} = dC^\infty(X)^{\Sigma,\varepsilon}$$

holds if and only if

$$\mathcal{Z}_{1,\mathbb{C}} \cap C_\gamma^\infty(T_\mathbb{C}^*)^{\Sigma,\varepsilon} = dC_\gamma^\infty(X)^{\Sigma,\varepsilon},$$

for all $\gamma \in \Gamma$.

PROOF: Since the exterior differential operator d acting on $C^\infty(X)$ is an elliptic homogeneous differential operator, the assertions of the proposition follow from (2.15) and Propositions 2.2,(iii), 2.27 and 2.28.

Proposition 2.32 gives us the following criteria, which are analogous to the criteria for the Guillemin rigidity and the infinitesimal rigidity of X given by Proposition 2.31:

PROPOSITION 2.33. *Let (X,g) be a symmetric space of compact type. The following assertions are equivalent:*
(i) *A differential form of degree 1 on the space X satisfies the Guillemin (resp. the zero-energy) condition if and only if it is exact.*

(ii) *The equality*

$$\mathcal{N}_{1,\mathbb{C}} \cap C^\infty_\gamma(T^*_\mathbb{C}) = dC^\infty_\gamma(X)$$

holds (resp. the equality

$$\mathcal{Z}_{1,\mathbb{C}} \cap C^\infty_\gamma(T^*_\mathbb{C}) = dC^\infty_\gamma(X)$$

holds) for all $\gamma \in \Gamma$.

Suppose that Σ is equal to a finite group Λ of isometries of X which commute with the action of G and that ε is equal to $+1$. Assume that Λ acts without fixed points and that the quotient space $Y = X/\Lambda$ is a symmetric space; furthermore, assume that there is a subgroup K' of G containing K and a G-equivariant diffeomorphism $\varphi : Y \to G/K'$ satisfying properties (i) and (ii) of §4. If the vector bundle F is a sub-bundle of $S^pT^*_\mathbb{C}$ which is invariant under Λ, we consider the sub-bundle F_Y of $S^pT^*_{Y,\mathbb{C}}$ determined by F; then for $\gamma \in \Gamma$, the G-submodule

$$C^\infty_\gamma(F)^\Lambda = C^\infty_\gamma(F)^{\Sigma,\varepsilon}$$

of $C^\infty_\gamma(F)$ is the isotypic component of $C^\infty(F)^\Lambda$ corresponding to γ and the isomorphism (2.5) induces an isomorphism

(2.19) $$\varpi^* : C^\infty_\gamma(Y, F_Y) \to C^\infty_\gamma(F)^\Lambda$$

of G-modules. If Λ is the group $\{\text{id}, \tau\}$, where τ is an involutive isometry of X, then, for $\gamma \in \Gamma$, the mapping (2.19) gives us an isomorphism

$$\varpi^* : C^\infty_\gamma(Y, F_Y) \to C^\infty_\gamma(F)^{\text{ev}}$$

of G-modules.

We choose a Cartan subalgebra of the complexification $\mathfrak{g}$ of the Lie algebra of G and fix a system of positive roots of $\mathfrak{g}$. Let p be an integer equal to 1 or 2 and consider the corresponding homogeneous differential operator $D^{p-1} : S^{p-1}T^* \to S^pT^*$. We consider the following properties which the space X and the group Λ might possess:

(A_p) Let γ be an arbitrary element of Γ, and let u be an arbitrary highest weight vector of the G-module $C^\infty_\gamma(S^pT^*_\mathbb{C})^\Lambda$; if the section u satisfies the Guillemin condition, then u belongs to $D^{p-1}C^\infty(S^{p-1}T^*_\mathbb{C})^\Lambda$.

(B_p) Let γ be an arbitrary element of Γ, and let u be an arbitrary highest weight vector of the G-module $C^\infty_\gamma(S^pT^*_\mathbb{C})^\Lambda$; if the section u satisfies the zero-energy condition, then u belongs to $D^{p-1}C^\infty(S^{p-1}T^*_\mathbb{C})^\Lambda$.

According to the relation (1.4) and Propositions 2.30 and 2.32, we see that in order to prove that the equality

$$\mathcal{N}_{p,\mathbb{C}} \cap C^\infty(S^p T^*_\mathbb{C})^\Lambda = D^{p-1} C^\infty(S^{p-1} T^*_\mathbb{C})^\Lambda$$

holds (resp. the equality

$$\mathcal{Z}_{p,\mathbb{C}} \cap C^\infty(S^p T^*_\mathbb{C})^\Lambda = D^{p-1} C^\infty(S^{p-1} T^*_\mathbb{C})^\Lambda$$

holds), it suffices to verify that X and Λ possess property A_p (resp. property B_p).

Thus according to Proposition 2.18 (resp. Proposition 2.19) and the relation (1.4), we know that, if the space X and the group Λ possess property A_2 (resp. property B_2), the space Y is rigid in the sense of Guillemin (resp. is infinitesimally rigid). On the other hand, according to Proposition 2.20, if the space X and the group Λ possess property A_1 (resp. property B_1), then every differential form of degree 1 on the space Y, which satisfies the Guillemin (resp. the zero-energy) condition, is exact.

These methods for proving the rigidity of a symmetric space of compact type were first introduced in [14] in the case of the complex projective space (see §5, Chapter III). The analogous method for proving the injectivity of Radon transforms for functions described above was first used by Funk to prove Theorem 2.23 for the 2-sphere S^2 and the real projective plane $\mathbb{RP}^2$ (see also Proposition 3.17); it was also applied by Grinberg in [31] to other projective spaces. The methods described above will be applied to the real Grassmannian $G^{\mathbb{R}}_{2,n}$ of 2-planes in $\mathbb{R}^{n+2}$ and to the complex quadric Q_n of dimension n. In fact, they shall be used in Chapter VI to show that the real Grassmannian $G^{\mathbb{R}}_{2,3}$ is rigid in the sense of Guillemin and that the complex quadric Q_3 is infinitesimally rigid, and in §4, Chapter X to show that the real Grassmannian $G^{\mathbb{R}}_{2,2}$ is infinitesimally rigid. Also the criterion for the exactness of a differential form of degree 1, which we have just described, shall be used in the case of the real projective plane $\mathbb{RP}^2$ in Chapter III, in the case of the complex quadric Q_n and of the real Grassmannian $G^{\mathbb{R}}_{2,n}$, with $n \geq 3$, in §11, Chapter VI, and in the case of the real Grassmannian $G^{\mathbb{R}}_{2,2}$ in §4, Chapter X.

§6. Lie algebras

Let $\mathfrak{g}$ be a complex semi-simple Lie algebra. The Casimir element of $\mathfrak{g}$ operates by a scalar $c(V)$ on an irreducible finite-dimensional $\mathfrak{g}$-module V. If $\mathfrak{g}$ is simple, the Casimir element of $\mathfrak{g}$ acts on the irreducible $\mathfrak{g}$-module $V_{\mathfrak{g}}$ corresponding to the adjoint representation of $\mathfrak{g}$ by the identity mapping, and so $c(V_{\mathfrak{g}}) = 1$ (see Theorem 3.11.2 of [55]).

LEMMA 2.34. *Let $\mathfrak{g}$ be a complex semi-simple Lie algebra. Let V_1 and V_2 be irreducible finite-dimensional $\mathfrak{g}$-modules. Then the $\mathfrak{g}$-modules V_1 and V_2 are isomorphic if and only if $c(V_1) = c(V_2)$.*

PROOF: Choose a Cartan subalgebra $\mathfrak{h}$ of $\mathfrak{g}$ and fix a system of positive roots of $\mathfrak{g}$. Let λ_1 and λ_2 be the highest weights of the irreducible $\mathfrak{g}$-modules V_1 and V_2, respectively. If $c(V_1) = c(V_2)$, then the infinitesimal characters of these $\mathfrak{g}$-modules are equal; if δ is half the sum of the positive roots, by Harish-Chandra's theorem there exists an element w of the Weyl group W of $(\mathfrak{g}, \mathfrak{h})$ such that $w(\lambda_1 + \delta) = \lambda_2 + \delta$ (see §5 and Theorem 5.62 in Chapter V of [39]). Since the weights $\lambda_1 + \delta$ and $\lambda_2 + \delta$ are dominant integral, we know that w must be equal to the identity element of W (see Lemma 4.7.4 of [55]), and so we have $\lambda_1 = \lambda_2$; hence the modules V_1 and V_2 are isomorphic.

Let G be a compact connected, semi-simple Lie group, whose Lie algebra we denote by $\mathfrak{g}_0$. A complex G-module V can be viewed as a $\mathfrak{g}_0$-module and so the Casimir element of $\mathfrak{g}_0$ operates on V; if V is an irreducible G-module, the Casimir element of $\mathfrak{g}_0$ acts by the scalar $c(V)$ on V.

From Lemma 2.34, we obtain the following result:

LEMMA 2.35. *Let G be a compact connected, semi-simple Lie group. Let V_1 and V_2 be irreducible complex G-modules. Then the G-modules V_1 and V_2 are isomorphic if and only if $c(V_1) = c(V_2)$.*

§7. Irreducible symmetric spaces

We consider the symmetric space (X, g) of compact type of §4. We write X as the homogeneous space G/K, where G is a compact, connected semi-simple Lie group and K is a closed subgroup of G. We suppose that g is a G-invariant metric and that (G, K) is a Riemannian symmetric pair of compact type. We continue to denote by Γ the dual $\hat{G}$ of the group G. Let x_0 be the point of X corresponding to the coset of the identity element of G. If $\mathfrak{g}_0$ and $\mathfrak{k}_0$ are the Lie algebras of G and K, respectively, we consider the Cartan decomposition $\mathfrak{g}_0 = \mathfrak{k}_0 \oplus \mathfrak{p}_0$ corresponding to the Riemannian symmetric pair (G, K), where $\mathfrak{p}_0$ is a subspace of $\mathfrak{g}_0$. We identify $\mathfrak{p}_0$ with the tangent space to X at the point x_0. If B is the Killing form of the Lie algebra $\mathfrak{g}_0$ of G, then the restriction of $-B$ to $\mathfrak{p}_0$ induces a G-invariant Riemannian metric g_0 on X. The complexifications $\mathfrak{g}$ of $\mathfrak{g}_0$ and $\mathfrak{p}$ of $\mathfrak{p}_0$ are K-modules. We denote by $S^k\mathfrak{p}$ the k-th symmetric product of $\mathfrak{p}$ and by $S^2_0\mathfrak{p}$ the K-submodule of $S^2\mathfrak{p}$ consisting of those elements of $S^2\mathfrak{p}$ of trace zero with respect to the Killing form of $\mathfrak{g}$. The isotropy group K acts on T_{x_0}; in fact, the K-modules $T_{\mathbb{C},x_0}$, $S^k T^*_{\mathbb{C},x_0}$ and $S^2_0 T^*_{\mathbb{C},x_0}$ are isomorphic to the K-modules $\mathfrak{p}$, $S^k\mathfrak{p}$ and $S^2_0\mathfrak{p}$, respectively. If X is an irreducible Hermitian symmetric space, then we have

(2.20) $\dim_{\mathbb{C}} \operatorname{Hom}_K(\mathfrak{g}, \mathbb{C}) = 1,$ $\dim_{\mathbb{C}} \operatorname{Hom}_K(\mathfrak{g}, \mathfrak{p}) = 2.$

If X is an irreducible symmetric space which is not Hermitian, then we have

(2.21) $\operatorname{Hom}_K(\mathfrak{g}, \mathbb{C}) = \{0\},$ $\dim_{\mathbb{C}} \operatorname{Hom}_K(\mathfrak{g}, \mathfrak{p}) = 1.$

The G-module

$$\mathcal{K} = \{\, \xi \in C^\infty(T) \mid D_0 \xi = 0 \,\}$$

of all Killing vector fields on X is isomorphic to $\mathfrak{g}_0$. We identify its complexification $\mathcal{K}_{\mathbb{C}}$ with the G-module

$$\{\, \xi \in C^\infty(T_{\mathbb{C}}) \mid D_0 \xi = 0 \,\}$$

of complex vector fields on X, which is isomorphic to $\mathfrak{g}$. We know that a Killing vector field ξ on X satisfies the relation $d^* g^\flat(\xi) = 0$; thus the subspaces $g^\flat(\mathcal{K}_{\mathbb{C}})$ and $dC^\infty(X)$ of $C^\infty(T_{\mathbb{C}}^*)$ are orthogonal.

The Lichnerowicz Laplacian Δ_g acting on $C^\infty(S^p T_{\mathbb{C}}^*)$ is self-adjoint and its eigenvalues are non-negative real numbers. Since the Laplacian Δ_g acting on $S^p T_{\mathbb{C}}^*$ is elliptic, the eigenspace

$$\{\, u \in C^\infty(S^p T_{\mathbb{C}}^*) \mid \Delta_g u = \mu u \,\}$$

of the Laplacian Δ_g corresponding to the eigenvalue $\mu \in \mathbb{R}$ is finite-dimensional. The Casimir element of $\mathfrak{g}_0$ acts by a scalar c_γ on an irreducible G-module which is a representative of $\gamma \in \Gamma$. According to [41, §5], the action of the Lichnerowicz Laplacian

$$\Delta_{g_0} : C^\infty(S^p T_{\mathbb{C}}^*) \to C^\infty(S^p T_{\mathbb{C}}^*)$$

corresponding to the metric g_0 on X on the G-module $C^\infty(S^p T_{\mathbb{C}}^*)$ coincides with the action of the Casimir element of $\mathfrak{g}_0$ on this G-module. Thus, for $\gamma \in \Gamma$, we see that $C^\infty_\gamma(S^p T_{\mathbb{C}}^*)$ is an eigenspace of the Lichnerowicz Laplacian Δ_{g_0} with eigenvalue c_γ. Since the operator Δ_{g_0} acting on $S^p T_{\mathbb{C}}^*$ is elliptic and real-analytic, the elements of $C^\infty_\gamma(S^p T_{\mathbb{C}}^*)$ are real-analytic sections of $S^p T_{\mathbb{C}}^*$. If F is a complex sub-bundle of $S^p T_{\mathbb{C}}^*$ invariant under the group G, then the Laplacian Δ_{g_0} preserves the space $C^\infty(F)$.

We suppose throughout the remainder of this section that X is an irreducible symmetric space. According to Lemma 1.21, the metric g is a positive multiple of g_0 and is an Einstein metric. In fact, by formula (1.65), we have Ric $= \lambda g$, where λ is a positive real number, and $g_0 = 2\lambda g$; moreover, the Lichnerowicz Laplacian Δ corresponding to the metric g is equal to $2\lambda\Delta_{g_0}$. Let F be a complex sub-bundle of $S^p T^*_{\mathbb{C}}$ invariant under the group G. If $\gamma \in \Gamma$, from Lemma 2.35 and the above remarks concerning the Laplacian Δ_{g_0} we infer that the G-submodule $C^\infty_\gamma(F)$ is equal to the eigenspace of Δ acting on $C^\infty(F)$ associated with the eigenvalue $\lambda_\gamma = 2\lambda c_\gamma$. Moreover, by Proposition 2.1 and the preceding remark, for $\mu \in \mathbb{R}$ we see that, if the eigenspace

$$\{\, u \in C^\infty(F) \mid \Delta u = \mu u \,\}$$

is non-zero, it is equal to the G-submodule $C^\infty_\gamma(F)$ of $C^\infty(F)$ and that μ is equal to λ_γ, for some appropriate element γ of Γ.

We denote by $\bar{u}$ the complex conjugate of an element of $C^\infty(S^p T^*_{\mathbb{C}})$. We consider the set Λ of all eigenvalues of the Laplacian Δ acting on $C^\infty(X)$ and the eigenspace $C^\infty(X)^\lambda$ of Δ associated with the eigenvalue $\lambda \in \Lambda$. For $\gamma \in \Gamma_0$, the irreducible G-module $C^\infty_\gamma(X)$ is equal to the eigenspace $C^\infty(X)^{\lambda_\gamma}$, where $\lambda_\gamma = 2\lambda c_\gamma$, and so is invariant under conjugation. Moreover, the mapping $\Gamma_0 \to \Lambda$, sending $\gamma \in \Gamma_0$ into λ_γ, is bijective. Hence by Proposition 2.1, the orthogonal direct sum

$$\bigoplus_{\gamma \in \Gamma_0} C^\infty_\gamma(X) = \bigoplus_{\lambda \in \Lambda} C^\infty(X)^\lambda$$

is a dense submodule of $C^\infty(X)$.

LEMMA 2.36. *Let (X, g) be an irreducible symmetric space of compact type. The subspaces $\mathcal{N}_{\mathbb{C}}$ and $\mathcal{Z}_{\mathbb{C}}$ of $C^\infty(S^p T^*_{\mathbb{C}})$ are invariant under the Lichnerowicz Laplacian Δ.*

PROOF: Let u be an element of $\mathcal{N}_{\mathbb{C}}$ (resp. of $\mathcal{Z}_{\mathbb{C}}$). Since Δ is a homogeneous differential operator, by (2.1) we have

$$P_\gamma \Delta u = \Delta P_\gamma u = \lambda_\gamma P_\gamma u,$$

for all $\gamma \in \Gamma$. If γ is an element of Γ, according to (2.12) (resp. to (2.14)), we know that $P_\gamma u$ belongs to $\mathcal{N}_{p,\mathbb{C}}$ (resp. to $\mathcal{Z}_{p,\mathbb{C}}$); therefore so does $P_\gamma \Delta u$. Since $\mathcal{N}_{\mathbb{C}}$ (resp. $\mathcal{Z}_{\mathbb{C}}$) is a closed subspace of $C^\infty(S^p T^*_{\mathbb{C}})$, we see that Δu also belongs to this subspace.

PROPOSITION 2.37. *Let (X, g) be an irreducible symmetric space of compact type. Let E be a G-invariant sub-bundle of $S^2 T^*$ and let h be a symmetric 2-form on X. Assume that there is a real number μ such that*

(2.22) $$\Delta h - \mu h \in C^\infty(E).$$

(i) *If μ is not an eigenvalue of the Laplacian Δ acting on $C^\infty(S^2 T_{\mathbb{C}}^*)$, then h is a section of E.*

(ii) *Assume that μ is an eigenvalue of the Laplacian Δ acting on $C^\infty(S^2 T_{\mathbb{C}}^*)$ and suppose that h satisfies $\operatorname{div} h = 0$ and $\operatorname{Tr} h = 0$. Then we can write $h = h_1 + h_2$, where h_1 is a section of $S^2 T^*$ and h_2 is a section of E satisfying*

$$\Delta h_1 = \mu h_1, \qquad \operatorname{div} h_1 = \operatorname{div} h_2 = 0, \qquad \operatorname{Tr} h_1 = \operatorname{Tr} h_2 = 0;$$

moreover, if h satisfies the Guillemin (resp. the zero-energy) condition, then we may require that h_1 and h_2 also satisfy the Guillemin (resp. the zero-energy) condition.

PROOF: Here we shall use Lemma 2.35 and the above remarks concerning the eigenspaces of the Laplacian Δ. For $\gamma \in \Gamma$, we infer from (2.22) that $P_\gamma(\Delta h - \mu h)$ belongs to $C^\infty(E_{\mathbb{C}})$; by (2.1), we have

$$(2.23) \qquad P_\gamma(\Delta h - \mu h) = \Delta P_\gamma h - \mu P_\gamma h = (\lambda_\gamma - \mu) P_\gamma h.$$

If μ is not an eigenvalue of Δ, then either the space $C_\gamma^\infty(S^2 T_{\mathbb{C}}^*)$ vanishes or we have the inequality $\lambda_\gamma \neq \mu$. Under this assumption on μ, we deduce that $P_\gamma h$ is an element of $C^\infty(E_{\mathbb{C}})$; it follows that h is a section of E. If μ is an eigenvalue of Δ, there is a unique element γ' of Γ satisfying $\lambda_{\gamma'} = \mu$. We set $h_1 = P_{\gamma'} h$; then $h_2 = h - h_2$ is orthogonal to the subspace $C_{\gamma'}^\infty(S^2 T_{\mathbb{C}}^*)$. If $\gamma \in \Gamma$ is not equal to γ', we have $\lambda_\gamma \neq \mu$ and, by (2.23), we see that $P_\gamma h_2 = P_\gamma h$ is an element of $C^\infty(E_{\mathbb{C}})$. Since $P_{\gamma'} h_2 = 0$, it follows that h_2 is a section of E. If h satisfies $\operatorname{Tr} h = 0$ and $\operatorname{div} h = 0$, by (2.1) we obtain the equalities $\operatorname{Tr} h_j = 0$ and $\operatorname{div} h_j = 0$, for $j = 1, 2$. If h satisfies the Guillemin (resp. the zero-energy) condition, then according to (2.12) (resp. to (2.14)) the section h_1 also satisfies the Guillemin (resp. the zero-energy) condition.

We shall require the following lemma in §8, Chapter V.

LEMMA 2.38. *Let (X, g) be an irreducible symmetric space of compact type. Let γ, γ' be elements of Γ. Let h be an element of $C_\gamma^\infty(S_0^2 T^*)$ and let f be an element of $C^\infty(X)$. Assume that the section fh of $S^2 T_{\mathbb{C}}^*$ is an element of $C_{\gamma'}^\infty(S^2 T_{\mathbb{C}}^*)$ and suppose that the sections h and fh belong to the space $D_0 C^\infty(T_{\mathbb{C}})$. Then we have*

$$(2.24) \qquad \tfrac{1}{2} df \cdot \operatorname{div} h - D^1((df)^\sharp \lrcorner h) + \tfrac{1}{2}(\lambda_\gamma - \lambda_{\gamma'}) fh = 0.$$

PROOF: The sections h and fh of $S_0^2 T_{\mathbb{C}}^*$ satisfy

$$\Delta h = \lambda_\gamma h, \qquad \Delta(fh) = \lambda_{\gamma'} fh.$$

The desired result is a direct consequence of formulas (1.39) and (1.10).

We now further assume that X is of type I, i.e. is not equal to a simple Lie group (see [36, p. 439]). We may suppose that the Lie group G is simple; then the complexification $\mathfrak{g}$ of the Lie algebra $\mathfrak{g}_0$ is simple. Let γ_1 be the element of Γ which is the equivalence class of the irreducible G-module $\mathfrak{g}$. We know that $c_{\gamma_1} = 1$, and hence we have $\lambda_{\gamma_1} = 2\lambda$. This observation and the above remarks concerning the Lichnerowicz Laplacians, together with the Frobenius reciprocity theorem, give us the following result:

LEMMA 2.39. *Let (X, g) be an irreducible symmetric space of compact type which is not equal to a simple Lie group. Let γ_1 be the element of Γ which is the equivalence class of the irreducible G-module $\mathfrak{g}$ and let F be a G-invariant complex sub-bundle of $S^p T_{\mathbb{C}}^*$. If λ is the positive real number satisfying $\mathrm{Ric} = \lambda g$, then the G-module $C_{\gamma_1}^\infty(F)$ is equal to the eigenspace*

$$\{\, u \in C^\infty(F) \mid \Delta u = 2\lambda u \,\}$$

of the Lichnerowicz Laplacian Δ, and the multiplicity of this G-module is equal to the dimension of the complex vector space $\mathrm{Hom}_K(\mathfrak{g}, F_{x_0})$.

If $E(X)_{\mathbb{C}}$ denotes the complexification of the space $E(X)$, Lemma 2.39 gives us the equalities

$$(2.25) \qquad \begin{aligned} E(X) &= \{\, h \in C_{\gamma_1}^\infty(S_0^2 T_{\mathbb{C}}^*) \mid h = \bar{h},\ \mathrm{div}\, h = 0 \,\}, \\ E(X)_{\mathbb{C}} &= \{\, h \in C_{\gamma_1}^\infty(S_0^2 T_{\mathbb{C}}^*) \mid \mathrm{div}\, h = 0 \,\}. \end{aligned}$$

Since $E(X)_{\mathbb{C}}$ is G-submodule of $C_{\gamma_1}^\infty(S_0^2 T_{\mathbb{C}}^*)$, we know that $E(X)_{\mathbb{C}}$ is equal to the direct sum of k copies of the irreducible G-module $\mathfrak{g}$, where k is the integer $\mathrm{Mult}\, E(X)_{\mathbb{C}}$; it follows that the G-module $E(X)$ is isomorphic to the direct sum of k copies of $\mathfrak{g}_0$. Moreover, we infer that the vanishing of the space $\dim_{\mathbb{C}} \mathrm{Hom}_K(\mathfrak{g}, S_0^2\mathfrak{p})$ implies that the space $E(X)$ vanishes.

Since the G-module $\mathcal{K}_{\mathbb{C}}$ is isomorphic to $\mathfrak{g}$, we see that

$$(2.26) \qquad \mathcal{K}_{\mathbb{C}} \subset C_{\gamma_1}^\infty(T_{\mathbb{C}});$$

this inclusion can also be obtained from Lemmas 1.5,(i) and 2.39. If X is not a Hermitian symmetric space, according to the second equality of (2.21) and the Frobenius reciprocity theorem we see that the relation

$$(2.27) \qquad C_{\gamma_1}^\infty(T_{\mathbb{C}}) = \mathcal{K}_{\mathbb{C}}$$

holds. If X is a Hermitian symmetric space, by (2.20) we see that $C_{\gamma_1}^\infty(X)$ is an irreducible G-module; since the decomposition of $T_{\mathbb{C}}^*$ given by (1.69) is G-invariant, by (2.20) we obtain the orthogonal decompositions

$$(2.28) \qquad C_{\gamma_1}^\infty(T_{\mathbb{C}}^*) = g^\flat(\mathcal{K}_{\mathbb{C}}) \oplus d C_{\gamma_1}^\infty(X) = \partial C_{\gamma_1}^\infty(X) \oplus \bar{\partial} C_{\gamma_1}^\infty(X),$$

whose components $g^\flat(\mathcal{K}_\mathbb{C})$, $dC^\infty_{\gamma_1}(X)$, $\partial C^\infty_{\gamma_1}(X)$ and $\bar{\partial} C^\infty_{\gamma_1}(X)$ are irreducible G-modules isomorphic to $\mathfrak{g}$.

PROPOSITION 2.40. *Let (X, g) be an irreducible symmetric space of compact type, which is not equal to a simple Lie group or to the sphere S^2. Then the space $E(X)$ of infinitesimal Einstein deformations of X is a G-module isomorphic to the direct sum of k copies of the irreducible G-module $\mathfrak{g}_0$ and its multiplicity k is equal to*

$$\dim_\mathbb{C} \operatorname{Hom}_K(\mathfrak{g}, S_0^2 \mathfrak{p}) - \dim_\mathbb{C} \operatorname{Hom}_K(\mathfrak{g}, \mathfrak{p}) + 1.$$

If X is a Hermitian symmetric space, this multiplicity k is equal to

$$\dim_\mathbb{C} \operatorname{Hom}_K(\mathfrak{g}, S_0^2 \mathfrak{p}) - 1.$$

If X is not a Hermitian symmetric space, this multiplicity k is equal to

$$\dim_\mathbb{C} \operatorname{Hom}_K(\mathfrak{g}, S_0^2 \mathfrak{p}),$$

and we have

$$(2.29) \quad E(X)_\mathbb{C} = C^\infty_{\gamma_1}(S_0^2 T_\mathbb{C}^*), \qquad E(X) = \{ h \in C^\infty_{\gamma_1}(S_0^2 T_\mathbb{C}^*) \mid h = \bar{h} \}.$$

PROOF: Since the differential operator $D_0^c : T \to S_0^2 T^*$ is homogeneous, from the decomposition (1.12) and the relations (2.1) and (2.25), we obtain the orthogonal decomposition

$$(2.30) \quad C^\infty_{\gamma_1}(S_0^2 T_\mathbb{C}^*) = D_0^c C^\infty_{\gamma_1}(T_\mathbb{C}) \oplus E(X)_\mathbb{C}.$$

We write

$$\mathcal{W} = \{ \xi \in C^\infty_{\gamma_1}(T_\mathbb{C}) \mid D_0^c \xi = 0 \}.$$

By (2.30), we have the equality

$$(2.31) \quad \operatorname{Mult} E(X)_\mathbb{C} = \operatorname{Mult} C^\infty_{\gamma_1}(S_0^2 T_\mathbb{C}^*) - \operatorname{Mult} C^\infty_{\gamma_1}(T_\mathbb{C}) + \operatorname{Mult} \mathcal{W}.$$

By Lemma 1.5,(ii), Proposition 1.6 and Lemma 2.39, we see that

$$\mathcal{W} = \mathcal{K}_\mathbb{C}.$$

Hence the equality (2.31) becomes

$$\operatorname{Mult} E(X)_\mathbb{C} = \operatorname{Mult} C^\infty_{\gamma_1}(S_0^2 T_\mathbb{C}^*) - \operatorname{Mult} C^\infty_{\gamma_1}(T_\mathbb{C}) + \operatorname{Mult} \mathcal{K}_\mathbb{C}.$$

Since Mult $\mathcal{K}_{\mathbb{C}}$ is equal to one, the preceding equality together with the Frobenius reciprocity theorem gives us the first assertion of the proposition. The other assertions of the proposition then follow from the second equalities of (2.20) and (2.21).

In [42], Koiso also showed that the assertions of the previous proposition also hold when the irreducible space X is a simple Lie group. The following lemma is stated without proof by Koiso (see Lemma 5.5 of [42]); for the irreducible symmetric spaces

$$SO(p+q)/S(O(p) \times O(q)), \qquad SU(p+q)/S(U(p) \times U(q)),$$

with $p, q \geq 2$, we shall verify the results of this lemma in Chapter IV (see Lemma 4.1), Chapter V (see Lemma 5.15) and §3, Chapter VIII.

LEMMA 2.41. *Let (X, g) be a simply-connected irreducible symmetric space of compact type which is not equal to a simple Lie group. If X is Hermitian, then the space $\mathrm{Hom}_K(\mathfrak{g}, S_0^2\mathfrak{p})$ is one-dimensional and if X is not Hermitian the space $\mathrm{Hom}_K(\mathfrak{g}, S_0^2\mathfrak{p})$ vanishes, unless X is one of the spaces appearing in the following table which gives the dimension of the space $\mathrm{Hom}_K(\mathfrak{g}, S_0^2\mathfrak{p})$:*

X	$\dim_{\mathbb{C}} \mathrm{Hom}_K(\mathfrak{g}, S_0^2\mathfrak{p})$
$SU(2)/S(U(1) \times U(1))$	0
$SU(p+q)/S(U(p) \times U(q))$, with $p, q \geq 2$	2
$SU(n)/SO(n)$, with $n \geq 3$	1
$SU(2n)/Sp(n)$, with $n \geq 3$	1
E_6/F_4	1

The first two spaces X of this table are Hermitian, while the last three are not Hermitian.

Since the space $SU(2)/S(U(1) \times U(1))$ is isometric to the sphere S^2, by Lemma 2.41 and the equalities (2.25) we know that the space $E(X)$ vanishes when X is the sphere S^2; we also proved this result directly in §3, Chapter I. Therefore from Proposition 2.40 and Lemma 2.41, we obtain the results of Theorem 1.22 when the space X of this theorem is not equal to a simple Lie group; moreover, when X is equal to one of the last four spaces of the table of Lemma 2.41, we see that the G-module $E(X)$ is isomorphic to $\mathfrak{g}_0$.

Thus according to Proposition 2.40 and Lemma 2.41, if X is an irreducible symmetric space of compact type, which is not equal to a simple Lie group, the space $E(X)$ either vanishes or is isomorphic to the G-module $\mathfrak{g}_0$.

From the relations (2.17), (2.18) and (2.26), we obtain the following result:

PROPOSITION 2.42. *Let (X, g) be an irreducible symmetric space of compact type, which is not equal to a simple Lie group, and let γ_1 be the element of Γ which is the equivalence class of the irreducible G-module $\mathfrak{g}$. Let Σ be a finite set of isometries of X which commute with the action of G on X and let ε be a real number equal to ± 1.*

(i) *For $\gamma \in \Gamma$, the equality*

$$\mathcal{N}_{2,\mathbb{C}} \cap C_\gamma^\infty (S^2 T_\mathbb{C}^*)^{\Sigma,\varepsilon} = D_0 C_\gamma^\infty (T_\mathbb{C})^{\Sigma,\varepsilon}$$

is equivalent to

$$\mathrm{Mult}\, \left(\mathcal{N}_{2,\mathbb{C}} \cap C_\gamma^\infty (S^2 T_\mathbb{C}^*)^{\Sigma,\varepsilon} \right) \leq \mathrm{Mult}\, C_\gamma^\infty (T_\mathbb{C})^{\Sigma,\varepsilon}$$

when $\gamma \neq \gamma_1$, or to

$$\mathrm{Mult}\, \left(\mathcal{N}_{2,\mathbb{C}} \cap C_{\gamma_1}^\infty (S^2 T_\mathbb{C}^*)^{\Sigma,\varepsilon} \right) \leq \mathrm{Mult}\, C_{\gamma_1}^\infty (T_\mathbb{C})^{\Sigma,\varepsilon} - \mathrm{Mult}\, \left(\mathcal{K}_\mathbb{C} \cap C^\infty (T_\mathbb{C})^{\Sigma,\varepsilon} \right)$$

when $\gamma = \gamma_1$.

(ii) *For $\gamma \in \Gamma$, the equality*

$$\mathcal{Z}_{2,\mathbb{C}} \cap C_\gamma^\infty (S^2 T_\mathbb{C}^*)^{\Sigma,\varepsilon} = D_0 C_\gamma^\infty (T_\mathbb{C})^{\Sigma,\varepsilon}$$

is equivalent to

$$\mathrm{Mult}\, \left(\mathcal{Z}_{2,\mathbb{C}} \cap C_\gamma^\infty (S^2 T_\mathbb{C}^*)^{\Sigma,\varepsilon} \right) \leq \mathrm{Mult}\, C_\gamma^\infty (T_\mathbb{C})^{\Sigma,\varepsilon}$$

when $\gamma \neq \gamma_1$, or to

$$\mathrm{Mult}\, \left(\mathcal{Z}_{2,\mathbb{C}} \cap C_{\gamma_1}^\infty (S^2 T_\mathbb{C}^*)^{\Sigma,\varepsilon} \right) \leq \mathrm{Mult}\, C_{\gamma_1}^\infty (T_\mathbb{C})^{\Sigma,\varepsilon} - \mathrm{Mult}\, \left(\mathcal{K}_\mathbb{C} \cap C^\infty (T_\mathbb{C})^{\Sigma,\varepsilon} \right)$$

when $\gamma = \gamma_1$.

Let d be the integer which is equal to 1 when X is a Hermitian symmetric space and equal to 0 otherwise. According to (2.27) and (2.28), we know that the multiplicity of the G-module $C_\gamma^\infty (T_\mathbb{C})$ is equal to $d+1$. Since its G-submodule $\mathcal{K}_\mathbb{C}$ is isomorphic to $\mathfrak{g}$, from Propositions 2.31 and 2.42 we deduce the following criteria for Guillemin rigidity and infinitesimal rigidity:

PROPOSITION 2.43. *Let (X, g) be an irreducible symmetric space of compact type, which is not equal to a simple Lie group, and let γ_1 be the element of Γ which is the equivalence class of the irreducible G-module $\mathfrak{g}$.*

(i) *If the inequality*

$$\mathrm{Mult}\, \left(\mathcal{N}_{2,\mathbb{C}} \cap C_\gamma^\infty (S^2 T_\mathbb{C}^*) \right) \leq \mathrm{Mult}\, C_\gamma^\infty (T_\mathbb{C})$$

holds for all $\gamma \in \Gamma$, with $\gamma \neq \gamma_1$, and if the inequality

$$\mathrm{Mult}\left(\mathcal{N}_{2,\mathbb{C}} \cap C_{\gamma_1}^\infty(S^2 T_\mathbb{C}^*)\right) \leq d$$

holds, then the symmetric space X is rigid in the sense of Guillemin.

(ii) If the inequality

$$\mathrm{Mult}\left(\mathcal{Z}_{2,\mathbb{C}} \cap C_\gamma^\infty(S^2 T_\mathbb{C}^*)\right) \leq \mathrm{Mult}\, C_\gamma^\infty(T_\mathbb{C})$$

holds for all $\gamma \in \Gamma$, with $\gamma \neq \gamma_1$, and if the inequality

$$\mathrm{Mult}\left(\mathcal{Z}_{2,\mathbb{C}} \cap C_{\gamma_1}^\infty(S^2 T_\mathbb{C}^*)\right) \leq d$$

holds, then the symmetric space X is infinitesimally rigid.

We choose a Cartan subalgebra of $\mathfrak{g}$ and fix a system of positive roots of $\mathfrak{g}$. If W is a G-submodule of $C_\gamma^\infty(S^p T_\mathbb{C}^*)$, with $\gamma \in \Gamma$, the dimension of its weight subspace, corresponding to the highest weight of $C_\gamma^\infty(S^p T_\mathbb{C}^*)$, is equal to the multiplicity of W. Thus according to Proposition 2.43,(i), to prove the Guillemin rigidity of X, it suffices to successively carry out the following steps:

(i) For all $\gamma \in \Gamma$, determine the multiplicities of the G-modules $C_\gamma^\infty(T_\mathbb{C})$ and $C_\gamma^\infty(S^2 T_\mathbb{C}^*)$.

(ii) For all $\gamma \in \Gamma$, describe an explicit basis for the weight subspace W_γ of the G-module $C_\gamma^\infty(S^2 T_\mathbb{C}^*)$ corresponding to its highest weight.

(iii) For $\gamma \in \Gamma$, consider the action of the Radon transform I_2 on the vectors of this basis for W_γ and prove that the inequality

$$\dim\left(\mathcal{N}_{2,\mathbb{C}} \cap W_\gamma\right) \leq \mathrm{Mult}\, C_\gamma^\infty(T_\mathbb{C})$$

holds whenever $\gamma \neq \gamma_1$, and that

$$\dim\left(\mathcal{N}_{2,\mathbb{C}} \cap W_{\gamma_1}\right) \leq d.$$

According to Proposition 2.43,(ii), to prove the infinitesimal rigidity of X, it suffices to carry out the steps (i) and (ii) given above and then the following step:

(iv) For $\gamma \in \Gamma$, prove that the inequality

$$\dim\left(\mathcal{Z}_{2,\mathbb{C}} \cap W_\gamma\right) \leq \mathrm{Mult}\, C_\gamma^\infty(T_\mathbb{C})$$

holds whenever $\gamma \neq \gamma_1$, and that

$$\dim\left(\mathcal{Z}_{2,\mathbb{C}} \cap W_{\gamma_1}\right) \leq d.$$

These methods for proving the rigidity of an irreducible symmetric space of compact type implement the criteria described at the end of §5. They were first used in [14] to show that the complex projective space is infinitesimally rigid (see §5, Chapter III).

§8. Criteria for the rigidity of an irreducible symmetric space

We consider the symmetric space (X, g) of compact type of §§4 and 7 and continue to view X as the homogeneous space G/K. We recall that a closed connected totally geodesic submanifold Y of X is a symmetric space; moreover, if x is a point of Y and the tangent space to Y at x is equal to the subspace V of T_x, then the submanifold Y is equal to the subset $\mathrm{Exp}_x V$ of X (see §7 in Chapter IV of [36]).

Let $\mathcal{F}$ be a family of closed connected totally geodesic surfaces of X which is invariant under the group G. Then the set $N_{\mathcal{F}}$ (resp. $N'_{\mathcal{F}}$) consisting of those elements of B (resp. of $\bigwedge^2 T^* \otimes \bigwedge^2 T^*$), which vanish when restricted to the submanifolds belonging to $\mathcal{F}$, is a sub-bundle of B (resp. of $\bigwedge^2 T^* \otimes \bigwedge^2 T^*$). Clearly, we have $N_{\mathcal{F}} \subset N'_{\mathcal{F}}$. According to formula (1.56), we see that

$$\tilde{B} \subset N_{\mathcal{F}};$$

we shall identify $N_{\mathcal{F}}/\tilde{B}$ with a sub-bundle of $B/\tilde{B}$. If $\beta : B/\tilde{B} \to B/N_{\mathcal{F}}$ is the canonical projection, we consider the differential operator

$$D_{1,\mathcal{F}} = \beta D_1 : S^2 T^* \to \mathcal{B}/\mathcal{N}_{\mathcal{F}}.$$

Let $\mathcal{F}'$ be a family of closed connected totally geodesic submanifolds of X. We denote by $\mathcal{L}(\mathcal{F}')$ the subspace of $C^\infty(S^2 T^*)$ consisting of all symmetric 2-forms h which satisfy the following condition: for all submanifolds $Z \in \mathcal{F}'$, the restriction of h to Z is a Lie derivative of the metric of Z induced by g. By Lemma 1.1, we know that $D_0 C^\infty(T)$ is a subspace of $\mathcal{L}(\mathcal{F}')$. We consider the following properties which the family $\mathcal{F}'$ might possess:

(I) If a section of $S^2 T^*$ over X satisfies the Guillemin condition, then its restriction to an arbitrary submanifold of X belonging to the family $\mathcal{F}'$ satisfies the Guillemin condition.

(II) Every submanifold of X belonging to $\mathcal{F}'$ is rigid in the sense of Guillemin.

(III) Every submanifold of X belonging to $\mathcal{F}'$ is infinitesimally rigid.

If the family $\mathcal{F}'$ possesses properties (I) and (II), then we see that

$$\mathcal{N}_2 \subset \mathcal{L}(\mathcal{F}').$$

On the other hand, the restriction of an element of $\mathcal{Z}_2$ to an arbitrary submanifold of X belonging to the family $\mathcal{F}'$ satisfies the zero-energy condition;

hence if the family $\mathcal{F}'$ possesses property (III), we have the inclusion

$$\mathcal{Z}_2 \subset \mathcal{L}(\mathcal{F}').$$

From Lemma 1.16, we obtain:

PROPOSITION 2.44. *Let (X, g) be a symmetric space of compact type. Let $\mathcal{F}$ be a family of closed connected totally geodesic surfaces of X which is invariant under the group G, and let $\mathcal{F}'$ be a family of closed connected totally geodesic submanifolds of X. Assume that each surface of X belonging to $\mathcal{F}$ is contained in a submanifold of X belonging to $\mathcal{F}'$. A symmetric 2-form h on X belonging to $\mathcal{L}(\mathcal{F}')$ satisfies the relation $D_{1,\mathcal{F}}h = 0$.*

PROPOSITION 2.45. *Let (X, g) be a symmetric space of compact type. Let $\mathcal{F}$ be a G-invariant family of closed connected totally geodesic surfaces of X with positive constant curvature. Let h be an element of $C^\infty(S^2T^*)$. Then the following assertions are equivalent:*

  (i) *The symmetric 2-form h belongs to $\mathcal{L}(\mathcal{F})$.*
  (ii) *The section $D_g h$ of $\bigwedge^2 T^* \otimes \bigwedge^2 T^*$ is an element of $C^\infty(N'_{\mathcal{F}})$.*
  (iii) *The symmetric 2-form h satisfies $D_{1,\mathcal{F}}h = 0$.*

PROOF: By Lemma 1.15, we know that assertion (i) implies (ii). Now suppose that assertion (ii) holds. Let Y be a totally geodesic submanifold of X belonging to the family $\mathcal{F}$ and let $i : Y \to X$ be the natural imbedding. Then we have $i^* D_g h = 0$. If g_Y is the Riemannian metric on Y induced by g, by Proposition 1.14,(i) the restriction i^*h of h to the manifold Y satisfies $D_{g_Y} i^* h = 0$. Theorem 1.18 gives us the exactness of the sequence (1.51) corresponding to the Riemannian manifold (Y, g_Y) with positive constant curvature; therefore the form i^*h on Y is a Lie derivative of the metric g_Y. Thus we know that h belongs to $\mathcal{L}(\mathcal{F})$, and so assertion (ii) implies (i). Since $\tilde{B} \subset N_{\mathcal{F}}$, the equivalence of assertions (ii) and (iii) is a consequence of Proposition 1.14,(ii).

PROPOSITION 2.46. *Let (X, g) be a symmetric space of compact type. Let $\mathcal{F}'$ be a family of closed connected totally geodesic submanifolds of X.*

  (i) *Suppose that each closed geodesic of X is contained in a submanifold of X belonging to the family $\mathcal{F}'$. Then we have the inclusion*

$$\mathcal{L}(\mathcal{F}') \subset \mathcal{Z}_2.$$

  (ii) *Suppose that the sequence (1.24), corresponding to an arbitrary submanifold of X belonging to the family $\mathcal{F}'$, is exact. Let h be an element of $C^\infty(S^2T^*)$ satisfying the relation $D_1 h = 0$. Then h belongs to $\mathcal{L}(\mathcal{F}')$.*

  (iii) *Suppose that the hypothesis of (i) and of (ii) hold, and that the space X is infinitesimally rigid. Then the sequence (1.24) is exact.*

PROOF: Let h be an element of $C^\infty(S^2T^*)$. First, suppose that h belongs to $\mathcal{L}(\mathcal{F}')$ and that the hypothesis of (i) holds. Let γ be an arbitrary closed geodesic of X; then there is a submanifold Y of X belonging to the family $\mathcal{F}'$ containing γ. Let $i : Y \to X$ be the natural inclusion. Since the symmetric 2-form i^*h on Y is a Lie derivative of the metric of Y, the integral of h over γ vanishes; thus the symmetric 2-form h satisfies the zero-energy condition and assertion (i) holds. Next, let Y be an arbitrary submanifold of X belonging to the family $\mathcal{F}'$ and let $i : Y \to X$ be the natural inclusion. If $D_{1,Y}$ is the differential operator on the symmetric space Y defined in §1, Chapter I, according to formula (1.58) of Proposition 1.14 the relation $D_1 h = 0$ implies that $D_{1,Y} i^* h = 0$. If $D_1 h = 0$ and if the sequence (1.24) for Y is exact, it follows that $i^* h$ is a Lie derivative of the metric of Y. Thus assertion (ii) is true. Finally, assertion (iii) is a direct consequence of (i) and (ii).

THEOREM 2.47. *Let (X, g) be a symmetric space of compact type. Let $\mathcal{F}$ be a family of closed connected totally geodesic surfaces of X which is invariant under the group G, and let $\mathcal{F}'$ be a family of closed connected totally geodesic submanifolds of X. Assume that each surface of X belonging to $\mathcal{F}$ is contained in a submanifold of X belonging to $\mathcal{F}'$. Suppose that the relation (1.48) and the equality*

$$(2.32) \qquad\qquad N_{\mathcal{F}} = \tilde{B}$$

hold.

    (i) *A symmetric 2-form h on X belonging to $\mathcal{L}(\mathcal{F}')$ is a Lie derivative of the metric g.*

    (ii) *If the family $\mathcal{F}'$ possesses properties (I) and (II), then the symmetric space X is rigid in the sense of Guillemin.*

    (iii) *If the family $\mathcal{F}'$ possesses property (III), then the symmetric space X is infinitesimally rigid.*

PROOF: First, let h be a symmetric 2-form h on X belonging to $\mathcal{L}(\mathcal{F}')$. By Proposition 2.44, we see that $D_{1,\mathcal{F}} h = 0$. According to the equality (2.32), we therefore know that $D_1 h = 0$. By the relation (1.48) and Theorem 1.18, the sequence (1.24) is exact, and so we see that h is a Lie derivative of the metric g. Thus we have proved assertion (i). Now assume that the family $\mathcal{F}'$ satisfies the hypothesis of (i) (resp. of (ii)). Then we know that the space $\mathcal{N}_2$ (resp. the space $\mathcal{Z}_2$) is contained in $\mathcal{L}(\mathcal{F}')$. Assertion (ii) (resp. (iii)) is a consequence of (i).

We now assume that (X, g) is an irreducible symmetric space of compact type; then we have $\mathrm{Ric} = \lambda g$, where λ is a positive real number.

THEOREM 2.48. *Let (X, g) be an irreducible symmetric space of compact type. Let $\mathcal{F}$ be a family of closed connected totally geodesic surfaces of X which is invariant under the group G, and let $\mathcal{F}'$ be a family of closed* connected totally geodesic submanifolds of X. Let E be a G-invariant sub-bundle of $S_0^2 T^*$. Assume that each surface of X belonging to $\mathcal{F}$ is contained in a submanifold of X belonging to $\mathcal{F}'$, and suppose that the relation

$$(2.33) \qquad\qquad \operatorname{Tr} N_{\mathcal{F}} \subset E$$

holds.

    (i) Let h be a symmetric 2-form on X satisfying $\operatorname{div} h = 0$. If h belongs to $\mathcal{L}(\mathcal{F}')$ or satisfies the relation $D_{1,\mathcal{F}} h = 0$, then we may write

$$h = h_1 + h_2,$$

where h_1 is an element of $E(X)$ and h_2 is a section of E; moreover, if h also satisfies the Guillemin (resp. the zero-energy) condition, we may require that h_1 and h_2 satisfy the Guillemin (resp. the zero-energy) condition.

    (ii) Let Λ be a finite group of isometries of X, and assume that the vector bundle E is invariant under Λ. Suppose that

$$(2.34) \qquad\qquad C^\infty(E)^\Lambda \cap \mathcal{L}(\mathcal{F}') = \{0\},$$

and that the equality

$$(2.35) \qquad\qquad \mathcal{N}_2 \cap E(X) = \{0\}$$

holds (resp. the equality

$$(2.36) \qquad\qquad \mathcal{Z}_2 \cap E(X) = \{0\}$$

holds). Then a Λ-invariant symmetric 2-form on X belonging to $\mathcal{L}(\mathcal{F}')$ satisfies the Guillemin (resp. the zero-energy) condition if and only if it is a Lie derivative of the metric g.

    (iii) If the relations

$$(2.37) \qquad\qquad C^\infty(E) \cap \mathcal{L}(\mathcal{F}') = \{0\}$$

and $E(X) \subset C^\infty(E)$ hold, then we have

$$\mathcal{L}(\mathcal{F}') = D_0 C^\infty(T).$$

    PROOF: In view of Proposition 2.44, to prove assertion (i) it suffices to consider a symmetric 2-form h on X satisfying the relations $\operatorname{div} h = 0$ and $D_{1,\mathcal{F}} h = 0$. Since $\operatorname{Tr} E = \{0\}$, by Lemma 1.10 and (2.33) we infer that $\operatorname{Tr} h = 0$ and that

$$\Delta h - 2\lambda h \in C^\infty(E).$$

Assertion (i) is now a consequence of Proposition 2.37,(ii), with $\mu = 2\lambda$. Next, let k be a symmetric 2-form on X belonging to $\mathcal{L}(\mathcal{F}')$. According to the decomposition (1.11), we may write k as

$$k = h + D_0\xi,$$

where h is an element of $C^\infty(S^2T^*)$ satisfying $\operatorname{div} h = 0$, which is uniquely determined by k, and where ξ is a vector field on X. If k is invariant under a finite group Λ of isometries of X, clearly h is also Λ-invariant. Since $D_0\xi$ is an element of $\mathcal{L}(\mathcal{F}')$, the 2-form h also belongs to $\mathcal{L}(\mathcal{F}')$. According to (i), we may write $h = h_1 + h_2$, where h_1 is an element of $E(X)$ and h_2 is a section of E. If k satisfies the Guillemin (resp. the zero-energy) condition, according to Lemma 2.10 (resp. Lemma 2.6) so does h, and we may suppose that h_1 also satisfies the Guillemin (resp. the zero-energy) condition. First, if $E(X) \subset C^\infty(E)$, then h_1 and h are also sections of E; if moreover the equality (2.37) holds, then h vanishes and so k is equal to $D_0\xi$. Next, under the hypotheses of (ii), if k is Λ-invariant and satisfies the Guillemin (resp. the zero-energy) condition and if the equality (2.35) (resp. the equality (2.36)) holds, then h is a Λ-invariant section of E; according to (2.34), we infer that h vanishes, and so k is equal to $D_0\xi$. We have thus verified both assertions (ii) and (iii).

Since the differential operator $D_{1,\mathcal{F}}$ corresponding to the family $\mathcal{F}$ of Theorem 2.48 is homogeneous, according to the proof of Proposition 2.37 the sections h_1 and h_2 given by Theorem 2.48,(i) satisfy the relations

$$D_{1,\mathcal{F}}h_1 = D_{1,\mathcal{F}}h_2 = 0.$$

The following theorem gives criteria for the Guillemin rigidity or the infinitesimal rigidity of an irreducible symmetric space of compact type.

THEOREM 2.49. *Let (X, g) be an irreducible symmetric space of compact type. Let $\mathcal{F}$ be a family of closed connected totally geodesic surfaces of X which is invariant under the group G, and let $\mathcal{F}'$ be a family of closed connected totally geodesic submanifolds of X. Let E be a G-invariant sub-bundle of $S_0^2T^*$. Assume that each surface of X belonging to $\mathcal{F}$ is contained in a submanifold of X belonging to $\mathcal{F}'$, and suppose that the relations (2.33) and (2.37) hold.*

(i) If the family $\mathcal{F}'$ possesses properties (I) and (II) and if the equality (2.35) holds, then the symmetric space X is rigid in the sense of Guillemin.

(ii) If the family $\mathcal{F}'$ possesses property (III) and if the equality (2.36) holds, then the symmetric space X is infinitesimally rigid.

PROOF: Under the hypotheses of (i) (resp. of (ii)), a symmetric 2-form h on X satisfying the Guillemin (resp. the zero-energy) condition belongs to $\mathcal{L}(\mathcal{F}')$; by Theorem 2.48,(ii), with $\Lambda = \{id\}$, we see that h is a Lie derivative of the metric g.

According to Proposition 2.13, we know that the equality (2.35) (resp. the equality (2.36)) is a necessary condition for the Guillemin rigidity (resp. the infinitesimal rigidity) of X.

If we take $E = \{0\}$ in Theorem 2.49, we obtain the following corollary of Theorem 2.49:

THEOREM 2.50. *Let (X, g) be an irreducible symmetric space of compact type. Let $\mathcal{F}$ be a family of closed connected totally geodesic surfaces of X which is invariant under the group G, and let $\mathcal{F}'$ be a family of closed connected totally geodesic submanifolds of X. Assume that each surface of X belonging to $\mathcal{F}$ is contained in a submanifold of X belonging to $\mathcal{F}'$. Suppose that the equality*

$$\mathrm{Tr}\, N_{\mathcal{F}} = \{0\}$$

holds. Then assertions (i) and (ii) of Theorem 2.49 hold.

Thus according to Theorem 2.50, when X is an irreducible space, in Theorem 2.47 in order to obtain assertion (ii) (resp. assertion (iii)) of the latter theorem we may replace the hypothesis that the relation (1.48) holds by the hypothesis that the equality (2.35) (resp. the equality (2.36)) holds.

We again assume that X is an arbitrary symmetric space of compact type. We consider the following properties which the family $\mathcal{F}'$ might possess:

(IV) If a one-form over X satisfies the Guillemin condition, then its restriction to an arbitrary submanifold of X belonging to the family $\mathcal{F}'$ satisfies the Guillemin condition.

(V) If Y is an arbitrary submanifold of X belonging to the family $\mathcal{F}'$, every form of degree one on Y satisfying the Guillemin is exact.

(VI) If Y is an arbitrary submanifold of X belonging to the family $\mathcal{F}'$, every form of degree one on Y satisfying the zero-energy is exact.

We consider the subset $C_{\mathcal{F}'}$ of $\bigwedge^2 T^*$ consisting of those elements of $\bigwedge^2 T^*$, which vanish when restricted to the submanifolds belonging to the family $\mathcal{F}'$; if the family $\mathcal{F}'$ is invariant under the group G, then $C_{\mathcal{F}'}$ is a sub-bundle of $\bigwedge^2 T^*$.

THEOREM 2.51. *Let (X, g) be a symmetric space of compact type. Let $\mathcal{F}$ and $\mathcal{F}'$ be two families of closed connected totally geodesic submanifolds of X. Assume that each submanifold of X belonging to $\mathcal{F}$ is contained in a submanifold of X belonging to $\mathcal{F}'$, and suppose that*

(2.38) $C_{\mathcal{F}} = \{0\}.$

(i) *If the family $\mathcal{F}'$ possesses properties (IV) and (V), then a differential form of degree one on X satisfies the Guillemin condition if and only if it is exact.*

(ii) *If the family $\mathcal{F}'$ possesses property (VI), then a differential form of degree one on X satisfies the zero-energy condition if and only if it is exact.*

PROOF: Suppose that the family $\mathcal{F}'$ possesses properties (IV) and (V) (resp. possesses property (VI)). Let θ be a 1-form on X satisfying the Guillemin (resp. the zero-energy) condition. Consider a submanifold Y of X belonging to the family $\mathcal{F}'$. According to our hypotheses, the restriction θ' of θ to Y satisfies the Guillemin (resp. the zero-energy) condition; it follows that the 1-form θ' on Y is closed. Hence the restriction of the 2-form $d\theta$ to Y vanishes, and so the restriction of $d\theta$ to an arbitrary submanifold of X belonging to the family $\mathcal{F}$ vanishes. From the relation (2.38), we infer that θ is closed. Since the cohomology group $H^1(X, \mathbb{R})$ vanishes, the form θ is exact.

CHAPTER III

SYMMETRIC SPACES OF RANK ONE

§1. Flat tori

Let (X, g) be a flat Riemannian manifold of dimension n. We first suppose that X is the circle S^1 of length L endowed with the Riemannian metric $g = dt \otimes dt$, where t is the canonical coordinate of S^1 defined modulo L. It is easily seen that this space X is infinitesimally rigid and that a 1-form on X satisfies the zero-energy condition if and only if it is exact.

In this section, we henceforth suppose that $n \geq 2$. We recall that $\tilde{B} = \{0\}$, that the operator D_1 is equal to D_g, and that the sequence (1.50) is exact. Let h be a section of $S^2 T^*$ over an open subset of X. According to formulas (1.20) and (1.21), we see that $D_g h$ is equal to the section $\mathcal{R}'_g h$ of B and that

$$
\begin{aligned}
(3.1) \qquad (D_g h)&(\xi_1, \xi_2, \xi_3, \xi_4) \\
&= \tfrac{1}{2}\{(\nabla^2 h)(\xi_1, \xi_3, \xi_2, \xi_4) + (\nabla^2 h)(\xi_2, \xi_4, \xi_1, \xi_3) \\
&\quad - (\nabla^2 h)(\xi_1, \xi_4, \xi_2, \xi_3) - (\nabla^2 h)(\xi_2, \xi_3, \xi_1, \xi_4)\},
\end{aligned}
$$

for $\xi_1, \xi_2, \xi_3, \xi_4 \in T$. From Proposition 1.8, with $\lambda = 0$, and formula (1.31), we obtain the equality

$$
(3.2) \qquad -\mathrm{Tr}\, D_g h = \tfrac{1}{2}(\Delta h - \mathrm{Hess}\,\mathrm{Tr}\, h) - D^1 \mathrm{div}\, h,
$$

which can also be deduced directly from formula (3.1). By (3.2) and (1.30), or by formula (1.41), we have

$$
(3.3) \qquad -\mathrm{Tr} \cdot \mathrm{Tr}\, D_g h = \Delta \mathrm{Tr}\, h + d^* \mathrm{div}\, h.
$$

In the remainder of this section, we suppose that (X, g) is a flat torus of dimension ≥ 2. We may consider X as the quotient of the space $\mathbb{R}^n$ endowed with the Euclidean metric g_0. In fact, there is a lattice Λ generated by a basis $\{\varepsilon_1, \ldots, \varepsilon_n\}$ of $\mathbb{R}^n$ such that X is equal to the quotient $\mathbb{R}^n / \Lambda$. We shall identify a tensor on X with the Λ-invariant tensor on $\mathbb{R}^n$ which it determines. Clearly, a tensor on $\mathbb{R}^n$ which is invariant under the group of all translations of $\mathbb{R}^n$ induces a tensor on X. Let $(x_1, \ldots, x_n)$ be the standard coordinate system of $\mathbb{R}^n$. In particular, for $1 \leq j \leq n$, the vector field $\partial/\partial x_j$ and the 1-form dx_j on $\mathbb{R}^n$ are invariant under the group of all translations of $\mathbb{R}^n$ and therefore induce a parallel vector field and a parallel 1-form on X, which we shall denote by ξ_j and α_j, respectively. Clearly,

$\{\xi_1, \ldots, \xi_n\}$ is a basis for the space of parallel vector fields on X, while $\{\alpha_1, \ldots, \alpha_n\}$ is a basis for the space of parallel 1-forms on X; in fact, we have

$$\langle \xi_j, \alpha_k \rangle = \delta_{jk},$$

for $1 \le j, k \le n$. Thus for all $x \in X$, the mapping

$$\{ \xi \in C^\infty(T) \mid \nabla \xi = 0 \} \to T_x,$$

sending a vector field ξ into the value $\xi(x)$ of ξ at x, is an isomorphism; moreover for $k, p \ge 0$, the mappings

$$\{ \theta \in C^\infty(\textstyle\bigwedge^k T^*) \mid \nabla \theta = 0 \} \to \textstyle\bigwedge^k T_x^*,$$
$$\{ u \in C^\infty(S^p T^*) \mid \nabla u = 0 \} \to S^p T_x^*,$$

sending a differential form θ of degree k into the value $\theta(x)$ of θ at x and a symmetric p-form u into the value $u(x)$ of u at x, respectively, are also isomorphisms. In fact, a parallel section θ of $\bigwedge^k T^*$ over X can be written in the form

$$\theta = \sum_{j_1,\ldots,j_k=1}^{n} c_{j_1\ldots j_k} \alpha_{j_1} \wedge \cdots \wedge \alpha_{j_k}$$

and a parallel section u of $S^p T^*$ over X can be written in the form

$$u = \sum_{j_1,\ldots,j_p=1}^{n} c'_{j_1\ldots j_p} \alpha_{j_1} \cdot \ldots \cdot \alpha_{j_p},$$

where the coefficients $c_{j_1\ldots j_p}$ and $c'_{j_1\ldots j_k}$ are real numbers. Hence the space of all parallel sections of $\bigwedge^k T^*$ (resp. of $S^p T^*$) over X is isomorphic to the space of all differential forms of degree k (resp. of all symmetric p-forms) on $\mathbb{R}^n$ with constant coefficients. In particular, the metric g is equal to the symmetric 2-form $\sum_{j,k=1}^{n} \alpha_j \otimes \alpha_k$. Since the cohomology group $H^k(X, \mathbb{R})$ is isomorphic to $\bigwedge^k T_x^*$, where x is an arbitrary point of X, we know that the space of harmonic forms of degree k on X is equal to the space of all parallel sections of $\bigwedge^k T^*$ over X.

PROPOSITION 3.1. *Let X be a flat torus of dimension ≥ 2. Let ξ be a vector field on X. Then ξ is a Killing vector field if and only if $\nabla \xi = 0$.*

PROOF: We consider the 1-form $\alpha = g^\flat(\xi)$ on X. First, we suppose that X is a flat torus of dimension 2 and we consider the volume form $\omega = \alpha_1 \wedge \alpha_2$ of X. According to formula (1.4), the 1-form α satisfies the relations $D^1\alpha = 0$ and $d^*\alpha = 0$. Since a harmonic 1-form on X is parallel, we may write

$$\alpha = \beta + d^*(f\omega),$$

where β is a 1-form satisfying $\nabla\beta = 0$ and f is a real-valued function on X. Then the 1-form $d^*(f\omega)$ on X is induced by the 1-form

$$\frac{\partial f}{\partial x_2}\, dx_1 - \frac{\partial f}{\partial x_1}\, dx_2$$

on $\mathbb{R}^2$, where f is viewed as a Λ-invariant function on $\mathbb{R}^2$. From the relation $D^1\alpha = 0$, we obtain

$$\frac{\partial^2 f}{\partial x_1 \partial x_2} = 0, \qquad \left(\frac{\partial^2}{\partial x_1^2} - \frac{\partial^2}{\partial x_2^2}\right)f = 0.$$

From the first of the preceding equations, we infer that the function f on $\mathbb{R}^2$ can be written in the form $f = f_1 + f_2$, where f_j is a real-valued function on $\mathbb{R}^2$ depending only on x_j. Then the second equality tells us that the function $f_1 - f_2$ on $\mathbb{R}^2$ is harmonic and so is constant. It follows that the two functions f_1 and f_2 are also constant, and so the 1-form α and the vector field ξ are parallel. Now, we assume that the dimension of X is ≥ 2. Let x_0 be a point of X and let ζ_1 and ζ_2 be parallel vector fields on X; then there is a totally geodesic flat 2-torus Y of X containing x_0 such that the vectors $\zeta_1(x_0)$ and $\zeta_2(x_0)$ are tangent to Y. We consider the vector field η on Y, whose value at $x \in Y$ is equal to the orthogonal projection of $\xi(x)$ onto the subspace $T_{Y,x}$ of T_x. If $i : Y \to X$ is the natural imbedding and $g_Y = i^*g$ is the Riemannian metric on Y induced by g, according to Lemma 1.1 we have $i^*\alpha = g_Y^\flat(\eta)$ and we know that η is a Killing vector field on Y. Therefore the 1-form $g_Y^\flat(\eta)$ on Y is parallel. Since i is a totally geodesic imbedding, it follows that

$$(\nabla\alpha)(\zeta_1,\zeta_2) = (\nabla_Y i^*\alpha)(\zeta_1,\zeta_2) = 0$$

at the point x_0. Thus we have shown that the 1-form α and the vector field ξ on X are parallel.

We recall that the cohomology of the complex

$$C^\infty(T) \xrightarrow{D_0} C^\infty(S^2 T^*) \xrightarrow{D_g} C^\infty(B)$$

is isomorphic to the space

$$H(X) = \{\, h \in C^\infty(S^2 T^*) \mid \operatorname{div} h = 0,\ D_g h = 0 \,\}.$$

PROPOSITION 3.2. *Let X be a flat torus of dimension ≥ 2. Let h be a symmetric 2-form on X. Then the following assertions are equivalent:*
  (i) *We have $\Delta h = 0$.*
  (ii) *We have $\nabla h = 0$.*
  (iii) *The section h belongs to $H(X)$.*

PROOF: Let ξ, η be parallel vector fields on X. If f is the real-valued function on X equal to $h(\xi, \eta)$, according to formula (1.52), we obtain the relation

$$\Delta f = (\Delta h)(\xi, \eta).$$

Therefore if Δh vanishes, the function f is constant; since the parallel vector fields ξ and η are arbitrary, we see that ∇h vanishes. If h is a parallel section of $S^2 T^*$, according to (3.1) we see that $D_g h = 0$, and so h belongs to $H(X)$. Finally, suppose that (iii) holds. Then according to (3.3), we see that $\Delta \operatorname{Tr} h = 0$; therefore $\operatorname{Tr} h$ is constant. Formula (3.2) now tells us that $\Delta h = 0$.

From Proposition 3.2, it follows that the space $H(X)$ is equal to the space

$$\{ h \in C^\infty(S^2 T^*) \mid \nabla h = 0 \}$$

of all parallel sections of $S^2 T^*$. According to remarks made in §3, Chapter I, we know that the cohomology group $H^1(X, \Theta)$ is isomorphic to this space, and therefore also to the vector space $S^2 T_x^*$, where x is an arbitrary point of X; thus the dimension of this cohomology group is equal to $n(n+1)/2$. Other proofs of these results are given in [2] and [15] (see Proposition 17.1 of [15]). From Proposition 3.2 and the decomposition (1.11), it follows that an element h of $C^\infty(S^2 T^*)$ satisfying $D_g h = 0$ can be written in the form

(3.4) $$h = \mathcal{L}_\xi g + h_0,$$

where ξ is a vector field on X and h_0 is a parallel section of $S^2 T^*$ over X.

LEMMA 3.3. *Let u be a parallel symmetric p-form on a flat torus X of dimension ≥ 2. If the form u satisfies the zero-energy condition, then it vanishes.*

PROOF: Let $\gamma : [0, L] \to X$ be a closed geodesic of X parametrized by its arc-length s. If $\dot\gamma(s)$ is the tangent vector to the geodesic γ at the point $\gamma(s)$, then we write $\varphi(s) = u(\dot\gamma(s), \ldots, \dot\gamma(s))$ and we have

$$(\nabla u)(\dot\gamma(s), \dot\gamma(s), \ldots, \dot\gamma(s)) = \frac{d}{ds}\varphi(s),$$

for all $0 \leq s \leq L$. Our hypothesis tells us that the function φ on $[0, L]$ is constant. If u satisfies the zero-energy condition, the integral of the function φ

over the interval $[0, L]$ vanishes, and hence the expression $u(\dot{\gamma}(s), \ldots, \dot{\gamma}(s))$ vanishes for all $0 \leq s \leq L$. If x is a point of X, we know that the set C_x of vectors ξ of $T_x - \{0\}$, for which $\mathrm{Exp}_x \mathbb{R}\xi$ is a closed geodesic of X, is a dense subset of T_x. From these last two observations, we obtain the desired result.

LEMMA 3.4. *Let h be a symmetric 2-form and u be a 1-form on a flat torus X of dimension ≥ 2 which satisfy the zero-energy condition.*
(i) *If the symmetric 2-form h satisfies the relation $D_g h = 0$, then it is a Lie derivative of the metric.*
(ii) *If the 1-form u satisfies the relation $du = 0$, then it is exact.*

PROOF: We first suppose that the relation $D_g h = 0$ holds. As we saw above, we have the decomposition (3.4), where ξ is a vector field on X and h_0 is a parallel symmetric 2-form on X. According to Lemma 2.6, the form h_0 also satisfies the zero-energy condition. From Lemma 3.3 with $p = 2$, we infer that h_0 vanishes, and so the equality $h = \mathcal{L}_\xi g$ holds. If the 1-form θ is closed, then we may write

$$\theta = df + \theta',$$

where f is a real-valued function on X and θ' is a harmonic 1-form on X. Clearly, θ' also satisfies the zero-energy condition. We saw above that θ' is parallel; hence by Lemma 3.3 with $p = 1$, we see that θ is equal to df.

Let $\{\varepsilon_1, \ldots, \varepsilon_n\}$ be a basis of $\mathbb{R}^n$ which generates the lattice Λ, and let $(y_1, \ldots, y_n)$ be the coordinates of $\mathbb{R}^2$ associated with this basis of $\mathbb{R}^n$. Let $\{\alpha'_1, \ldots, \alpha'_n\}$ be the parallel 1-forms on $\mathbb{R}^n$ determined by

$$\langle \partial/\partial y_j, \alpha'_k \rangle = \delta_{jk},$$

for $1 \leq j, k \leq n$. For $1 \leq j \leq n$, let ζ_j be the parallel vector field on X induced by the vector field $\partial/\partial y_j$.

We now suppose that $n = 2$. We fix a pair of integers (p_1, p_2), with $p_1 \neq 0$. If u is an arbitrary real number, we consider the line segment $\gamma_u : [0, 1] \to \mathbb{R}^2$ defined in terms of the coordinates (y_1, y_2) by

$$\gamma_u(t) = (u + p_2 t, -p_1 t),$$

for $0 \leq t \leq 1$; clearly its image in the torus X is a closed geodesic also denoted by γ_u. The parallel vector field

$$\xi = p_2 \frac{\partial}{\partial y_1} - p_1 \frac{\partial}{\partial y_2}$$

on $\mathbb{R}^2$ is tangent to the line segment γ_u. If θ is a symmetric p-form on X and $\hat{\theta}$ is the Λ-invariant symmetric p-form determined by θ, then the integral of θ over the closed geodesic γ_u of X is given by

$$(3.5) \qquad \int_{\gamma_u} \theta = c_p \int_0^1 \hat{\theta}(\xi, \ldots, \xi)(\gamma_u(t)) \, dt,$$

where the constant $c_p \in \mathbb{R}$ depends only on the integers p, p_1 and p_2 and the basis $\{\varepsilon_1, \varepsilon_2\}$ of $\mathbb{R}^2$.

The following proposition is due to Michel [46].

PROPOSITION 3.5. *The X-ray transform for functions on a flat torus of dimension $n \geq 2$ is injective.*

PROOF: We first suppose that $n = 2$. Let f be a complex-valued function on X satisfying the zero-energy condition; we also denote by f the Λ-invariant function on $\mathbb{R}^2$ which it determines. We consider the Fourier series

$$f(y_1, y_2) = \sum_{q_1, q_2 \in \mathbb{Z}} a_{q_1 q_2} e^{2i\pi(q_1 y_1 + q_2 y_2)}$$

of the function f on $\mathbb{R}^2$, where the $a_{q_1 q_2}$ are its Fourier coefficients. We now fix a pair of integers (p_1, p_2); we suppose that $p_1 \neq 0$ and we consider the closed geodesic γ_u of X, which we associated above with the integers p_1 and p_2 and with the real number u. We then consider the function ψ on $\mathbb{R}$ whose value at $u \in \mathbb{R}$ is equal to the integral of f/c_0 over the closed geodesic γ_u, where c_0 is the constant appearing in the equality (3.5), with $p = 0$; according to our hypothesis, the function ψ vanishes identically. Therefore the sum

$$\sum_{q_1, q_2 \in \mathbb{Z}} \int_0^1 a_{q_1 q_2} e^{2i\pi(q_1 u + (p_2 q_1 - p_1 q_2)t)} \, dt$$

vanishes for all $u \in \mathbb{R}$. The Fourier series of the function ψ is given by

$$\psi(u) = \sum_{q_1 \in \mathbb{Z}} e^{2i\pi q_1 u} \sum_{q_2 \in \mathbb{Z}} a_{q_1 q_2} \int_0^1 e^{2i\pi(p_2 q_1 - p_1 q_2)t} \, dt$$

for $u \in \mathbb{R}$. Since the Fourier coefficient of ψ corresponding to the integer p_1 vanishes, we see that $a_{p_1 p_2} = 0$. A similar argument shows that the coefficient $a_{p_1 p_2}$ vanishes when $p_1 = 0$ and $p_2 \neq 0$. Thus the function f is constant and therefore vanishes. Since an arbitrary point of a flat torus of dimension ≥ 2 is contained in a totally geodesic flat torus of dimension 2, we obtain the desired result in all cases.

PROPOSITION 3.6. *Suppose that X is a flat torus of dimension 2. Let h be a symmetric 2-form and θ be a 1-form on X; suppose that these two forms satisfy the zero-energy condition. Then we have the relations $D_g h = 0$ and $d\theta = 0$.*

PROOF: We consider the Λ-invariant symmetric forms

$$h = a\alpha_1' \otimes \alpha_1' + b\alpha_1' \cdot \alpha_2' + c\alpha_2' \otimes \alpha_2', \qquad \theta = a^1\alpha_1' + a^2\alpha_2'$$

on $\mathbb{R}^2$ determined by h and θ, where a, b, c, a^1, a^2 are Λ-invariant functions on $\mathbb{R}^2$. We now fix a pair of integers (p_1, p_2); we suppose that $p_1 \neq 0$ and we consider the closed geodesic γ_u of X which we associated above with the integers p_1 and p_2 and with the real number u. We consider the functions ψ_1 and ψ_2 on $\mathbb{R}$ whose values at $u \in \mathbb{R}$ are equal to the integrals of θ/c_1 and h/c_2 over the closed geodesic γ_u, respectively; here c_p is the constant appearing in the equality (3.5). According to our hypotheses, the functions ψ_1 and ψ_2 vanish identically. The Fourier series of the function ψ_2 is given by

$$\sum_{q_1 \in \mathbb{Z}} e^{2i\pi q_1 u} \sum_{q_2 \in \mathbb{Z}} (p_2^2 a_{q_1 q_2} - 2p_1 p_2 b_{q_1 q_2} + p_1^2 c_{q_1 q_2}) \int_0^1 e^{2i\pi(p_2 q_1 - p_1 q_2)t} \, dt,$$

where $a_{q_1 q_2}$, $b_{q_1 q_2}$ and $c_{q_1 q_2}$ are the Fourier coefficients of the functions a, b and c, respectively; on the other hand, the Fourier series of the function ψ_1 is given by

$$\sum_{q_1 \in \mathbb{Z}} e^{2i\pi q_1 u} \sum_{q_2 \in \mathbb{Z}} (p_2 a_{q_1 q_2}^1 - p_1 a_{q_1 q_2}^2) \int_0^1 e^{2i\pi(p_2 q_1 - p_1 q_2)t} \, dt,$$

where $a_{q_1 q_2}^1$ and $a_{q_1 q_2}^2$ are the Fourier coefficients of the functions a^1 and a^2, respectively. Since the Fourier coefficients of ψ_1 and ψ_2 corresponding to the integer p_1 vanish, we see that

$$p_2^2 a_{p_1 p_2} - 2p_1 p_2 b_{p_1 p_2} + p_1^2 c_{p_1 p_2} = 0, \qquad p_2 a_{p_1 p_2}^1 - p_1 a_{p_1 p_2}^2 = 0.$$

A similar argument shows that these two relations also hold when $p_1 = 0$ and $p_2 \neq 0$. It follows that

$$\frac{\partial^2 a}{\partial y_2^2} - 2\frac{\partial^2 b}{\partial y_1 \partial y_2} + \frac{\partial^2 c}{\partial y_1^2} = 0, \qquad \frac{\partial a^1}{\partial y_2} - \frac{\partial a^2}{\partial y_1} = 0.$$

By (3.1), these last relations are equivalent to the equalities

$$(D_g h)(\zeta_1, \zeta_2, \zeta_1, \zeta_2) = 0, \qquad (d\theta)(\zeta_1, \zeta_2) = 0,$$

respectively. Since $\{\zeta_1, \zeta_2\}$ is a frame for the tangent bundle of $\mathbb{R}^2$, we see that $D_g h = 0$ and $d\theta = 0$.

THEOREM 3.7. *A flat torus of dimension ≥ 2 is infinitesimally rigid.*

THEOREM 3.8. *A differential form of degree 1 on a flat torus of dimension ≥ 2 satisfies the zero-energy condition if and only if it is exact.*

We now simultaneously prove Theorems 3.7 and 3.8. Let h be a symmetric 2-form and θ be a 1-form on the flat torus X, both of which satisfy the zero-energy condition. Let x be an arbitrary point of X and $\{\xi, \eta\}$ be an orthonormal set of vectors of T_x. If F is the subspace $\mathbb{R}\xi \oplus \mathbb{R}\eta$ of T_x, then $Y = \operatorname{Exp}_x F$ is a closed totally geodesic submanifold of X isometric to a flat 2-torus. Let $i : Y \to X$ be the natural imbedding; the forms i^*h and $i^*\theta$ on Y satisfy the zero-energy condition. If g_Y is the metric on Y induced by g, Proposition 3.6 tells us that $D_{g_Y} i^*h = 0$ and $di^*\theta = 0$. According to Proposition 1.14,(i), the restriction $i^* D_g h$ of the section $D_g h$ of B to Y vanishes. Hence we have

$$(D_g h)(\xi, \eta, \xi, \eta) = 0, \qquad (d\theta)(\xi, \eta) = 0.$$

Thus these equalities holds for all $\xi, \eta \in T$, and we see that $D_g h = 0$ and $d\theta = 0$. According to Lemma 3.4, the symmetric form h on X is Lie derivative of the metric and the 1-form θ is exact.

Theorems 3.7 and 3.8 are due to Michel [46]; our proofs of these theorems are essentially the same as those given by Estezet [12]. The next theorem, which generalizes both of these theorems, was proved by Michel [46] when the integer p is equal to 0, 1 or an odd integer and by Estezet [12] in all the other cases.

THEOREM 3.9. *Let X be a flat torus of dimension ≥ 2. For all $p \geq 0$, the space $\mathcal{Z}_{p+1}$ of all sections of $C^\infty(S^{p+1}T^*)$ satisfying the zero-energy condition is equal to $D^p C^\infty(S^p T^*)$.*

The next result is given by Proposition 1 of [20].

PROPOSITION 3.10. *Let (X, g) be a Riemannian manifold. Let Y be a totally geodesic submanifold of X isometric to a flat torus of dimension ≥ 2. Let $i : Y \to X$ be the natural imbedding and $g_Y = i^*g$ be the Riemannian metric on Y induced by g. Let ξ be a Killing vector field on X.*
(i) *We have*
$$i^* dg^\flat(\xi) = 0.$$

(ii) *If the 1-form $g^\flat(\xi)$ satisfies the zero-energy condition, then we have*

$$i^* g^\flat(\xi) = 0.$$

PROOF: We consider the vector field η on Y, whose value at $x \in Y$ is equal to the orthogonal projection of $\xi(x)$ onto the subspace $T_{Y,x}$ of T_x.

According to Lemma 1.1, we have $g_Y^\flat(\eta) = i^* g^\flat(\xi)$ and we know that η is a Killing vector field on Y. Therefore by Proposition 3.1, we see that $\nabla^Y g_Y^\flat(\eta) = 0$, and so the 1-form $g_Y^\flat(\eta)$ on Y is harmonic. Thus we have $i^* dg^\flat(\xi) = dg_Y^\flat(\eta) = 0$. If $g^\flat(\xi)$ satisfies the zero-energy condition, then so does the 1-form $g_Y^\flat(\eta)$ on Y; by Theorem 3.8, the 1-form $g_Y^\flat(\eta)$ on the flat torus Y is exact and therefore vanishes.

§2. The projective spaces

In the remainder of this chapter, we shall consider the symmetric spaces of compact type of rank one. They are the spheres (S^n, g_0), the real projective spaces $(\mathbb{RP}^n, g_0)$, the complex projective spaces $\mathbb{CP}^n$, the quaternionic projective spaces $\mathbb{HP}^n$, with $n \geq 2$, and the Cayley plane. The following two theorems are consequences of Theorems 3.20, 3.26, 3.39, 3.40, 3.44 and 3.45, which appear below in this chapter.

THEOREM 3.11. *A symmetric space of rank one, which is not isometric to a sphere, is infinitesimally rigid.*

THEOREM 3.12. *Let X be a symmetric space of rank one, which is not isometric to a sphere. A differential form of degree 1 on X satisfies the zero-energy condition if and only if it is exact.*

Most of the results described in the remainder of this section are to be found in Chapter 3 of [5]. Let n be an integer ≥ 1. Let $\mathbb{K}$ be one of the fields $\mathbb{R}$, $\mathbb{C}$ or $\mathbb{H}$. We set $a = \dim_{\mathbb{R}} \mathbb{K}$. We endow $\mathbb{K}^{n+1}$ with its right vector space structure over $\mathbb{K}$, with the Hermitian scalar product

$$(3.6) \qquad \langle x, y \rangle = \begin{cases} \sum_{j=0}^{n} x_j \bar{y}_j & \text{if } \mathbb{K} = \mathbb{R} \text{ or } \mathbb{C}, \\ \sum_{j=0}^{n} \bar{x}_j y_j & \text{if } \mathbb{K} = \mathbb{H}, \end{cases}$$

where $x = (x_0, x_1, \ldots, x_n)$ and $y = (y_0, y_1, \ldots, y_n)$ are vectors of $\mathbb{K}^{n+1}$, and with the real scalar product

$$(3.7) \qquad \langle x, y \rangle_{\mathbb{R}} = \operatorname{Re} \langle x, y \rangle,$$

for $x, y \in \mathbb{K}^{n+1}$.

The projective space $\mathbb{KP}^n$ is the orbit space of the space $\mathbb{K}^{n+1} - \{0\}$ under the action of the group $\mathbb{K}^* = \mathbb{K} - \{0\}$; it is a manifold of dimension na. Let

$$\pi : \mathbb{K}^{n+1} - \{0\} \to \mathbb{KP}^n$$

be the natural projection sending $x \in \mathbb{K}^{n+1} - \{0\}$ into its orbit $\pi(x)$; then two non-zero vectors x, y of $\mathbb{K}^{n+1}$ have the same image under π if and only if there exists $\lambda \in \mathbb{K}^*$ such that $x = y \cdot \lambda$.

Let
$$SK^{n+1} = \{\, x \in K^{n+1} \mid \langle x, x \rangle = 1 \,\}$$

be the unit sphere in K^{n+1}; its dimension is $na + a - 1$. Then $SK = SK^1$ is a subgroup of K^*, and the restriction

(3.8) $$\pi : SK^{n+1} \to KP^n$$

of the mapping π is a principal bundle with structure group SK.

Let x be a point of SK^{n+1}. The tangent space $T_x(SK^{n+1})$ of the sphere SK^{n+1} at x is identified with the space

$$\{\, u \in K^{n+1} \mid \langle x, u \rangle_{\mathbb{R}} = 0 \,\}.$$

We denote by $T_{\pi(x)}(KP^n)$ the tangent space of KP^n at the point $\pi(x)$. If $u \in K^{n+1}$ satisfies $\langle x, u \rangle_{\mathbb{R}} = 0$, we denote by (x, u) the tangent vector belonging to $T_x(SK^{n+1})$ corresponding to u, and we consider its image $\pi_*(x, u)$ in $T_{\pi(x)}(KP^n)$ under the mapping π_*; in fact, we have

$$(x, u) = \frac{d}{dt}\,(x + tu)_{|t=0}, \qquad \pi_*(x, u) = \frac{d}{dt}\,\pi(x + tu)_{|t=0};$$

moreover if $\lambda \in K$, we see that

$$\pi_*(x\lambda, u\lambda) = \pi_*(x, u).$$

The subspace $V_x(SK^{n+1})$ of $T_x(SK^{n+1})$ consisting of the vectors tangent to the fibers of the projection π is equal to

$$\{\, (x, x\lambda) \mid \lambda \in \mathbb{R}, \ \text{with} \ \operatorname{Re} \lambda = 0 \,\}.$$

We also consider the subspace

$$H_x(SK^{n+1}) = \{\, (x, u) \mid u \in K^{n+1}, \ \text{with} \ \langle x, u \rangle = 0 \,\}$$

of $T_x(SK^{n+1})$. Then the decomposition

$$T_x(SK^{n+1}) = H_x(SK^{n+1}) \oplus V_x(SK^{n+1})$$

is orthogonal with respect to the scalar product (3.6). The projection

(3.9) $$\pi_* : H_x(SK^{n+1}) \to T_{\pi(x)}(KP^n)$$

is an isomorphism.

We endow the sphere $S\mathbb{K}^{n+1}$ with the Riemannian metric g_0 induced by the scalar product (3.7), and the projective space $\mathbb{K}P^n$ with the Riemannian metric g determined by

$$g(\pi_*(x, u), \pi_*(x, v)) = \langle u, v \rangle_{\mathbb{R}},$$

where $x \in S\mathbb{K}^{n+1}$ and $u, v \in \mathbb{K}^{n+1}$ satisfy $\langle x, u \rangle = \langle x, v \rangle = 0$. In fact, since the equality

$$\langle u, v \rangle_{\mathbb{R}} = \langle u\lambda, v\lambda \rangle_{\mathbb{R}}$$

holds for $u, v \in \mathbb{K}^{n+1}$ and $\lambda \in S\mathbb{K}$, the metric g on $\mathbb{K}P^n$ is well-defined. We see that the projection (3.8) is a Riemannian submersion and that the isomorphism (3.9) is an isometry.

Let x be a point of $S\mathbb{K}^{n+1}$ and let $u, v \in \mathbb{K}^{n+1}$, with $\langle x, u \rangle = \langle x, v \rangle = 0$. When $\mathbb{K} = \mathbb{C}$, the tangent space $T_{\pi(x)}(\mathbb{C}P^n)$ admits a structure of a complex vector space; in fact, we have

$$\pi_*(x, u) \cdot \lambda = \pi_*(x, u \cdot \lambda),$$

for $\lambda \in \mathbb{C}$. Clearly, by construction $\mathbb{C}P^n$ has the structure of a complex manifold, and its complex structure J is determined by

$$J\pi_*(x, u) = \pi_*(x, ui).$$

In general, if the vectors u, v are non-zero, we say that the two vectors $\pi_*(x, u)$ and $\pi_*(x, v)$ are $\mathbb{K}$-dependent if there exists an element $\lambda \in \mathbb{K}^*$ such that $u = v \cdot \lambda$; this relation among non-zero vectors of $T_{\pi(x)}(\mathbb{K}P^n)$ is easily seen to be well-defined. If the vector u is non-zero and $\xi = \pi_*(x, u)$ is the tangent vector of $T_{\pi(x)}(\mathbb{K}P^n)$ corresponding to u, the subset

$$\xi\mathbb{K} = \{\, \pi_*(x, u \cdot \lambda) \mid \lambda \in \mathbb{K}^* \,\} \cup \{\pi_*(x, 0)\}$$

is a real subspace of $T_{\pi(x)}(\mathbb{K}P^n)$ of dimension (over $\mathbb{R}$) equal to $\dim_{\mathbb{R}} \mathbb{K}$. Moreover, if $\mathbb{K} = \mathbb{C}$, then $\xi\mathbb{K}$ is equal to the real subspace $\mathbb{C}\xi$ of $T_{\pi(x)}(\mathbb{C}P^n)$ generated by ξ and $J\xi$.

The following lemma is proved in [5, §3.9].

LEMMA 3.13. *Let x be a point of $S\mathbb{K}^{n+1}$ and u, v be non-zero vectors of $\mathbb{K}^{n+1}$, with $\langle x, u \rangle = \langle x, v \rangle = 0$. If $\xi = \pi_*(x, u)$ and $\eta = \pi_*(x, v)$ are the tangent vectors of $T_{\pi(x)}(\mathbb{K}P^n)$ corresponding to u and v, the following assertions are equivalent:*

(i) *We have $\langle u, v \rangle = 0$.*

(ii) *The vector ξ is orthogonal to the subspace $\eta\mathbb{K}$ with respect to the metric g.*

(iii) *The subspaces $\xi\mathbb{K}$ and $\eta\mathbb{K}$ are orthogonal with respect to the metric g.*

If $\xi_1, \ldots, \xi_n$ are non-zero vectors of $T_{\pi(x)}(\mathbb{KP}^n)$ such that the subspaces $\xi_j\mathbb{K}$ are mutually orthogonal, then we have the decomposition

$$T_{\pi(x)}(\mathbb{KP}^n) = \xi_1\mathbb{K} \oplus \cdots \oplus \xi_n\mathbb{K}.$$

We consider the subgroup $U(n+1, \mathbb{K})$ of $GL(n+1, \mathbb{K})$ consisting of all elements $A \in GL(n+1, \mathbb{K})$ satisfying

$$\langle Ax, Ay \rangle = \langle x, y \rangle,$$

for all $x, y \in \mathbb{K}^{n+1}$. When $\mathbb{K} = \mathbb{R}$ or $\mathbb{C}$, let $SU(n+1, \mathbb{K})$ be the subgroup of $U(n+1, \mathbb{K})$ consisting of all elements $A \in U(n+1, \mathbb{K})$ satisfying $\det A = 1$. Then $SU(n+1, \mathbb{R})$ is equal to $SO(n+1)$ and $SU(n+1, \mathbb{C})$ is equal to $SU(n+1)$, while $U(n+1, \mathbb{H})$ is equal to $Sp(n+1)$. When $\mathbb{K}$ is either $\mathbb{R}$ or $\mathbb{C}$, let G denote the group $SU(n+1, \mathbb{K})$; when $\mathbb{K}$ is equal to $\mathbb{H}$, let G be the group $U(n+1, \mathbb{K}) = Sp(n+1)$.

The group G acts transitively on $S\mathbb{K}^{n+1}$ and on $\mathbb{KP}^n$ by isometries. Let $\{e_0, e_1, \ldots, e_n\}$ be the standard basis of $\mathbb{K}^{n+1}$. Let K be the subgroup of G leaving the point $\pi(e_0)$ fixed, and let K' be the subgroup of K consisting of those elements of G leaving the point e_0 fixed. Then we have diffeomorphism

$$\varphi : G/K \to \mathbb{KP}^n, \qquad \varphi' : G/K' \to S\mathbb{K}^{n+1}$$

defined by

$$\varphi(\phi \cdot K) = \pi\phi(e_0), \qquad \varphi'(\phi \cdot K') = \phi(e_0),$$

for $\phi \in G$. If $\phi \in G$ belongs to K, then we have

$$\phi(e_0) = e_0 \cdot \lambda,$$

where $\lambda \in S\mathbb{K}$; hence the element a of G can be written in the form

(3.10)
$$\begin{pmatrix} \lambda & 0 \\ 0 & B \end{pmatrix},$$

where B belongs to $U(n, \mathbb{K})$. If K_1 denotes the subgroup $U(1, \mathbb{K}) \times U(n, \mathbb{K})$ of $U(n+1, \mathbb{K})$, the group K is therefore equal to $G \cap K_1$. The subgroup K' consists of the elements (3.10) of K with $\lambda = 1$. When $\mathbb{K} = \mathbb{R}$, we identify K' with the group $SO(n)$.

When $\mathbb{K} = \mathbb{R}$, in the remainder of this section we suppose that $n \geq 2$. If I_n denotes the unit matrix of order n, the element

$$s = \begin{pmatrix} -1 & 0 \\ 0 & I_n \end{pmatrix}$$

of $U(n+1, \mathbb{K})$ determines an involution σ of G which sends $a \in G$ into sas^{-1}. Then K is equal to the set of fixed points of σ; thus (G, K) is a Riemannian symmetric pair. When $\mathbb{K} = \mathbb{R}$, the subgroup K' coincides with the identity component of the group K, and (G, K') is also a Riemannian symmetric pair in this case. The Cartan decomposition of the Lie algebra $\mathfrak{g}_0$ of G corresponding to σ is

$$\mathfrak{g}_0 = \mathfrak{k}_0 \oplus \mathfrak{p}_0;$$

here $\mathfrak{k}_0$ is the Lie algebra of K and $\mathfrak{p}_0$ is the space of all matrices

(3.11)
$$\begin{pmatrix} 0 & -{}^t\bar{Z} \\ Z & 0 \end{pmatrix}$$

of $\mathfrak{g}_0$, where $Z \in \mathbb{K}^n$ is viewed as a column vector and ${}^t\bar{Z}$ is its conjugate transpose. We identify $\mathfrak{p}_0$ with the vector space $\mathbb{K}^n$ and, in particular, the element (3.11) of $\mathfrak{p}_0$ with the vector Z of $\mathbb{K}^n$. The adjoint action of K on $\mathfrak{p}_0$ is expressed by

$$\mathrm{Ad}\,\phi \cdot Z = B \cdot Z \cdot \bar{\lambda},$$

where ϕ is the element (3.10) of K and $Z \in \mathbb{K}^n$. When $\mathbb{K} = \mathbb{R}$, we know that $\mathfrak{k}_0$ is also the Lie algebra of K'.

We identify $\mathfrak{p}_0$ with the tangent space of G/K (resp. of G/K' when $\mathbb{K} = \mathbb{R}$) at the coset of the identity element of G. Via the above identification of $\mathfrak{p}_0$ with the vector space $\mathbb{K}^n$, we transfer the scalar product on $\mathbb{K}^n$ given by (3.7) to $\mathfrak{p}_0$ and we note that this scalar product on $\mathfrak{p}_0$ is invariant under the adjoint action of K and therefore induces G-invariant metrics on the homogeneous space G/K and, when $\mathbb{K} = \mathbb{R}$, on the space G/K', which we denote by g_1. Endowed with this metric g_1, the manifold G/K (resp. the manifold G/K' when $\mathbb{K} = \mathbb{R}$) is an irreducible symmetric space of compact type. Then the diffeomorphism φ from $(G/K, g_1)$ to $(\mathbb{K}\mathbb{P}^n, g)$ and, when $\mathbb{K} = \mathbb{R}$, the diffeomorphism φ from $(G/K', g_1)$ to $(S\mathbb{R}^{n+1}, g_0)$ are isometries (see §C in Chapter 3 of [5] and §10 in Chapter XI of [40]). It follows that the space $(\mathbb{R}\mathbb{P}^n, g)$, with $n \geq 2$, has constant curvature 1.

Since the complex structure of the manifold $\mathbb{C}\mathbb{P}^n$ is invariant under the group $SU(n+1, \mathbb{C})$, we see that $(\mathbb{C}\mathbb{P}^n, g)$ is a Hermitian symmetric space (see Proposition 4.2 in Chapter VIII of [36]). In fact, the metric g on complex manifold $\mathbb{C}\mathbb{P}^n$ is the Fubini-Study metric of constant holomorphic curvature 4 (see Example 10.5 in Chapter XI of [40]). On the other hand, the projective lines $\mathbb{CP}^1$ and $\mathbb{HP}^1$ are isometric to the spheres S^2 and S^4, respectively, endowed with their metrics of constant curvature 4. In §4, we shall verify this last assertion in the case of the complex projective line.

The following result is proved in §D of [5, Chapter 3].

PROPOSITION 3.14. *Let X be the projective space $\mathbb{KP}^n$, with $n \geq 2$. Let x be a point of X and $\xi_1, \ldots, \xi_q \in T_x$ be unitary tangent vectors.*

   (i) *If the subspaces $\xi_1\mathbb{K}, \ldots, \xi_q\mathbb{K}$ of T_x are mutually orthogonal, then $\mathrm{Exp}_x(\mathbb{R}\xi_1 \oplus \cdots \oplus \mathbb{R}\xi_q)$ is a closed totally geodesic submanifold of X isometric to the real projective space $\mathbb{RP}^q$ endowed with its metric of constant curvature 1.*

   (ii) *If the subspaces $\xi_1\mathbb{K}, \ldots, \xi_q\mathbb{K}$ of T_x are mutually orthogonal, then $\mathrm{Exp}_x(\xi_1\mathbb{K} \oplus \cdots \oplus \xi_q\mathbb{K})$ is a closed totally geodesic submanifold of X isometric to the projective space $\mathbb{KP}^q$.*

   (iii) *Suppose that $\mathbb{K}$ is equal to $\mathbb{H}$ and that $q = 2r$, where $r \geq 1$. Suppose that the subspaces $\xi_1\mathbb{H}, \ldots, \xi_r\mathbb{H}$ of T_x are mutually orthogonal and that there are vectors $u_j \in H_x(S\mathbb{K}^{n+1})$ and $\lambda \in \mathbb{H}$ satisfying $\mathrm{Re}\,\lambda = 0$ and*

$$\xi_j = \pi_*(x, u_j), \qquad \xi_{j+r} = \pi_*(x, u_j\lambda),$$

for $1 \leq j \leq r$. Then $\mathrm{Exp}_x(\mathbb{R}\xi_1 \oplus \cdots \oplus \mathbb{R}\xi_q)$ is a closed totally geodesic submanifold of X isometric to the complex projective space $\mathbb{CP}^r$ endowed with its metric of constant holomorphic curvature 4.

The following result is a direct consequence of Proposition 3.14.

PROPOSITION 3.15. *Suppose that $\mathbb{K}$ is equal to $\mathbb{C}$ or $\mathbb{H}$. Let X be the projective space $\mathbb{KP}^n$, with $n \geq 2$. Let x be a point of X and let ξ_1 and ξ_2 be linearly independent unitary tangent vectors of T_x. If ξ_2 belongs to $\xi_1\mathbb{K}$, then $\mathrm{Exp}_x(\mathbb{R}\xi_1 \oplus \mathbb{R}\xi_2)$ is a closed totally geodesic submanifold of X isometric to the complex projective line $\mathbb{CP}^1$ endowed with its metric of constant curvature 4.*

PROPOSITION 3.16. *The projective space $X = \mathbb{KP}^n$, with $n \geq 2$, is a C_π-manifold.*

PROOF: By Proposition 3.14,(i), with $q = 1$, if x is a point of X and ξ is a unitary vector of T_x, we see that $\mathrm{Exp}_x\mathbb{R}\xi$ is isometric to $\mathbb{RP}^1$ and is therefore a closed geodesic of length π.

Clearly a C_L-manifold cannot contain a totally geodesic flat torus of dimension ≥ 2. Therefore, from Proposition 3.15 it follows that the symmetric spaces $\mathbb{KP}^n$, with $n \geq 2$, have rank one.

In §G of [5, Chapter 3], the structure of symmetric space of rank one is defined on the Cayley plane X. An analogue of Proposition 3.14 holds for the Cayley plane. In fact, the inclusion of the quaternions $\mathbb{H}$ into the Cayley algebra gives rise to closed totally geodesic submanifolds of X isometric to $\mathbb{HP}^2$.

§3. The real projective space

We consider the sphere $S^n = S\mathbb{R}^{n+1}$, with $n \geq 2$, endowed with its metric g_0 of constant curvature 1, which is equal to the metric introduced in §2. The anti-podal involution τ of the sphere S^n is an isometry. The real projective space $(\mathbb{RP}^n, g_0)$ is the quotient of S^n by the group $\Lambda = \{\mathrm{id}, \tau\}$. When $\mathbb{K} = \mathbb{R}$, we observed in §2 that the group $S\mathbb{K} = \{+1, -1\}$ acts on $\mathbb{K}^{n+1}$ and on the unit sphere $S^n = S\mathbb{K}^{n+1}$; in fact, its action is equal to that of the group Λ. Therefore the metric g_0 on $\mathbb{RP}^n$ is equal to the metric g of §2. As in §4, Chapter II, we say that a function f on S^n is even (resp. odd) if $\tau^* f = \varepsilon f$, where $\varepsilon = +1$ (resp. $\varepsilon = -1$). If $X = S^n$, we denote by $C^\infty(X)^{\mathrm{ev}}$ (resp. $C^\infty(X)^{\mathrm{odd}}$) the subspace of $C^\infty(X)$ consisting of all even (resp. odd) functions.

Let $\mathfrak{g} = \mathfrak{so}(n + 1, \mathbb{C})$ be the complexification of the Lie algebra of $G = SO(n+1)$ and let Γ be the dual of the group G. Let γ_0 and γ_1 be the elements of Γ which are the equivalence classes of the trivial G-module $\mathbb{C}$ and of the irreducible G-module $\mathfrak{g}$, respectively.

We view the sphere $X = S^n = S\mathbb{R}^{n+1}$ as the irreducible symmetric space $SO(n+1)/SO(n)$. The set of eigenvalues of the Laplacian Δ acting on $C^\infty(X)$ consists of all the integers $\lambda_k = k(n + k - 1)$, where k is an integer ≥ 0. The eigenspace $\mathcal{H}_k$ of Δ associated with the eigenvalue λ_k consists of all the complex-valued functions on X which are restrictions to S^n of harmonic polynomials of degree k on $\mathbb{R}^{n+1}$. According to observations made in §7, Chapter II and by Proposition 2.1, we know that $\mathcal{H}_k$ is an irreducible $SO(n + 1)$-submodule of $C^\infty(X)$ and that the direct sum $\bigoplus_{k \geq 0} \mathcal{H}_k$ is a dense submodule of $C^\infty(X)$ (see §C.I in Chapter III of [4]). For $k \geq 2$, let γ_k be the element of Γ corresponding to the irreducible G-module $\mathcal{H}_k$; then we have the equalities

$$C^\infty_{\gamma_k}(X) = \mathcal{H}_k,$$

for all $k \geq 0$. The homogeneous polynomial f_k of degree k on $\mathbb{R}^{n+1}$ defined by

$$f_k(x_1, \ldots, x_{n+1}) = \mathrm{Re}\,(x_1 + ix_2)^k,$$

for $(x_1, \ldots, x_{n+1}) \in \mathbb{R}^{n+1}$, is harmonic; its restriction $\tilde{f}_k$ to S^n therefore belongs to $\mathcal{H}_k$. Since $\tilde{f}_k$ is an even (resp. odd) function on S^n when the integer k is even (resp. odd), we obtain the inclusions

$$\mathcal{H}_{2k} \subset C^\infty(X)^{\mathrm{ev}}, \qquad \mathcal{H}_{2k+1} \subset C^\infty(X)^{\mathrm{odd}},$$

for all $k \geq 0$.

Let Y be the real projective space $\mathbb{RP}^n$ endowed with the metric g_0, which we view as an irreducible symmetric space. According to §7, Chapter II and the isomorphism (2.19), we see that the set of eigenvalues of the Laplacian Δ_Y acting on $C^\infty(Y)$ consists of all the integers λ_{2k}, with $k \geq 0$; moreover if k is an even integer, the G-module $C^\infty_{\gamma_k}(Y)$ is equal to the space of functions on Y obtained from $\mathcal{H}_k$ by passage to the quotient. Therefore $\bigoplus_{k \geq 0} C^\infty_{\gamma_{2k}}(Y)$ is a dense subspace of $C^\infty(Y)$, and the first non-zero eigenvalue of the Laplacian Δ_Y acting on $C^\infty(Y)$ is equal to $2(n+1)$.

We consider the closed geodesic γ of S^n, which is the great circle defined by

$$\gamma(s) = (\cos s, 0, \ldots, 0, \sin s),$$

with $0 \leq s \leq 2\pi$. Then we have the relations

$$I_0(\tilde{f}_{2k})(\gamma) = \int_\gamma \tilde{f}_{2k} = \int_0^{2\pi} \cos^{2k} t \, dt > 0.$$

The following result is a consequence of the above observations and Proposition 2.29, with $\Sigma = \{\tau\}$ and $\varepsilon = +1$.

PROPOSITION 3.17. *An even function on the sphere S^n, with $n \geq 2$, whose X-ray transform vanishes, vanishes identically.*

In §4, Chapter II, we noted that the preceding proposition is equivalent to assertion (i) of Theorem 2.23 and also that it implies that the X-ray transform for functions is injective on the real projective space $\mathbb{RP}^n$, with $n \geq 2$.

THEOREM 3.18. *The X-ray transform for functions on a symmetric space of compact type of rank one, which is not isometric to a sphere, is injective.*

PROOF: According to Proposition 3.14 and the discussion which follows this proposition, each point of such a projective space X is contained in a closed totally geodesic submanifold of X isometric to the projective plane $\mathbb{RP}^2$. The desired result is then a consequence of Theorem 2.23,(ii).

Theorem 3.18, together with Proposition 2.26, implies that the X-ray transform for functions on a symmetric space X of compact type is injective if and only if X is not isometric to a sphere. Theorem 3.18 is also a consequence of Theorem 2.24.

We shall now establish the infinitesimal rigidity of the real projective space $\mathbb{RP}^n$, a result due to Michel [45]. We first consider the case of the real projective plane.

PROPOSITION 3.19. *The real projective plane $X = \mathbb{RP}^2$ is infinitesimally rigid.*

PROOF: Let h be a symmetric 2-form on X satisfying the zero-energy condition. We know that the relation (1.64) holds on the sphere (S^2, g_0), which is a covering space of X; therefore we may write

$$h = D_0\xi + fg,$$

where ξ is a vector field and f is a real-valued function on X. If γ is closed geodesic of X, we have

$$\int_\gamma f = \int_\gamma fg.$$

Since the Lie derivative $\mathcal{L}_\xi g$ satisfies the zero-energy condition, we see that the function f also satisfies the zero-energy condition. From Theorem 2.23,(ii), with $n = 2$, we deduce that the function f vanishes and so we have $h = D_0\xi$.

THEOREM 3.20. *The real projective space $X = \mathbb{RP}^n$, with $n \geq 2$, is infinitesimally rigid.*

PROOF: Let h be a symmetric 2-form on X satisfying the zero-energy condition. Let x be an arbitrary point of X and $\{\xi, \eta\}$ be an orthonormal set of vectors of T_x; we consider the subspace $F = \mathbb{R}\xi \oplus \mathbb{R}\eta$ of T_x. According to Proposition 3.14,(ii), we know that $Y = \mathrm{Exp}_x F$ is a totally geodesic submanifold of X isometric to $\mathbb{RP}^2$. If $i : Y \to X$ is the natural imbedding, by Proposition 3.19 we know that the restriction i^*h of h to Y is a Lie derivative of the metric g_Y on Y induced by g. According to Proposition 1.14,(i), we see that the restriction $i^*D_g h$ of the section $D_g h$ of B to Y is equal to $D_{g_Y} i^*h$. Therefore by the relation (1.49), the section $i^*D_g h$ vanishes and so we have

$$(D_g h)(\xi, \eta, \xi, \eta) = 0.$$

Thus this last equality holds for all $\xi, \eta \in T$, and we see that $D_g h = 0$. According to Theorem 1.18, the sequence (1.51) is exact, and so h is a Lie derivative of the metric.

The proof of Proposition 3.19 given above is due to Bourguignon and our proof of Theorem 3.20 is inspired by the one given in Chapter 5 of [5]. We now present a variant of the version given in [30, §2] of Michel's original proof of Proposition 3.19 (see [45]).

We suppose that (X, g) is the real projective plane $(\mathbb{RP}^2, g_0)$. Let $\gamma : [0, \pi] \to X$ be an arbitrary closed geodesic parametrized by its arclength. We set $\gamma(0) = x$, and let $e_1(t) = \dot{\gamma}(t)$ be the tangent vector to the geodesic at $\gamma(t)$, for $0 \leq t \leq \pi$. We choose a unit vector $e_2 \in T_x$ orthogonal to $e_1(0)$ and consider the family of tangent vectors $e_2(t) \in T_{\gamma(t)}$, with $0 \leq t \leq \pi$, obtained by parallel transport of the vector e_2 along γ. Clearly, if u is an element of $C^\infty(S^p T^*)$, we have

$$\int_\gamma u = \int_0^\pi u(e_1(t), e_1(t), \ldots, e_1(t)) \, dt;$$

on the other hand, if v is an element of $C^\infty(\bigotimes^p T^*)$ and $\xi_j(t)$ is a vector field along $\gamma(t)$ equal to either $e_1(t)$ or $e_2(t)$ for $1 \leq j \leq p-1$, we have

$$(\nabla v)(e_1(t), \xi_1(t), \ldots, \xi_{p-1}(t)) = \frac{d}{dt} v(\xi_1(t), \ldots, \xi_{p-1}(t)),$$

and so

(3.12)
$$\int_0^\pi (\nabla v)(e_1(t), \xi_1(t), \ldots, \xi_{p-1}(t)) \, dt = 0.$$

We consider the space $\mathcal{Z}_2$ of symmetric 2-forms on $X = \mathbb{RP}^2$ satisfying the zero-energy condition. The following result given by Lemma 2.36 is proved in [45] by considering the Jacobi fields along the closed geodesics of X.

LEMMA 3.21. *The space $\mathcal{Z}_2$ of symmetric 2-forms on $X = \mathbb{RP}^2$ is invariant under the Lichnerowicz Laplacian Δ.*

LEMMA 3.22. *Let X be the real projective plane $\mathbb{RP}^2$. If $h \in \mathcal{Z}_2$ satisfies $\operatorname{div} h = 0$, then we have*

$$\int_0^\pi (\nabla^2 h)(e_2(t), e_2(t), e_2(t), e_2(t)) \, dt = \int_\gamma \operatorname{Tr} h.$$

PROOF: We have

$$(\nabla^2 h)(e_2(t), e_2(t), e_2(t), e_2(t)) = -(\nabla^2 h)(e_2(t), e_1(t), e_1(t), e_2(t))$$
$$= -(\nabla^2 h)(e_1(t), e_2(t), e_1(t), e_2(t))$$
$$+ h(e_2(t), e_2(t)) - h(e_1(t), e_1(t)),$$

for $0 \leq t \leq \pi$; the first equality holds because $\operatorname{div} h = 0$, while the second one is obtained using the expression for the curvature of (X, g). The lemma is now a consequence of (3.12).

LEMMA 3.23. *Let X be the real projective plane $\mathbb{RP}^2$. If $h \in \mathcal{Z}_2$ satisfies $\operatorname{div} h = 0$, then*

$$\int_\gamma (\Delta \operatorname{Tr} h - \operatorname{Tr} h) = 0.$$

PROOF: According to (1.52), we have

$$(\Delta h)(e_2(t), e_2(t)) = -(\nabla^2 h)(e_1(t), e_1(t), e_2(t), e_2(t))$$
$$- (\nabla^2 h)(e_2(t), e_2(t), e_2(t), e_2(t))$$
$$+ 4h(e_2(t), e_2(t)) - 2(\operatorname{Tr} h)(\gamma(t)),$$

for $0 \le t \le \pi$. By (1.30), (3.12), Lemmas 3.21 and 3.22, and the preceding equality, we see that

$$\int_\gamma \Delta \operatorname{Tr} h = \int_\gamma \operatorname{Tr} \Delta h = \int_0^\pi (\Delta h)(e_2(t), e_2(t))\, dt = \int_\gamma \operatorname{Tr} h.$$

LEMMA 3.24. *Let X be the real projective plane $\mathbb{RP}^2$. If $h \in \mathcal{Z}_2$ satisfies $\operatorname{div} h = 0$, then $\operatorname{Tr} h$ vanishes.*

PROOF: According to Theorem 2.23,(ii), with $n = 2$, and Lemma 3.23, we see that $\Delta \operatorname{Tr} h = \operatorname{Tr} h$. As the first non-zero eigenvalue of the Laplacian Δ acting on $C^\infty(X)$ is equal to 6, we see that $\operatorname{Tr} h = 0$.

LEMMA 3.25. *Let X be the real projective plane $\mathbb{RP}^2$. An element h of $\mathcal{Z}_2$ satisfying $\operatorname{div} h = 0$ vanishes.*

PROOF: Let h be an element of $\mathcal{Z}_2$ satisfying $\operatorname{div} h = 0$. According to Lemma 3.24 and the equality (1.53), the symmetric 2-form h belongs to $H(X)$. Then Proposition 1.20 tells us that h vanishes.

Now Proposition 3.19 is a direct consequence of Proposition 2.13 and Lemma 3.25.

Our approach to the rigidity questions, which led us to the criteria of Theorem 2.49 and the methods introduced in [22] for the study of the complex quadrics, were partially inspired by the proof of the infinitesimal rigidity of the real projective plane which we have just completed. The correspondence between the arguments given here in the case of the real projective plane and those used in the case of the complex quadric is pointed out in [30].

The following result is due to Michel [47].

THEOREM 3.26. *A differential form of degree 1 on the real projective space* $\mathbb{RP}^n$, *with* $n \geq 2$, *satisfies the zero-energy condition if and only if it is exact.*

In the case of the real projective plane $\mathbb{RP}^2$, according to Proposition 2.20, we see that Theorem 3.26 is a consequence of Proposition 3.29,(ii), which is proved below; on the other hand, the proof of this theorem given in [47] for $\mathbb{RP}^2$ is elementary and requires only Stokes's theorem for functions in the plane. In fact, the result given by Theorem 3.26 for the real projective plane implies the result in the general case. Let X be the real projective space $\mathbb{RP}^n$, with $n \geq 2$; by Proposition 3.14, we easily see that an element of $\bigwedge^2 T^*$, which vanishes when restricted to the totally geodesic surfaces of X isometric to the real projective plane, must be equal to 0. Then the desired result for X is a consequence of Proposition 2.51,(ii).

The following result due to Bailey and Eastwood [1] generalizes both Theorems 3.20 and 3.26.

THEOREM 3.27. *Let* X *be the real projective space* $\mathbb{RP}^n$, *with* $n \geq 2$. *For all* $p \geq 0$, *the space* $\mathcal{Z}_{p+1}$ *of all sections of* $C^\infty(S^{p+1}T^*)$ *satisfying the zero-energy condition is equal to* $D^p C^\infty(S^p T^*)$.

In the case $p = 2$, the assertion of Theorem 3.27 was first established by Estezet (see [12] and [29]).

§4. The complex projective space

In this section, we suppose that X is the complex projective space $\mathbb{CP}^n$, with $n \geq 1$, endowed with the metric g of §2. We have seen that g is the Fubini-Study metric of constant holomorphic curvature 4. We denote by J the complex structure of X. As in §2, we identify X with the Hermitian symmetric space G/K, where G is the group $SU(n+1)$ and K is its subgroup $S(U(n) \times U(1))$.

The curvature tensor $\tilde{R}$ of X is given by

$$
(3.13) \qquad \begin{aligned}
\tilde{R}(\xi, \eta)\zeta = {} & g(\eta, \zeta)\xi - g(\xi, \zeta)\eta + g(J\eta, \zeta)J\xi - g(J\xi, \zeta)J\eta \\
& - 2g(J\xi, \eta)J\zeta,
\end{aligned}
$$

for all $\xi, \eta, \zeta \in T$; it follows that

$$
(3.14) \qquad\qquad \mathrm{Ric} = 2(n+1)g.
$$

Let $\zeta = (\zeta_0, \zeta_1, \ldots, \zeta_n)$ be the standard complex coordinate system of $\mathbb{C}^{n+1}$ and consider the complex vector field

$$
\zeta^0 = \sum_{j=0}^{n} \zeta_j \frac{\partial}{\partial \zeta_j}
$$

on $\mathbb{C}^{n+1}$. We consider the objects associated with $\mathbb{K} = \mathbb{C}$ in §2 and the natural projection

(3.15) $$\pi : \mathbb{C}^{n+1} - \{0\} \to \mathbb{CP}^n.$$

The group $U(1) = S\mathbb{K}$ acts by multiplication on $\mathbb{C}^{n+1} = \mathbb{K}^{n+1}$ and on the unit sphere $S^{2n+1} = S\mathbb{K}^{n+1}$ of dimension $2n + 1$ in $\mathbb{C}^{n+1}$. The restriction

$$\pi : S^{2n+1} \to \mathbb{CP}^n$$

of the mapping π (which is also given by (3.8), with $\mathbb{K} = \mathbb{C}$) is a principal bundle with structure group $U(1)$.

If z is a point of the unit sphere S^{2n+1}, the subspace $V_z(S^{2n+1})$ of $T_z(S^{2n+1})$ is spanned by the tangent vector

(3.16) $$(z, iz) = i(\zeta^0 - \overline{\zeta^0})(z).$$

If $H_z(S^{2n+1})$ is the complement of $V_z(S^{2n+1})$ in $T_z(S^{2n+1})$ defined in §2, we know that the induced mapping

$$\pi_* : H_z(S^{2n+1}) \to T_z(\mathbb{CP}^n)$$

is an isometry.

Let h be a complex symmetric 2-form on $\mathbb{C}^{n+2}$, which is $U(1)$-invariant and which satisfies

$$(\zeta^0 - \overline{\zeta^0}) \lrcorner \, h = 0.$$

According to (3.16) and the preceding observations, by restricting h to the sphere S^{2n+1} and passing to the quotient, we obtain a complex symmetric 2-form h' on $\mathbb{CP}^n$, which is uniquely characterized by the relation

$$\pi^* h' = h_{|S^{2n+1}},$$

where π is the mapping (3.15).

Let u, u' be vectors of S^{2n+1} satisfying $\langle u, u' \rangle = 0$. We consider the paths σ and $\dot{\sigma}$ in S^{2n+1} defined by

(3.17)
$$\sigma(t) = \cos t \cdot u + \sin t \cdot u',$$
$$\dot{\sigma}(t) = -\sin t \cdot u + \cos t \cdot u',$$

for $t \in \mathbb{R}$; the unit tangent vector

$$\alpha(t) = (\sigma(t), \dot{\sigma}(t))$$

of $T_{\sigma(t)}(S^{2n+1})$ belongs to the subspace $H_{\sigma(t)}(S^{2n+1})$ and is tangent to the path σ. Then the path $\gamma = \gamma_{u,u'}$ defined by $\gamma(t) = (\pi \circ \sigma)(t)$, for $0 \leq t \leq \pi$, is a closed geodesic of $\mathbb{CP}^n$ parametrized by its arc-length t, and we see that $\pi_* \alpha(t) = \dot{\gamma}(t)$, for all $0 \leq t \leq \pi$.

Let f be a complex-valued function on $\mathbb{C}^{n+1}$, whose restriction to the sphere S^{2n+1} is invariant under the group $U(1)$; then the restriction of f to S^{2n+1} induces by passage to the quotient a function on $\mathbb{CP}^n$, which we denote by $\tilde{f}$. In fact, if π is the mapping (3.15), this function $\tilde{f}$ satisfies the relation

$$\pi^* \tilde{f} = f_{|S^{2n+1}}.$$

If z is a point of S^{2n+1} and ξ is a vector of $H_z(S^{2n+1})$, then we easily see that the equalities

$$(3.18) \quad \langle df, \xi \rangle = \langle d\tilde{f}, \pi_* \xi \rangle, \quad \langle \partial f, \xi \rangle = \langle \partial \tilde{f}, \pi_* \xi \rangle, \quad \langle \bar{\partial} f, \xi \rangle = \langle \bar{\partial} \tilde{f}, \pi_* \xi \rangle$$

hold. In particular, if we consider the path σ in S^{2n+1} and the closed geodesic $\gamma = \gamma_{u,u'}$ of $\mathbb{CP}^n$, by (3.18) we have

$$(3.19) \qquad \langle d\tilde{f}, \dot{\gamma}(t) \rangle = \langle df, \alpha(t) \rangle = \frac{d}{dt} f(\sigma(t)),$$

$$\langle \partial \tilde{f}, \dot{\gamma}(t) \rangle = \langle \partial f, \alpha(t) \rangle, \qquad \langle \bar{\partial} \tilde{f}, \dot{\gamma}(t) \rangle = \langle \bar{\partial} f, \alpha(t) \rangle,$$

for all $0 \leq t \leq \pi$.

Let $\mathfrak{g} = \mathfrak{sl}(n+1, \mathbb{C})$ be the complexification of the Lie algebra of G and let Γ be the dual of the group G. Let γ_0 and γ_1 be the elements of Γ which are the equivalence classes of the trivial G-module $\mathbb{C}$ and of the irreducible G-module $\mathfrak{g}$, respectively. As in §7, Chapter II, we identify the complexification $\mathcal{K}_{\mathbb{C}}$ of the space $\mathcal{K}$ of all Killing vector fields on X with the G-submodule

$$\operatorname{Ker} D_0 = \{ \xi \in C^\infty(T_{\mathbb{C}}) \mid D_0 \xi = 0 \}$$

of $C^\infty_{\gamma_1}(T_{\mathbb{C}})$ isomorphic to $\mathfrak{g}$.

Let $\mathcal{A}_k$ be the G-module consisting of all $U(1)$-invariant homogeneous complex polynomials on $\mathbb{C}^{n+1}$ of degree $2k$ in the variables ζ and $\bar{\zeta}$. Clearly, a homogeneous polynomial belongs to $\mathcal{A}_k$ if and only if it is homogeneous of degree k in ζ and of degree k in $\bar{\zeta}$. Let $\mathcal{P}_k$ be the G-submodule of $\mathcal{A}_k$ consisting of all elements of $\mathcal{A}_k$ which are harmonic, and let $\mathcal{H}_k$ be the G-module of all functions on X which are induced by the restrictions of the polynomials of $\mathcal{P}_k$ to the sphere S^{2n+1}.

For $0 \leq j, k \leq n$, we consider the $U(1)$-invariant homogeneous complex polynomial f_{jk} on $\mathbb{C}^{n+1}$ defined by

$$f_{jk}(\zeta) = \zeta_j \bar{\zeta}_k,$$

for $\zeta \in \mathbb{C}^{n+1}$, and the function $\tilde{f}_{jk}$ on X induced by f_{jk}. The space $\mathcal{A}_1$ is generated by the functions $\{f_{jk}\}$, with $0 \leq j, k \leq n$, while the space $\mathcal{P}_1$ is generated by the functions $\{f_{jk}\}$ and $\{f_{jj} - f_{kk}\}$, with $0 \leq j, k \leq n$ and $j \neq k$. For $0 \leq j, k \leq n$, let $E_{jk} = (b_{rs})$ be the element of $\mathfrak{gl}(n+1, \mathbb{C})$ determined by $b_{jk} = 1$ and $b_{rs} = 0$ when $(r, s) \neq (j, k)$. The mapping

$$\varphi : \mathcal{A}_1 \to \mathfrak{gl}(n+1, \mathbb{C}),$$

which sends f_{jk} into E_{kj}, is an isomorphism of G-modules; it induces by restriction an isomorphism of G-modules $\varphi : \mathcal{P}_1 \to \mathfrak{g}$.

The set of eigenvalues of the Laplacian Δ acting on $C^\infty(X)$ consists of all the integers $4k(n+k)$, where k is an integer ≥ 0. If $k \geq 0$ is an integer, the G-module $\mathcal{H}_k$ is equal to the eigenspace of the Laplacian Δ acting on $C^\infty(X)$ associated with the eigenvalue $4k(n+k)$. According to observations made in §7, Chapter II and by Proposition 2.1, we know that $\mathcal{H}_k$ is an irreducible G-submodule of $C^\infty(X)$ and that the direct sum $\bigoplus_{k \geq 0} \mathcal{H}_k$ is a dense submodule of $C^\infty(X)$ (see §§C.I and C.III in Chapter III of [4]). Then $\mathcal{H}_0 = C^\infty_{\gamma_0}(X)$ is the space of all constant functions on X; on the other hand, we know that $\mathcal{H}_1$ is isomorphic to the G-module $\mathfrak{g}$, and hence $\mathcal{H}_1$ is equal to $C^\infty_{\gamma_1}(X)$. For $k \geq 2$, let γ_k be the element of Γ corresponding to the irreducible G-module $\mathcal{H}_k$. Then in fact, we have the equalities

$$C^\infty_{\gamma_k}(X) = \mathcal{H}_k,$$

for all $k \geq 0$; moreover, if γ is an element of Γ which cannot be written in the form γ_k, for some integer $k \geq 0$, we know that $C^\infty_\gamma(X) = \{0\}$.

The function $f_k = (f_{n0})^k$ is a $U(1)$-invariant homogeneous complex polynomial on $\mathbb{C}^{n+1}$ of degree $2k$ which is easily seen to be harmonic; thus the function $\tilde{f}_k$ on X induced by f_k is non-zero and belongs to $\mathcal{H}_k$. Since the differential operators ∂ and $\bar{\partial}$ are homogeneous and X is a Kähler manifold, it follows that, for $k \geq 1$, the sections $\partial \tilde{f}_k$ of $T^{1,0}$ and $\bar{\partial} \tilde{f}_k$ of $T^{0,1}$ are both non-zero and belong to the G-modules $C^\infty_{\gamma_k}(T^{1,0})$ and $C^\infty_{\gamma_k}(T^{0,1})$, respectively.

If we consider the path σ in S^{2n+1} and the closed geodesic $\gamma = \gamma_{u,u'}$ of $\mathbb{CP}^n$, and if we write $u = (u_0, u_1, \ldots, u_n)$ and $u' = (u'_0, u'_1, \ldots, u'_n)$, from the formulas (3.19) we deduce that

$$\langle \partial \tilde{f}_1, \dot{\gamma}(t) \rangle = \bar{\zeta}_0(\sigma(t)) \cdot \frac{d}{dt} \zeta_n(\sigma(t))$$

$$= (\cos t \cdot \bar{u}_0 + \sin t \cdot \bar{u}'_0)(-\sin t \cdot u_n + \cos t \cdot u'_n),$$

(3.20)

$$\langle \bar{\partial} \tilde{f}_1, \dot{\gamma}(t) \rangle = \zeta_n(\sigma(t)) \cdot \frac{d}{dt} \bar{\zeta}_0(\sigma(t))$$

$$= (\cos t \cdot u_n + \sin t \cdot u'_n)(-\sin t \cdot \bar{u}_0 + \cos t \cdot \bar{u}'_0),$$

for all $0 \le t \le \pi$; these relations imply that

$$(3.21) \qquad \langle (\partial - \bar{\partial})\tilde{f}_1, \dot{\gamma}(t) \rangle = u'_n \bar{u}_0 - u_n \bar{u}'_0,$$

for all $0 \le t \le \pi$.

Let $\mathfrak{k}$ be the complexification of the Lie algebra of K. The group of all diagonal matrices of G is a maximal torus of G and of K. The complexification $\mathfrak{t}$ of the Lie algebra $\mathfrak{t}_0$ of this torus is a Cartan subalgebra of the semi-simple Lie algebra $\mathfrak{g}$ and also of the reductive Lie algebra $\mathfrak{k}$. For $0 \le j \le n$, the linear form $\lambda_j : \mathfrak{t} \to \mathbb{C}$, sending the diagonal matrix with $a_0, a_1, \ldots, a_n \in \mathbb{C}$ as its diagonal entries into a_j, is purely imaginary on $\mathfrak{t}_0$. Then

$$\Delta = \{ \lambda_i - \lambda_j \mid 0 \le i, j \le n \text{ and } i \ne j \}$$

is the system of roots of $\mathfrak{g}$ with respect to $\mathfrak{t}$, and

$$\Delta' = \{ \lambda_i - \lambda_j \mid 1 \le i, j \le n \text{ and } i \ne j \}$$

is the system of roots of $\mathfrak{k}$ with respect to $\mathfrak{t}$. We fix the positive system

$$\Delta^+ = \{ \lambda_i - \lambda_j \mid 0 \le i < j \le n \}$$

for the roots of $\mathfrak{g}$, and the positive system $\Delta'^+ = \Delta' \cap \Delta^+$ for the roots of $\mathfrak{k}$.

Then we see that highest weight of the irreducible G-module $\mathcal{H}_k$ is equal to $k\lambda_0 - k\lambda_n$, and that $\tilde{f}_k$ is a highest weight vector of this module. Clearly, we have

$$(3.22) \qquad C^\infty_{\gamma_k}(\{g\}) = \mathcal{H}_k \cdot g.$$

The fibers of the homogeneous vector bundles $T^{1,0}$ and $T^{0,1}$ at the point $\pi(e_0)$ of X considered in §2 are irreducible K-modules of highest weights equal to $\lambda_0 - \lambda_n$ and $-\lambda_0 + \lambda_1$, respectively. According to the branching law for G and K (see Proposition 3.1 of [14]), from the results of [14, §4] we obtain the following:

PROPOSITION 3.28. *Let X be the complex projective space $\mathbb{CP}^n$, with $n \ge 1$.*

(i) *The G-modules $C^\infty_{\gamma_k}(T^{1,0})$ and $C^\infty_{\gamma_k}(T^{0,1})$ vanish when $k = 0$ and are irreducible when $k > 0$.*

(ii) *Suppose that $n = 1$. Then for $\gamma \in \Gamma$, the G-modules $C^\infty_\gamma(T^{1,0})$ and $C^\infty_\gamma(T^{0,1})$ vanish unless γ is equal to γ_k, for some integer $k > 0$.*

(iii) *We have*

$$C^\infty_{\gamma_1}((S^2 T^*)^-_\mathbb{C}) = \{0\}.$$

We now consider the formalism of Kähler geometry on $X = \mathbb{CP}^n$. We consider the open subset

$$V = \pi\left(\{\,(\zeta_0, \ldots, \zeta_n) \in \mathbb{C}^{n+1} \mid \zeta_0 \neq 0\,\}\right)$$

of $X = \mathbb{CP}^n$, the point $a = \pi(1, 0, \ldots, 0)$ of V, and the holomorphic coordinate $z = (z_1, \ldots, z_n)$ on V, where z_j is the function which satisfies $\pi^* z_j = \zeta_j / \zeta_0$ on $\mathbb{C}^* \times \mathbb{C}^n$, with $1 \leq j \leq n$. We set

$$|z| = (|z_1|^2 + \cdots + |z_n|^2)^{\frac{1}{2}}.$$

The Fubini-Study metric g of X is given on V by

$$g_{j\bar{k}} = \frac{1}{2}\left(\frac{\delta_{jk}}{1 + |z|^2} - \frac{\bar{z}_j z_k}{(1 + |z|^2)^2}\right),$$

for $1 \leq j, k \leq n$. We recall that the Christoffel symbols of the Levi-Civita connection ∇ of g are determined on V by

$$\Gamma^l_{jk} = \overline{\Gamma^{\bar{l}}_{\bar{j}\bar{k}}} = -\frac{\bar{z}_j \delta_{kl} + \bar{z}_k \delta_{jl}}{1 + |z|^2}.$$

If f is a complex-valued function on V, we have

$$\Delta f = -2 \sum_{j,k=1}^n g^{j\bar{k}} \frac{\partial^2 f}{\partial z_j \partial \bar{z}_k};$$

in particular, we see that

$$(3.23) \qquad (\Delta f)(a) = -4 \sum_{j,k=1}^n \frac{\partial^2 f}{\partial z_j \partial \bar{z}_k}(a).$$

On the open subset V of X, we see that

$$\tilde{f}_{jk} = \frac{z_j \bar{z}_k}{1 + |z|^2}, \qquad \tilde{f}_{0k} = \frac{\bar{z}_k}{1 + |z|^2},$$

$$\tilde{f}_{j0} = \frac{z_j}{1 + |z|^2}, \qquad \tilde{f}_{00} = \frac{1}{1 + |z|^2},$$

for $1 \leq j, k \leq n$. For $0 \leq j, k \leq n$, with $j \neq k$, and $1 \leq l \leq n$, from formula (3.23) we obtain the equalities

$$(3.24) \qquad (\Delta \tilde{f}_{jk})(a) = 0, \qquad (\Delta \tilde{f}_{ll})(a) = -4, \qquad (\Delta \tilde{f}_{00})(a) = 4n.$$

For the remainder of this section, we suppose that $n = 1$ and that $X = \mathbb{CP}^1$. We consider the sphere $S^2 = S\mathbb{R}^3$; the mapping

$$\varphi : S^2 \to \mathbb{CP}^1$$

is well-defined by

$$\varphi(x_1, x_2, x_3) = \begin{cases} \pi(1 - x_3, x_1 + ix_2) & \text{if } (x_1, x_2, x_3) \neq (0, 0, 1), \\ \pi(x_1 - ix_2, 1 + x_3) & \text{if } (x_1, x_2, x_3) \neq (0, 0, -1), \end{cases}$$

where $(x_1, x_2, x_3) \in \mathbb{R}^3$, with $x_1^2 + x_2^2 + x_3^2 = 1$. The two expressions for φ correspond to the stereographic projections whose poles are the points $(0, 0, 1)$ and $(0, 0, -1)$, respectively, and so we know that φ is a diffeomorphism. We also consider the involution Ψ of $\mathbb{CP}^1$ which sends the point $\pi(u)$ of $\mathbb{CP}^1$, where u is a non-zero vector of $\mathbb{C}^2$, into the point $\pi(v)$, where v is a non-zero vector of $\mathbb{C}^2$ orthogonal to u. If τ is the anti-podal involution of the sphere S^2, it is easily verified that the diagram

$$
\begin{array}{ccc}
S^2 & \stackrel{\tau}{\longrightarrow} & S^2 \\
\downarrow{\scriptstyle \varphi} & & \downarrow{\scriptstyle \varphi} \\
\mathbb{CP}^1 & \stackrel{\Psi}{\longrightarrow} & \mathbb{CP}^1
\end{array}
$$

(3.25)

is commutative.

The mapping $\Psi : \mathbb{C}^2 \to \mathbb{C}^2$ defined by

$$\Psi(u_0, u_1) = (-\bar{u}_1, \bar{u}_0),$$

for $u_0, u_1 \in \mathbb{C}$, is an automorphism of $\mathbb{C}^2$ (over $\mathbb{R}$) satisfying $\Psi^2 = -\text{id}$; it induces by restriction a mapping $\Psi : S\mathbb{C}^2 \to S\mathbb{C}^2$ such that the diagram

$$
\begin{array}{ccc}
S\mathbb{C}^2 & \stackrel{\Psi}{\longrightarrow} & S\mathbb{C}^2 \\
\downarrow{\scriptstyle \pi} & & \downarrow{\scriptstyle \pi} \\
\mathbb{CP}^1 & \stackrel{\Psi}{\longrightarrow} & \mathbb{CP}^1
\end{array}
$$

(3.26)

commutes. It is easily seen that we have the equality

$$\Psi^* f_{10} = -f_{10}$$

among functions on $\mathbb{C}^2$; therefore the function $\tilde{f}_k$ on $\mathbb{CP}^1$ satisfies

(3.27) $$\Psi^* \tilde{f}_k = (-1)^k \tilde{f}_k.$$

If J is the complex structure of the vector space $\mathbb{C}^2$, we see that the automorphisms Ψ and J of $\mathbb{C}^2$ satisfy

$$J \circ \Psi = -\Psi \circ J.$$

Thus if J is the complex structure of $\mathbb{CP}^1$, it follows that

$$\Psi_* \cdot J = -J \cdot \Psi_*,$$

as mappings acting on the tangent bundle of $X = \mathbb{CP}^1$; from the preceding relation, we infer that

(3.28)
$$\Psi^*\partial = \bar{\partial}\Psi^*$$

on $\bigwedge^p T_{\mathbb{C}}^*$.

We now verify that φ is an isometry from (S^2, g_0) to $(\mathbb{CP}^1, g')$, where $g' = 4g$ is the metric of constant curvature 1. In fact, we consider the holomorphic coordinate $z = z_1$ on the open subset V of $\mathbb{CP}^1$. We saw above that the metric g' of $\mathbb{CP}^1$ is given on V by

$$g' = \frac{2}{(1 + |z|^2)^2} \, dz \cdot d\bar{z}.$$

Let U be the subset of $\mathbb{R}^3$ consisting of all points $(x_1, x_2, x_3) \in \mathbb{R}^3$, with $x_3 \neq 1$, and let $\varphi' : U \to V$ be the mapping determined by

$$(z \circ \varphi')(x_1, x_2, x_3) = \frac{x_1 + ix_2}{1 - x_3},$$

for $(x_1, x_2, x_3) \in \mathbb{R}^3$ belonging to U. Then the restrictions of φ and φ' to the open subset

$$U_0 = S^2 \cap U = S^2 - \{(0, 0, 1)\}$$

of S^2 are equal, and we know that φ is a diffeomorphism from U_0 to V. If i is the natural inclusion of U_0 into U, we then easily verify that

$$i^*\varphi'^* g' = i^*(dx_1 \otimes dx_1 + dx_2 \otimes dx_2 + dx_3 \otimes dx_3)$$

on U_0. It follows that $\varphi^* g' = g_0$ on the open subset U_0 of S^2 and hence on all of S^2.

We say that a symmetric p-form θ on X is even (resp. is odd) if and only if $\Psi^*\theta = \varepsilon u$, where $\varepsilon = 1$ (resp. $\varepsilon = -1$). According to the commutativity of diagram (3.25), a symmetric p-form θ on X is even (resp. is odd) if and only if the symmetric p-form $\varphi^*\theta$ on S^2 is even (resp. is odd).

Since $\mathcal{H}_k$ is an irreducible G-module which contains the function $\tilde{f}_k$, by (3.27) we see that

$$(3.29) \qquad \begin{aligned} C_{\gamma_{2k}}^\infty(X) &= C_{\gamma_{2k}}^\infty(X)^{\mathrm{ev}} = \mathcal{H}_{2k}, \\ C_{\gamma_{2k+1}}^\infty(X) &= C_{\gamma_{2k+1}}^\infty(X)^{\mathrm{odd}} = \mathcal{H}_{2k+1}, \end{aligned}$$

for $k \geq 0$. Now let k be a given positive integer. Since the differential operators d and $\partial - \bar{\partial}$ from $C^\infty(X)$ to $C^\infty(T_{\mathbb{C}}^*)$ are homogeneous and since the sections $\partial \tilde{f}_k$ of $T^{1,0}$ and $\bar{\partial}\tilde{f}_k$ of $T^{0,1}$ are non-zero, from the relations (3.28) and (3.29) and Proposition 3.28, we obtain the following equalities among irreducible G-modules

$$(3.30) \qquad \begin{aligned} C_{\gamma_k}^\infty(T_{\mathbb{C}}^*)^{\mathrm{ev}} &= \begin{cases} d\mathcal{H}_k & \text{if } k \text{ is even,} \\ (\partial - \bar{\partial})\mathcal{H}_k & \text{if } k \text{ is odd,} \end{cases} \\[2mm] C_{\gamma_k}^\infty(T_{\mathbb{C}}^*)^{\mathrm{odd}} &= \begin{cases} (\partial - \bar{\partial})\mathcal{H}_k & \text{if } k \text{ is even,} \\ d\mathcal{H}_k & \text{if } k \text{ is odd.} \end{cases} \end{aligned}$$

Moreover, the sections $d\tilde{f}_k$ and $\tilde{f}_{k-1}d\tilde{f}_1$ are highest weight vectors of the irreducible G-module $d\mathcal{H}_k$, while $(\partial - \bar{\partial})\tilde{f}_k$ and $\tilde{f}_{k-1}(\partial - \bar{\partial})\tilde{f}_1$ are highest weight vectors of the irreducible G-module $(\partial - \bar{\partial})\mathcal{H}_k$. From the relations (3.30), with $k = 1$, and (2.28) we obtain the equalities

$$(3.31) \qquad C_{\gamma_1}^\infty(T_{\mathbb{C}}^*)^{\mathrm{ev}} = (\partial - \bar{\partial})\mathcal{H}_1 = g^\flat(\mathcal{K}_{\mathbb{C}})$$

of irreducible G-modules.

According to the commutativity of diagram (3.25), the first assertion of the next proposition is equivalent to the result concerning the sphere S^2 given by Proposition 3.17; moreover according to Proposition 2.20, the second assertion of the next proposition implies the result concerning the real projective plane stated in Theorem 3.26.

PROPOSITION 3.29. *Let X be the complex projective space $\mathbb{CP}^1$.*

(i) *An even function on X, whose X-ray transform vanishes, vanishes identically.*

(ii) *An even differential form of degree 1 on X satisfies the zero-energy condition if and only if it is exact.*

PROOF: Let $\delta : [0, \pi] \to X$ be the closed geodesic $\gamma_{u,u'}$ of X corresponding to the pair of unit vectors $u = (1,0)$ and $u' = (0,1)$. For all $0 \leq t \leq \pi$, by (3.20) and (3.21) we easily verify that

$$\tilde{f}_1(\delta(t)) = \sin t \cdot \cos t, \qquad \langle (\partial - \bar{\partial})\tilde{f}_1, \dot{\delta}_1(t) \rangle = 1.$$

When $k \geq 0$ is an even integer, it follows that

$$\int_{\delta} \tilde{f}_k = \int_{\delta} \tilde{f}_k \, (\partial - \bar{\partial}) \tilde{f}_1 = \int_0^{\pi} (\cos t \cdot \sin t)^k \, dt > 0.$$

Therefore the function $\tilde{f}_k$ and the 1-form $\tilde{f}_k \, (\partial - \bar{\partial}) \tilde{f}_1$ on X do not satisfy the zero-energy condition. When $k \geq 0$ is an even integer, by (3.30) we know that the sections $\tilde{f}_k$ and $\tilde{f}_k \, (\partial - \bar{\partial}) \tilde{f}_1$ are highest weight vectors of the irreducible G-modules $C^{\infty}_{\gamma_k}(X)^{\mathrm{ev}}$ and $C^{\infty}_{\gamma_{k+1}}(T^*_{\mathbb{C}})^{\mathrm{ev}}$, respectively; thus we have proved the relation

$$\mathcal{Z}_{0,\mathbb{C}} \cap C^{\infty}_{\gamma_k}(X)^{\mathrm{ev}} = \{0\}$$

when $k \geq 0$ is an even integer, and the relation

$$\mathcal{Z}_{1,\mathbb{C}} \cap C^{\infty}_{\gamma_k}(T^*_{\mathbb{C}})^{\mathrm{ev}} = \{0\}$$

when $k \geq 1$ is an odd integer. From Proposition 3.28,(ii), the equalities (3.29) and (3.30), and the previous relations, we now deduce the equalities

$$\mathcal{Z}_{0,\mathbb{C}} \cap C^{\infty}_{\gamma}(X)^{\mathrm{ev}} = \{0\}, \qquad \mathcal{Z}_{1,\mathbb{C}} \cap C^{\infty}_{\gamma}(T^*_{\mathbb{C}})^{\mathrm{ev}} = dC^{\infty}_{\gamma}(X),$$

for all $\gamma \in \Gamma$. Then by Propositions 2.29,(ii) and 2.32,(ii), with $\Sigma = \{\Psi\}$ and $\varepsilon = +1$, we see that the restriction of the X-ray transform to $C^{\infty}(X)^{\mathrm{ev}}$ is injective and we obtain the equality

$$\mathcal{Z}_{1,\mathbb{C}} \cap C^{\infty}(T^*_{\mathbb{C}})^{\mathrm{ev}} = dC^{\infty}(X);$$

these results imply the two assertions of the proposition.

The bundle $(S^2 T^*)^+_{\mathbb{C}}$ has rank one and is generated by its section g; thus in this case, from Proposition 3.28,(iii) and the relation (3.22) we obtain the equalities

$$(3.32) \qquad C^{\infty}_{\gamma_1}(S^2 T^*_{\mathbb{C}}) = C^{\infty}_{\gamma_1}(\{g\}_{\mathbb{C}}) = \mathcal{H}_1 \cdot g$$

of irreducible G-modules. Since $C^{\infty}_{\gamma_1}(X) = \mathcal{H}_1$ is the eigenspace of the Laplacian associated with the eigenvalue 8 and since the differential operator Hess is homogeneous, from the relation (3.32) we deduce that

$$C^{\infty}_{\gamma_1}(S^2 T^*_{\mathbb{C}}) = \mathrm{Hess} \, \mathcal{H}_1$$

and that

$$(3.33) \qquad \mathrm{Hess} \, f = -4fg,$$

for all $f \in \mathcal{H}_1$. If f and f' are elements of $C^\infty(X)$, by (1.5) we see that

$$D_0(f'(df)^\sharp) = f' D_0(df)^\sharp + df \cdot df' = 2f \operatorname{Hess} f' + df \cdot df';$$

in particular, when f is an element of $\mathcal{H}_1$, by (3.33) we obtain the relation

(3.34) $$D_0(f'(df)^\sharp) = -8ff'g + df \cdot df'.$$

§5. The rigidity of the complex projective space

From formula (3.13), we deduce that

$$\tilde{B}^+ = \rho(g_1^+)R = \{0\}$$

and, by (1.72), we obtain the equality

$$\rho(\iota(\beta))R = -2\psi(\check{\beta}),$$

for $\beta \in (\textstyle\bigwedge^2 T^*)^-$. Thus we have

(3.35) $$\tilde{B} = \tilde{B}^- = \psi((\textstyle\bigwedge^2 T^*)^-)$$

and, by Lemma 1.24, the morphism

$$\psi : (\textstyle\bigwedge^2 T^*)^- \to \tilde{B}$$

is an isomorphism.

We now introduce various families of closed connected totally geodesic submanifolds of X. Let x be a point of X, and let $\mathcal{F}_{1,x}$ be the family of all closed connected totally geodesic surfaces of X passing through x of the form $\operatorname{Exp}_x F$, where F is the subspace of the tangent space T_x generated by an orthonormal set of vectors $\{\xi, \eta\}$ of T_x satisfying $\mathbb{C}\xi \subset (\mathbb{C}\eta)^\perp$. Let $\mathcal{F}_{2,x}$ be the family of all closed connected totally geodesic surfaces of X passing through x of the form $\operatorname{Exp}_x F$, where F is the subspace $\mathbb{C}\xi$ of the tangent space T_x determined by a unitary vector ξ of T_x. We consider the G-invariant families

$$\mathcal{F}_1 = \bigcup_{x \in X} \mathcal{F}_{1,x}, \qquad \mathcal{F}_2 = \bigcup_{x \in X} \mathcal{F}_{2,x},$$
$$\mathcal{F}_3 = \mathcal{F}_1 \cup \mathcal{F}_2$$

of closed connected totally geodesic surfaces of X.

According to Proposition 3.14, a surface belonging to the family $\mathcal{F}_1$ is isometric to the real projective plane with its metric of constant curvature 1, while a surface belonging to the family $\mathcal{F}_2$ is isometric to the complex projective line with its metric of constant curvature 4.

For $j = 1, 2, 3$, we consider the sub-bundle $N_j = N_{\mathcal{F}_j}$ of B consisting of those elements of B which vanish when restricted to the submanifolds of $\mathcal{F}_j$. An element u of B belongs to N_1 if and only if the relation

$$u(\xi, \eta, \xi, \eta) = 0$$

holds for all vectors $\xi, \eta \in T$ satisfying $\mathbb{C}\xi \subset (\mathbb{C}\eta)^\perp$; moreover, an element u of B belongs to N_2 if and only if the relation

$$u(\xi, J\xi, \xi, J\xi) = 0$$

holds for all vectors $\xi \in T$. We set

$$N_1^+ = N_1 \cap B^+, \qquad N_1^- = N_1 \cap B^-;$$

we easily see that the sub-bundle N_1 of B is stable under the involution J, and so we have

$$N_1 = N_1^+ \oplus N_1^-.$$

An elementary computation shows that the rank of N_1 is $\leq n(2n-1)$ (see [18, Proposition 4.1]). Moreover, it is easily verified that

$$\psi(\wedge^2 T^*) \subset N_1.$$

From these observations and the relations (3.35), we obtain the following:

PROPOSITION 3.30. *We have*

$$N_1^+ = \psi(T_{\mathbb{R}}^{1,1}), \qquad N_1^- = \tilde{B} = \psi((\wedge^2 T^*)^-),$$
$$N_1 = \psi(\wedge^2 T^*) = \tilde{B} \oplus \psi(T_{\mathbb{R}}^{1,1}).$$

Let β be an element of $\wedge^2 T^*$ and ξ be a unit vector of T; then we see that

$$\psi(\beta)(\xi, J\xi, \xi, J\xi) = 6\beta(\xi, J\xi).$$

Thus if β belongs to $(\wedge^2 T^*)^-$, we have

$$\psi(\beta)(\xi, J\xi, \xi, J\xi) = 0.$$

On the other hand, if β belongs to $T_{\mathbb{R}}^{1,1}$ and is equal to $\check{h}$, with $h \in (S^2 T^*)^+$, we have

$$\psi(\beta)(\xi, J\xi, \xi, J\xi) = 6h(\xi, \xi).$$

From these remarks, we obtain

$$\psi((\textstyle\bigwedge^2 T^*)^-) \subset N_2, \qquad \psi(T_{\mathbb{R}}^{1,1}) \cap N_2 = \{0\}.$$

From Proposition 3.30 and these relations, we deduce the following result, which is also given by Theorem 8.2 of [13].

PROPOSITION 3.31. *For $n \geq 2$, we have*

$$N_3 = \tilde{B}.$$

In [13, Lemma 8.7], we also show that

(3.36) $$H \cap (T^* \otimes N_3) = \{0\}$$

and then deduce the relation (1.48) for the complex projective of dimension ≥ 2 from Proposition 3.31; thus we have the following:

PROPOSITION 3.32. *For $n \geq 2$, we have*

$$H \cap (T^* \otimes \tilde{B}) = \{0\}.$$

We consider the space $\mathcal{Z}_2$ of symmetric 2-forms on X satisfying the zero-energy condition. Since the space X is simply connected, from Proposition 3.32 and Theorem 1.18 it follows that the sequence (1.24) is exact.

LEMMA 3.33. *We have*

$$\mathcal{L}(\mathcal{F}_2) \subset \mathcal{L}(\mathcal{F}_1) = \mathcal{Z}_2.$$

PROOF: By Proposition 3.19, every submanifold of X belonging to $\mathcal{F}_1$ is infinitesimally rigid, and so we obtain the inclusion $\mathcal{Z}_2 \subset \mathcal{L}(\mathcal{F}_1)$. For $j = 1$ or 2, we see that every closed geodesic of X is contained in a submanifold of X belonging to the family $\mathcal{F}_j$; then from Proposition 2.46,(i), we obtain the inclusion $\mathcal{L}(\mathcal{F}_j) \subset \mathcal{Z}_2$.

For $j = 1$ or 3, we consider the differential operator $D_{1,\mathcal{F}_j}$ of §8, Chapter II corresponding to the family $\mathcal{F}_j$.

LEMMA 3.34. *Let h be an element of $C^\infty(S^2 T^*)$ belonging to $\mathcal{L}(\mathcal{F}_2)$. Then we have $D_1 h = 0$.*

PROOF: By Lemma 3.33, we know that h also belongs to $\mathcal{L}(\mathcal{F}_1)$. Hence by Proposition 2.45, with $\mathcal{F} = \mathcal{F}_3$, we see that $D_{1,\mathcal{F}_3} h = 0$. By Proposition 3.31, we therefore know that $D_1 h = 0$.

The equivalence of assertions (i) and (iii) of the following theorem is originally due to Michel [45]. We now provide an alternate proof of Michel's result following [13, §8].

THEOREM 3.35. *Let h be a symmetric 2-form on $X = \mathbb{CP}^n$, with $n \geq 2$. The following assertions are equivalent:*
  (i) *The symmetric 2-form h belongs to $\mathcal{L}(\mathcal{F}_2)$.*
  (ii) *We have $D_1 h = 0$.*
  (iii) *The symmetric 2-form h is a Lie derivative of the metric g.*

PROOF: From the exactness of sequence (1.24), we obtain the equivalence of assertions (ii) and (iii). The implication (i) $\Rightarrow$ (ii) is given by Lemma 3.34, and the implication (iii) $\Rightarrow$ (i) is a consequence of Lemma 1.1.

Let N_1' be the sub-bundle of $\bigwedge^2 T^* \otimes \bigwedge^2 T^*$ consisting of those elements of $\bigwedge^2 T^* \otimes \bigwedge^2 T^*$ which vanish when restricted to the submanifolds of $\mathcal{F}_1$; clearly, we have $N_1 \subset N_1'$. We consider the quotient bundle

$$E = (\textstyle\bigwedge^2 T^* \otimes \bigwedge^2 T^*)/N_1'$$

and the natural projection

$$\beta : \textstyle\bigwedge^2 T^* \otimes \bigwedge^2 T^* \to E.$$

Let

$$D_g' : S^2 T^* \to \mathcal{E}$$

be the differential operator equal to the composition $\beta \circ D_g$. According to Lemma 1.1 and Proposition 2.45, with $\mathcal{F} = \mathcal{F}_1$, the sequence

$$(3.37) \qquad C^\infty(T) \xrightarrow{\;D_0\;} C^\infty(S^2 T^*) \xrightarrow{\;D_g'\;} C^\infty(E)$$

is a complex. From Proposition 2.45, with $\mathcal{F} = \mathcal{F}_1$, and Lemma 3.33, we obtain the following:

LEMMA 3.36. *Let h be an element of $C^\infty(S^2 T^*)$. The following assertions are equivalent:*
  (i) *The symmetric 2-form h satisfies the zero-energy condition.*
  (ii) *We have $D_g' h = 0$.*
  (iii) *We have $D_{1,\mathcal{F}_1} h = 0$.*

Consequently, we have:

LEMMA 3.37. *Let X be the complex projective space $\mathbb{CP}^n$, with $n \geq 2$. The following assertions are equivalent:*
  (i) *The space X is infinitesimally rigid.*
  (ii) *The complex (3.37) is exact.*

LEMMA 3.38. *A real-valued function f on X satisfies $D_g'(fg) = 0$ if and only if it vanishes identically.*

PROOF: Let f be a real-valued function f on X satisfying $D_g'(fg) = 0$. According to Lemma 3.36, the symmetric 2-form fg satisfies the zero-energy condition, and so the X-ray transform $\check{f}$ of f vanishes. From Theorem 3.18, we obtain the vanishing of f.

Lemma 4.7 of [14] provides us with a more direct proof of Lemma 3.38. In fact, let f is a real-valued function on X satisfying $D'_g(fg) = 0$. Let x be a point of X; then there is a set $\{\xi_1, \ldots, \xi_n\}$ of vectors of T_x such that $\{\xi_1, \ldots, \xi_n, J\xi_1, \ldots, J\xi_n\}$ is an orthonormal basis of T_x. According to the second formula of (1.22) and (3.13), we have

$$D_g(fg)(\xi_j, \xi_k, \xi_j, \xi_k) = \frac{1}{2}\{(\text{Hess } f)(\xi_j, \xi_j) + (\text{Hess } f)(\xi_k, \xi_k)\} + f(x),$$

for $1 \leq j < k \leq n$. From the preceding equality, we infer that

$$\sum_{1 \leq j < k \leq n} \{(\text{Hess } f)(\xi_j, \xi_j) + (\text{Hess } f)(\xi_k, \xi_k)\} = -n(n-1)f(x),$$

and so

$$\sum_{1 \leq j \leq n} (\text{Hess } f)(\xi_j, \xi_j) = -nf(x).$$

It follows that

$$\Delta f = 2nf.$$

Since $2n$ is not an eigenvalue of Δ, it follows that f vanishes identically. In fact, according to Lichnerowicz's theorem (see [43, p. 135] or Theorem D.I.1 in Chapter III of [4]) and (3.14), we see that the first non-zero eigenvalue of Δ is $> 2(n+1)$; we have also seen that the first non-zero eigenvalue of Δ is equal to $4(n+1)$.

The following theorem gives us the infinitesimal rigidity of the complex projective spaces of dimension ≥ 2.

THEOREM 3.39. *The complex projective space $X = \mathbb{CP}^n$, with $n \geq 2$, is infinitesimally rigid.*

The infinitesimal rigidity of the complex projective spaces of dimension ≥ 2 was first proved by Tsukamoto [53]; in fact, he first proved directly the infinitesimal rigidity of $\mathbb{CP}^2$, and then used the above-mentioned result of Michel given by Theorem 3.35 to derive the rigidity of the complex projective spaces of dimension > 2. Other proofs of Theorem 3.39 may be found in [14] and [18].

We can also obtain the infinitesimal rigidity of the complex projective spaces of dimension ≥ 2 from the rigidity of the complex projective plane by means of Theorem 2.47. In fact, we apply this theorem to the family $\mathcal{F}$ equal to $\mathcal{F}_3$ and to the family $\mathcal{F}'$ consisting of all closed totally geodesic submanifolds of X isometric to $\mathbb{CP}^2$; according to Propositions 3.14, 3.31 and 3.32, we know that the hypotheses of Theorem 2.47 hold.

We remark that the equivalence of assertions (i) and (iii) of Theorem 3.35 may be obtained from Theorem 3.39 and Lemma 3.33 without requiring Propositions 3.31 and 3.32.

We now present an outline of the proof of Theorem 3.39 given in [14]. Since the differential operator D'_g is homogeneous and the differential operator D_0 is elliptic, according to Lemma 3.37 and Proposition 2.3 we see that the space X is infinitesimally rigid if and only if the complex

$$(3.38) \qquad C^\infty_\gamma(T_{\mathbb{C}}) \xrightarrow{D_0} C^\infty_\gamma(S^2 T^*_{\mathbb{C}}) \xrightarrow{D'_g} C^\infty_\gamma(E_{\mathbb{C}})$$

is exact for all $\gamma \in \Gamma$.

We choose a Cartan subalgebra of $\mathfrak{g}$ and fix a system of positive roots of $\mathfrak{g}$. If γ is an arbitrary element of Γ, we determine the multiplicities of the G-modules $C^\infty_\gamma(T_{\mathbb{C}})$ and $C^\infty_\gamma(S^2 T^*_{\mathbb{C}})$ and then describe an explicit basis for the weight subspace W_γ of $C^\infty_\gamma(S^2 T^*_{\mathbb{C}})$ corresponding to its highest weight in terms of elements of the eigenspaces $\mathcal{H}_k$; either the multiplicity of $C^\infty_\gamma(S^2 T^*_{\mathbb{C}})$ is equal to 4 and the multiplicity of $C^\infty_\gamma(T^*_{\mathbb{C}})$ is equal to 2, or these two multiplicities are ≤ 2. Since X is an irreducible Hermitian symmetric space, according to (2.28) the multiplicity of the G-module $C^\infty_{\gamma_1}(T_{\mathbb{C}})$ is equal to 2. The multiplicity of $C^\infty_{\gamma_1}(S^2 T^*_{\mathbb{C}})$ is also equal to 2; in fact, we show that

$$(3.39) \qquad C^\infty_{\gamma_0}(S^2 T^*_{\mathbb{C}}) = \mathcal{H}_0 \cdot g, \qquad C^\infty_{\gamma_1}(S^2 T^*_{\mathbb{C}}) = \mathcal{H}_1 \cdot g + \operatorname{Hess} \mathcal{H}_1.$$

We also recall that the space $\operatorname{Ker} D_0$ of all complex Killing vector fields on X is an irreducible G-submodule of $C^\infty_{\gamma_1}(T_{\mathbb{C}})$.

According to these observations, in order to prove the exactness of the sequence (3.38) corresponding to $\gamma \in \Gamma$, it suffices to consider the action of the differential operator D'_g on the vectors of our basis for W_γ and to prove that

$$\dim_{\mathbb{C}} D'_g W_\gamma \geq \operatorname{Mult} C^\infty_\gamma(S^2 T^*_{\mathbb{C}}) - \operatorname{Mult} C^\infty_\gamma(T_{\mathbb{C}})$$

if $\gamma \neq \gamma_1$, or that $D'_g W_\gamma \neq 0$ when $\gamma = \gamma_1$. In fact, to obtain this result we need to verify that $D'_g W_\gamma \neq 0$ or that

$$\dim_{\mathbb{C}} D'_g W_\gamma \geq 2,$$

as the case may be. This last step is carried out in [14, §5]; in fact, to prove the first inequality, we choose an element h of W_γ, a point $x \in X$ and vectors $\xi, \eta \in T_x$ satisfying $\mathbb{C}\xi \subset (\mathbb{C}\eta)^\perp$ and

$$(D_g h)(\xi, \eta, \xi, \eta) \neq 0.$$

Before proceeding to the description of another proof of Theorem 3.39 given in [18], we shall prove the following result which first appeared in [17].

THEOREM 3.40. *A differential form of degree 1 on* $X = \mathbb{CP}^n$, *with* $n \geq 2$, *satisfies the zero-energy condition if and only if it is exact.*

PROOF: Let F be the sub-bundle of $\bigwedge^2 T^*$ consisting of those elements of $\bigwedge^2 T^*$ which vanish when restricted to the submanifolds of $\mathcal{F}_1$. An element α of $\bigwedge^2 T^*$ belongs to F if and only if the relation

$$\alpha(\xi, \eta) = 0$$

holds for all vectors $\xi, \eta \in T$ satisfying $\mathbb{C}\xi \subset (\mathbb{C}\eta)^\perp$. An elementary algebraic computation shows that F is the line bundle generated by the Kähler form ω of X. Let θ be a differential form of degree 1 on X. Since any closed geodesic of X is contained in a submanifold belonging to the family $\mathcal{F}_1$, we see that the 1-form θ satisfies the zero-energy condition if and only if the restrictions of θ to the submanifolds belonging to the family $\mathcal{F}_1$ satisfy the zero-energy condition. According to Theorem 3.26, the latter property of θ holds if and only if $d\theta$ is a section of F. We now suppose that the 1-form θ satisfies the zero-energy condition; our previous observations imply that θ satisfies the relation

$$d\theta = f\omega,$$

where f is a real-valued function on X. From this equality, we infer that

$$df \wedge \omega = 0,$$

and so f is constant. Since the Kähler ω is harmonic, the function f vanishes and so the form θ is closed; hence θ is exact.

The simplicity of the preceding proof, which is based on a remark of Demailly (see [18]), rests on the correct interpretation of the bundle F. This observation led us to a new proof of the infinitesimal rigidity of $\mathbb{CP}^n$, which can be found in [18] and which requires a minimal amount of harmonic analysis. For symmetric 2-forms, the analogue of the bundle F of the preceding proof is the bundle N_1, and its interpretation is given by Proposition 3.30.

We now describe the proof of Theorem 3.39 given in [18]. We consider the first-order differential operator D_1' introduced in §3, Chapter I. Clearly, if u is a section of B satisfying $\nabla u = 0$, from the definition of D_1' we infer that $D_1' \alpha u = 0$. The following two results are proved in [18].

LEMMA 3.41. *Let* β *be a section of the bundle* $T_{\mathbb{R}}^{1,1}$ *over an open subset of* X *satisfying*

(3.40) $$D_1' \alpha \psi(\beta) = 0.$$

Then the 2-form β *is closed.*

LEMMA 3.42. *We have*

$$\{ f \in C^\infty(X) \mid D_1'\alpha\psi(\partial\bar\partial f) = 0 \} = \mathcal{H}_0 \oplus \mathcal{H}_1.$$

Algebraic computations, the first Bianchi identity and certain properties of the Killing vector fields on X are the ingredients which enter into the proof of Lemma 3.41. In fact, in [18] it is shown that a Killing vector field ξ on X satisfies the relation

(3.41) $$D_1'\alpha\psi(dg^\flat(\xi)) = 0.$$

By (2.28), we know that

$$dC_{\gamma_1}^\infty(T_{\mathbb{C}}^*) = dg^\flat(\mathcal{K}_{\mathbb{C}}) = \partial\bar\partial\mathcal{H}_1.$$

Thus from (3.41) and these equalities, we infer that

$$D_1'\alpha\psi(\partial\bar\partial\mathcal{H}_1) = 0.$$

Since the differential operator $D_1'\alpha\psi\partial\bar\partial$ is homogeneous and $\tilde f_k$ is a non-zero element of the irreducible G-module $\mathcal{H}_k$, according to Proposition 2.1, in order to prove Lemma 3.42 it therefore suffices to show that

$$D_1'\alpha\psi(\partial\bar\partial\tilde f_k) \neq 0,$$

for all integers $k > 1$. This inequality is verified in the appendix of [18].

Let h be a symmetric 2-form on X satisfying the zero-energy condition. From Proposition 3.19, it follows that h belongs to the space $\mathcal{L}(\mathcal{F}_1)$. By Proposition 2.44 or Lemma 1.15, we see that $D_1 h$ is a section of $N_1/\tilde B$. Thus by Proposition 3.30, there exists a section β of $T_{\mathbb{R}}^{1,1}$ over X such that

$$D_1 h = \alpha\psi(\beta).$$

Then Lemma 1.17 tells us that the equality (3.40) holds; hence according to Lemma 3.41, the 2-form β is closed. Therefore there exists a constant $c \in \mathbb{R}$ and a real-valued function f on X such that

(3.42) $$\beta = c\omega + \partial\bar\partial f.$$

Since ω is parallel, we know that $D_1'\alpha\psi(\omega) = 0$. Hence the equality (3.40) implies that $D_1'\alpha\psi(\partial\bar\partial f) = 0$. By Lemma 3.42, the function f belongs to $\mathcal{H}_0 \oplus \mathcal{H}_1$. Therefore without loss of generality, we may assume that the function f belongs to $\mathcal{H}_1$.

We consider the subspaces

$$V = C^\infty_{\gamma_0}(S^2 T^*_\mathbb{C}) \oplus C^\infty_{\gamma_1}(S^2 T^*_\mathbb{C}), \qquad W = C^\infty_{\gamma_0}((B/\tilde{B})_\mathbb{C}) \oplus C^\infty_{\gamma_1}((B/\tilde{B})_\mathbb{C})$$

of $C^\infty(S^2 T^*_\mathbb{C})$ and $C^\infty((B/\tilde{B})_\mathbb{C})$, and the element $h' = P_{\gamma_0} h + P_{\gamma_1} h$ of V. Then $h'' = h - h'$ is an element of $C^\infty(S^2 T^*_\mathbb{C})$ orthogonal to the subspace V. Since D_1 is a homogeneous differential operator, by (2.1) we see that $D_1 h'$ belongs to W and that $D_1 h''$ is orthogonal to W. Since the operator $\partial\bar{\partial}$ is homogeneous and ω is a G-invariant form, according to (2.1) and (3.42) we know that β belongs to $C^\infty_{\gamma_0}(\bigwedge^2 T^*_\mathbb{C}) \oplus C^\infty_{\gamma_1}(\bigwedge^2 T^*_\mathbb{C})$; as the morphism $\alpha\psi$ is G-equivariant, we see that $\alpha\psi(\beta)$ is an element of the subspace W. It follows that

$$D_1 h' = \alpha\psi(\beta), \qquad D_1 h'' = 0.$$

By (3.39), we may write

$$h' = f_1 g + \mathrm{Hess}\, f_2,$$

where f_1, f_2 are real-valued functions on X. According to formulas (1.27) and (1.28), we have

$$D_1 h' = D_1(f_1 g) = \alpha D_g(f_1 g).$$

Thus by Proposition 3.30, we see that $D_g(f_1 g)$ is a section of N_1. Then Lemma 3.38 tells us that the function f_1 vanishes identically. Therefore $D_1 h'$ also vanishes, and so $D_1 h = 0$. By Proposition 3.32 and Theorem 1.18, the complex (1.24) is exact, and so h is a Lie derivative of the metric.

§6. The other projective spaces

Let X be a projective space equal either to the quaternionic projective space $\mathbb{HP}^n$, with $n \geq 2$, or to the Cayley plane. Let $\mathcal{F}_1$ be the family of all closed connected totally geodesic surfaces of X which are isometric either to the real projective plane with its metric of constant curvature 1 or to the sphere S^2 with its metric of constant curvature 4. Let $\mathcal{F}_2$ be the family of all closed connected totally geodesic submanifolds of X isometric to the projective plane $\mathbb{CP}^2$. We verify that every surface belonging to the family $\mathcal{F}_1$ is contained in a submanifold belonging to the family $\mathcal{F}_2$ (see Proposition 3.14, [45, §3.2] and [17, §3]).

We consider the sub-bundle $N = N_{\mathcal{F}_1}$ of B consisting of those elements of B which vanish when restricted to the submanifolds of $\mathcal{F}_1$. The following proposition can be proved by means of computations similar to the ones used in [14] to prove Proposition 3.31 and the relation (3.36) for the complex projective spaces.

PROPOSITION 3.43. *Let X be equal to the quaternionic projective space $\mathbb{HP}^n$, with $n \geq 2$, or to the Cayley plane. Then we have*

$$N = \tilde{B}, \qquad H \cap (T^* \otimes N) = \{0\}.$$

We remark that the equalities of the preceding proposition imply that the relation (1.48) holds for the space X. Thus we may apply Theorem 2.47 to the families $\mathcal{F}_1$ and $\mathcal{F}_2$ in order to obtain the following result from the infinitesimal rigidity of $\mathbb{CP}^2$.

THEOREM 3.44. *The quaternionic projective space $\mathbb{HP}^n$, with $n \geq 2$, and the Cayley plane are infinitesimally rigid.*

THEOREM 3.45. *Let X be equal to the quaternionic projective space $\mathbb{HP}^n$, with $n \geq 2$, or to the Cayley plane. A differential form of degree 1 on X satisfies the zero-energy condition if and only if it is exact.*

PROOF: According to Propositions 3.14 and 3.15 (see Corollary 3.26 of [5]) when X is equal to a quaternionic projective space, or by observations made in [17, §3] when X is the Cayley plane, we easily see that

$$C_{\mathcal{F}_1} = \{0\}.$$

By Theorem 3.40, we know that the hypotheses of Theorem 2.51 are satisfied. The latter theorem now gives us the desired result.

Let $\mathcal{F}'$ be the family of all closed connected totally geodesic submanifolds of X isometric to the projective line $\mathbb{HP}^1$ (which is a sphere of dimension 4), or to the projective line over the Cayley algebra (which is a sphere of dimension 8), as the case may be. By means of the methods which we used in §5 to prove Theorem 3.35 for the complex projective spaces, we may also derive the following theorem, which is weaker than Theorem 3.44.

THEOREM 3.46. *Let X be equal to the quaternionic projective space $\mathbb{HP}^n$, with $n \geq 2$, or to the Cayley plane. If h is a symmetric 2-form on X, the following assertions are equivalent:*
   (i) *The symmetric 2-form h belongs to $\mathcal{L}(\mathcal{F}')$.*
   (ii) *We have $D_1 h = 0$.*
   (iii) *The symmetric 2-form h is a Lie derivative of the metric g.*

The equivalence of assertions (i) and (iii) of Theorem 3.46 were first proved by Michel (see [45]). In [53], Tsukamoto deduced the infinitesimal rigidity of X from this result of Michel, the infinitesimal rigidity of $\mathbb{CP}^2$ and the exactness of sequence (1.51) for the sphere of dimension ≥ 2; he requires the equality (1.57) of Proposition 1.14,(i) and uses an argument similar to the one appearing in the proof of Theorem 3.20.

CHAPTER IV

THE REAL GRASSMANNIANS

§1. The real Grassmannians

Let $m \geq 1$, $n \geq 0$ be given integers and let F be a real vector space of dimension $m + n$ endowed with a positive definite scalar product. Let X be the real Grassmannian $\widetilde{G}_m^{\mathbb{R}}(F)$ of all oriented m-planes in F.

Let $V = V_X$ be the canonical vector bundle (of rank m) over X whose fiber at $x \in X$ is the subspace of F determined by the oriented m-plane x. We denote by $W = W_X$ the vector bundle of rank n over X whose fiber at $x \in X$ is the orthogonal complement W_x of V_x in F. Then we have a natural isomorphism of vector bundles

$$(4.1) \qquad\qquad V^* \otimes W \to T$$

over X. We may view X as a submanifold of $\bigwedge^m F$. In fact, the point $x \in X$ corresponds to the vector $v_1 \wedge \cdots \wedge v_m$ of $\bigwedge^m F$, where $\{v_1, \ldots, v_m\}$ is a positively oriented orthonormal basis of the oriented m-plane x. The isomorphism (4.1) sends an element $\theta \in (V^* \otimes W)_x$ into the tangent vector $dx_t/dt|_{t=0}$ to X at x, where x_t is the point of X corresponding to the vector

$$(v_1 + t\theta(v_1)) \wedge \cdots \wedge (v_m + t\theta(v_m))$$

of $\bigwedge^m F$, for $t \in \mathbb{R}$.

Since the vector bundles V and W are sub-bundles of the trivial vector bundle over X whose fiber is F, the scalar product on F induces by restriction positive definite scalar products g_1 and g_2 on the vector bundles V and W, respectively. If we identify the vector bundle V^* with V by means of the scalar product g_1, the isomorphism (4.1) gives rise to a natural isomorphism

$$(4.2) \qquad\qquad V \otimes W \to T$$

of vector bundles over X, which allows us to identify these two vector bundles and the vector bundle $\bigotimes^2 T^*$ with $\bigotimes^2 V^* \otimes \bigotimes^2 W^*$. In fact, if $\theta_1 \in \bigotimes^2 V^*$, $\theta_2 \in \bigotimes^2 W^*$, we identify the element $\theta_1 \otimes \theta_2$ of $\bigotimes^2 V^* \otimes \bigotimes^2 W^*$ with the element u of $\bigotimes^2 T^*$ determined by

$$u(v_1 \otimes w_1, v_2 \otimes w_2) = \theta_1(v_1, v_2) \cdot \theta_2(w_1, w_2),$$

for $v_1, v_2 \in V$ and $w_1, w_2 \in W$. The scalar product g on T induced by the scalar product $g_1 \otimes g_2$ on $V \otimes W$ is a Riemannian metric on X.

The involution τ of X, corresponding to the change of orientation of an m-plane of F, is an isometry of X. The group Λ of isometries of X generated by τ, which is of order 2, acts freely on X and we may consider the Riemannian manifold $Y = X/\Lambda$ endowed with the Riemannian metric g_Y induced by g. The natural projection $\varpi : X \to Y$ is a two-fold covering. We identify Y with the real Grassmannian $G_m^{\mathbb{R}}(F)$ of all m-planes in F. When $m = 1$, the Grassmannian $G_1^{\mathbb{R}}(F)$ is the projective space of F.

Let V_Y and W_Y be the vector bundles over Y whose fibers at the point $y \in Y$ are equal to V_x and W_x, respectively, where x is one of the points of X satisfying $\varpi(x) = y$. Then the tangent space $T_{Y,y}$ of Y at $y \in Y$ is identified with $(V_Y \otimes W_Y)_y$.

For $x \in X$, the tangent space $T_{\tau(x)}$ is equal to $(V \otimes W)_x$; it is easily verified that the mapping $\tau_* : T_x \to T_{\tau(x)}$ is equal to the identity mapping of $(V \otimes W)_x$. A vector field ξ on X is even (resp. odd) with respect to the involution τ if $\tau_*\xi = \xi$ (resp. $\tau_*\xi = -\xi$). We say that a symmetric p-form u on X is even (resp. odd) with respect to τ if $\tau^*u = \varepsilon u$, where $\varepsilon = 1$ (resp. $\varepsilon = -1$). Such a form u is even if and only if we can write $u = \varpi^*u'$, where u' is a symmetric p-form on Y. If E is a sub-bundle of S^pT^* invariant under the isometry τ, there exists a unique sub-bundle E_Y of $S^pT_Y^*$ such that, for all $x \in X$, the isomorphism $\varpi^* : S^pT_{Y,y}^* \to S^pT_x^*$, where $y = \varpi(x)$, induces by restriction an isomorphism $\varpi^* : E_{Y,y} \to E_x$. A symmetric p-form u on Y is a section of E_Y if and only if the even symmetric p-form ϖ^*u on X is a section of E.

Throughout the remainder of this section, we suppose that $n \geq 1$. The curvature R of the Riemannian manifold (X, g) is determined by

$$R(v_1 \otimes w_1, v_2 \otimes w_2, v_3 \otimes w_3, v_4 \otimes w_4)$$
$$= (\langle v_1, v_4 \rangle \langle v_2, v_3 \rangle - \langle v_1, v_3 \rangle \langle v_2, v_4 \rangle)\langle w_1, w_2 \rangle \langle w_3, w_4 \rangle$$
$$+ (\langle w_1, w_4 \rangle \langle w_2, w_3 \rangle - \langle w_1, w_3 \rangle \langle w_2, w_4 \rangle)\langle v_1, v_2 \rangle \langle v_3, v_4 \rangle,$$

for $v_j \in V$, $w_j \in W$, with $1 \leq j \leq 4$, where $\langle v_j, v_k \rangle = g_1(v_j, v_k)$ and $\langle w_j, w_k \rangle = g_2(w_j, w_k)$. It follows that g is an Einstein metric; in fact, its Ricci tensor is given by

$$(4.3) \qquad\qquad \mathrm{Ric} = (m + n - 2)g.$$

The sub-bundles S^2T^* and $\bigwedge^2 T^*$ of $\bigotimes^2 T^*$ admit decompositions

$$(4.4) \qquad S^2T^* = (S^2V^* \otimes S^2W^*) \oplus (\textstyle\bigwedge^2 V^* \otimes \bigwedge^2 W^*),$$

$$(4.5) \qquad \textstyle\bigwedge^2 T^* = (S^2V^* \otimes \bigwedge^2 W^*) \oplus (\bigwedge^2 V^* \otimes S^2W^*).$$

We denote by $S_0^2V^*$ the sub-bundle of S^2V^* which is the orthogonal complement, with respect to the scalar product induced by g_1, of the line bundle $\{g_1\}$ generated by the section g_1 of S^2V^*. Similarly, we denote by $S_0^2W^*$ the sub-bundle of S^2W^* which is the orthogonal complement of the line bundle $\{g_2\}$ generated by the section g_2. We consider the sub-bundles

$$E_1 = \{g_1\} \otimes S_0^2W^*, \quad E_2 = S_0^2V^* \otimes \{g_2\}, \quad E_3 = S_0^2V^* \otimes S_0^2W^*$$

of S^2T^*. If E_0 is the line bundle $\{g\}$ generated by the section $g = g_1 \otimes g_2$ of S^2T^*, from the equality (4.4) we obtain the orthogonal decomposition

$$(4.6) \qquad\qquad S^2T^* = \bigoplus_{j=0}^{3} E_j \oplus (\wedge^2 V^* \otimes \wedge^2 W^*).$$

We consider the sub-bundle $E = E_X$ of S^2T^* consisting of all elements h of S^2T^* which satisfy

$$(4.7) \qquad\qquad h(\xi, \xi) = 0,$$

for all elements ξ of $V \otimes W$ of rank one. The sub-bundle E is invariant under the isometry τ and we shall also consider the corresponding sub-bundle E_Y of $S^2T_Y^*$ induced by E; if $x \in X$, the fiber of E_Y at the point $y = \varpi(x)$ consists of all elements h of $S^2T_{Y,y}^*$ satisfying the relation (4.7) for all elements ξ of $T_{Y,y} = (V \otimes W)_x$ of rank one.

We easily see that an element h of E satisfies the relation

$$h(v_1 \otimes w_1, v_2 \otimes w_2) + h(v_1 \otimes w_2, v_2 \otimes w_1) = 0,$$

for all $v_1, v_2 \in V$, $w_1, w_2 \in W$. Clearly, the vector bundle $\wedge^2 V^* \otimes \wedge^2 W^*$ is a sub-bundle of E; then from the decomposition (4.4), we obtain the equality

$$E = \wedge^2 V^* \otimes \wedge^2 W^*.$$

This last relation implies that

$$(4.8) \qquad\qquad \operatorname{Tr} E = \{0\}, \qquad \operatorname{Tr} E_Y = \{0\}.$$

We also consider the Grassmannians $X' = \tilde{G}_n^{\mathbb{R}}(F)$ and $Y' = G_n^{\mathbb{R}}(F)$, the natural projection $\varpi : X' \to Y'$ and the involution τ of X'. Let V' be the canonical vector bundle of rank n over X' whose fiber at $a \in X'$ is the subspace of F determined by the oriented n-plane a, and let W' be the vector bundle of rank m over X' whose fiber over $a \in X'$ is the orthogonal complement W_a' of V_a' in F. As above, we identify the tangent bundle of X' with the bundle $V' \otimes W'$; the scalar product on F induces Riemannian metrics on X' and Y'. There is a natural diffeomorphism

$$\Psi : G_m^{\mathbb{R}}(F) \to G_n^{\mathbb{R}}(F),$$

sending an m-plane of F into its orthogonal complement. When $m = n$, this mapping is an involution of $Y = G_n^{\mathbb{R}}(F)$; in this case, we say that a symmetric p-form u on Y is even (resp. odd) if $\Psi^* u = \varepsilon u$, where $\varepsilon = 1$ (resp. $\varepsilon = -1$).

Now suppose that the vector space F is oriented and let Ω be a unit vector of $\bigwedge^{m+n} F$ which is positive with respect to the orientation of F. The oriented m-plane $x \in X$ gives us an orientation of V_x, which in turn induces an orientation of W_x: if $\{v_1, \ldots, v_m\}$ is a positively oriented orthonormal basis of V_x, then the orientation of W_x is determined by an orthonormal basis $\{w_1, \ldots, w_n\}$ of W_x satisfying

$$v_1 \wedge \cdots \wedge v_m \wedge w_1 \wedge \cdots \wedge w_n = \Omega.$$

Then there is a natural diffeomorphism

$$\Psi : \widetilde{G}_m^{\mathbb{R}}(F) \to \widetilde{G}_n^{\mathbb{R}}(F),$$

sending $x \in \widetilde{G}_m^{\mathbb{R}}(F)$ into the n-plane W_x endowed with the orientation described above. For $x \in X$, we have $V'_{\Psi(x)} = W_x$ and $W'_{\Psi(x)} = V_x$. It is easily verified that the induced mapping $\Psi_* : (V \otimes W)_x \to (V' \otimes W')_{\Psi(x)}$ sends $v \otimes w$ into $-w \otimes v$, where $v \in V_x$ and $w \in W_x$; therefore Ψ is an isometry. Clearly the diagram

(4.9)
$$
\begin{array}{ccc}
\widetilde{G}_m^{\mathbb{R}}(F) & \xrightarrow{\Psi} & \widetilde{G}_n^{\mathbb{R}}(F) \\
\downarrow{\varpi} & & \downarrow{\varpi} \\
G_m^{\mathbb{R}}(F) & \xrightarrow{\Psi} & G_n^{\mathbb{R}}(F)
\end{array}
$$

commutes. It follows that the mapping $\Psi : G_m^{\mathbb{R}}(F) \to G_n^{\mathbb{R}}(F)$ is also an isometry. Also the diagram

$$
\begin{array}{ccc}
\widetilde{G}_m^{\mathbb{R}}(F) & \xrightarrow{\Psi} & \widetilde{G}_n^{\mathbb{R}}(F) \\
\downarrow{\tau} & & \downarrow{\tau} \\
\widetilde{G}_m^{\mathbb{R}}(F) & \xrightarrow{\Psi} & \widetilde{G}_n^{\mathbb{R}}(F)
\end{array}
$$

is easily seen to commute.

Let $\{e_1, \ldots, e_{m+n}\}$ be the standard basis of $\mathbb{R}^{m+n}$. We henceforth suppose that F is the vector space $\mathbb{R}^{m+n}$ endowed with the standard Euclidean scalar product. We now consider the real Grassmannians

$$X = \widetilde{G}_{m,n}^{\mathbb{R}} = \widetilde{G}_m^{\mathbb{R}}(\mathbb{R}^{m+n}), \qquad Y = G_{m,n}^{\mathbb{R}} = G_m^{\mathbb{R}}(\mathbb{R}^{m+n}),$$

endowed with the Riemannian metrics g and g_Y induced by the standard Euclidean scalar product of $\mathbb{R}^{m+n}$. Throughout the remainder of this section, we also suppose that $m + n \geq 3$.

The group $SO(m + n)$ acting on $\mathbb{R}^{m+n}$ sends every oriented m-plane into another oriented m-plane. This gives rise to an action of the group $SO(m + n)$ on X. In fact, the group $SO(m + n)$ acts transitively on the Riemannian manifold (X, g) by isometries. The isotropy group of the point x_0 of X corresponding to the vector $e_1 \wedge \cdots \wedge e_m$ of $\bigwedge^m \mathbb{R}^{m+n}$ is the subgroup $K = SO(m) \times SO(n)$ of $SO(m + n)$ consisting of the matrices

$$(4.10) \qquad\qquad\qquad \phi = \begin{pmatrix} A & 0 \\ 0 & B \end{pmatrix},$$

where $A \in SO(m)$ and $B \in SO(n)$. The diffeomorphism

$$\Phi : SO(m + n)/K \to X,$$

which sends the class $\phi \cdot K$, where $\phi \in SO(m+n)$, into the oriented m-plane of $\mathbb{R}^{m+n}$ corresponding to the vector $\phi(e_1) \wedge \cdots \wedge \phi(e_m)$, is compatible with the actions of $SO(m + n)$ on $SO(m + n)/K$ and X.

If I_p denotes the unit matrix of order p, the element

$$s = \begin{pmatrix} -I_m & 0 \\ 0 & I_n \end{pmatrix}$$

of $O(m + n)$ determines an involution σ of $SO(m + n)$ which sends $\phi \in SO(m+n)$ into $s\phi s^{-1}$. Then K is equal to the identity component of the set of fixed points of σ, and $(SO(m + n), K)$ is a Riemannian symmetric pair. The Cartan decomposition of the Lie algebra $\mathfrak{g}_0$ of $SO(m+n)$ corresponding to σ is

$$\mathfrak{g}_0 = \mathfrak{k}_0 \oplus \mathfrak{p}_0;$$

here $\mathfrak{k}_0$ is the Lie algebra of K and $\mathfrak{p}_0$ is the space of all matrices

$$(4.11) \qquad\qquad\qquad \begin{pmatrix} 0 & -{}^t Z \\ Z & 0 \end{pmatrix}$$

of $\mathfrak{g}_0$, where Z is a real $n \times m$ matrix and ${}^t Z$ is its transpose. We identify $\mathfrak{p}_0$ with the vector space $M_{n,m}$ of all real $n \times m$ matrices and, in particular, the element (4.11) of $\mathfrak{p}_0$ with the matrix Z of $M_{n,m}$. The adjoint action of K on $\mathfrak{p}_0$ is expressed by

$$\mathrm{Ad}\,\phi \cdot Z = B \cdot Z \cdot A^{-1},$$

where ϕ is the element (4.10) of K and $Z \in M_{n,m}$.

We identify $\mathfrak{p}_0$ with the tangent space of $SO(m+n)/K$ at the coset of the identity element of $SO(m+n)$. Clearly the diffeomorphism Φ sends this coset into the point x_0 of X. Since V_{x_0} is the subspace of $\mathbb{R}^{m+n}$ generated by $\{e_1, \ldots, e_m\}$, clearly W_{x_0} is the subspace generated by $\{e_{m+1}, \ldots, e_n\}$. Then it is easily verified that the isomorphism $\Phi_* : \mathfrak{p}_0 \to (V \otimes W)_{x_0}$ sends the element (4.11) of $\mathfrak{p}_0$ corresponding to the matrix $Z = (z_{jk})$ of $M_{n,m}$, with $1 \le j \le n$ and $1 \le k \le m$ into the vector

$$\sum_{\substack{1 \le j \le n \\ 1 \le k \le m}} z_{jk} e_k \otimes e_{j+m}$$

of $(V \otimes W)_{x_0}$.

If B is the Killing form of $\mathfrak{g}_0$, the restriction to $\mathfrak{p}_0$ of the scalar product $-B$ is invariant under the adjoint action of K and therefore induces an $SO(m+n)$-invariant metric g_0 on the homogeneous space $SO(m+n)/K$. Endowed with this metric g_0, the manifold $SO(m+n)/K$ is a symmetric space of compact type of rank $\min(m,n)$. It is easily verified that $g_0 = 2(m+n-2)\Phi^*g$. Thus Φ is an isometry from the symmetric space $SO(m+n)/K$, endowed with the metric

$$\frac{1}{2(m+n-2)} \, g_0,$$

to X; henceforth, we shall identify these Riemannian manifolds by means of this $SO(m+n)$-equivariant isometry. From Lemma 1.21, we again obtain the equality (4.3). Moreover, the symmetric space $SO(m+n)/K$ is irreducible unless $m = n = 2$. On the other hand, we shall see below that the Grassmannian $\widetilde{G}^{\mathbb{R}}_{2,2}$ is not irreducible and is in fact isometric to a product of 2-spheres (see Proposition 4.3).

The vector bundles V and W are homogeneous sub-bundles of the trivial vector bundle over X whose fiber is $\mathbb{R}^{m+n}$; it is easily seen that (4.2) is an isomorphism of homogeneous $SO(m+n)$-bundles over X. All the vector bundles appearing in the decomposition (4.6) and the bundle E are homogeneous sub-bundles of $S^2 T^*$; hence the fibers at x_0 of these vector bundles are K-submodules of $S^2 T^*_{x_0}$.

When $m = 1$, we easily see that $\widetilde{G}^{\mathbb{R}}_{1,n}$ endowed with the metric g is isometric to the sphere (S^n, g_0) and that $G^{\mathbb{R}}_{1,n}$ endowed with the metric g_Y is isometric to the real projective space $(\mathbb{RP}^n, g_0)$. In fact, the mapping $\tilde{\varphi} : S^n \to \widetilde{G}^{\mathbb{R}}_{1,n}$, sending the unit vector $u \in S^n$ into the point of $\widetilde{G}^{\mathbb{R}}_{1,n}$ corresponding to the oriented basis $\{u\}$ of the line generated by u is easily seen to be an isometry; moreover, by passage to the quotient, $\tilde{\varphi}$ induces an isometry $\varphi : \mathbb{RP}^n \to G^{\mathbb{R}}_{1,n}$. When $m = 1$, we also know that $E = \{0\}$.

The Grassmannian $\widetilde{G}^{\mathbb{R}}_{n,m}$ is also a homogeneous space of the group $SO(m+n)$. We shall always choose the orientation of $\mathbb{R}^{m+n}$ induced by the unit vector $e_1 \wedge \cdots \wedge e_{m+n}$ of $\bigwedge^{m+n}\mathbb{R}^{m+n}$. It is easily verified that the isometry

$$\Psi : \widetilde{G}^{\mathbb{R}}_{m,n} \to \widetilde{G}^{\mathbb{R}}_{n,m}$$

determined by this orientation satisfies

$$\Psi \circ \phi = \phi \circ \Psi,$$

for all $\phi \in SO(m+n)$. Thus $\widetilde{G}^{\mathbb{R}}_{m,n}$ and $\widetilde{G}^{\mathbb{R}}_{n,m}$ are isometric as symmetric spaces. For $x \in X$, we shall always consider the orientation of the space W_x induced by the orientation of its orthogonal complement V_x (with respect to our orientation of $\mathbb{R}^{m+n}$).

The involutive isometry τ of X satisfies

(4.12) $$\tau \cdot \phi(x) = \phi \cdot \tau(x),$$

for all $\phi \in SO(m+n)$ and $x \in X$. From (4.12), we see that the action of the group $SO(m+n)$ on X passes to the quotient Y. In fact, the group $SO(m+n)$ acts transitively on Y and, if $K_1 = O(m) \times O(n)$ is the subgroup of $O(m+n)$ consisting of the matrices (4.10), where $A \in O(m)$ and $B \in O(n)$, it is easily verified that the isotropy group of the point $\varpi(x_0)$ is equal to the subgroup $K' = SO(m+n) \cap K_1$ of $SO(m+n)$. In fact, K' is equal to the set of fixed points of the involution σ of $SO(m+n)$. Therefore $(SO(m+n), K')$ is a Riemannian symmetric pair and $SO(m+n)/K'$ is a symmetric space of compact type of rank $\min(m,n)$; we identify this space with Y by means of the isometry sending $\phi \cdot K'$ into the point $\phi(\varpi(x_0))$, for $\phi \in SO(m+n)$. Then by (4.12), we see that the projection $\varpi : X \to Y$ is identified with the natural submersion $SO(m+n)/K \to SO(m+n)/K'$ of symmetric spaces. The space Y is irreducible unless $m = n = 2$; moreover, Y is the adjoint space of the symmetric space X whenever $m \neq n$.

From the commutativity of the diagram (4.9), with $F = \mathbb{R}^{m+n}$, we infer that $G^{\mathbb{R}}_{m,n}$ and $G^{\mathbb{R}}_{n,m}$ are isometric as symmetric spaces.

The notion of even or odd tensor on X (with respect to the involutive isometry τ) defined here coincides with the one considered in §4, Chapter II. If F is an $SO(m+n)$-invariant sub-bundle of $T_{\mathbb{C}}$ or of $S^p T_{\mathbb{C}}^*$, which is also invariant under the isometry τ, the space $C^\infty(F)^{\mathrm{ev}}$ (resp. $C^\infty(F)^{\mathrm{odd}}$) consisting of all even (resp. odd) sections of F over X is an $SO(m+n)$-submodule of $C^\infty(F)$, and we then have the decomposition (2.8) of the $SO(m+n)$-module $C^\infty(F)$. Also if the bundle F is a sub-bundle of $S^p T_{\mathbb{C}}^*$, the sub-bundle F_Y of $S^p T_{Y,\mathbb{C}}^*$ induced by F is invariant under the group $SO(m+n)$ and coincides with the one considered in §4, Chapter II; moreover, the mapping ϖ^* induces an isomorphism from the $SO(m+n)$-module $C^\infty(Y, F_Y)$ of all sections of F_Y over Y to the $SO(m+n)$-module $C^\infty(F)^{\mathrm{ev}}$.

If F is an arbitrary vector space of dimension $m + n$ endowed with a positive definite scalar product, an isometry $\varphi : \mathbb{R}^{m+n} \to F$ induces isometries $\varphi : \widetilde{G}_{m,n}^{\mathbb{R}} \to \widetilde{G}_m^{\mathbb{R}}(F)$ and $\varphi : G_{m,n}^{\mathbb{R}} \to G_m^{\mathbb{R}}(F)$. Thus the Riemannian manifolds $\widetilde{G}_m^{\mathbb{R}}(F)$ and $G_m^{\mathbb{R}}(F)$ are symmetric spaces. If we write $X = \widetilde{G}_{m,n}^{\mathbb{R}}$ and $X' = \widetilde{G}_m^{\mathbb{R}}(F)$, for $x \in X$ the isomorphism $\varphi : \mathbb{R}^{m+n} \to F$ induces by restriction isomorphisms

$$\varphi : V_{X,x} \to V_{X',\varphi(x)}, \qquad \varphi : W_{X,x} \to W_{X',\varphi(x)};$$

hence the isomorphism $\varphi_* : T_{X,x} \to T_{X',\varphi(x)}$ is equal to the natural mapping

$$\varphi \otimes \varphi : V_{X,x} \otimes W_{X,x} \to V_{X',\varphi(x)} \otimes W_{X',\varphi(x)}.$$

It follows that

$$\varphi^* E_{X',\varphi(x)} = E_{X,x},$$

for all $x \in X$. Moreover, since $\tau \circ \varphi = \varphi \circ \tau$ as mappings from X to X', we see that, if u is an even (resp. odd) symmetric form on X', then $\varphi^* u$ is an even (resp. odd) symmetric form on X. When $m = n$, since $\Psi \circ \varphi = \varphi \circ \Psi$ as mappings from $G_{n,n}^{\mathbb{R}}$ to $G_n^{\mathbb{R}}(F)$, we see that, if u is an even (resp. odd) symmetric form on $G_n^{\mathbb{R}}(F)$, then $\varphi^* u$ is an even (resp. odd) symmetric form on $G_{n,n}^{\mathbb{R}}$.

The orientations of the spaces V_x, with $x \in X$, considered above, together with the scalar product g_1 on V, give us a Hodge operator

$$* : \textstyle\bigwedge^p V^* \to \bigwedge^{m-p} V^*.$$

On the other hand, the orientations of the spaces W_x induced by the orientations of the spaces V_x, for $x \in X$, together with the scalar product g_2 on W, determine a Hodge operator

$$* : \textstyle\bigwedge^p W^* \to \bigwedge^{n-p} W^*.$$

If $m = 4$, we consider the eigenbundles $\bigwedge^+ V^*$ and $\bigwedge^- V^*$ of the Hodge operator $* : \bigwedge^2 V^* \to \bigwedge^2 V^*$ corresponding to the eigenvalues $+1$ and -1, respectively; since this operator is an involution, we obtain the decomposition

(4.13) $$\textstyle\bigwedge^2 V^* = \bigwedge^+ V^* \oplus \bigwedge^- V^*.$$

When $n = 4$, the Hodge operator $* : \bigwedge^2 W^* \to \bigwedge^2 W^*$ gives us an analogous decomposition

(4.14) $$\textstyle\bigwedge^2 W^* = \bigwedge^+ W^* \oplus \bigwedge^- W^*,$$

where $\bigwedge^+ W^*$ and $\bigwedge^- W^*$ are the eigenbundles of this Hodge operator corresponding to the eigenvalues $+1$ and -1, respectively. If $m = 4$ (resp. $n = 4$), the vector bundles appearing in the decomposition (4.13) (resp. (4.14)) are homogeneous sub-bundles of $\bigwedge^2 V^*$ (resp. $\bigwedge^2 W^*$) and their fibers at x_0 are K-submodules of $\bigwedge^2 V_{x_0}^*$ (resp. $\bigwedge^2 W_{x_0}^*$).

We now suppose that $m, n \geq 2$. The complexification $\mathfrak{g}$ of the Lie algebra $\mathfrak{g}_0$ is equal to $\mathfrak{so}(m + n, \mathbb{C})$, and the complexification $\mathfrak{k}$ of the Lie algebra $\mathfrak{k}_0$ admits the decomposition

$$\mathfrak{k} = \mathfrak{k}_1 \oplus \mathfrak{k}_2,$$

where $\mathfrak{k}_1$ and $\mathfrak{k}_2$ are subalgebras of $\mathfrak{k}$ isomorphic to $\mathfrak{so}(m, \mathbb{C})$ and $\mathfrak{so}(n, \mathbb{C})$, respectively. If $\mathfrak{p}$ denotes the subspace of $\mathfrak{g}$ generated by $\mathfrak{p}_0$, the Lie algebra $\mathfrak{g}$ admits the decomposition

$$\mathfrak{g} = \mathfrak{k}_1 \oplus \mathfrak{k}_2 \oplus \mathfrak{p}$$

into K-modules. In fact, the K-module $\mathfrak{k}_1$ is isomorphic to $\bigwedge^2 V_{\mathbb{C},x_0}^*$, the K-module $\mathfrak{k}_2$ is isomorphic to $\bigwedge^2 W_{\mathbb{C},x_0}^*$ and $\mathfrak{p}$ is isomorphic to the K-module $(V \otimes W)_{\mathbb{C},x_0}$; in fact, $\mathfrak{p}$ is irreducible when $m, n \geq 3$.

The fibers at x_0 of the vector bundles $E_{j,\mathbb{C}}$ are irreducible K-modules. In fact, $E_{0,\mathbb{C},x_0}$ is isomorphic to the trivial K-module, while $E_{1,\mathbb{C},x_0}$ is isomorphic to the irreducible K-module $(S_0^2 W^*)_{\mathbb{C},x_0}$ and $E_{2,\mathbb{C},x_0}$ is isomorphic to the irreducible K-module $(S_0^2 V^*)_{\mathbb{C},x_0}$. Therefore $E_{3,\mathbb{C},x_0}$ is isomorphic to the irreducible K-module $E_{1,\mathbb{C},x_0} \otimes E_{2,\mathbb{C},x_0}$. If $m \neq 4$ (resp. $n \neq 4$), the K-module $\bigwedge^2 V_{\mathbb{C},x_0}^*$ (resp. $\bigwedge^2 W_{\mathbb{C},x_0}^*$) is irreducible; on the other hand, when $m = 4$ (resp. $n = 4$), the K-submodules $\bigwedge^+ V_{\mathbb{C},x_0}^*$ and $\bigwedge^- V_{\mathbb{C},x_0}^*$ of $\bigwedge^2 V_{\mathbb{C},x_0}^*$ (resp. $\bigwedge^+ W_{\mathbb{C},x_0}^*$ and $\bigwedge^- W_{\mathbb{C},x_0}^*$ of $\bigwedge^2 W_{\mathbb{C},x_0}^*$) are irreducible. If $m, n \geq 3$, it is easily seen that none of these three irreducible K-modules $E_{j,\mathbb{C},x_0}$ is isomorphic to one of the K-modules $\bigwedge^2 V_{\mathbb{C},x_0}^*$, $\bigwedge^2 W_{\mathbb{C},x_0}^*$, or $(V \otimes W)_{\mathbb{C},x_0}$, or to any one of their irreducible components.

If $m, n \neq 4$, clearly $(\bigwedge^2 V^* \otimes \bigwedge^2 W^*)_{\mathbb{C},x_0}$ is an irreducible K-module. When either m or n is equal to 4, the bundle $\bigwedge^2 V^* \otimes \bigwedge^2 W^*$ admits an $SO(n+m)$-invariant decomposition arising from the decompositions (4.13) and (4.14); the fibers at x_0 of the components of the corresponding decomposition of $(\bigwedge^2 V^* \otimes \bigwedge^2 W^*)_{\mathbb{C}}$ are irreducible K-modules. If $m, n \geq 3$, the K-module $(\bigwedge^2 V^* \otimes \bigwedge^2 W^*)_{\mathbb{C},x_0}$ does not contain a K-submodule isomorphic to any one of the K-modules $\bigwedge^2 V_{\mathbb{C},x_0}^*$, $\bigwedge^2 W_{\mathbb{C},x_0}^*$, or to any one of their irreducible components. If $m, n \geq 3$ and $m + n \geq 7$, we easily see that the K-module $(\bigwedge^2 V^* \otimes \bigwedge^2 W^*)_{\mathbb{C},x_0}$ does not contain a submodule isomorphic to the irreducible K-module $(V \otimes W)_{\mathbb{C},x_0}$ or to $\mathfrak{p}$.

We now suppose that $m = n = 3$. If we identify the vector bundle V^* with V and the vector bundle W^* with W by means of the scalar products g_1 on V and g_2 on W, the Hodge operators

$$* : V^* \to \bigwedge^2 V^*, \qquad * : W^* \to \bigwedge^2 W^*$$

determine isomorphisms

$$* : V \to \bigwedge^2 V^*, \qquad * : W \to \bigwedge^2 W^*.$$

In turn, via the identification (4.2) the isomorphism

$$* \otimes * : V \otimes W \to \bigwedge^2 V^* \otimes \bigwedge^2 W^*$$

gives rise to an $SO(6)$-equivariant isomorphism of vector bundles

(4.15) $$* : T \to \bigwedge^2 V^* \otimes \bigwedge^2 W^*.$$

In this case, the irreducible K-module $(\bigwedge^2 V^* \otimes \bigwedge^2 W^*)_{\mathbb{C},x_0}$ is isomorphic to $(V \otimes W)_{\mathbb{C},x_0}$ and hence to $\mathfrak{p}$.

From the above remarks, we obtain the following results:

LEMMA 4.1. *Let X be the real Grassmannian $\widetilde{G}^{\mathbb{R}}_{m,n}$, with $m, n \geq 3$.*
(i) *We have*
$$\mathrm{Hom}_K(\mathfrak{g}, E_{j,\mathbb{C},x_0}) = \{0\},$$

for $j = 0, 1, 2, 3$.
(ii) *If $m + n \geq 7$, we have*

$$\mathrm{Hom}_K(\mathfrak{g}, (\bigwedge^2 V^* \otimes \bigwedge^2 W^*)_{\mathbb{C},x_0}) = \{0\}.$$

(iii) *If $m = n = 3$, we have*

$$\dim \mathrm{Hom}_K(\mathfrak{g}, (\bigwedge^2 V^* \otimes \bigwedge^2 W^*)_{\mathbb{C},x_0}) = 1.$$

We now suppose that $m, n \geq 3$. From Lemma 4.1 and the decomposition (4.6), we deduce that

(4.16) $$\mathrm{Hom}_K(\mathfrak{g}, S^2 T^*_{\mathbb{C},x_0}) = \{0\}$$

when $m, n \geq 3$ and $m + n \geq 7$, and that

(4.17) $$\dim \mathrm{Hom}_K(\mathfrak{g}, S^2 T^*_{\mathbb{C},x_0}) = \dim \mathrm{Hom}_K(\mathfrak{g}, S^2_0 T^*_{\mathbb{C},x_0}) = 1$$

when $m = n = 3$. We recall that the Grassmannian $\widetilde{G}_{3,3}^{\mathbb{R}}$ is isometric to the symmetric space $SU(4)/SO(4)$ (see [36, p. 519]). Thus by Lemma 2.41, we know that the space $\mathrm{Hom}_K(\mathfrak{g}, S_0^2 T_{\mathbb{C},x_0}^*)$ vanishes for the space $\widetilde{G}_{m,n}^{\mathbb{R}}$, when $m, n \geq 3$ and $m + n \geq 7$, and that $\mathrm{Hom}_K(\mathfrak{g}, S_0^2 T_{\mathbb{C},x_0}^*)$ is one-dimensional for the space $\widetilde{G}_{3,3}^{\mathbb{R}}$; as we have just seen, both of these assertions are also consequences of Lemma 4.1.

Let γ_1 be the element of the dual of the group $SO(m + n)$ which is the equivalence class of the irreducible $SO(m + n)$-module $\mathfrak{g}$. We denote by $\mathcal{K}$ the $SO(m + n)$-module of all Killing vector fields on X and by $\mathcal{K}_{\mathbb{C}}$ its complexification. The irreducible symmetric space X is not equal to a simple Lie group. Thus according to (2.27), we know that the $SO(m + n)$-module $C_{\gamma_1}^\infty(T_{\mathbb{C}}^*)$ is irreducible and is equal to $\mathcal{K}_{\mathbb{C}}$. When $m + n \geq 7$, by (4.16) the Frobenius reciprocity theorem tells us that

$$(4.18) \qquad C_{\gamma_1}^\infty(S^2 T_{\mathbb{C}}^*) = \{0\},$$

and the equality (2.25) then says that $E(X) = \{0\}$. We now again consider the case when $m = n = 3$. The isomorphism (4.15) provides us with an isomorphism

$$* : C_{\gamma_1}^\infty(T_{\mathbb{C}}) \to C_{\gamma_1}^\infty((\wedge^2 V^* \otimes \wedge^2 W^*)_{\mathbb{C}})$$

of $SO(6)$-modules; hence by (2.27), we have

$$C_{\gamma_1}^\infty((\wedge^2 V^* \otimes \wedge^2 W^*)_{\mathbb{C}}) = * \mathcal{K}_{\mathbb{C}}.$$

By (4.17) and Lemma 4.1, the Frobenius reciprocity theorem gives us the equalities

$$(4.19) \qquad C_{\gamma_1}^\infty(S^2 T_{\mathbb{C}}^*) = C_{\gamma_1}^\infty(S_0^2 T_{\mathbb{C}}^*) = C_{\gamma_1}^\infty((\wedge^2 V^* \otimes \wedge^2 W^*)_{\mathbb{C}}) = * \mathcal{K}_{\mathbb{C}}.$$

By Proposition 2.40 and the equality (4.17), we see that $E(X)$ is an irreducible $SO(6)$-module isomorphic to the Lie algebra $\mathfrak{g}_0$; moreover by (4.19) and (2.29), we have

$$(4.20) \qquad E(X)_{\mathbb{C}} = C_{\gamma_1}^\infty((\wedge^2 V^* \otimes \wedge^2 W^*)_{\mathbb{C}}) = * \mathcal{K}_{\mathbb{C}}.$$

Since $\wedge^2 V^* \otimes \wedge^2 W^*$ is a sub-bundle of E, the above discussion gives us the following result:

PROPOSITION 4.2. Let X be the Grassmannian $\widetilde{G}_{m,n}^{\mathbb{R}}$, with $m, n \geq 3$.
(i) If $m + n \geq 7$, then we have

$$E(X) = \{0\}.$$

(ii) *If $m = n = 3$, then $E(X)$ is an irreducible $SO(6)$-module isomorphic to the Lie algebra $\mathfrak{g}_0 = \mathfrak{so}(6)$, and is equal to the $SO(6)$-submodule*

$$\{\, h \in C^\infty_{\gamma_1}((\wedge^2 V^* \otimes \wedge^2 W^*)_{\mathbb{C}}) \mid h = \bar{h} \,\} = *\mathcal{K}$$

of $C^\infty(E)$.

When $m + n \geq 7$, the vanishing of the space $E(X)$ is also given by Theorem 1.22 (see Koiso [41] and [42]).

For the remainder of this section, we suppose that $m = n \geq 2$. Then the isometry Ψ of the Grassmannian $Y = G^{\mathbb{R}}_{n,n}$ is an involution. The group Λ of isometries of Y generated by Ψ, which is of order 2, acts freely on Y and we may consider the Riemannian manifold $\bar{Y} = \bar{G}^{\mathbb{R}}_{n,n}$ equal to the quotient Y/Λ endowed with the Riemannian metric $g_{\bar{Y}}$ induced by g. The natural projections $\varpi' : Y \to \bar{Y}$ and $\varpi'' : X \to \bar{Y}$ are two-fold and four-fold coverings, respectively. The action of the group $SO(2n)$ on Y passes to the quotient $\bar{Y}$. In fact, $SO(2n)$ acts transitively on $\bar{Y}$ and it is easily verified that the isotropy group of the point $\varpi''(x_0)$ is equal to the subgroup of $SO(2n)$ generated by K' and the matrix

$$\begin{pmatrix} 0 & -I_n \\ I_n & 0 \end{pmatrix}$$

of $SO(2n)$. In fact, $\bar{G}^{\mathbb{R}}_{n,n}$ is a symmetric space of compact type of rank n. When $n \geq 3$, it is irreducible and equal to the adjoint space of X and of Y. On the other hand, when $n = 2$, it is not irreducible, and we have the following result, whose proof appears below in §9, Chapter V:

PROPOSITION 4.3. *The symmetric space $S^2 \times S^2$ (resp. $\mathbb{RP}^2 \times \mathbb{RP}^2$) endowed with the Riemannian metric which is the product of the metrics of constant curvature 1 on each factor is isometric to the Grassmannian $\bar{G}^{\mathbb{R}}_{2,2}$ (resp. $\bar{G}^{\mathbb{R}}_{2,2}$) endowed with the Riemannian metric $2g$ (resp. $2g_{\bar{Y}}$).*

The following proposition is a direct consequence of Propositions 4.3 and 10.2, and of Theorem 2.23,(ii).

PROPOSITION 4.4. *The maximal flat Radon transform for functions on the symmetric space $\bar{G}^{\mathbb{R}}_{2,2}$ is injective.*

The notion of even or odd tensor on Y (with respect to the involutive isometry Ψ) defined here coincides with the one considered in §4, Chapter II. In fact, a section u of $S^p T^*_Y$ over Y is even if and only if we can write $u = \varpi'^* u'$, where u' is a symmetric p-form on $\bar{Y}$. Lemma 2.17 gives us the following result:

LEMMA 4.5. *A symmetric p-form u on $\bar{G}^{\mathbb{R}}_{n,n}$ satisfies the Guillemin condition if and only if the even symmetric p-form $\varpi'^* u$ on $G^{\mathbb{R}}_{n,n}$ satisfies the Guillemin condition.*

§2. The Guillemin condition on the real Grassmannians

Let $m, n \geq 1$ be given integers. In this section, we again consider the real Grassmannians $X = \widetilde{G}^{\mathbb{R}}_{m,n}$ and $Y = G^{\mathbb{R}}_{m,n}$, endowed with the metrics g and g_Y, and the natural Riemannian submersion $\varpi : X \to Y$, and continue to identify the tangent bundle T of X with the vector bundle $V \otimes W$ as in §1.

Let $V_1, \ldots, V_r$ be mutually orthogonal subspaces of $\mathbb{R}^{m+n}$ and let $p_1, \ldots, p_r$ be given integers, with $1 \leq p_j \leq \dim V_j$ and $p_1 + \cdots + p_r = m$. For $1 \leq j \leq r$, the space V_j is endowed with the scalar product induced by the Euclidean scalar product of $\mathbb{R}^{m+n}$; in turn, this scalar product induces a Riemannian metric on $\widetilde{G}^{\mathbb{R}}_{p_1}(V_j)$. Then there is a natural totally geodesic imbedding

$$\iota : \widetilde{G}^{\mathbb{R}}_{p_1}(V_1) \times \cdots \times \widetilde{G}^{\mathbb{R}}_{p_r}(V_r) \to \widetilde{G}^{\mathbb{R}}_{m,n}$$

which is defined as follows. For $1 \leq j \leq r$, let x_j be a point of $\widetilde{G}^{\mathbb{R}}_{p_j}(V_j)$ corresponding to the vector $v^j_1 \wedge \cdots \wedge v^j_{p_j}$ of $\bigwedge^{p_j} V_j$, where $\{v^j_1, \ldots, v^j_{p_j}\}$ is a positively oriented orthonormal basis of the oriented p_j-plane x_j. The mapping ι sends $(x_1, \ldots, x_r)$ into the point x of $\widetilde{G}^{\mathbb{R}}_{m,n}$ corresponding to the vector

$$v^1_1 \wedge \cdots \wedge v^1_{p_1} \wedge \cdots \wedge v^r_1 \wedge \cdots \wedge v^r_{p_r}$$

of $\bigwedge^m \mathbb{R}^{m+n}$. Moreover there is a unique totally geodesic imbedding

$$\bar{\iota} : G^{\mathbb{R}}_{p_1}(V_1) \times \cdots \times G^{\mathbb{R}}_{p_r}(V_r) \to G^{\mathbb{R}}_{m,n}$$

such that the diagram

$$
\begin{array}{ccc}
\widetilde{G}^{\mathbb{R}}_{p_1}(V_1) \times \cdots \times \widetilde{G}^{\mathbb{R}}_{p_r}(V_r) & \xrightarrow{\iota} & \widetilde{G}^{\mathbb{R}}_{m,n} \\
\downarrow{\scriptstyle \varpi \times \cdots \times \varpi} & & \downarrow{\scriptstyle \varpi} \\
G^{\mathbb{R}}_{p_1}(V_1) \times \cdots \times G^{\mathbb{R}}_{p_r}(V_r) & \xrightarrow{\bar{\iota}} & G^{\mathbb{R}}_{m,n}
\end{array}
$$

commutes.

If V'_j is the p_j-dimensional subspace of V_j spanned by the vectors $\{v^j_1, \ldots, v^j_{p_j}\}$ of V_j and if W_j is the orthogonal complement of V'_j in V_j, the tangent space of $\widetilde{G}^{\mathbb{R}}_{p_j}(V_{p_j})$ at x_j is identified with $V'_j \otimes W_j$, for $1 \leq j \leq r$. Then $V'_1 \oplus \cdots \oplus V'_r$ is equal to V_x, while $W_1 \oplus \cdots \oplus W_r$ is a subspace of W_x. It is easily seen that the mapping $\iota_{*(x_1, \ldots, x_r)}$ from the tangent space of

$\widetilde{G}_{p_1}^{\mathbb{R}}(V_1) \times \cdots \times \widetilde{G}_{p_r}^{\mathbb{R}}(V_r)$ at $(x_1, \ldots, x_r)$ to the tangent space of $\widetilde{G}_{m,n}^{\mathbb{R}}$ at x induced by ι is identified with the mapping

$$(V_1' \otimes W_1) \oplus \cdots \oplus (V_r' \otimes W_r) \to (V \otimes W)_x$$

sending $(\theta_1, \ldots, \theta_r)$ into $\theta_1 + \cdots + \theta_r$, where $\theta_j \in V_j' \otimes W_j$.

We consider the totally geodesic imbeddings $\varphi : \widetilde{G}_{p_1}^{\mathbb{R}}(V_1) \to X$ and $\bar{\varphi} : G_{p_1}^{\mathbb{R}}(V_1) \to Y$ defined by

$$\varphi(z) = \iota(z, x_2, \ldots, x_r), \qquad \bar{\varphi}(\varpi(z)) = \bar{\iota}(\varpi(z), \varpi(x_2), \ldots, \varpi(x_r)),$$

for $z \in \widetilde{G}_{p_1}^{\mathbb{R}}(V_1)$; then we have $\varphi(x_1) = x$. We write $Z = \widetilde{G}_{p_1}^{\mathbb{R}}(V_1)$; then the mapping $\varphi_* : (V_Z \otimes W_Z)_z \to (V \otimes W)_{\varphi(z)}$ is the natural inclusion. Therefore if h is a section of the vector bundle E over X, we see that the symmetric 2-form $\varphi^* h$ is a section of the vector bundle E_Z over Z.

We have the equality

$$(4.21) \qquad \operatorname{Exp}_x V_1' \otimes W_1 = \varphi(\widetilde{G}_{p_1}^{\mathbb{R}}(V_1))$$

of closed totally geodesic submanifolds of $\widetilde{G}_{m,n}^{\mathbb{R}}$. Indeed, using the above description of the mapping ι_* at $(x_1, \ldots, x_r)$, we see that the tangent spaces of these two submanifolds of $\widetilde{G}_{m,n}^{\mathbb{R}}$ at x are equal. From the formula for the curvature of $\widetilde{G}_{m,n}^{\mathbb{R}}$, we infer that $\operatorname{Exp}_x V_1' \otimes W_1$ is a totally geodesic submanifold of $\widetilde{G}_{m,n}^{\mathbb{R}}$ and a globally symmetric space. Clearly, the submanifold $\varphi(\widetilde{G}_{p_1}^{\mathbb{R}}(V_1))$ possesses these same properties. In fact, the subgroup $SO(m+n, V_1)$ of $SO(m+n)$ consisting of all elements of $SO(m+n)$ which preserve the subspace V_1 and which are the identity on the orthogonal complement of V_1 acts transitively on these submanifolds by isometries. These various observations yield the relation (4.21), which in turn gives us the equality

$$(4.22) \qquad \operatorname{Exp}_{\varpi(x)} V_1' \otimes W_1 = \bar{\varphi}(G_{p_1}^{\mathbb{R}}(V_1))$$

of closed totally geodesic submanifolds of $G_{m,n}^{\mathbb{R}}$.

It is easily verified that the diagram

$$(4.23) \qquad
\begin{array}{ccc}
\widetilde{G}_{p_1}^{\mathbb{R}}(V_1) \times \widetilde{G}_{p_2}^{\mathbb{R}}(V_2) \times \cdots \times \widetilde{G}_{p_r}^{\mathbb{R}}(V_r) & \xrightarrow{\iota} & \widetilde{G}_{m,n}^{\mathbb{R}} \\
\downarrow{\tau \times \mathrm{id} \times \cdots \times \mathrm{id}} & & \downarrow{\tau} \\
\widetilde{G}_{p_1}^{\mathbb{R}}(V_1) \times \widetilde{G}_{p_2}^{\mathbb{R}}(V_2) \times \cdots \times \widetilde{G}_{p_r}^{\mathbb{R}}(V_r) & \xrightarrow{\iota} & \widetilde{G}_{m,n}^{\mathbb{R}}
\end{array}$$

is commutative. It follows that, if u is an even (resp. odd) symmetric p-form on $\widetilde{G}_{m,n}^{\mathbb{R}}$, the symmetric p-form $\varphi^* u$ on Z is even (resp. odd).

From the above observations and the equalities (4.21) and (4.22), we obtain:

LEMMA 4.6. *Let X be the real Grassmannian $\widetilde{G}_{m,n}^{\mathbb{R}}$, with $m, n \geq 2$. Let x be a point of X and let V' and W' be non-zero subspaces of V_x and W_x of dimension p and q, respectively. Then $X' = \mathrm{Exp}_x V' \otimes W'$ is a closed totally geodesic submanifold of X invariant under the involution τ, which is isometric to the Grassmannian $\widetilde{G}_{p,q}^{\mathbb{R}}$, and $Y' = \mathrm{Exp}_{\varpi(x)} V' \otimes W'$ is a closed totally geodesic submanifold of $Y = G_{m,n}^{\mathbb{R}}$, which is isometric to the Grassmannian $G_{p,q}^{\mathbb{R}}$, with $\varpi^{-1}(Y') = X'$. Moreover, if F is the subspace $V' \oplus W'$ of $\mathbb{C}^{m+n}$ of dimension $p + q$ and Z is the Grassmannian $\widetilde{G}_p^{\mathbb{R}}(F)$, there are isometric imbeddings $i : Z \to X$ and $\bar{\imath} : G_p^{\mathbb{R}}(F) \to Y$ whose images are equal to X' and Y', respectively, satisfying $\varpi \circ i = \bar{\imath} \circ \varpi$ and which possess the following properties:*

(i) *if z is the unique point of Z satisfying $i(z) = x$, we have*

$$V_{Z,z} = V', \qquad W_{Z,z} = W',$$

and the mapping $i_* : (V_Z \otimes W_Z)_z \to (V \otimes W)_x$ induced by i is the natural inclusion;

(ii) *if u is an even (resp. odd) symmetric form on X, the form i^*u on Z is even (resp. odd);*

(iii) *if h is a section of the sub-bundle E of S^2T^* over X, then i^*h is a section of the sub-bundle E_Z of $S^2T_Z^*$.*

If the subspaces V_j are all 2-dimensional and the integers p_j are all equal to 1, then the images of the mappings ι and $\bar{\iota}$ are totally geodesic flat r-tori of $\widetilde{G}_{m,n}^{\mathbb{R}}$ and $G_{m,n}^{\mathbb{R}}$. In particular, when $m \leq n$ and $r = m$, these images are maximal flat totally geodesic tori of $\widetilde{G}_{m,n}^{\mathbb{R}}$ and $G_{m,n}^{\mathbb{R}}$, and all maximal flat totally geodesic tori of $\widetilde{G}_{m,n}^{\mathbb{R}}$ and $G_{m,n}^{\mathbb{R}}$ arise in this way. On the other hand when $n < m$ and $r = n + 1$, if the subspaces V_j are 2-dimensional and $p_j = 1$ for $1 \leq j \leq n$, and if $p_{n+1} = m - n$, then the images

$$\iota(\widetilde{G}_1^{\mathbb{R}}(V_1) \times \cdots \times \widetilde{G}_1^{\mathbb{R}}(V_n) \times \{x_{n+1}\})$$

and

$$\bar{\iota}(G_1^{\mathbb{R}}(V_1) \times \cdots \times G_1^{\mathbb{R}}(V_n) \times G_{m-n}^{\mathbb{R}}(V_{n+1}))$$

are maximal flat totally geodesic n-tori of $\widetilde{G}_{m,n}^{\mathbb{R}}$ and $G_{m,n}^{\mathbb{R}}$, and all maximal flat totally geodesic tori of $\widetilde{G}_{m,n}^{\mathbb{R}}$ and $G_{m,n}^{\mathbb{R}}$ arise in this way. In this case, $G_{m-n}^{\mathbb{R}}(V_{n+1})$ consists of the single point $\varpi(x_{n+1})$.

From Lemma 2.17, or from the commutativity of diagram (4.23) and the above remarks concerning totally geodesic flat tori of the Grassmannians, we obtain the following result:

LEMMA 4.7. *A symmetric p-form u on Y satisfies the Guillemin condition if and only if the even symmetric p-form $\varpi^* u$ on X satisfies the Guillemin condition.*

LEMMA 4.8. *Let X be the real Grassmannian $\widetilde{G}^{\mathbb{R}}_{m,n}$, with $m, n \geq 2$. Let $2 \leq p \leq m$ and $2 \leq q \leq n$ be given integers. Let $\mathcal{F}$ be the family of all totally geodesic submanifolds of X passing through a point $x \in X$ which can be written in the form $\mathrm{Exp}_x V' \otimes W'$, where V' is a p-dimensional subspace of V_x and W' is a q-dimensional subspace of W_x. Let h be an element of $S^2 T_x^*$, with $x \in X$. If the restriction of h to an arbitrary submanifold of the family $\mathcal{F}$ vanishes, then h vanishes.*

PROOF: Let $\{v_1, \ldots, v_m\}$ and $\{w_1, \ldots, w_n\}$ be orthonormal bases of the spaces V_x and W_x, respectively. If $1 \leq i, k \leq m$ and $1 \leq j, l \leq n$ are given integers, the two vectors $\xi = v_i \otimes w_j$ and $\eta = v_k \otimes w_l$ of T_x are tangent to a submanifold of X belonging to the family $\mathcal{F}$; thus we have $h(\xi, \eta) = 0$. It follows that h vanishes.

LEMMA 4.9. *Let Y be the real Grassmannian $G^{\mathbb{R}}_{m,n}$, with $2 \leq m < n$. Let F_1 and F_2 be orthogonal subspaces of $\mathbb{R}^{m+n}$ of dimension $2m - 2$ and ≥ 3, respectively, and let*

$$(4.24) \qquad \varphi : G^{\mathbb{R}}_{m-1}(F_1) \times G^{\mathbb{R}}_1(F_2) \to Y$$

be the totally geodesic imbedding corresponding to the orthogonal subspaces F_1 and F_2. Let $x_0 \in G^{\mathbb{R}}_1(F_2)$ and let Y' be the totally geodesic submanifold

$$\varphi(G^{\mathbb{R}}_{m-1}(F_1) \times \{x_0\})$$

of Y, which is isometric to $G^{\mathbb{R}}_{m-1,m-1}$. If u is a symmetric p-form on Y satisfying the Guillemin condition, then the restriction of u to Y' satisfies the Guillemin condition.

PROOF: The rank of the symmetric space $G^{\mathbb{R}}_{m-1}(F_1)$ is equal to $m-1$. Let Z be a maximal flat totally geodesic torus of $G^{\mathbb{R}}_{m-1}(F_1)$ of dimension $m - 1$ and ξ be a unitary parallel vector field on Z. We define a unitary vector field ξ' on the submanifold $\varphi(Z \times G^{\mathbb{R}}_1(F_2))$ of Y by

$$\xi'(\varphi(z, x)) = \varphi_{*(z,x)}(\xi(z), 0),$$

for $z \in Z$, $x \in G^{\mathbb{R}}_1(F_2)$. For $x \in G^{\mathbb{R}}_1(F_2)$, we consider the totally geodesic flat torus $Z_x = \varphi(Z \times \{x\})$ of Y of dimension $m - 1$. Let u be a symmetric p-form on Y. We define a real-valued function f on $G^{\mathbb{R}}_1(F_2)$ by

$$f(x) = \int_{Z_x} u(\xi', \ldots, \xi') \, dZ_x,$$

for $x \in G_1^{\mathbb{R}}(F_2)$. If $i : Y' \to Y$ is the inclusion mapping, we have the equality

$$f(x_0) = \int_Z (i^* u)(\xi, \ldots, \xi) \, dZ.$$

Let $\gamma : [0, L] \to G_1^{\mathbb{R}}(F_2)$ be a closed geodesic of the real projective space $G_1^{\mathbb{R}}(F_2)$. Since the rank of the symmetric space Y is equal to m, we see that $Z_\gamma = \varphi(Z \times \gamma[0, L])$ is a maximal flat totally geodesic torus of Y. Clearly the equality

$$\int_\gamma f = \int_{Z_\gamma} u(\xi', \ldots, \xi') \, dZ_\gamma$$

holds. Now we suppose that u satisfies the Guillemin condition; then the above integral vanishes, and so the function f on the real projective space $G_1^{\mathbb{R}}(F_2)$ of dimension ≥ 2 satisfies the zero-energy condition. The injectivity of the X-ray transform for functions on the real projective space of dimension ≥ 2, given by Theorem 2.23,(ii), tells us that the function f vanishes. From the equality $f(x_0) = 0$, we infer that the restriction of u to Y' satisfies the Guillemin condition.

The following proposition is a generalization of Lemma 5.3 of [23].

PROPOSITION 4.10. *Let y be a point of the real Grassmannian $Y = G_{m,n}^{\mathbb{R}}$, with $2 \leq m < n$. Let Y' be a closed totally geodesic submanifold of Y isometric to the real Grassmannian $G_{m-1,n}^{\mathbb{R}}$ which can be written in the form $\operatorname{Exp}_y V' \otimes W_{Y,y}$, where V' is an $(m-1)$-dimensional subspace of $V_{Y,y}$. If u is a symmetric p-form on Y satisfying the Guillemin condition, then the restriction of u to Y' satisfies the Guillemin condition.*

PROOF: We consider the subspace $V_1 = V' \oplus W_{Y,y}$ of $\mathbb{R}^{m+n}$ of dimension $m + n - 1$ and its orthogonal complement V_2 in $\mathbb{R}^{m+n}$. Since $G_1^{\mathbb{R}}(V_2)$ consists of a single point x_0, from the equality (4.22) we see that the image of the totally geodesic imbedding

$$\bar{\iota} : G_{m-1}^{\mathbb{R}}(V_1) \times G_1^{\mathbb{R}}(V_2) \to Y,$$

corresponding to the subspaces V_1 and V_2 of $\mathbb{R}^{m+n}$, is equal to Y'. We choose a maximal flat totally geodesic torus Z of $G_{m-1}^{\mathbb{R}}(V_1)$ of dimension $m - 1$. We know that there exists a subspace F_1 of V_1 of dimension $2m-2$ such that Z is a totally geodesic submanifold of the submanifold $G_{m-1}^{\mathbb{R}}(F_1)$ of $G_{m-1}^{\mathbb{R}}(V_1)$. Since $n \geq m+1$, the orthogonal complement F_2 of F_1 in $\mathbb{R}^{m+n}$ is of dimension $n-m+2 \geq 3$ and contains V_2. Then x_0 belongs to the real projective space $G_1^{\mathbb{R}}(F_2)$. We now consider the totally geodesic imbedding (4.24) corresponding to the orthogonal subspaces F_1 and F_2 of $\mathbb{R}^{m+n}$. Let u be a symmetric p-form on Y satisfying the Guillemin condition. According to Lemma 4.9, the restrictions of u to the totally geodesic submanifold $\varphi(G^{\mathbb{R}}_{m-1}(F_1) \times \{x_0\})$ or to its maximal flat torus $\varphi(Z \times \{x_0\})$ satisfy the Guillemin condition. Since $\bar{\iota}(Z \times \{x_0\}) = \varphi(Z \times \{x_0\})$ is a maximal flat totally geodesic torus of the submanifold Y' and since all such tori of Y' arise in this way, we infer that the restriction of u to Y' also satisfies the Guillemin condition.

PROPOSITION 4.11. *Let y be a point of the real Grassmannian $Y = G^{\mathbb{R}}_{m,n}$, with $2 \le m < n$. Let Y' be a closed totally geodesic submanifold of Y isometric to the real Grassmannian $G^{\mathbb{R}}_{m,m-1}$ which can be written in the form $\mathrm{Exp}_y V_{Y,y} \otimes W'$, where W' is an $(m-1)$-dimensional subspace of $W_{Y,y}$. If u is a symmetric p-form on Y satisfying the Guillemin condition, then the restriction of u to Y' satisfies the Guillemin condition.*

PROOF: We consider the subspace $F = V_{Y,y} \oplus W'$ of $\mathbb{R}^{m+n}$ of dimension $2m-1$. From the equality (4.22), we see that Y' is equal to the image of the natural totally geodesic imbedding $G^{\mathbb{R}}_m(F) \to Y$, which sends the m-plane of F into the m-plane of $\mathbb{R}^{m+n}$ which it determines. We choose a maximal flat totally geodesic torus Z' of Y' of dimension $m-1$. We know that there exists a subspace F_1 of F of dimension $2m-2$ such that Z' is contained in the image of the totally geodesic imbedding

$$\bar{\iota} : G^{\mathbb{R}}_{m-1}(F_1) \times G^{\mathbb{R}}_1(F_1') \to Y,$$

where F_1' is the orthogonal complement of F_1 in F and $G^{\mathbb{R}}_1(F_1')$ consists of a single point x_0. Thus we may write $Z' = \bar{\iota}(Z \times \{x_0\})$, where Z is a maximal flat totally geodesic torus of $G^{\mathbb{R}}_{m-1}(F_1)$. Since $n \ge m+1$, the orthogonal complement F_2 of F_1 in $\mathbb{R}^{m+n}$ is of dimension $n-m+2 \ge 3$ and contains F_1'. Then x_0 belongs to the real projective space $G^{\mathbb{R}}_1(F_2)$. We now consider the totally geodesic imbedding (4.24) corresponding to the orthogonal subspaces F_1 and F_2 of $\mathbb{R}^{m+n}$. Let u be a symmetric p-form on Y satisfying the Guillemin condition. According to Lemma 4.9, the restrictions of u to the totally geodesic submanifold $\varphi(G^{\mathbb{R}}_{m-1}(F_1) \times \{x_0\})$ or to its maximal flat torus $Z' = \varphi(Z \times \{x_0\})$ satisfy the Guillemin condition. Since Z' is an arbitrary maximal flat geodesic torus of Y', we infer that the restriction of u to Y' also satisfies the Guillemin condition.

From Proposition 4.10 and Lemma 4.6, by means of the isometries $\Psi : G^{\mathbb{R}}_{p,q} \to G^{\mathbb{R}}_{q,p}$ we deduce the following:

PROPOSITION 4.12. *Let y be a point of the real Grassmannian $Y = G^{\mathbb{R}}_{m,n}$. Let Y' be a closed totally geodesic submanifold of Y isometric to the real Grassmannian $G^{\mathbb{R}}_{q,r}$ which can be written in the form $\mathrm{Exp}_y V' \otimes W'$, where V' is a q-dimensional subspace of $V_{Y,y}$ and W' is an r-dimensional subspace of $W_{Y,y}$. Assume either that $2 \le m < n$ and $r = n$, or that $2 \le n < m$ and $q = m$. If u is a symmetric p-form on Y satisfying the*

Guillemin condition, then the restriction of u to Y' satisfies the Guillemin condition.

When $1 \le q \le m - 1$, the following proposition is a consequence of Propositions 4.11 and 4.12 and Lemma 4.6. If $m \le q \le n$ in the following proposition, the submanifold Y' of Y considered there has the same rank as Y, and in this case the conclusion of the proposition is immediate.

PROPOSITION 4.13. *Let y be a point of the real Grassmannian $Y = G_{m,n}^{\mathbb{R}}$, with $2 \le m < n$. Let Y' be a closed totally geodesic submanifold of Y isometric to the real Grassmannian $G_{m,q}^{\mathbb{R}}$ which can be written in the form $\mathrm{Exp}_y V_{Y,y} \otimes W'$, where W' is a q-dimensional subspace of $W_{Y,y}$. If u is a symmetric p-form on Y satisfying the Guillemin condition, then the restriction of u to Y' satisfies the Guillemin condition.*

From Proposition 4.12 and the injectivity of the X-ray transform for functions on a real projective space, we now obtain the following proposition, which is also given by Theorem 2.24.

PROPOSITION 4.14. *For $m, n \ge 2$, with $m \ne n$, the maximal flat Radon transform for functions on the real Grassmannian $G_{m,n}^{\mathbb{R}}$ is injective.*

PROOF: Without any loss of generality, we may suppose that $m < n$. Let y be an arbitrary point of $G_{m,n}^{\mathbb{R}}$; we consider a submanifold Y' of $Y = G_{m,n}^{\mathbb{R}}$ which is of the form $\mathrm{Exp}_y V' \otimes W_{Y,y}$, where V' is a one-dimensional subspace of $V_{Y,y}$. Let f be a function on Y satisfying the Guillemin condition. Since Y' is isometric to the real projective space $\mathbb{RP}^n$ of dimension $n \ge 3$, by Proposition 4.12, with $q = 1$ and $r = n$, we see that the restriction f' of f to Y' satisfies the zero-energy condition. The injectivity of the X-ray transform for functions on the real projective space $\mathbb{RP}^n$, given by Theorem 2.23,(ii), now tells us that f' vanishes. Hence the function f vanishes at y.

The following proposition is a generalization of Proposition 5.2 of [21].

PROPOSITION 4.15. *Let y be a point of the real Grassmannian $Y = G_{m,n}^{\mathbb{R}}$, with $2 \le m < n$, and let $2 \le r \le n$ be a given integer. Let Y' be a closed totally geodesic submanifold of Y isometric to the real Grassmannian $G_{1,r}^{\mathbb{R}}$ which can be written in the form $\mathrm{Exp}_y v \otimes W'$, where v is a unit vector of $V_{Y,y}$ and W' is an r-dimensional subspace of $W_{Y,y}$. If θ is a 1-form on Y satisfying the Guillemin condition, then the restriction of θ to Y' is exact.*

PROOF: Let Z be the closed totally geodesic submanifold of Y equal to $\mathrm{Exp}_y v \otimes W_{Y,y}$, which is isometric to the real Grassmannian $G_{1,n}^{\mathbb{R}}$; clearly Y' is a totally geodesic submanifold of Z. Let θ be a 1-form on Y satisfying the Guillemin condition. By Proposition 4.12, the restriction of θ to the submanifold Z satisfies the zero-energy condition; therefore so does the restriction θ' of θ to the submanifold Y'. By Theorem 3.26, we know that θ' is exact.

CHAPTER V

THE COMPLEX QUADRIC

§1. Outline

This chapter is devoted to the complex quadric which plays a central
role in the rigidity problems. In §§2 and 3, we describe the differential geometry of the quadric Q_n viewed as a complex hypersurface of the complex
projective space $\mathbb{CP}^{n+1}$. We show that Q_n is a Hermitian symmetric space
and a homogeneous space of the group $SO(n+2)$. The involutions of the
tangent spaces of Q_n, which arise from the second fundamental form of the
quadric, allow us to introduce various objects and vector bundles on Q_n.
In particular, we decompose the bundle of symmetric 2-forms on Q_n into
irreducible $SO(n+2)$-invariant sub-bundles; one of these bundles L, which
is of rank 2, was first introduced in [18]. In §4, we develop the local formalism of Kähler geometry on the complex quadric following [22]; we wish to
point out that auspicious choices lead to remarkably simple formulas. The
identification of the quadric Q_n with the Grassmannian $\widetilde{G}_{2,n}^{\mathbb{R}}$ of oriented
2-planes in $\mathbb{R}^{n+2}$ given in §5 allows us to relate the geometries of these two
manifolds and to define the objects introduced in §3 in an intrinsic man-
ner. In the next section, we describe the tangent spaces of various families
of totally geodesic submanifolds of Q_n and present results concerning the
spaces of tensors of curvature type which vanish when restricted to some of
these families. In §7, we determine explicitly the space of infinitesimal Einstein deformations of Q_n and, from the point of view of harmonic analysis
on homogeneous spaces, we compute the multiplicities of a class of isotypic
components of the $SO(n+2)$-module of complex symmetric 2-forms on Q_n
and establish properties of these components. Finally, §8 is devoted to results concerning sections of the sub-bundle L of S^2T^*; in §9, we prove that
the complex quadric Q_2 is isometric to the product of spheres $S^2 \times S^2$.

§2. The complex quadric viewed as a symmetric space

Let n be an integer ≥ 2. In this chapter, we suppose that X is the complex quadric Q_n, which is the complex hypersurface of complex projective
space $\mathbb{CP}^{n+1}$ defined by the homogeneous equation

$$\zeta_0^2 + \zeta_1^2 + \cdots + \zeta_{n+1}^2 = 0,$$

where $\zeta = (\zeta_0, \zeta_1, \ldots, \zeta_{n+1})$ is the standard complex coordinate system
of $\mathbb{C}^{n+2}$. Let g be the Kähler metric on X induced by the Fubini-Study
metric $\tilde{g}$ on $\mathbb{CP}^{n+1}$ of constant holomorphic curvature 4 considered in Chapter III. We denote by J the complex structure of X or of $\mathbb{CP}^{n+1}$.

We consider the natural projection

$$\pi : \mathbb{C}^{n+2} - \{0\} \rightarrow \mathbb{C}\mathbb{P}^{n+1}$$

and the unit sphere S^{2n+3} of $\mathbb{C}^{n+2}$ endowed with the Riemannian metric induced by the real scalar product (3.7) on $\mathbb{C}^{n+2}$. For $z \in S^{2n+3}$, we consider the space

$$H_z(S^{2n+3}) = \{ (z, u) \mid u \in \mathbb{C}^{n+2}, \, \langle z, u \rangle = 0 \}$$

of §2, Chapter III defined in terms of the Hermitian scalar product (3.6), with $\mathbb{K} = \mathbb{C}$, and we view it as a subspace of the tangent space $T_z(S^{2n+3})$ of S^{2n+3} at z; we denote by $T_{\pi(z)}(\mathbb{C}\mathbb{P}^{n+1})$ the tangent space of $\mathbb{C}\mathbb{P}^{n+1}$ at $\pi(z)$. We recall that the restriction

$$\pi : S^{2n+3} \rightarrow \mathbb{C}\mathbb{P}^{n+1}$$

of the mapping π is a Riemannian submersion and that, for $z \in S^{2n+3}$, the induced mapping

$$\pi_* : H_z(S^{2n+3}) \rightarrow T_{\pi(z)}(\mathbb{C}\mathbb{P}^{n+1})$$

is an isometry.

We endow $\mathbb{C}^{n+2}$ with the complex bilinear form h defined by

$$h(z, w) = \sum_{j=0}^{n+1} z_j w_j,$$

where $z = (z_0, z_1, \ldots, z_{n+1})$ and $w = (w_0, w_1, \ldots, w_{n+1})$ are elements of $\mathbb{C}^{n+2}$. In fact, we have

$$Q_n = \{ \pi(z) \mid z \in \mathbb{C}^{n+2} - \{0\}, \, h(z, z) = 0 \}$$
$$= \{ \pi(z) \mid z \in S^{2n+3}, \, h(z, z) = 0 \}.$$

If $z \in S^{2n+3}$ satisfies $h(z, z) = 0$, we consider the subspace

$$H_z'(S^{2n+3}) = \{ (z, u) \mid u \in \mathbb{C}^{n+2}, \, \langle z, u \rangle = 0, \, h(z, u) = 0 \}$$

of $H_z(S^{2n+3})$; then the mapping

$$\pi_* : H_z'(S^{2n+3}) \rightarrow T_{\pi(z)}$$

is an isometry (see Example 10.6 in Chapter XI of [40]).

Let $\{e_0, e_1, \ldots, e_{n+1}\}$ be the standard basis of $\mathbb{C}^{n+2}$. Let $\tilde{b}$ be the point $(e_0 + ie_1)/\sqrt{2}$ of S^{2n+3}; then $h(\tilde{b}, \tilde{b}) = 0$ and $b = \pi(\tilde{b})$ is a point of Q_n. We consider the vectors

$$\tilde{\nu} = \left(\tilde{b}, (-e_0 + ie_1)/\sqrt{2}\right), \qquad \tilde{\nu}' = \left(\tilde{b}, (ie_0 + e_1)/\sqrt{2}\right)$$

of $H_{\tilde{b}}(S^{2n+3})$; clearly, we have $\pi_* \tilde{\nu}' = -J\pi_* \tilde{\nu}$. Then

$$\left\{ (\tilde{b}, e_2), \ldots, (\tilde{b}, e_{n+1}), (\tilde{b}, ie_2), \ldots, (\tilde{b}, ie_{n+1}) \right\}$$

is an orthonormal basis of $H'_{\tilde{b}}(S^{2n+3})$ and $\{\tilde{\nu}, \tilde{\nu}'\}$ is an orthonormal basis for the orthogonal complement of $H'_{\tilde{b}}(S^{2n+3})$ in $H_{\tilde{b}}(S^{2n+3})$.

The group $SU(n+2)$ acts on $\mathbb{C}^{n+2}$ and $\mathbb{CP}^{n+1}$ by holomorphic isometries. Its subgroup $G = SO(n+2)$ leaves the submanifold X of $\mathbb{CP}^{n+1}$ invariant; in fact, the group G acts transitively and effectively on the Riemannian manifold (X, g) by holomorphic isometries. The isotropy group of the point b is equal to the subgroup $K = SO(2) \times SO(n)$ of G consisting of the matrices

$$\begin{pmatrix} A & 0 \\ 0 & B \end{pmatrix},$$

where $A \in SO(2)$ and $B \in SO(n)$. For $\theta \in \mathbb{R}$, we denote by $R(\theta)$ the element

$$\begin{pmatrix} \cos\theta & -\sin\theta \\ \sin\theta & \cos\theta \end{pmatrix}$$

of $SO(2)$ and by $R'(\theta)$ the element

$$\begin{pmatrix} R(\theta) & 0 \\ 0 & I \end{pmatrix}$$

of K, where I is the identity element of $SO(n)$. Since $R'(\theta)\tilde{b} = e^{-i\theta}\tilde{b}$, we see that

(5.1) $\qquad R'(\theta)_* \pi_* (\tilde{b}, e_j) = \pi_* (e^{-i\theta}\tilde{b}, e_j) = \pi_* (\tilde{b}, e^{i\theta} e_j),$

for $2 \leq j \leq n+1$.

Let $\{e'_1, \ldots, e'_n\}$ be the standard basis of $\mathbb{C}^n$ and let

$$\psi : T_b \to \mathbb{C}^n$$

be the isomorphism of real vector spaces determined by

$$\psi\pi_*(\tilde{b}, e_j) = e'_{j-1}, \qquad \psi\pi_*(\tilde{b}, ie_j) = ie'_{j-1},$$

for $2 \leq j \leq n+1$. If we identify T_b with $\mathbb{C}^n$ by means of this isomorphism ψ, since $J\pi_*(\tilde{b}, e_j) = \pi_*(\tilde{b}, ie_j)$, for $2 \leq j \leq n+1$, the complex structure of T_b is the one determined by the multiplication by i on $\mathbb{C}^n$, and the Kähler metric g at b is the one obtained from the standard Hermitian scalar product on $\mathbb{C}^n$, given by (3.6). Moreover, by (5.1) we see that the action of the element

$$(5.2) \qquad \phi = \begin{pmatrix} R(\theta) & 0 \\ 0 & B \end{pmatrix}$$

of K, with $B \in SO(n)$, $\theta \in \mathbb{R}$, on $T_b = \mathbb{C}^n$ is given by

$$(5.3) \qquad \phi_*\zeta = e^{i\theta} B\zeta,$$

for $\zeta \in \mathbb{C}^n$, where $SO(n)$ is considered as a subgroup of $SU(n)$.

Since G acts transitively on X, the mapping

$$\Psi : G/K \to X,$$

which sends the class $\phi \cdot K$, where $\phi \in G$, into the point $\phi(b)$, is a diffeomorphism compatible with the actions of G on G/K and X.

The element $j = R'(\pi/2)$ of K belongs to the center of K and is of order 4. The element $s = j^2$ of G determines an involution σ of G which sends $\phi \in G$ into $s\phi s^{-1}$. Then K is equal to the identity component of the set of fixed points of σ and (G, K) is a Riemannian symmetric pair. The Cartan decomposition of the Lie algebra $\mathfrak{g}_0$ of G corresponding to σ is

$$\mathfrak{g}_0 = \mathfrak{k}_0 \oplus \mathfrak{p}_0,$$

where $\mathfrak{k}_0$ is the Lie algebra of K and $\mathfrak{p}_0$ is the space of all matrices

$$(5.4) \qquad \begin{pmatrix} 0 & 0 & -^t\xi \\ 0 & 0 & -^t\eta \\ \xi & \eta & 0 \end{pmatrix}$$

of $\mathfrak{g}_0$, where ξ, η are vectors of $\mathbb{R}^n$ considered as column vectors. We identify $\mathfrak{p}_0$ with the tangent space of G/K at the coset of the identity element of G and also with the vector space $\mathbb{R}^n \oplus \mathbb{R}^n$; in particular, the matrix (5.4) of $\mathfrak{p}_0$ is identified with the vector $(\xi, \eta) \in \mathbb{R}^n \oplus \mathbb{R}^n$.

If B is the Killing form of $\mathfrak{g}_0$, the restriction to $\mathfrak{p}_0$ of the scalar product $-B$ is invariant under the adjoint action of K and therefore induces a G-invariant Riemannian metric g_0 on the homogeneous space G/K. The restriction of $\mathrm{Ad}\, j$ to $\mathfrak{p}_0$ is a K-invariant complex structure on $\mathfrak{p}_0$ and so gives rise to a G-invariant almost complex structure on G/K. According to Proposition 4.2 in Chapter VIII of [36], this almost complex structure is integrable and the manifold G/K, endowed with the corresponding complex structure and the metric g_0, is a Hermitian symmetric space. The space G/K is of compact type and of rank 2; when $n \geq 3$, it is irreducible.

It is easily verified that the isomorphism $\Psi_* : \mathfrak{p}_0 \to T_b$ sends $(\xi, \eta) \in \mathfrak{p}_0$, with $\xi, \eta \in \mathbb{R}^n$, into the element $(\xi + i\eta)/\sqrt{2}$ of $\mathbb{C}^n$. The group K acts on T_b and, for $\phi \in K$, we have

$$\Psi_* \circ \operatorname{Ad} \phi = \phi \circ \Psi_*$$

as mappings from $\mathfrak{p}_0$ to T_b. We also see that $\Psi_* \circ \operatorname{Ad} j = J \circ \Psi_*$ and that

$$g_0 = 4n \Psi^* g.$$

Thus Ψ is a holomorphic isometry from the Hermitian symmetric space G/K, endowed with the metric $(1/4n) \cdot g_0$, to X; henceforth, we shall identify these two Kähler manifolds by means of this isometry. According to formula (1.65), it follows that X is an Einstein manifold and that its Ricci tensor Ric is given by

$$(5.5) \hspace{4cm} \operatorname{Ric} = 2ng.$$

§3. The complex quadric viewed as a complex hypersurface

We begin by recalling some results of Smyth [49] (see also [21]). The second fundamental form C of the complex hypersurface X of $\mathbb{CP}^{n+1}$ is a symmetric 2-form with values in the normal bundle of X in $\mathbb{CP}^{n+1}$. We denote by S the bundle of unit vectors of this normal bundle.

Let x be a point of X and ν be an element of S_x. We consider the element h_ν of $S^2 T_x^*$ defined by

$$h_\nu(\xi, \eta) = \tilde{g}(C(\xi, \eta), \nu),$$

for all $\xi, \eta \in T_x$. Since $\{\nu, J\nu\}$ is an orthonormal basis for the fiber of the normal bundle of X in $\mathbb{CP}^{n+1}$ at the point x, we see that

$$C(\xi, \eta) = h_\nu(\xi, \eta)\nu + h_{J\nu}(\xi, \eta)J\nu,$$

for all $\xi, \eta \in T_x$. If μ is another element of S_x, we have

$$(5.6) \hspace{4cm} \mu = \cos\theta \cdot \nu + \sin\theta \cdot J\nu,$$

with $\theta \in \mathbb{R}$. We consider the symmetric endomorphism K_ν of T_x determined by

$$h_\nu(\xi, \eta) = g(K_\nu \xi, \eta),$$

for all $\xi, \eta \in T_x$. Since our manifolds are Kähler, we have

$$C(\xi, J\eta) = JC(\xi, \eta),$$

for all $\xi, \eta \in T_x$; from this relation, we deduce the equalities

(5.7) $$K_{J\nu} = JK_\nu = -K_\nu J.$$

It follows that h_ν and $h_{J\nu}$ are linearly independent. By (5.7), we see that h_ν belongs to $(S^2 T^*)^-$ and that

(5.8) $$\check{h}_\nu = -h_{J\nu}.$$

Then if μ is the element of S_x given by (5.6), it is easily verified that

(5.9) $$K_\mu = \cos\theta \cdot K_\nu + \sin\theta \cdot JK_\nu.$$

The Gauss equation gives us an expression for the Riemann curvature tensor R of (X, g) in terms of the Riemann curvature tensor R_0 of $\mathbb{CP}^{n+1}$ (endowed with the metric $\tilde{g}$) and the second fundamental form C; in fact, we have

$$R(\xi_1, \xi_2, \xi_3, \xi_4) = R_0(\xi_1, \xi_2, \xi_3, \xi_4) + \tilde{g}(C(\xi_1, \xi_4), C(\xi_2, \xi_3))$$
$$- \tilde{g}(C(\xi_1, \xi_3), C(\xi_2, \xi_4)),$$

for all $\xi_1, \xi_2, \xi_3, \xi_4 \in T$. Using formulas (3.13) and (5.7), from the above relation we obtain the equality

(5.10)
$$\tilde{R}(\xi, \eta)\zeta = g(\eta, \zeta)\xi - g(\xi, \zeta)\eta + g(J\eta, \zeta)J\xi - g(J\xi, \zeta)J\eta$$
$$- 2g(J\xi, \eta)J\zeta + g(K_\nu\eta, \zeta)K_\nu\xi - g(K_\nu\xi, \zeta)K_\nu\eta$$
$$+ g(JK_\nu\eta, \zeta)JK_\nu\xi - g(JK_\nu\xi, \zeta)JK_\nu\eta,$$

for all $\xi, \eta, \zeta \in T_x$. From (5.7), we infer that the trace of the endomorphism K_ν of T_x vanishes. According to this last remark and formulas (5.10) and (5.7), we see that

$$\mathrm{Ric}(\xi, \eta) = -2g(K_\nu^2\xi, \eta) + 2(n+1)g(\xi, \eta),$$

for all $\xi, \eta \in T_x$. From (5.5), it follows that K_ν is an involution. We call K_ν the *real structure* of the quadric associated to the unit normal ν.

We denote by T_ν^+ and T_ν^- the eigenspaces of K_ν corresponding to the eigenvalues $+1$ and -1, respectively. Then by (5.7), we infer that J induces isomorphisms of T_ν^+ onto T_ν^- and of T_ν^- onto T_ν^+, and that

(5.11) $$T_x = T_\nu^+ \oplus T_\nu^-$$

is an orthogonal decomposition. If μ is the unit normal given by (5.6), then according to (5.7) we easily see that

$$(5.12) \qquad T_\mu^+ = \{\, \cos\theta' \cdot \xi + \sin\theta' \cdot J\xi \mid \xi \in T_\nu^+ \,\},$$

where $\theta' = \theta/2$. In particular, if μ is the unit normal $-\nu$, we obtain the equalities

$$(5.13) \qquad T_\mu^+ = JT_\nu^+ = T_\nu^-.$$

If ϕ is an isometry of $\mathbb{CP}^{n+1}$ which preserves X, we have

$$C(\phi_*\xi, \phi_*\eta) = \phi_*C(\xi,\eta),$$

for all $\xi, \eta \in T$. Thus, if μ is the tangent vector $\phi_*\nu$ belonging to $S_{\phi(x)}$, we see that

$$h_\mu(\phi_*\xi, \phi_*\eta) = h_\nu(\xi, \eta),$$

for all $\xi, \eta \in T_x$, and hence that

$$(5.14) \qquad K_\mu\phi_* = \phi_*K_\nu$$

on T_x. Therefore ϕ induces isomorphisms

$$\phi_* : T_\nu^+ \to T_\mu^+, \qquad \phi_* : T_\nu^- \to T_\mu^-.$$

Now let ν be a section of S over an open subset U of X. Let h_ν be the section of S^2T^* over U corresponding to the unit normal field ν defined by

$$(h_\nu)(x) = h_{\nu(x)},$$

for $x \in U$; we then consider the symmetric endomorphism K_ν of $T_{|U}$ determined by

$$K_\nu(x) = K_{\nu(x)},$$

for $x \in U$. We also consider the sub-bundles T_ν^+ and T_ν^- of $T_{|U}$, which are the eigenbundles of K_ν corresponding to the eigenvalues $+1$ and -1, respectively; we have $T_{\nu,x}^+ = T_{\nu(x)}^+$ and $T_{\nu,x}^- = T_{\nu(x)}^-$, for $x \in U$. If $\widetilde{\nabla}$ denotes the Levi-Civita connection of the metric $\tilde{g}$ on $\mathbb{CP}^{n+1}$, we consider the 1-form φ_ν on U defined by

$$\langle \xi, \varphi_\nu \rangle = \tilde{g}(\widetilde{\nabla}_\xi \nu, J\nu),$$

for all $\xi \in T_{|U}$. Then we have the equality

$$(5.15) \qquad \widetilde{\nabla}_\xi \nu = -K_\nu\xi + \langle \xi, \varphi_\nu \rangle J\nu,$$

for all $\xi \in T_{|U}$. Now let ξ, η be vectors of $T_{|U}$. According to (5.15) and the definition of the curvature tensor $\widetilde{R}_0$ of $\mathbb{CP}^{n+1}$, we easily see that the equality

$$
(5.16) \quad \begin{aligned} \widetilde{R}_0(\xi, \eta)\nu &= \langle \xi \wedge \eta, d\varphi_\nu \rangle J\nu - (\widetilde{\nabla}K_\nu)(\xi, \eta) + (\widetilde{\nabla}K_\nu)(\eta, \xi) \\ &\quad + \langle \xi, \varphi_\nu \rangle JK_\nu\eta - \langle \eta, \varphi_\nu \rangle JK_\nu\xi \end{aligned}
$$

holds. On the other hand, formula (3.13) tells us that

$$
(5.17) \quad \widetilde{R}_0(\xi, \eta)\nu = -2\omega(\xi, \eta)J\nu.
$$

Since K_ν is an involution of $T_{|U}$, by (5.7) the definition of the second fundamental form C gives us the relation

$$
(5.18) \quad \begin{aligned} (\widetilde{\nabla}K_\nu)(\xi, \eta) - (\widetilde{\nabla}K_\nu)(\eta, \xi) &= (\nabla K_\nu)(\xi, \eta) - (\nabla K_\nu)(\eta, \xi) \\ &\quad - 2\omega(\xi, \eta)J\nu. \end{aligned}
$$

We equate the normal and tangential components of the right-hand sides of (5.16) and (5.17); using (5.18), we then find that

$$
(5.19) \quad d\varphi_\nu = -4\omega
$$

and that the Gauss-Codazzi equation

$$
(5.20) \quad (\nabla K_\nu)(\xi, \eta) - (\nabla K_\nu)(\eta, \xi) = \langle \xi, \varphi_\nu \rangle JK_\nu\eta - \langle \eta, \varphi_\nu \rangle JK_\nu\xi
$$

holds. Since K_ν is an involution, we know that $\nabla K_\nu^2 = 0$; hence if η is an element of T_ν^+ (resp. of T_ν^-), we see that $(\nabla K_\nu)(\xi, \eta)$ belongs to T_ν^- (resp. to T_ν^+). From this last remark, the equality (5.20) and the fact that J induces isomorphisms $J : T_\nu^+ \to T_\nu^-$ and $J : T_\nu^- \to T_\nu^+$, we deduce that the relation

$$
(5.21) \quad (\nabla K_\nu)(\xi, \eta) = \langle \xi, \varphi_\nu \rangle JK_\nu\eta
$$

holds whenever ξ belongs to T_ν^+ (resp. to T_ν^-) and η is an element of T_ν^- (resp. of T_ν^+). Since X is a Kähler manifold, by (5.7) we see that

$$
(\nabla K_\nu)(\xi, J\eta) = -J(\nabla K_\nu)(\xi, \eta).
$$

If ξ and η belong to the same eigenbundle of K_ν, by (5.21) and (5.7) we have

$$
(\nabla K_\nu)(\xi, J\eta) = \langle \xi, \varphi_\nu \rangle K_\nu\eta;
$$

from the last two equalities, we infer that the relation (5.21) also holds in this case. Thus we have shown that

$$(5.22) \qquad\qquad \nabla K_\nu = \varphi_\nu \otimes J K_\nu.$$

The remainder of this section is devoted to results of [21] and [23]. We consider the sub-bundle L of $(S^2T^*)^-$ introduced in [18], whose fiber at $x \in X$ is equal to

$$L_x = \{\, h_\mu \mid \mu \in S_x \,\};$$

according to (5.9), if $\nu \in S_x$, this fiber L_x is generated by the elements h_ν and $h_{J\nu}$ and so the sub-bundle L of $(S^2T^*)^-$ is of rank 2. We denote by $(S^2T^*)^{-\perp}$ the orthogonal complement of L in $(S^2T^*)^-$. By (5.8), we see that L is stable under the endomorphism (1.68) of $(S^2T^*)^-$; since the automorphism J of T is an isometry, the orthogonal complement $(S^2T^*)^{-\perp}$ of L in $(S^2T^*)^-$ is also stable under this endomorphism. We denote by L', L'', $(S^{2,0}T^*)^\perp$ and $(S^{0,2}T^*)^\perp$ the eigenbundles corresponding to the eigenvalues $+i$ and $-i$ of the endomorphism of $(S^2T^*)^-_{\mathbb{C}}$ induced by the mapping (1.68). In fact, we have the equalities

$$L' = L_{\mathbb{C}} \cap S^{2,0}T^*, \qquad L'' = L_{\mathbb{C}} \cap S^{0,2}T^*,$$
$$(S^{2,0}T^*)^\perp = (S^2T^*)^{-\perp}_{\mathbb{C}} \cap S^{2,0}T^*, \qquad (S^{0,2}T^*)^\perp = (S^2T^*)^{-\perp}_{\mathbb{C}} \cap S^{0,2}T^*$$

and the decompositions

$$L_{\mathbb{C}} = L' \oplus L'', \qquad (S^2T^*)^{-\perp}_{\mathbb{C}} = (S^{2,0}T^*)^\perp \oplus (S^{0,2}T^*)^\perp.$$

By (5.8), if $x \in X$ and $\nu \in S_x$, we infer that $h_\nu + ih_{J\nu}$ generates L'_x and that $h_\nu - ih_{J\nu}$ generates L''_x. Clearly, we have the equalities

$$L'' = \overline{L'}, \qquad (S^{0,2}T^*)^\perp = \overline{(S^{2,0}T^*)^\perp}$$

and the orthogonal decompositions

$$(5.23) \qquad S^{2,0}T^* = L' \oplus (S^{2,0}T^*)^\perp, \qquad S^{0,2}T^* = L'' \oplus (S^{0,2}T^*)^\perp.$$

Now let x a point of X and ν be an element of S_x. For $\beta \in T^{1,1}_{\mathbb{R},x}$ and $h \in (S^2T^*)^+_x$, we define elements $K_\nu\beta$ of $\bigwedge^2 T^*_x$ and $K_\nu h$ of $S^2T^*_x$ by

$$(K_\nu\beta)(\xi, \eta) = \beta(K_\nu\xi, K_\nu\eta), \qquad (K_\nu h)(\xi, \eta) = h(K_\nu\xi, K_\nu\eta),$$

for all $\xi, \eta \in T_x$. Using (5.7) and (5.9), we see that $K_\nu\beta$ and $K_\nu h$ belong to $T^{1,1}_{\mathbb{R}}$ and $(S^2T^*)^+$, respectively, and do not depend on the choice of the unit normal ν. Using (5.7) and (5.9), we see that $K_\nu\beta$ and $K_\nu h$ belong to $T_{\mathbb{R}}^{1,1}$ and $(S^2T^*)^+$, respectively, and do not depend on the choice of the unit normal ν. We thus obtain canonical involutions of $T_{\mathbb{R}}^{1,1}$ and $(S^2T^*)^+$ over all of X, which give us the orthogonal decompositions

$$T_{\mathbb{R}}^{1,1} = (T_{\mathbb{R}}^{1,1})^+ \oplus (T_{\mathbb{R}}^{1,1})^-,$$
$$(S^2T^*)^+ = (S^2T^*)^{++} \oplus (S^2T^*)^{+-}$$

into the direct sum of the eigenbundles $(T_{\mathbb{R}}^{1,1})^+$, $(T_{\mathbb{R}}^{1,1})^-$, $(S^2T^*)^{++}$ and $(S^2T^*)^{+-}$ corresponding to the eigenvalues $+1$ and -1, respectively, of these involutions. We easily see that

$$
\begin{aligned}
(S^2T^*)_x^{++} &= \{\, h \in (S^2T^*)_x^+ \mid h(\xi, J\eta) = 0, \text{ for all } \xi, \eta \in T_\nu^+ \,\}, \\
(S^2T^*)_x^{+-} &= \{\, h \in (S^2T^*)_x^+ \mid h(\xi, \eta) = 0, \text{ for all } \xi, \eta \in T_\nu^+ \,\}, \\
(T_{\mathbb{R}}^{1,1})_x^+ &= \{\, \beta \in (T_{\mathbb{R}}^{1,1})_x \mid \beta(\xi, J\eta) = 0, \text{ for all } \xi, \eta \in T_\nu^+ \,\}, \\
(T_{\mathbb{R}}^{1,1})_x^- &= \{\, \beta \in (T_{\mathbb{R}}^{1,1})_x \mid \beta(\xi, \eta) = 0, \text{ for all } \xi, \eta \in T_\nu^+ \,\}.
\end{aligned}
$$

(5.24)

By (5.7), the morphism (1.67) induces by restriction isomorphisms

(5.25) $$(S^2T^*)^{++} \to (T_{\mathbb{R}}^{1,1})^-, \qquad (S^2T^*)^{+-} \to (T_{\mathbb{R}}^{1,1})^+.$$

Using the equalities (5.24), we find that

$$\operatorname{rank}(S^2T^*)^{++} = \frac{n(n+1)}{2}, \qquad \operatorname{rank}(S^2T^*)^{+-} = \frac{n(n-1)}{2}.$$

The metric g is a section of $(S^2T^*)^{++}$ and generates a line bundle $\{g\}$, whose orthogonal complement in $(S^2T^*)^{++}$ is the sub-bundle $(S^2T^*)_0^{++}$ consisting of the traceless symmetric tensors of $(S^2T^*)^{++}$. The Kähler form ω of X is the image under the morphism (1.67), or under the first isomorphism of (5.25), and is therefore a section of $(T_{\mathbb{R}}^{1,1})^-$. We denote by $(T_{\mathbb{R}}^{1,1})_0^-$ the orthogonal complement in $(T_{\mathbb{R}}^{1,1})^-$ of the line bundle $\{\omega\}$ generated by the section ω; the vector bundle $(T_{\mathbb{R}}^{1,1})_0^-$ is the image of $(S^2T^*)_0^{++}$ under the morphism (1.67). We thus obtain the orthogonal decompositions

(5.26) $$S^2T^* = L \oplus (S^2T^*)^{-\perp} \oplus \{g\} \oplus (S^2T^*)_0^{++} \oplus (S^2T^*)^{+-},$$

(5.27) $$S^2T_{\mathbb{C}}^* = L' \oplus L'' \oplus (S^{2,0}T^*)^\perp \oplus (S^{0,2}T^*)^\perp$$
$$\oplus \{g\}_{\mathbb{C}} \oplus (S^2T^*)_{0\mathbb{C}}^{++} \oplus (S^2T^*)_{\mathbb{C}}^{+-},$$

(5.28) $$T_{\mathbb{R}}^{1,1} = \{\omega\} \oplus (T_{\mathbb{R}}^{1,1})_0^- \oplus (T_{\mathbb{R}}^{1,1})^+.$$

Using the relation (5.14), we easily see that the decompositions (5.26)–(5.28) are G-invariant.

Let

$$\pi_{++} : S^2T^* \to (S^2T^*)^{++}, \qquad \pi^0_{++} : S^2T^* \to (S^2T^*)^{++}_0,$$

$$\pi_{+-} : S^2T^* \to (S^2T^*)^{+-}, \qquad \rho_+ : T^{1,1}_{\mathbb{R}} \to (T^{1,1}_{\mathbb{R}})^+$$

be the orthogonal projections. Clearly, we have

$$\pi^0_{++}h = \pi_{++}h - \frac{1}{n}\,(\mathrm{Tr}\,h) \cdot g,$$

for $h \in S^2T^*$.

Now let ν be a section of S over an open subset U of X and consider the symmetric endomorphism K_ν of $T_{|U}$ and the corresponding sub-bundle T^+_ν of $T_{|U}$. If $\beta \in T^{1,1}_{\mathbb{R},x}$ and $h \in (S^2T^*)^+_x$, with $x \in U$, we have

$$\pi_{+-}h = \tfrac{1}{2}(h - K_\nu h), \qquad \rho_+\beta = \tfrac{1}{2}(\beta + K_\nu\beta).$$

Let h be an element of $(S^2T^*)^+$; if k is the element $\pi_{+-}h$, then by (5.7) we see that

$$(5.29) \qquad\qquad \check{k} = \rho_+\check{h}.$$

Let f be a real-valued function on X; if k is the section $\pi_{+-}\mathrm{Hess}\,f$ of $(S^2T^*)^{+-}$, according to (5.29) and Lemma 1.25, we see that

$$(5.30) \qquad\qquad \check{k} = i\rho_+\partial\bar{\partial}f.$$

If u is a section of $(S^2T^*)^+$ or of $(T^{1,1}_{\mathbb{R}})^+$ over U, by (5.22) we have

$$
\begin{aligned}
(\nabla K_\nu u)(\xi, \eta, \zeta) &= (\nabla u)(\xi, K_\nu\eta, K_\nu\zeta) \\
&\quad + u((\nabla_\xi K_\nu)\eta, K_\nu\zeta) + u(K_\nu\eta, (\nabla_\xi K_\nu)\zeta), \\
&= (\nabla u)(\xi, K_\nu\eta, K_\nu\zeta) \\
&\quad + \langle \xi, \varphi_\nu\rangle(u(JK_\nu\eta, K_\nu\zeta) + u(K_\nu\eta, JK_\nu\zeta)) \\
&= (\nabla u)(\xi, K_\nu\eta, K_\nu\zeta),
\end{aligned}
$$

for all $\xi, \eta, \zeta \in T_{|U}$. If h is a section of $(S^2T^*)^+$ over X, from the preceding relations we infer that

$$(5.31) \qquad \nabla_\xi\pi_{++}h = \pi_{++}\nabla_\xi h, \qquad \nabla_\xi\pi_{+-}h = \pi_{+-}\nabla_\xi h,$$

for all $\xi \in T$; moreover if β is a section of $(T_{\mathbb{R}}^{1,1})^+$ over U, we obtain the relation

$$(5.32) \qquad (d\rho_+\beta)(\xi, \eta, \zeta) = (d\beta)(\xi, \eta, \zeta),$$

for all $\xi, \eta, \zeta \in T_{\nu}^+$.

The complex conjugation of $\mathbb{C}^{n+2}$ induces an involutive isometry τ of $\mathbb{CP}^{n+1}$ satisfying

$$(5.33) \qquad \tau \cdot \phi(z) = \phi \cdot \tau(z),$$

for all $\phi \in G$ and $z \in \mathbb{CP}^{n+1}$; moreover τ preserves the submanifold X of $\mathbb{CP}^{n+1}$. The group Λ of isometries of X generated by τ, which is of order 2, acts freely on X and we may therefore consider the Riemannian manifold $Y = X/\Lambda$, with the metric g_Y induced by g, and the natural projection $\varpi : X \to Y$, which is a two-fold covering. By (5.33), we see that the action of the group G on X passes to the quotient Y; in fact, the group G acts transitively on Y. If $K_1 = O(2) \times O(n)$ is the subgroup of $O(n+2)$ consisting of the matrices

$$\begin{pmatrix} A & 0 \\ 0 & B \end{pmatrix},$$

where $A \in O(2)$ and $B \in O(n)$, it is easily verified that the isotropy group of the point $\varpi(b)$ is equal to the subgroup $K' = G \cap K_1$ of G. We know that G/K' is a symmetric space of compact type of rank 2, which we may identify with Y by means of the isometry

$$\bar{\Psi} : G/K' \to Y$$

sending $\phi \cdot K'$ into the point $\phi(\varpi(b))$, for $\phi \in G$. Then by (5.33), we see that the projection $\varpi : X \to Y$ is identified with the natural submersion $G/K \to G/K'$ of symmetric spaces.

Clearly, we have

$$(5.34) \qquad \tau_* \cdot J = -J \cdot \tau_*,$$

as mappings acting on the tangent bundle of $\mathbb{CP}^{n+1}$. Since τ is an isometry of $\mathbb{CP}^{n+1}$ preserving X, the tangent vector $\mu = \tau_* \nu$ belongs to $S_{\tau(x)}$ and by (5.14) we have

$$\tau_* K_\nu = K_\mu \tau_*$$

on T_x. It follows that τ^* preserves all the sub-bundles of $S^2 T^*$ appearing in the decomposition (5.26), and hence also $(S^2 T^*)^{++}$; moreover, we have

$$(5.35) \qquad \tau^* \pi_{++} = \pi_{++} \tau^*, \qquad \tau^* \pi_{+-} = \pi_{+-} \tau^*.$$

By (5.34), we see that

(5.36) $$\tau^* \pi' = \pi'' \tau^*$$

on $S^2 T^*_{\mathbb{C}}$, and

(5.37) $$\tau^* \partial = \bar{\partial} \tau^*$$

on $\bigwedge^p T^*_{\mathbb{C}}$.

We consider the sub-bundle $\{g_Y\}$ of $S^2 T^*_Y$ generated by its section g_Y. The bundles of the decomposition (5.26) are invariant under the group Λ. Thus if F is one of the vector bundles appearing in the right-hand side of the decomposition (5.26), we consider the sub-bundle F_Y of $S^2 T^*_Y$ which it determines; then the mapping $\varpi^* : F_{Y,\varpi(b)} \to F_b$ is an isomorphism of K-modules. We denote by $(S^2 T^*_Y)^{-\perp}$, $(S^2 T^*_Y)^{++}_0$ and $(S^2 T^*_Y)^{+-}$ the sub-bundles of $S^2 T^*_Y$ determined by the sub-bundles $(S^2 T^*)^{-\perp}$, $(S^2 T^*)^{++}_0$ and $(S^2 T^*)^{+-}$ of $S^2 T^*$, respectively. Then from (5.26) we obtain the decomposition

(5.38) $$S^2 T^*_Y = L_Y \oplus (S^2 T^*_Y)^{\perp} \oplus \{g_Y\} \oplus (S^2 T^*_Y)^{++}_0 \oplus (S^2 T^*_Y)^{+-}$$

over Y.

All the fibers at b of the G-invariant vector bundles appearing in the right-hand side of the decompositions (5.26) and (5.27) are K-submodules of $S^2 T^*_b$ or $S^2 T^*_{b,\mathbb{C}}$. The fiber $S^2 T^*_{Y,\varpi(b)}$ is a K'-module and all the fibers at $\varpi(b)$ of the vector bundles appearing in the right-hand side of the decomposition (5.38) are K'-submodules of $S^2 T^*_{Y,\varpi(b)}$.

§4. Local Kähler geometry of the complex quadric

We now introduce the formalism of Kähler geometry on the complex quadric $X = Q_n$, with $n \geq 2$, developed in [22, §4].

Let $\zeta = (\zeta_0, \zeta_1, \ldots, \zeta_{n+1})$ be the standard complex coordinate system of $\mathbb{C}^{n+2}$. We consider the natural projection $\pi : \mathbb{C}^{n+2} - \{0\} \to \mathbb{CP}^{n+1}$, the open subset

$$V = \pi \left(\{ (\zeta_0, \ldots, \zeta_{n+1}) \in \mathbb{C}^{n+2} \mid \zeta_0 \neq 0 \} \right)$$

of $\mathbb{CP}^{n+1}$ and the holomorphic coordinate $z = (z_1, \ldots, z_{n+1})$ on V, where z_j is the function which satisfies $\pi^* z_j = \zeta_j / \zeta_0$ on $\mathbb{C}^* \times \mathbb{C}^{n+1}$. We set

$$|z| = (|z_1|^2 + \cdots + |z_{n+1}|^2)^{\frac{1}{2}}.$$

The Fubini-Study metric $\tilde{g}$ of $\mathbb{CP}^{n+1}$ is given on V by

$$\tilde{g}_{j\bar{k}} = \frac{1}{2} \left(\frac{\delta_{jk}}{1 + |z|^2} - \frac{\bar{z}_j z_k}{(1 + |z|^2)^2} \right), \quad \text{for } 1 \leq j, k \leq n+1.$$

We recall that the Christoffel symbols of the Levi-Civita connection $\widetilde{\nabla}$ of $\tilde{g}$ are determined on V by

$$\widetilde{\Gamma}^l_{jk} = \overline{\widetilde{\Gamma}^{\bar{l}}_{\bar{j}\bar{k}}} = -\frac{\bar{z}_j \delta_{kl} + \bar{z}_k \delta_{jl}}{1 + |z|^2}.$$

The intersection $X \cap V$ is equal to the hypersurface of V given by the equation

(5.39)
$$z_1^2 + \cdots + z_{n+1}^2 = -1.$$

We consider the open subset

$$V' = \{ z = (z_1, \ldots, z_{n+1}) \mid \operatorname{Im} z_{n+1} \neq 0 \}$$

of V and the corresponding open subset $U = X \cap V'$ of X. The holomorphic vector fields

$$\xi_j = \frac{\partial}{\partial z_j} - \frac{z_j}{z_{n+1}} \frac{\partial}{\partial z_{n+1}},$$

with $1 \leq j \leq n$, on V' are tangent to X, and over U they constitute a frame for the bundle T' of tangent vectors of type $(1,0)$ of X. On V', we consider the vector fields

$$\nu' = \sum_{l=1}^{n+1} \frac{(z_l - \bar{z}_l)}{1 + |z|^2} \frac{\partial}{\partial z_l},$$

$$\zeta = \sum_{k=1}^{n} \frac{(z_k - \bar{z}_k)}{1 + |z|^2} \xi_k, \qquad \eta_j = \xi_j + z_j \zeta$$

and the 1-forms

$$\omega^j = dz_j - \frac{z_j - \bar{z}_j}{z_{n+1} - \bar{z}_{n+1}} dz_{n+1},$$

with $1 \leq j \leq n$; we set $\nu'' = \overline{\nu'}$.

The real vector field

(5.40)
$$\nu = \sum_{l=1}^{n+1} (z_l - \bar{z}_l) \left(\frac{\partial}{\partial z_l} - \frac{\partial}{\partial \bar{z}_l} \right)$$

on V' satisfies

$$\frac{\nu}{1 + |z|^2} = \nu' + \nu''.$$

It is easily seen that the restriction of ν to the open subset U of X is a section of the bundle S of unit normals over U, and that

(5.41)
$$\langle \eta_k, \omega^j \rangle = \delta_k^j, \qquad \langle \nu, \omega^j \rangle = 0$$

on U, for $1 \le i, j \le n$. From the first relation of (5.41), it follows that $\{\eta_1, \ldots, \eta_n\}$ is a frame for T' over U and that

$$\{\operatorname{Re}\eta_1, \ldots, \operatorname{Re}\eta_n, \operatorname{Im}\eta_1, \ldots, \operatorname{Im}\eta_n\}$$

is a frame for T over U.

By means of the equation (5.39), which defines the complex quadric, we see that the decomposition of the tangent vector $\partial/\partial z_{n+1}$ into its tangential and normal components is given by

$$(5.42) \qquad \frac{1}{z_{n+1}} \frac{\partial}{\partial z_{n+1}} = \zeta - \nu'$$

on U. It follows that

$$(5.43) \qquad \eta_j = \frac{\partial}{\partial z_j} + z_j \nu'$$

on U, for $1 \le j \le n$. By means of (5.42), we obtain the equality

$$(5.44) \qquad \widetilde{\nabla}_{\xi_j} \xi_k = \nabla_{\xi_j} \xi_k + \left(\delta_{jk} + \frac{z_j z_k}{z_{n+1}^2} \right) \nu',$$

where

$$(5.45) \qquad \begin{aligned} \nabla_{\xi_j} \xi_k = {} & \frac{z_j \bar{z}_{n+1} - \bar{z}_j z_{n+1}}{z_{n+1}(1 + |z|^2)^2} \xi_k + \frac{z_k \bar{z}_{n+1} - \bar{z}_j z_{n+1}}{z_{n+1}(1 + |z|^2)^2} \xi_j \\ & - \left(\delta_{jk} + \frac{z_j z_k}{z_{n+1}^2} \right) \zeta, \end{aligned}$$

for $1 \le j, k \le n$. From (5.45), we easily deduce that

$$(5.46) \qquad \nabla_{\eta_j} \eta_k = -\frac{\bar{z}_k}{1 + |z|^2} \eta_j - \frac{z_j + \bar{z}_j}{1 + |z|^2} \eta_k.$$

Since $\nabla_{\xi_j} \bar{\xi}_k = 0$, we have

$$(5.47) \qquad \nabla_{\eta_j} \bar{\eta}_k = -\frac{\bar{z}_k}{1 + |z|^2} \bar{\eta}_j, \qquad \nabla_{\bar{\eta}_j} \eta_k = -\frac{z_k}{1 + |z|^2} \eta_j.$$

The Hermitian metric g of X is determined on U by

$$(5.48) \qquad g_{j\bar{k}} = g(\eta_j, \bar{\eta}_k) = \frac{1}{2} \left(\frac{\delta_{jk}}{1 + |z|^2} - \frac{z_j \bar{z}_k + \bar{z}_j z_k}{(1 + |z|^2)^2} \right),$$

and the inverse matrix $(g^{j\bar{k}})$ of $(g_{j\bar{k}})$ is given by the formula

(5.49)
$$g^{j\bar{k}} = 2(1+|z|^2)\left\{\delta_{jk} - \frac{1+|z_{n+1}|^2}{(z_{n+1}-\bar{z}_{n+1})^2}(z_j\bar{z}_k + \bar{z}_j z_k) \right.$$
$$\left. + \frac{(1+\bar{z}_{n+1}^2)z_j z_k + (1+z_{n+1}^2)\bar{z}_j\bar{z}_k}{(z_{n+1}-\bar{z}_{n+1})^2}\right\},$$

for $1 \le j, k \le n$. The image a of the point $(1,0,\ldots,0,i)$ of $\mathbb{C}^{n+2}$ under the natural projection $\pi : \mathbb{C}^{n+2} - \{0\} \to \mathbb{CP}^{n+1}$ belongs to the subset U of X; then a is the point of U with coordinates $(0,\ldots,0,i)$. By (5.48), we see that

(5.50)
$$g_{j\bar{k}}(a) = g(\eta_j, \bar{\eta}_k)(a) = \frac{1}{4}.$$

We consider the section h_ν over U and the symmetric endomorphism K_ν of $T_{|U}$ corresponding to the unit normal field ν. For $1 \le j \le n$, we verify directly that

$$\tilde{\nabla}_{\xi_j}\nu = -\frac{\partial}{\partial\bar{z}_j} + \frac{z_j}{z_{n+1}}\frac{\partial}{\partial\bar{z}_{n+1}} + \frac{z_j\bar{z}_{n+1} - \bar{z}_j z_{n+1}}{z_{n+1}}\nu'$$

on U. From this last relation and the decomposition (5.42), we infer that

$$\tilde{\nabla}_{\xi_j}\nu = -\bar{\xi}_j + \frac{z_j\bar{z}_{n+1} - \bar{z}_j z_{n+1}}{z_{n+1}}(\bar{\zeta} + \nu' - \nu'')$$

on U. The equality (5.15) implies that

$$K_\nu\xi_j = \bar{\xi}_j - \frac{z_j\bar{z}_{n+1} - \bar{z}_j z_{n+1}}{z_{n+1}}\bar{\zeta}.$$

It follows that

$$K_\nu\zeta = \frac{\bar{z}_{n+1}}{z_{n+1}}\bar{\zeta},$$

and thus we obtain the relation

(5.51)
$$K_\nu\eta_j = \bar{\eta}_j.$$

By (5.51), (5.46) and (5.47), we have

$$(\nabla K_\nu)(\eta_j, \eta_k) = \frac{z_j + \bar{z}_j}{1+|z|^2}\bar{\eta}_k,$$

for $1 \le k \le n$; according to (5.22), we then see that

$$(5.52) \qquad \langle \eta_j, \varphi_\nu \rangle = i \, \frac{z_j + \bar{z}_j}{1 + |z|^2}.$$

For $1 \le j \le n$, we consider the vector fields

$$v_j = \sqrt{2}\,(\eta_j + \bar{\eta}_j), \qquad w_j = J v_j = i\sqrt{2}\,(\eta_j - \bar{\eta}_j)$$

on U. By (5.51), we see that $\{v_1, \ldots, v_n\}$ and $\{w_1, \ldots, w_n\}$ are frames for T_ν^+ and T_ν^-, respectively, over U. By (5.50), we see that $\{v_1(a), \ldots, v_n(a)\}$ is an orthonormal basis of $T_{\nu(a)}^+ = T_{\nu,a}^+$. By (5.51) and (5.7), we obtain the relations

$$(5.53) \qquad \begin{aligned} h_\nu(\eta_j, \eta_k) &= g_{j\bar{k}} = h_\nu(\bar{\eta}_j, \bar{\eta}_k), \\ h_{J\nu}(\eta_j, \eta_k) &= -i\, g_{j\bar{k}} = -h_{J\nu}(\bar{\eta}_j, \bar{\eta}_k), \\ h_\nu(\eta_j, \bar{\eta}_k) &= h_{J\nu}(\eta_j, \bar{\eta}_k) = 0, \end{aligned}$$

for $1 \le j, k \le n$.

The relations (5.46) and (5.47) tell us that $\nabla v_j = \nabla w_j$ at the point a, for $1 \le j \le n$. Therefore there are orthonormal frames $\{v_1', \ldots, v_n'\}$ for T_ν^+ and $\{w_1', \ldots, w_n'\}$ for T_ν^- over U satisfying

$$(\nabla v_j)(a) = (\nabla w_j)(a) = 0$$

and $v_j'(a) = v_j(a)$, $w_j'(a) = w_j(a)$, for $1 \le j \le n$. Thus since the group $G = SO(n+2)$ acts transitively on X, if x is a given point of X, from the preceding remark and (5.14) we infer that there exist a section μ of S over a neighborhood U' of x and an orthonormal frame $\{\zeta_1, \ldots, \zeta_n\}$ for the vector bundle T_μ^+ over U' satisfying $(\nabla \zeta_j)(x) = 0$, for $1 \le j \le n$.

If $\tilde{a}$ is the point $(e_0 + ie_{n+1})/\sqrt{2}$ of S^{2n+3}, then we note that $\pi(\tilde{a}) = a$; moreover,

$$\{(\tilde{a}, e_1), \ldots, (\tilde{a}, e_n), (\tilde{a}, ie_1), \ldots, (\tilde{a}, ie_n)\}$$

is an orthonormal basis of $H_{\tilde{a}}'(S^{2n+1})$, and the unit vector

$$\mu = \big(\tilde{a}, (-e_0 + ie_{n+1})/\sqrt{2}\big)$$

of $H_{\tilde{a}}(S^{2n+1})$ is orthogonal to $H_{\tilde{a}}'(S^{2n+3})$. We easily verify that

$$\pi_*(\tilde{a}, e_j) = v_j(a), \qquad \pi_*(\tilde{a}, ie_j) = w_j(a), \qquad \pi_* \mu = \nu(a),$$

for $1 \le j \le n$. Since $K_\nu v_j = v_j$ and $K_\nu w_j = -w_j$, we see that

$$K_\nu \pi_*(\tilde{a}, e_j) = \pi_*(\tilde{a}, e_j), \qquad K_\nu \pi_*(\tilde{a}, i e_j) = -\pi_*(\tilde{a}, i e_j),$$

for $1 \le j \le n$.

We know that the vectors $\pi_* \tilde{\nu}$ and $\pi_* \tilde{\nu}' = -J \pi_* \tilde{\nu}$ introduced in §2 belong to S_b. By interchanging the roles of the coordinates ζ_1 and ζ_{n+1}, from the preceding relations we obtain the equalities

$$(5.54) \qquad K_{\pi_* \tilde{\nu}} \pi_*(\tilde{b}, e_j) = \pi_*(\tilde{b}, e_j), \qquad K_{\pi_* \tilde{\nu}} \pi_*(\tilde{b}, i e_j) = -\pi_*(\tilde{b}, i e_j),$$

for $2 \le j \le n+1$.

Let f be a complex-valued function on $\mathbb{CP}^{n+1}$ and $\tilde{f}$ be its restriction to X. For $1 \le j, k \le n$, let f_j, $f_{\bar{j}}$, $f_{j\bar{k}}$, f_{jk} and $f_{\bar{j}\bar{k}}$ be the functions on V' defined by

$$f_j = \eta_j f, \qquad f_{\bar{j}} = \bar{\eta}_j f,$$

$$f_{j\bar{k}} = \eta_j f_{\bar{k}} + \frac{\bar{z}_k}{1 + |z|^2} f_{\bar{j}},$$

$$f_{jk} = \eta_j f_k + \frac{\bar{z}_k}{1 + |z|^2} f_j + \frac{z_j + \bar{z}_j}{1 + |z|^2} f_k,$$

$$f_{\bar{j}\bar{k}} = \bar{\eta}_j f_{\bar{k}} + \frac{z_k}{1 + |z|^2} f_{\bar{j}} + \frac{z_j + \bar{z}_j}{1 + |z|^2} f_{\bar{k}}.$$

By formulas (5.46) and (5.47), for $1 \le j, k \le n$ we have

$$(\text{Hess } \tilde{f})(\eta_j, \bar{\eta}_k) = f_{j\bar{k}}, \qquad (\text{Hess } \tilde{f})(\eta_j, \eta_k) = f_{jk},$$

$$(5.55)$$

$$(\text{Hess } \tilde{f})(\bar{\eta}_j, \bar{\eta}_k) = f_{\bar{j}\bar{k}}$$

on U. For $1 \le j, k \le n$, from the preceding formulas, we obtain the equalities

$$(\text{Hess } \tilde{f})(\eta_j, \bar{\eta}_k) = \frac{\partial^2 f}{\partial z_j \partial \bar{z}_k},$$

$$(5.56) \qquad (\text{Hess } \tilde{f})(\eta_j, \eta_k) = \frac{\partial^2 f}{\partial z_j \partial z_k} + i\, \delta_{jk} \frac{\partial f}{\partial z_{n+1}},$$

$$(\text{Hess } \tilde{f})(\bar{\eta}_j, \bar{\eta}_k) = \frac{\partial^2 f}{\partial \bar{z}_j \partial \bar{z}_k} - i\, \delta_{jk} \frac{\partial f}{\partial \bar{z}_{n+1}}$$

at the point a; moreover when $j \ne k$, by (5.46) and (5.47) we see that the relations

$$(\nabla \operatorname{Hess} \tilde{f})(\bar{\eta}_j, \eta_j, \eta_k) = \frac{\partial^3 f}{\partial z_j \partial \bar{z}_j \partial z_k} + \frac{1}{2} \frac{\partial f}{\partial z_k},$$

$$(\nabla \operatorname{Hess} \tilde{f})(\eta_j, \bar{\eta}_j, \bar{\eta}_k) = \frac{\partial^3 f}{\partial z_j \partial \bar{z}_j \partial \bar{z}_k} + \frac{1}{2} \frac{\partial f}{\partial \bar{z}_k},$$

(5.57)

$$(\nabla \operatorname{Hess} \tilde{f})(\eta_j, \eta_j, \eta_k) = \frac{\partial^3 f}{\partial z_j^2 \partial z_k} + \frac{1}{2} \frac{\partial f}{\partial z_k} + i \frac{\partial^2 f}{\partial z_k \partial z_{n+1}},$$

$$(\nabla \operatorname{Hess} \tilde{f})(\eta_j, \eta_j, \eta_j) = \frac{\partial^3 f}{\partial z_j^3} + \frac{3}{2} \frac{\partial f}{\partial z_j} + 3i \frac{\partial^2 f}{\partial z_j \partial z_{n+1}}$$

hold at the point a. By (5.55) and (5.51), for $1 \le j, k \le n$, we obtain the equality

(5.58)

$$(\pi_{++} \operatorname{Hess} \tilde{f})(\eta_j, \bar{\eta}_k) = \frac{1}{2} \left(\frac{\partial^2 f}{\partial z_j \partial \bar{z}_k} + \frac{\partial^2 f}{\partial z_k \partial \bar{z}_j} \right),$$

$$(\pi_{+-} \operatorname{Hess} \tilde{f})(\eta_j, \bar{\eta}_k) = \frac{1}{2} \left(\frac{\partial^2 f}{\partial z_j \partial \bar{z}_k} - \frac{\partial^2 f}{\partial z_k \partial \bar{z}_j} \right)$$

at the point a.

§5. The complex quadric and the real Grassmannians

We also consider $\{e_0, e_1, \ldots, e_{n+1}\}$ as the standard basis of $\mathbb{R}^{n+2}$. We consider the real Grassmannian $\widetilde{G}_{2,n}^{\mathbb{R}}$ of oriented 2-planes in $\mathbb{R}^{n+2}$, which is a homogeneous space of $G = SO(n+2)$, endowed with the Riemannian metric g' defined in §1, Chapter IV and denoted there by g; we also consider the homogeneous vector bundles V and W over $\widetilde{G}_{2,n}^{\mathbb{R}}$.

We define an almost complex structure J on $\widetilde{G}_{2,n}^{\mathbb{R}}$ as follows. If $x \in \widetilde{G}_{2,n}^{\mathbb{R}}$ and $\{v_1, v_2\}$ is a positively oriented orthonormal basis of the oriented 2-plane V_x, the endomorphism J of V_x, determined by

$$J v_1 = v_2, \qquad J v_2 = -v_1,$$

is independent of the choice of basis of V_x and we have $J^2 = -\operatorname{id}$. Clearly, the almost complex structure J of $\widetilde{G}_{2,n}^{\mathbb{R}}$, which is equal to $J \otimes \operatorname{id}$ on the tangent space $(V \otimes W)_x$ of $\widetilde{G}_{2,n}^{\mathbb{R}}$ at $x \in \widetilde{G}_{2,n}^{\mathbb{R}}$, is invariant under the group G. Since $\widetilde{G}_{2,n}^{\mathbb{R}}$ is a symmetric space, according to Proposition 4.2 in Chapter VIII of [36], this almost complex structure J is integrable and the manifold $\widetilde{G}_{2,n}^{\mathbb{R}}$, endowed with the corresponding complex structure and the metric g', is a Hermitian symmetric space.

As in §1, Chapter IV, we consider the diffeomorphism

$$\Phi : G/K \to \widetilde{G}^{\mathbb{R}}_{2,n},$$

which sends the class $\phi \cdot K$, where $\phi \in G$, into the oriented 2-plane of $\mathbb{R}^{n+2}$ corresponding to the vector $\phi(e_0) \wedge \phi(e_1)$, and which is compatible with the actions of G on G/K and X. If x_0 is the point of $\widetilde{G}^{\mathbb{R}}_{2,n}$ corresponding to the vector $e_0 \wedge e_1$, we see that V_{x_0} is generated by $\{e_0, e_1\}$ and W_{x_0} is generated by $\{e_2, \ldots, e_{n+1}\}$; moreover, the isomorphism $\Phi_* : \mathfrak{p}_0 \to (V \otimes W)_{x_0}$ induced by Φ satisfies

$$\Phi_* \circ \operatorname{Ad} j = J \circ \Phi_*.$$

The complex quadric $X = Q_n$ is endowed with the metric g of §2. It is easily verified that the diffeomorphism

$$\Theta = \Psi \circ \Phi^{-1} : \widetilde{G}^{\mathbb{R}}_{2,n} \to X$$

sends the oriented 2-plane of $\mathbb{R}^{n+2}$ determined by $v_1 \wedge v_2$, where $\{v_1, v_2\}$ is an orthonormal system of vectors of $\mathbb{R}^{n+2}$, into the point of X equal to $\pi(v_1 + iv_2)$. Clearly, we have $\Theta(x_0) = b$ and

$$\Theta \circ \phi = \phi \circ \Theta,$$

for all $\phi \in G$. Thus Φ is a holomorphic isometry from G/K to X. In §1, Chapter IV, we saw that $g_0 = 2n\Phi^* g'$; therefore we have

(5.59) $$\Theta^* g = \tfrac{1}{2} g'.$$

If τ_0 denotes the involution of $\widetilde{G}^{\mathbb{R}}_{2,n}$ corresponding to the change of orientation of a 2-plane of $\mathbb{R}^{n+2}$ (which is denoted by τ in §1, Chapter IV), we see that

(5.60) $$\tau \circ \Theta = \Theta \circ \tau_0.$$

Also we consider the diffeomorphisms $\bar{\Phi}$ from the homogeneous space G/K' to the real Grassmannian $G^{\mathbb{R}}_{2,n}$ of (unoriented) 2-planes in $\mathbb{R}^{n+2}$, which sends $\phi \cdot K'$, where $\phi \in G$, into the 2-plane of $\mathbb{R}^{n+2}$ spanned by $\phi(e_0)$ and $\phi(e_1)$, and the diffeomorphism

$$\bar{\Theta} = \bar{\Psi} \circ \bar{\Phi}^{-1} : G^{\mathbb{R}}_{2,n} \to Y.$$

Then by (5.60), it is easily seen that the diagram

(5.61)
$$
\begin{array}{ccc}
\widetilde{G}^{\mathbb{R}}_{2,n} & \xrightarrow{\ \Theta\ } & X \\
\Big\downarrow{\varpi} & & \Big\downarrow{\varpi} \\
G^{\mathbb{R}}_{2,n} & \xrightarrow{\ \bar{\Theta}\ } & Y
\end{array}
$$

commutes.

If $x \in \widetilde{G}_{2,n}^{\mathbb{R}}$, let $\mathcal{K}_x$ be the set of endomorphisms κ of V_x satisfying

$$J\kappa = -\kappa J, \qquad \kappa^2 = \mathrm{id}.$$

If $x \in X$ and $\{v_1, v_2\}$ is a positively oriented orthonormal basis of the oriented 2-plane V_x, the endomorphism κ_0 of V_x determined by

$$\kappa_0 v_1 = v_1 \qquad \kappa_0 v_2 = -v_2,$$

belongs to $\mathcal{K}_x$. It is easily seen that $\mathcal{K}_x$ consists precisely of all endomorphisms κ of V_x which can be written in the form

$$\kappa = \cos\theta \cdot \kappa_0 + \sin\theta \cdot J\kappa_0,$$

where $\theta \in \mathbb{R}$. For $\kappa \in \mathcal{K}_x$, we denote also by κ the endomorphism $\kappa \otimes \mathrm{id}$ of the tangent space $(V \otimes W)_x$ of $\widetilde{G}_{2,n}^{\mathbb{R}}$ at x.

We again consider the point x_0 of $\widetilde{G}_{2,n}^{\mathbb{R}}$ and its image $b = \Theta(x_0)$ in X. If $\nu' = \pi_* \tilde{\nu}'$ is the element of S_b of §2, using the relations (5.54) we easily verify that the endomorphism $K_{\nu'}$ of T_b is determined by

$$K_{\nu'}\Theta_*(e_0 \otimes e_j) = \Theta_*(e_0 \otimes e_j), \qquad K_{\nu'}\Theta_*(e_1 \otimes e_j) = -\Theta_*(e_1 \otimes e_j),$$

for $2 \le j \le n+1$. Thus if κ_0 is the endomorphism of V_{x_0} corresponding to the oriented orthonormal basis $\{e_0, e_1\}$ of the oriented 2-plane V_{x_0}, we have

$$K_{\nu'}\Theta_* = \Theta_* \kappa_0,$$

as mappings from the tangent space of $\widetilde{G}_{2,n}^{\mathbb{R}}$ at x_0 to the tangent space of X at b. It follows that the mapping Θ induces a bijective mapping $\Theta : \mathcal{K}_x \to S_x$, with $\Theta(\kappa_0) = \nu'$, such that, for all $\kappa \in \mathcal{K}_x$, we have the equality

$$(5.62) \qquad\qquad K_\nu \Theta_* = \Theta_* \kappa$$

of mappings from $(V \otimes W)_x$ to the tangent space of X at $\Theta(x)$, where ν is the element $\Theta(\kappa)$ of S_x.

We henceforth identify the real Grassmannian $\widetilde{G}_{2,n}^{\mathbb{R}}$ with the complex quadric $X = Q_n$ by means of the holomorphic diffeomorphism Θ, and the real Grassmannian $G_{2,n}^{\mathbb{R}}$ with the manifold Y by means of the diffeomorphism $\bar{\Theta}$. Then for $x \in X$, the tangent space T_x is identified with $(V \otimes W)_x$. According to the commutativity of the diagram (5.61), we see that the mapping $\varpi : X \to Y$ is then identified with the natural projection from $\widetilde{G}_{2,n}^{\mathbb{R}}$ to $G_{2,n}^{\mathbb{R}}$. The involutive isometry τ of Q_n defined in §3 is identified with the involution τ_0 of $\widetilde{G}_{2,n}^{\mathbb{R}}$ corresponding to the change of orientation of a 2-plane of $\mathbb{R}^{n+2}$. According to (5.59), the Kähler metric $2g$ on Q_n is identified with the metric g' of $\widetilde{G}_{2,n}^{\mathbb{R}}$. Following §1, Chapter IV, we identify the vector bundles S^2T^* and

$$(S^2V^* \otimes S^2W^*) \oplus (\wedge^2V^* \otimes \wedge^2W^*).$$

Thus we may identify the fiber of the latter vector bundle at the point x_0 with the fiber $S^2T_b^*$ (resp. the fiber $S^2T_{Y,\varpi(b)}^*$) as K-modules (resp. as K'-modules).

In §4, Chapter II and in §1, Chapter IV, we introduced the notion of even or odd tensor (with respect to the involutive isometry τ). We recall that, if F is a G-invariant sub-bundle of $T_{\mathbb{C}}$ or of $S^pT_{\mathbb{C}}^*$, which is also invariant under τ, the $SO(m+n)$-module $C^\infty(F)$ admits the decomposition (2.8), where $C^\infty(F)^{\mathrm{ev}}$ (resp. $C^\infty(F)^{\mathrm{odd}}$) is the $SO(m+n)$-submodule of $C^\infty(F)$ consisting of all even (resp. odd) sections of F over X. In particular, we have the decomposition

$$C^\infty(X) = C^\infty(X)^{\mathrm{ev}} \oplus C^\infty(X)^{\mathrm{odd}}$$

of the $SO(m+n)$-module $C^\infty(X)$.

If x is a point of X and ν is an element of S_x, according to the discussion appearing above, there exists a positively oriented basis $\{v_1, v_2\}$ of V_x such that

$$J(v_1 \otimes w) = v_2 \otimes w, \qquad J(v_2 \otimes w) = -v_1 \otimes w,$$

$$v_1 \otimes W_x = T_\nu^+, \qquad v_2 \otimes W_x = T_\nu^-,$$

for all $w \in W_x$. If $\{\alpha_1, \alpha_2\}$ is the basis of V_x^* dual to the basis $\{v_1, v_2\}$, by (5.7) we easily verify that the equalities

$$(5.63) \qquad h_\nu = \tfrac{1}{2}(\alpha_1^2 - \alpha_2^2) \otimes g_2, \qquad h_{J\nu} = \alpha_1 \cdot \alpha_2 \otimes g_2$$

hold. If F is a subspace of T_ν^+, there exists a subspace W_1 of W_x of the same dimension as F such that

$$F \oplus JF = V_x \otimes W_1.$$

If W' is a subspace of W_x of dimension $k \geq 2$, according to Lemma 4.6 we know that the closed totally geodesic submanifold $\mathrm{Exp}_x V_x \otimes W'$ of X is isometric to the quadric Q_k.

The sub-bundles $(S^2T^*)^{++}$ and $(S^2T^*)^{+-}$ of $(S^2T^*)^+$ can be defined directly in terms of the intrinsic structure of the real Grassmannian $\widetilde{G}_{2,n}^{\mathbb{R}}$, without having recourse to the imbedding of X as a complex hypersurface of $\mathbb{CP}^{n+1}$ by

$$(S^2 T^*)_x^{++} = \{ h \in (S^2 T^*)_x^+ \mid h(\kappa \xi, \kappa \eta) = h(\xi, \eta), \text{ for all } \xi, \eta \in T_x \},$$

$$(S^2 T^*)_x^{+-} = \{ h \in (S^2 T^*)_x^+ \mid h(\kappa \xi, \kappa \eta) = -h(\xi, \eta), \text{ for all } \xi, \eta \in T_x \},$$

for $x \in X$, where κ is an arbitrary element of $\mathcal{K}_x$. We shall sometimes write $(S^2 T_X^*)^{+-} = (S^2 T^*)^{+-}$.

Let $x \in X$ and $\{v_1, v_2\}$ be an orthonormal basis of V_x. An element $h \in S^2 T_x^*$ belongs to the sub-bundle $(S^2 T^*)^+$ (resp. $(S^2 T^*)^-$) if and only if

$$h(v_2 \otimes w_1, v_2 \otimes w_2) = \varepsilon h(v_1 \otimes w_1, v_1 \otimes w_2),$$

$$h(v_1 \otimes w_1, v_2 \otimes w_2) = -\varepsilon h(v_1 \otimes w_2, v_2 \otimes w_1),$$

for all $w_1, w_2 \in W_x$, where $\varepsilon = 1$ (resp. $\varepsilon = -1$). It is easily seen that an element h of $S^2 T_x^*$ belongs to the sub-bundle $(S^2 T^*)^{++}$ if and only if

$$h(v_2 \otimes w_1, v_2 \otimes w_2) = h(v_1 \otimes w_1, v_1 \otimes w_2), \quad h(v_1 \otimes w_1, v_2 \otimes w_2) = 0,$$

for all $w_1, w_2 \in W_x$; moreover, an element h of $S^2 T_x^*$ belongs to the sub-bundle $(S^2 T^*)^{+-}$ if and only if

$$h(v_j \otimes w_1, v_j \otimes w_2) = 0, \quad h(v_1 \otimes w_1, v_2 \otimes w_2) = -h(v_1 \otimes w_2, v_2 \otimes w_1),$$

for all $w_1, w_2 \in W_x$ and $j = 1, 2$.

We consider the sub-bundles $S_0^2 V^*$ of $S^2 V^*$ and $S_0^2 W^*$ of $S^2 W^*$, the sub-bundles E, E_j, with $j = 1, 2, 3$, of $S^2 T^*$ defined in §1, Chapter IV. The above observations concerning the sub-bundles $(S^2 T^*)^-$, $(S^2 T^*)^{++}$ and $(S^2 T^*)^{+-}$ lead us to the equality

$$(5.64) \qquad\qquad (S^2 T^*)^{+-} = E$$

and to the inclusions

$$E_1 \subset (S^2 T^*)_0^{++}, \qquad S_0^2 V^* \otimes S^2 W^* \subset (S^2 T^*)^-,$$

$$\textstyle\bigwedge^2 V^* \otimes \bigwedge^2 W^* \subset (S^2 T^*)^{+-}.$$

On the other hand, the equalities (5.63) tell us that

$$(5.65) \qquad\qquad E_2 = L.$$

From the above inclusions, the relations (5.64) and (5.65), and the decompositions (4.6) and (5.26), we obtain the equalities

(5.66)
$$E_1 = (S^2 T^*)_0^{++}, \qquad E_3 = (S^2 T^*)^{-\perp},$$
$$S_0^2 V^* \otimes S^2 W^* = (S^2 T^*)^-,$$
$$\wedge^2 V^* \otimes \wedge^2 W^* = E = (S^2 T^*)^{+-}.$$

We now suppose that n is even. In §1, Chapter IV, we saw that the oriented 2-plane $x \in X$ determines an orientation of the space W_x. Let x be a point of X and let ν be an element of S_x. We say that an orthonormal basis $\{\zeta_1, \ldots, \zeta_n\}$ of T_ν^+ is positively (resp. negatively) oriented if there is a positively oriented orthonormal basis $\{v_1, v_2\}$ of V_x and a positively (resp. negatively) oriented orthonormal basis $\{w_1, \ldots, w_n\}$ of W_x such that

$$\zeta_j = v_1 \otimes w_j,$$

for $1 \le j \le n$. Since n is even, it is easily seen that the notions of positively and negatively oriented orthonormal bases of T_ν^+ are well-defined. Also an arbitrary orthonormal basis of T_ν^+ is either positively or negatively oriented.

We now consider the case when $n = 4$. The orientations of the spaces W_a, with $a \in X$, and the scalar product g_2 give rise to a Hodge operator

$$* : \wedge^2 W^* \to \wedge^2 W^*.$$

In turn, this operator induces an involution $* = \mathrm{id} \otimes *$ of the vector bundle $\wedge^2 V^* \otimes \wedge^2 W^*$. Let x be a point of X and let $\{w_1, \ldots, w_4\}$ be a positively oriented orthonormal basis of W_x; according to the definition of the involution $*$ of $\wedge^2 V^* \otimes \wedge^2 W^*$, we easily see that

(5.67)
$$(* h)(v_1 \otimes w_1, v_2 \otimes w_2) = h(v_1 \otimes w_3, v_2 \otimes w_4),$$

for all $h \in (\wedge^2 V^* \otimes \wedge^2 W^*)_x$. Let ν be an element of S_x and let $\{\zeta_1, \ldots, \zeta_4\}$ be a positively oriented orthonormal basis of T_ν^+. By (5.67), we have

(5.68)
$$(* h)(\zeta_1, J\zeta_2) = h(\zeta_3, J\zeta_4),$$

for all $h \in (S^2 T^*)_x^{+-}$.

By formulas (3.6) of [21] and (5.68), we easily verify that this automorphism $*$ of the vector bundle $\wedge^2 V^* \otimes \wedge^2 W^*$ is equal to the involution $*$ of the vector bundle $(S^2 T^*)^{+-}$ defined in [21, §3] in terms of an appropriate orientation of the real structures of X. Thus the eigenbundles F^+ and F^- of this involution of $(S^2 T^*)^{+-}$ corresponding to the eigenvalues $+1$

and -1, which are considered in [21, §3], are equal to $\bigwedge^2 V^* \otimes \bigwedge^+ W^*$ and $\bigwedge^2 V^* \otimes \bigwedge^- W^*$, respectively. The decomposition

$$(5.69) \qquad (S^2 T^*)^{+-} = F^+ \oplus F^-$$

then gives rise to the equality

$$(5.70) \qquad (S^2 T^*)^{+-}_{\mathbb{C}} = F^+_{\mathbb{C}} \oplus F^-_{\mathbb{C}}.$$

Since the mapping $\tau_* : T_x \to T_{\tau(x)}$, with $x \in X$, is equal to the identity mapping of $(V \otimes W)_x$, we easily see that

$$(5.71) \qquad \tau^* \cdot * = - * \cdot \tau^*,$$

as mappings from $(S^2 T^*)^{+-}_{\tau(x)}$ to $(S^2 T^*)^{+-}_x$.

If h is a section of $(S^2 T^*)^{+-}$ over X, then the equality

$$(5.72) \qquad \nabla_\xi * h = * \nabla_\xi h$$

holds for all $\xi \in T$. Indeed, let x be an arbitrary point of X; in §4, we saw that there exists a section ν of S over a neighborhood U of X and an orthonormal frame $\{\zeta_1, \dots, \zeta_4\}$ for T^+_ν satisfying $(\nabla \zeta_j)(x) = 0$, for $1 \le j \le n$. Without loss of generality, we may suppose that, for each point $a \in U$, the orthonormal basis $\{\zeta_1(a), \dots, \zeta_4(a)\}$ of $T^+_{\nu(a)}$ is positively oriented. By means of this frame and the relation (5.68), we see that the equality (5.72) holds for all $\xi \in T_x$.

The following result is given by Lemma 4.1 of [21].

LEMMA 5.1. Let X be the quadric Q_4. Let $x \in X$ and $\nu \in S_x$, and let $\{\zeta_1, \zeta_2, \zeta_3, \zeta_4\}$ be a positively oriented orthonormal basis of $T^+_{\nu,x}$. Then for $h \in C^\infty((S^2 T^*)^{+-})$, we have

$$(\operatorname{div} * h)(J\zeta_1) = -(d\check{h})(\zeta_2, \zeta_3, \zeta_4).$$

PROOF: By the second equalities of (5.24) and (5.30), and by (5.68) and (5.72), we have

$$\begin{aligned}
-(\operatorname{div} * h)(J\zeta_1) &= (\nabla * h)(\zeta_2, \zeta_2, J\zeta_1) + (\nabla * h)(\zeta_3, \zeta_3, J\zeta_1) \\
&\quad + (\nabla * h)(\zeta_4, \zeta_4, J\zeta_1) \\
&= (* \nabla_{\zeta_2} h)(\zeta_2, J\zeta_1) + (* \nabla_{\zeta_3} h)(\zeta_3, J\zeta_1) \\
&\quad + (* \nabla_{\zeta_4} h)(\zeta_4, J\zeta_1) \\
&= (\nabla\check{h})(\zeta_2, \zeta_3, \zeta_4) + (\nabla\check{h})(\zeta_3, \zeta_4, \zeta_2) + (\nabla\check{h})(\zeta_4, \zeta_2, \zeta_3) \\
&= (d\check{h})(\zeta_2, \zeta_3, \zeta_4).
\end{aligned}$$

The following result is given by Lemma 3.2 of [22].

LEMMA 5.2. *Let X be the quadric Q_4. For all $f \in C^\infty(X)$, we have*

$$\operatorname{div} * \pi_{+-} \operatorname{Hess} f = 0.$$

PROOF: Let x be a point of X and ν be an element of S_x, and let $\{\zeta_1, \zeta_2, \zeta_3, \zeta_4\}$ be a positively oriented orthonormal basis of $T^+_{\nu,x}$. If f is an element of $C^\infty(X)$, by Lemma 5.1 and formulas (5.30) and (5.32) we see that

$$\begin{aligned}
(\operatorname{div} * \pi_{+-} \operatorname{Hess} f)(J\zeta_1) &= -(id\rho_+ \partial\bar\partial f)(\zeta_2, \zeta_3, \zeta_4) \\
&= -(id\partial\bar\partial f)(\zeta_2, \zeta_3, \zeta_4) \\
&= 0.
\end{aligned}$$

§6. Totally geodesic surfaces
and the infinitesimal orbit of the curvature

We begin by giving an explicit representation of the infinitesimal orbit of the curvature of the complex quadric $X = Q_n$, with $n \geq 3$.

We consider the morphism of vector bundles

$$\tau_B : S^2 T^* \otimes S^2 T^* \to B, \qquad \hat\tau_B : S^2 T^* \to B$$

of §1, Chapter I and the morphisms of vector bundles

$$\psi : \textstyle\bigwedge^2 T^* \to B, \qquad \check\psi : (\textstyle\bigwedge^2 T^*)^- \to B$$

of §4, Chapter I; we saw that the morphisms $\hat\tau_B$ and ψ are injective.

If x is a point of X and ν is an element of S_x, for $\beta \in \bigwedge^2 T^*_x$ we define an element β^{K_ν} of $S^2 T^*_x$ by

$$\beta^{K_\nu}(\xi, \eta) = \beta(K_\nu \xi, \eta) + \beta(\xi, K_\nu \eta),$$

for $\xi, \eta \in T_x$. We easily verify that the element

$$\beta^{K_\nu} \otimes h_\nu + \beta^{K_{J\nu}} \otimes h_{J\nu}$$

of $S^2 T^*_x \otimes S^2 T^*_x$ does not depend on the choice of the element ν of S_x. If we set

$$\chi(\beta) = \tau_B(\beta^{K_\nu} \otimes h_\nu + \beta^{K_{J\nu}} \otimes h_{J\nu}),$$

we then obtain a well-defined morphism

$$\chi : \textstyle\bigwedge^2 T^* \to B$$

of vector bundles over X. We easily see that

$$\rho(\iota(\beta))h_\nu = \beta^{K_\nu},$$

for all $\beta \in \bigwedge^2 T_x^*$. From formulas (1.72) and (5.10), it follows that

$$\rho(\iota(\beta))R = -2\chi(\beta), \qquad \text{for } \beta \in T_{\mathbb{R}}^{1,1},$$

$$\rho(\iota(\beta))R = -2(\check{\psi} + \chi)(\beta), \qquad \text{for } \beta \in (\bigwedge^2 T^*)^-.$$

In fact, if β is an element of $(T_{\mathbb{R}}^{1,1})^+$ or if $\beta = \omega$, by relation (1.78) we have

$$\rho(\iota(\beta))R = 0.$$

By means of the decomposition (5.28), Lemma 1.13 and the preceding formulas, we obtain the equalities

(5.73) $\tilde{B}^+ = \chi((T_{\mathbb{R}}^{1,1})_0^-), \qquad \tilde{B}^- = (\check{\psi} + \chi)((\bigwedge^2 T^*)^-).$

We now introduce various families of closed connected totally geodesic submanifolds of X. Let x be a point of X and ν be an element of S_x. If $\{\xi, \eta\}$ is an orthonormal set of vectors of T_ν^+, according to formula (5.10) we see that the set $\mathrm{Exp}_x F$ is a closed connected totally geodesic surface of X, whenever F is the subspace of T_x generated by one of following families of vectors:

(A$_1$) $\{\xi, J\eta\}$;
(A$_2$) $\{\xi + J\eta, J\xi - \eta\}$;
(A$_3$) $\{\xi, J\xi\}$;
(A$_4$) $\{\xi, \eta\}$.

According to [10], if F is generated by the family (A$_2$) (resp. the family (A$_3$)) of vectors, where $\{\xi, \eta\}$ is an orthonormal set of vectors of T_ν^+, the surface $\mathrm{Exp}_x F$ is isometric to the complex projective line $\mathbb{CP}^1$ with its metric of constant holomorphic curvature 4 (resp. curvature 2). Moreover, if F is generated by the family (A$_1$), where $\{\xi, \eta\}$ is an orthonormal set of vectors of T_ν^+, the surface $\mathrm{Exp}_x F$ is isometric to a flat torus.

For $1 \le j \le 4$, we denote by $\tilde{\mathcal{F}}^{j,\nu}$ the set of all closed totally geodesic surfaces of X which can be written in the form $\mathrm{Exp}_x F$, where F is a subspace of T_x generated by a family of vectors of type (A$_j$).

According to §5, there exists a unit vector v of V_x such that an arbitrary submanifold Z belonging to the family $\tilde{\mathcal{F}}^{4,\nu}$ can be written in the form $\mathrm{Exp}_x v \otimes W'$, where W' is a two-dimensional subspace of W_x. We consider the Riemannian metric g' on the Grassmannian $\tilde{G}_{1,2}^{\mathbb{R}}$ defined in Chapter IV; by Lemma 4.6 and the relation (5.59), we see that the submanifold Z is isometric to the Grassmannian $\tilde{G}_{1,2}^{\mathbb{R}}$ endowed with the Riemannian metric $\frac{1}{2}g'$. Therefore such a submanifold Z is isometric to a sphere of constant curvature 2 (see also [10]); moreover, by Lemma 4.6 we also see that the image of Z under the mapping $\varpi : X \to Y$ is a closed totally geodesic surface of Y isometric to the real projective plane endowed with its metric of constant curvature 2.

If ε is a number equal to ± 1 and ξ, η, ζ are unit vectors of T_ν^+ satisfying

$$g(\xi, \eta) = g(\xi, \zeta) = 3g(\eta, \zeta) = \varepsilon \frac{3}{5},$$

and if F is the subspace of T_x generated by the vectors

$$\{\xi + J\zeta, \eta + \varepsilon J(\xi - \eta) - J\zeta\},$$

according to (5.10) we also see that the set $\mathrm{Exp}_x F$ is a closed connected totally geodesic surface of X. Moreover, according to [10] this surface is isometric to a sphere of constant curvature $2/5$. We denote by $\tilde{\mathcal{F}}^{5,\nu}$ the set of all such closed totally geodesic surfaces of X.

If $\{\xi_1, \xi_2, \xi_3, \xi_4\}$ is an orthonormal set of vectors of T_ν^+ and if F is the subspace of T_x generated by the vectors

$$\{\xi_1 + J\xi_2, \xi_3 + J\xi_4\},$$

according to (5.10) we see that the set $\mathrm{Exp}_x F$ is a closed connected totally geodesic surface of X. Moreover, according to [10] this surface is isometric to the real projective plane $\mathbb{RP}^2$ of constant curvature 1. Clearly such submanifolds of X only occur when $n \geq 4$. We denote by $\tilde{\mathcal{F}}^{6,\nu}$ the set of all such closed totally geodesic surfaces of X.

If $\{\xi_1, \xi_2, \xi_3, \xi_4\}$ is an orthonormal set of vectors of T_ν^+ and if F is the subspace of T_x generated by the vectors

$$\{\xi_1 + J\xi_2, J\xi_1 - \xi_2, \xi_3 + J\xi_4, J\xi_3 - \xi_4\},$$

according to (5.10) we see that the set $\mathrm{Exp}_x F$ is a closed connected totally geodesic submanifold of X. Moreover, this submanifold is isometric to the complex projective plane $\mathbb{CP}^2$ of constant holomorphic curvature 4. Clearly such submanifolds of X only occur when $n \geq 4$. We denote by $\tilde{\mathcal{F}}^{7,\nu}$ the set of all such closed totally geodesic submanifolds of X.

When $n \geq 4$, clearly a surface belonging to the family $\tilde{\mathcal{F}}^{2,\nu}$ or to the family $\tilde{\mathcal{F}}^{6,\nu}$ is contained in a closed totally geodesic submanifold of X belonging to the family $\tilde{\mathcal{F}}^{7,\nu}$. In fact, the surfaces of the family $\tilde{\mathcal{F}}^{2,\nu}$ (resp. the family $\tilde{\mathcal{F}}^{6,\nu}$) correspond to complex lines (resp. to linearly imbedded real projective planes) of the submanifolds of X belonging to the family $\tilde{\mathcal{F}}^{7,\nu}$ viewed as complex projective planes.

Let Z be a surface belonging to the family $\tilde{\mathcal{F}}^{j,\nu}$, with $1 \leq j \leq 6$; we may write $Z = \mathrm{Exp}_x F$, where F is an appropriate subspace of T_x. Clearly, this space F is contained in a subspace of T_x which can be written in the form $F_1 \oplus JF_1$, where F_1 is a subspace of T_ν^+ of dimension k; we may suppose that this integer k is given by

$$k = \begin{cases} 2 & \text{when} \quad 1 \le j \le 4, \\ 3 & \text{when} \quad j = 5, \\ 4 & \text{when} \quad j = 6, 7. \end{cases}$$

According to observations made in §5, the surface $Z = \operatorname{Exp}_x F$ is contained in a closed totally geodesic submanifold $\operatorname{Exp}_x V_x \otimes W_1$ of X isometric to the quadric Q_k, where W_1 is a subspace of W_x of dimension k.

Let $\mathcal{F}'_x$ be the family of all closed connected totally geodesic submanifolds of X passing through x which can be written as $\operatorname{Exp}_x V_x \otimes W_1$, where W_1 is a subspace of W_x of dimension 3. We know that a submanifold of X belonging to $\mathcal{F}'_x$ is isometric to the quadric Q_3 of dimension 3.

For $1 \le j \le 7$, we consider the G-invariant families

$$\tilde{\mathcal{F}}^j_x = \bigcup_{\nu \in S_x} \tilde{\mathcal{F}}^{j,\nu}, \qquad \tilde{\mathcal{F}}^j = \bigcup_{x \in X} \tilde{\mathcal{F}}^j_x$$

of closed connected totally geodesic submanifolds of X. When $n \ge 4$, we know that a surface belonging to the family $\tilde{\mathcal{F}}^2$ is contained in a closed totally geodesic submanifold of X belonging to the family $\tilde{\mathcal{F}}^7$. We write

$$\mathcal{F}_1 = \tilde{\mathcal{F}}^1 \cup \tilde{\mathcal{F}}^3 \cup \tilde{\mathcal{F}}^4, \qquad \mathcal{F}_2 = \tilde{\mathcal{F}}^1 \cup \tilde{\mathcal{F}}^2 \cup \tilde{\mathcal{F}}^6,$$

$$\mathcal{F}_3 = \tilde{\mathcal{F}}^1 \cup \tilde{\mathcal{F}}^2 \cup \tilde{\mathcal{F}}^4 \cup \tilde{\mathcal{F}}^5.$$

We also consider the G-invariant family

$$\mathcal{F}' = \bigcup_{x \in X} \mathcal{F}'_x$$

of closed connected totally geodesic submanifolds of X isometric to Q_3. We have seen that a surface belonging to the family $\tilde{\mathcal{F}}^j$, with $1 \le j \le 5$, is contained in a closed totally geodesic submanifold of X belonging to the family $\mathcal{F}'$.

Since the group G acts transitively on the set Ξ of all maximal flat totally geodesic tori of X and also on a torus belonging to Ξ, and since a surface of $\tilde{\mathcal{F}}^1$ is a flat 2-torus, we see that, if Z is an element of Ξ and if x is a point of Z, there exists an element $\nu \in S_x$ and an orthonormal set of vectors $\{\xi, \eta\}$ of T_ν^+ such that

$$Z = \operatorname{Exp}_x(\mathbb{R}\xi \oplus \mathbb{R}J\eta).$$

It follows that the family $\tilde{\mathcal{F}}^1$ is equal to Ξ.

In [10], Dieng classified all closed connected totally geodesic surfaces of X and proved the following:

PROPOSITION 5.3. *If $n \geq 3$, then the family of all closed connected totally geodesic surfaces of X is equal to $\mathcal{F}_1 \cup \mathcal{F}_2 \cup \mathcal{F}_3$.*

If Z is a surface of X belonging to the family $\tilde{\mathcal{F}}^j$, with $1 \leq j \leq 6$, there is a subgroup of G which acts transitively on Z. Thus for $1 \leq j \leq 6$, we see that an element u of $\bigotimes^q T_x^*$, with $x \in X$, vanishes when restricted to an arbitrary surface belonging to the family $\tilde{\mathcal{F}}^j$ if and only if it vanishes when restricted to an arbitrary surface belonging to the family $\tilde{\mathcal{F}}_x^j$ of closed connected totally geodesic surfaces of X passing through x.

We now establish relationships between the families of closed totally geodesic surfaces of X introduced above, the G-invariant sub-bundles of $S^2 T^*$ and the infinitesimal orbit of the curvature $\tilde{B}$. If $\mathcal{F}$ is a G-invariant family of closed connected totally geodesic surfaces of X, we denote by $N_{\mathcal{F}}$ the sub-bundle of B consisting of those elements of B which vanish when restricted to the submanifolds of $\mathcal{F}$.

For $1 \leq k \leq 6$ and $j = 1, 2, 3$, we set

$$\tilde{N}_k = N_{\tilde{\mathcal{F}}^k}, \qquad N_j = N_{\mathcal{F}_j},$$

$$\tilde{N}_k^+ = \tilde{N}_k \cap B^+, \qquad \tilde{N}_k^- = \tilde{N}_k \cap B^-,$$

$$N_j^+ = N_j \cap B^+, \qquad N_j^- = N_j \cap B^-.$$

Using the relation (5.13), we easily verify that the sub-bundles $\tilde{N}_k$ and N_j of B are stable under the involution J; hence we have

$$(5.74) \qquad \tilde{N}_k = \tilde{N}_k^+ \oplus \tilde{N}_k^-, \qquad N_j = N_j^+ \oplus N_j^-,$$

for $1 \leq k \leq 6$ and $j = 1, 2, 3$. According to formula (1.56), we see that

$$(5.75) \qquad \tilde{B} \subset N_j,$$

for $j = 1, 2, 3$.

Let x be a point of X, and let ν be an element of S_x and $\{\xi, \eta\}$ be an orthonormal system of vectors of T_ν^+. Let u be an element of the vector bundle $\tilde{N}_1$; clearly we have

$$(5.76) \qquad u(\xi, J\eta, \xi, J\eta) = 0.$$

Since ν is an arbitrary element of S_x, from the relation (5.12) we easily infer that u also satisfies

$$(5.77) \qquad u(\xi + tJ\xi, J\eta - t\eta, \xi + tJ\xi, J\eta - t\eta) = 0,$$

for all $t \in \mathbb{R}$. Since the vectors $\zeta_1 = \xi + \eta$ and $\zeta_2 = \xi - \eta$ of T_ν^+ are orthogonal and have the same length, the equality (5.77), with $t = 1$, tells us that

$$u(\zeta_1 + J\zeta_1, J\zeta_2 - \zeta_2, \zeta_1 + J\zeta_1, J\zeta_2 - \zeta_2) = 0;$$

the preceding relation implies that the element u of $\tilde{N}_1$ satisfies

$$(5.78) \qquad u(\xi + J\eta, \eta + J\xi, \xi + J\eta, \eta + J\xi) = 0.$$

Clearly an element u of the vector bundle $\tilde{N}_2$ satisfies

$$(5.79) \qquad u(\xi + J\eta, J\xi - \eta, \xi + J\eta, J\xi - \eta) = 0,$$

while an element u of $\tilde{N}_3$ verifies the relation

$$(5.80) \qquad u(\xi, J\xi, \xi, J\xi) = 0.$$

On the other hand, an element u of $\tilde{N}_4$ satisfies

$$(5.81) \qquad u(\xi, \eta, \xi, \eta) = 0.$$

Finally, if $n \geq 4$ and u is an element of $\tilde{N}_6$, and if $\{\zeta, \zeta'\}$ is an orthonormal system of vectors of T_ν^+ orthogonal to the vectors ξ and η, we see that

$$(5.82) \qquad u(\xi + J\zeta, \eta + J\zeta', \xi + J\zeta, \eta + J\zeta') = 0.$$

We remark that an element u of B^- always satisfies the relation (5.80).

Clearly, an element u of B_x belongs to $N_{2,x}$ if and only if u satisfies the relations (5.76), (5.79) and (5.82), for all $\nu \in S_x$. It is easily verified that the vector bundles $\hat{\tau}_B(L)$ and $\psi((\wedge^2 T^*)^-)$ are sub-bundles of N_2^-. Using the formula (1.3), we see that

$$(5.83) \qquad \operatorname{Tr} \hat{\tau}_B(L) = L, \qquad \operatorname{Tr} \psi((\wedge^2 T^*)^-) = \{0\},$$

and so we obtain the inclusion

$$(5.84) \qquad L \subset \operatorname{Tr} N_2^-.$$

The following three lemmas are proved in [21].

LEMMA 5.4. *Suppose that $n \geq 3$. Let ν be an element of S_x and $\{\xi, \eta, \zeta\}$ be an orthonormal system of vectors of T_ν^+, and let u be an element of the vector bundle $\tilde{N}_1$. Then the following assertions hold:*
(i) *We have*

$$(5.85) \qquad u(\xi, J\zeta, \eta, J\zeta) = 0.$$

(ii) *If u belongs to $\tilde{N}_1^-$, we have*

$$(5.86) \qquad u(\xi, J\xi, \eta, J\xi) + u(\xi, J\eta, \eta, J\eta) = 0.$$

LEMMA 5.5. *Suppose that $n \geq 3$. Let ν be an element of S_x and $\{\xi, \eta\}$ be an orthonormal system of vectors of T_ν^+, and let u be an element of the vector bundle $\tilde{N}_1^+ \cap \tilde{N}_2^+$. Then we have*

$$(5.87) \qquad 2u(\xi, \eta, \xi, \eta) + u(\xi, J\xi, \xi, J\xi) + u(\eta, J\eta, \eta, J\eta) = 0.$$

LEMMA 5.6. *Suppose that $n \geq 4$. Let ν be an element of S_x and $\{\xi, \eta, \zeta, \zeta'\}$ be an orthonormal system of vectors of T_ν^+, and let u be an element of $\tilde{N}_1 \cap \tilde{N}_6$. Then the following assertions hold:*
  (i) *We have*

$$(5.88) \qquad u(\xi, \eta, \xi, \eta) + u(J\zeta, J\zeta', J\zeta, J\zeta') = 0,$$
$$(5.89) \qquad u(\xi, \eta, J\zeta, \eta) + u(\xi, J\zeta', J\zeta, J\zeta') = 0.$$

  (ii) *If u belongs to B^-, we have*

$$(5.90) \qquad u(\xi, \eta, \xi, \zeta) = -u(\zeta', \eta, \zeta', \zeta).$$

  (iii) *If $n \geq 5$, we have*

$$(5.91) \qquad u(\xi, \eta, \xi, \zeta) = 0.$$

  (iv) *If $n \geq 5$ and if u belongs to B^+, we have*

$$(5.92) \qquad u(\xi_1, \xi_2, \xi_3, \xi_4) = 0,$$

for all vectors $\xi_1, \xi_2, \xi_3, \xi_4 \in T_\nu^+$.

LEMMA 5.7. *Suppose that $n \geq 5$. Let ν be an element of S_x and $\{\xi, \eta\}$ be an orthonormal system of vectors of T_ν^+, and let u be an element of N_2^+. Then the equality (5.80) holds and we have*

$$(5.93) \qquad u(\xi, \eta, \xi, J\xi) = 0.$$

Let x be a point of X, and let ν be an element of S_x and $\{\xi, \eta\}$ be an orthonormal system of vectors of T_ν^+. We choose an orthonormal basis $\{\xi_1, \ldots \xi_n\}$ of T_ν^+, with $\xi_1 = \xi$ and $\xi_2 = \eta$; then $\{J\xi, \ldots J\xi_n\}$ is an orthonormal basis of T_ν^-. Then for $u \in B_x$ and $\zeta \in T_x$, we have

$$(\text{Tr}\, u)(\xi, \zeta) = u(\xi, \eta, \zeta, \eta) + u(\xi, J\xi, \zeta, J\xi) + u(\xi, J\eta, \zeta, J\eta)$$
$$+ \sum_{j=3}^{n} (u(\xi, \xi_j, \zeta, \xi_j) + u(\xi, J\xi_j, \zeta, J\xi_j)).$$

Thus if u is an element of $\tilde{N}_{1,x}$, by the relation (5.76) and Lemma 5.4,(i) we see that

$$(\operatorname{Tr} u)(\xi, \xi) = u(\xi, \eta, \xi, \eta) + u(\xi, J\xi, \xi, J\xi) + \sum_{j=3}^{n} u(\xi, \xi_j, \xi, \xi_j),$$

(5.94)

$$(\operatorname{Tr} u)(\xi, \eta) = u(\xi, J\xi, \eta, J\xi) + u(\xi, J\eta, \eta, J\eta) + \sum_{j=3}^{n} u(\xi, \xi_j, \eta, \xi_j)$$

and

$$(\operatorname{Tr} u)(\xi, J\eta) = u(\xi, \eta, J\eta, \eta) + u(\xi, J\xi, J\eta, J\xi)$$

(5.95)

$$+ \sum_{j=3}^{n} (u(\xi, \xi_j, J\eta, \xi_j) + u(\xi, J\xi_j, J\eta, J\xi_j)).$$

LEMMA 5.8. *For $n \geq 3$, we have*

$$\operatorname{Tr} N_1 \subset (S^2 T^*)^{+-}.$$

PROOF: Let x be a point of X and u be an element of $N_{1,x}$. Let ν be an element of S_x and ξ be a unit vector of T_ν^+. We choose an orthonormal basis $\{\xi_1, \ldots \xi_n\}$ of T_ν^+, with $\xi_1 = \xi$. According to the first formula of (5.95), with $\eta = \xi_2$, and the relations (5.80) and (5.81), the expression $(\operatorname{Tr} u)(\xi, \xi)$ vanishes. Hence, by polarization we obtain the equality

(5.96) $(\operatorname{Tr} u)(\xi, \eta) = 0,$

for all vectors η of T_ν^+. Now let η be a given vector of T_ν^+. Since ν is an arbitrary element of S_x, from the equalities (5.96) and (5.12) we infer that the function f defined by

$$f(s) = (\operatorname{Tr} u)(\xi + sJ\xi, \eta + sJ\eta),$$

for $s \in \mathbb{R}$, vanishes identically. The equality $f'(0) = 0$ gives the relation

(5.97) $(\operatorname{Tr} u)(\xi, J\eta) + (\operatorname{Tr} u)(J\xi, \eta) = 0.$

If u belongs to N_1^+, according to (5.96) and (5.24) we find that $\operatorname{Tr} u$ is an element of $(S^2 T^*)^{+-}$. If u belongs to N_1^-, from (5.96) and (5.97) it follows that $\operatorname{Tr} u = 0$. The desired result is now a consequence of the second equality of (5.74), with $j = 1$.

PROPOSITION 5.9. (i) *For $n \geq 4$, we have*

$$(5.98) \qquad \text{Tr}\, N_2^- = L, \qquad \text{Tr}\, N_2^+ \subset (S^2 T^*)^{+-}.$$

(ii) *For $n \geq 5$, we have*

$$(5.99) \qquad \text{Tr}\, N_2^+ = \{0\}.$$

PROOF: We suppose that $n \geq 4$. Let x be a point of X and u be an element of $N_{2,x}$. Let ν be an element of S_x and $\{\xi, \eta\}$ be an orthonormal system of vectors of T_ν^+. We choose an orthonormal basis $\{\xi_1, \ldots \xi_n\}$ of T_ν^+, with $\xi_1 = \xi$ and $\xi_2 = \eta$. First, suppose that u belongs to N_2^-. When $n \geq 5$, by Lemma 5.6,(iii) we see that

$$u(\xi, \xi_j, \eta, \xi_j) = 0,$$

for all $3 \leq j \leq n$; by the equality (5.90) of Lemma 5.6,(ii), we have

$$u(\xi, \xi_3, \eta, \xi_3) + u(\xi, \xi_4, \eta, \xi_4) = 0.$$

Therefore according to the second formula of (5.94) and (5.86), we obtain

$$(5.100) \qquad (\text{Tr}\, u)(\xi, \eta) = 0.$$

By the equality (5.88) of Lemma 5.6,(i), we see that

$$u(\xi, \xi_j, \xi, \xi_j) = u(\eta, \xi_k, \eta, \xi_k),$$

for $3 \leq j, k \leq n$, with $j \neq k$; hence we have the equality

$$\sum_{j=3}^{n} u(\xi, \xi_j, \xi, \xi_j) = \sum_{j=3}^{n} u(\eta, \xi_j, \eta, \xi_j).$$

As u belongs to B^-, we know that (5.80) holds, and so from the previous relation and the first formula of (5.94), we deduce that

$$(5.101) \qquad (\text{Tr}\, u)(\xi, \xi) = (\text{Tr}\, u)(\eta, \eta).$$

Since ν is an arbitrary element of S_x, from the relation (5.12) and the equalities (5.100) and (5.101), we easily infer that u also satisfies

$$(\text{Tr}\, u)(\xi + tJ\xi, \eta + tJ\eta) = 0,$$

$$(\text{Tr}\, u)(\xi + tJ\xi, \xi + tJ\xi) = (\text{Tr}\, u)(\eta + tJ\eta, \eta + tJ\eta),$$

for all $t \in \mathbb{R}$. Since $\operatorname{Tr} u$ belongs to $(S^2 T^*)^-$, we obtain the equalities

$$(5.102) \qquad (\operatorname{Tr} u)(\xi, J\eta) = 0, \qquad (\operatorname{Tr} u)(\xi, J\xi) = (\operatorname{Tr} u)(\eta, J\eta).$$

We set

$$a = (\operatorname{Tr} u)(\xi, \xi), \qquad b = (\operatorname{Tr} u)(\xi, J\xi).$$

By (5.7), (5.100), (5.101) and (5.102), it is easily seen that $\operatorname{Tr} u$ is equal to the element $ah_\nu + bh_{J\nu}$ of L_x, and so we obtain the inclusion

$$\operatorname{Tr} N_2^- \subset L.$$

This inclusion and (5.84) give us the first relation of (5.98). Now suppose that u belongs to N_2^+. When $n \geq 5$, by Lemma 5.7 we know that the relation (5.80) holds; then according to the first formula of (5.94) and Lemma 5.6,(iv), we see that the expression $(\operatorname{Tr} u)(\xi, \xi)$ vanishes. When $n = 4$, according to the first formula of (5.94) and the equalities (5.87) of Lemma 5.5, we have

$$(\operatorname{Tr} u)(\xi, \xi) + (\operatorname{Tr} u)(\eta, \eta) = u(\xi, \xi_3, \xi, \xi_3) + u(\xi, \xi_4, \xi, \xi_4)$$
$$+ u(\eta, \xi_3, \eta, \xi_3) + u(\eta, \xi_4, \eta, \xi_4);$$

now by the relation (5.88) of Lemma 5.6,(i), the right-hand side of the preceding equality vanishes. Therefore if $n = 4$ and ζ is a unit vector of T_ν^+ orthogonal to ξ and η, we see that

$$(\operatorname{Tr} u)(\xi, \xi) = -(\operatorname{Tr} u)(\eta, \eta) = (\operatorname{Tr} u)(\zeta, \zeta) = -(\operatorname{Tr} u)(\xi, \xi).$$

Hence the expression $(\operatorname{Tr} u)(\xi, \xi) = 0$ also vanishes when $n = 4$. By polarization, we see that the equality $(\operatorname{Tr} u)(\xi, \xi') = 0$ holds for all $\xi' \in T_\nu^+$; by (5.24), we find that $\operatorname{Tr} u$ belongs to $(S^2 T^*)^{+-}$. We now suppose that $n \geq 5$. By Lemma 5.7, we know that the equality (5.93) holds, and so we obtain

$$u(\xi, \eta, \eta, J\eta) = 0,$$
$$u(\xi, J\xi, J\eta, J\xi) = -u(J\xi, \xi, \eta, \xi) = 0.$$

By the equality (5.89) of Lemma 5.6,(i), we see that the sum

$$\sum_{j=3}^{n} (u(\xi, \xi_j, J\eta, \xi_j) + u(\xi, J\xi_j, J\eta, J\xi_j))$$

vanishes. Hence according to (5.95), the expression $(\operatorname{Tr} u)(\xi, J\eta)$ vanishes. Since $\operatorname{Tr} u$ belongs to $(S^2 T^*)^+$, we know that $(\operatorname{Tr} u)(\xi, J\xi) = 0$. Thus we have proved assertion (ii).

The following two propositions are direct consequences of Proposition 5.9 and the second equality of (5.74), with $j = 2$. In fact, Proposition 5.11 is given by Proposition 5.1 of [21].

PROPOSITION 5.10. *For $n \geq 5$, we have*

$$\mathrm{Tr}\, N_2 = L.$$

PROPOSITION 5.11. *For $n = 4$, we have*

$$\mathrm{Tr}\, N_2 \subset L \oplus (S^2 T^*)^{+-}.$$

In [18], we verified that the sum

$$\hat{\tau}_B(L) \oplus \psi((\wedge^2 T^*)^-) \oplus \tilde{B}^-$$

is direct; we also know that it is a sub-bundle of N_2^-. Using the relations (1.79) and (5.73), in [18] we were able to determine the ranks of the vector bundles $\tilde{B}^+$ and $\tilde{B}^-$. When $n \geq 5$, by means of Lemmas 5.4–5.7 and other analogous results, in [21] we found explicit bounds for the ranks of the vector bundles N_2^+ and N_2^-. From these results, the relation (1.79), the second equality of (5.74) and the inclusion (5.75), with $j = 2$, we obtain the following proposition (see [18, §5]):

PROPOSITION 5.12. *For $n \geq 5$, we have*

$$N_2^+ = \tilde{B}^+, \qquad N_2^- = \tilde{B}^- \oplus \psi((\wedge^2 T^*)^-) \oplus \hat{\tau}_B(L),$$
$$N_2 = \tilde{B} \oplus \psi((\wedge^2 T^*)^-) \oplus \hat{\tau}_B(L).$$

According to Lemma 1.7 and (5.83), we see that Proposition 5.10 may also be deduced from Proposition 5.12.

By methods similar to those used in [21] to prove Lemmas 5.4–5.7, Dieng [10] showed that $N_3 \subset \tilde{N}_3$ and proved the following result:

PROPOSITION 5.13. *For $n \geq 3$, we have*

$$N_3 = \tilde{B}.$$

When $n \geq 3$, Dieng [10] showed that

$$H \cap (T^* \otimes N_3) = \{0\},$$

and then deduced the relation (1.48) for the complex quadric X from Proposition 5.13; thus, we have the following result:

PROPOSITION 5.14. *For $n \geq 3$, we have*

$$H \cap (T^* \otimes \tilde{B}) = \{0\}.$$

From Proposition 5.14 and Theorem 1.18, we deduce the exactness of the sequence (1.24) for the complex quadric $X = Q_n$.

§7. Multiplicities

In this section, we shall suppose that $n \geq 3$. Let $\mathfrak{g}$ and $\mathfrak{k}$ denote the complexifications of the Lie algebras $\mathfrak{g}_0$ and $\mathfrak{k}_0$ of $G = SO(n+2)$ and its subgroup K, respectively. Let $\Gamma = \hat{G}$ and $\hat{K}$ be the duals of the groups G and K, respectively.

For $\mu \in \mathbb{C}$, we set

$$L(\mu) = \begin{pmatrix} 0 & -i\mu \\ i\mu & 0 \end{pmatrix}.$$

If $m \geq 1$, for $\theta_0, \theta_1, \ldots, \theta_m \in \mathbb{R}$, $\mu_0, \mu_1, \ldots, \mu_m \in \mathbb{C}$, we consider the $2(m+1) \times 2(m+1)$ matrices

$$R(\theta_0, \theta_1, \ldots, \theta_m) = \begin{pmatrix} R(\theta_0) & 0 & \cdots & 0 \\ 0 & R(\theta_1) & \cdots & 0 \\ \vdots & \vdots & \ddots & \vdots \\ 0 & 0 & \cdots & R(\theta_m) \end{pmatrix},$$

$$L(\mu_0, \mu_1, \ldots, \mu_m) = \begin{pmatrix} L(\mu_0) & 0 & \cdots & 0 \\ 0 & L(\mu_1) & \cdots & 0 \\ \vdots & \vdots & \ddots & \vdots \\ 0 & 0 & \cdots & L(\mu_m) \end{pmatrix}.$$

We define a subgroup $\mathbb{T}$ of G and a Lie subalgebra $\mathfrak{t}$ of $\mathfrak{g}$ as follows. If $n = 2m$, with $m \geq 2$, the subgroup $\mathbb{T}$ consists of all $(n+2) \times (n+2)$ matrices $R(\theta_0, \theta_1, \ldots, \theta_m)$, with $\theta_0, \theta_1, \ldots, \theta_m \in \mathbb{R}$, and the Lie algebra $\mathfrak{t}$ consists of all $(n+2) \times (n+2)$ matrices $L(\mu_0, \mu_1, \ldots, \mu_m)$, with $\mu_0, \mu_1, \ldots, \mu_m \in \mathbb{C}$. If $n = 2m+1$, with $m \geq 1$, the subgroup $\mathbb{T}$ consists of all $(n+2) \times (n+2)$ matrices

$$\begin{pmatrix} R(\mu_0, \mu_1, \ldots, \mu_m) & 0 \\ 0 & 1 \end{pmatrix},$$

with $\theta_0, \theta_1, \ldots, \theta_m \in \mathbb{R}$, and the Lie algebra $\mathfrak{t}$ consists of all $(n+2) \times (n+2)$ matrices

$$L'(\mu_0, \mu_1, \ldots, \mu_m) = \begin{pmatrix} L(\mu_0, \mu_1, \ldots, \mu_m) & 0 \\ 0 & 0 \end{pmatrix},$$

with $\mu_0, \mu_1, \ldots, \mu_m \in \mathbb{C}$.

The subgroup $\mathbb{T}$ is a maximal torus of G, and $\mathfrak{t}$ is the complexification of the Lie algebra $\mathfrak{t}_0$ of $\mathbb{T}$ and is a Cartan subalgebra of the semi-simple Lie algebras $\mathfrak{g}$ and $\mathfrak{k}$. If $n = 2m$ (resp. $n = 2m+1$), for $0 \leq j \leq m$, the linear form λ_j on $\mathfrak{t}$, which sends the element $L(\mu_0, \mu_1, \ldots, \mu_m)$ (resp. the element $L'(\mu_0, \mu_1, \ldots, \mu_m)$) of $\mathfrak{t}$, with $\mu_0, \mu_1, \ldots, \mu_m \in \mathbb{C}$, into μ_j, is purely imaginary on $\mathfrak{t}_0$; for $0 \leq i \leq m-1$, we write $\alpha_i = \lambda_i - \lambda_{i+1}$.

We first suppose that $n = 2m$, with $m \geq 2$. We set $\alpha_m = \lambda_{m-1} + \lambda_m$. We choose Weyl chambers of $(\mathfrak{g}, \mathfrak{t})$ and $(\mathfrak{k}, \mathfrak{t})$ for which the system of simple roots of $\mathfrak{g}$ and $\mathfrak{k}$ are equal to $\{\alpha_0, \alpha_1, \ldots, \alpha_m\}$ and $\{\alpha_1, \ldots, \alpha_m\}$, respectively. The highest weight of an irreducible G-module (resp. K-module) is a linear form

$$c_0 \lambda_0 + c_1 \lambda_1 + \cdots + \varepsilon c_m \lambda_m$$

on $\mathfrak{t}$, where $\varepsilon = \pm 1$ and $c_0, c_1, \ldots, c_m$ are integers satisfying

(5.103) $c_0 \geq c_1 \geq \cdots \geq c_m \geq 0$ (resp. $c_1 \geq \cdots \geq c_m \geq 0$).

The equivalence class of such a G-module (resp. K-module) is determined by this weight. In this case, we identify Γ (resp. $\hat{K}$) with the set of all such linear forms on $\mathfrak{t}$.

We next suppose that $n = 2m + 1$, with $m \geq 1$. We set $\alpha_m = \lambda_m$. We choose Weyl chambers of $(\mathfrak{g}, \mathfrak{t})$ and $(\mathfrak{k}, \mathfrak{t})$ for which the system of simple roots of $\mathfrak{g}$ and $\mathfrak{k}$ are equal to $\{\alpha_0, \alpha_1, \ldots, \alpha_m\}$ and $\{\alpha_1, \ldots, \alpha_m\}$, respectively. The highest weight of an irreducible G-module (resp. K-module) is a linear form

$$c_0 \lambda_0 + c_1 \lambda_1 + \cdots + c_m \lambda_m$$

on $\mathfrak{t}$, where $c_0, c_1, \ldots, c_m$ are integers satisfying the inequalities (5.103). The equivalence class of such a G-module (resp. K-module) is determined by this weight. In this case, we identify Γ (resp. $\hat{K}$) with the set of all such linear forms on $\mathfrak{t}$.

For $r, s \geq 0$, we consider the elements

$$\gamma_{r,s} = (2r + s)\lambda_0 + s\lambda_1, \qquad \gamma'_{r,s} = (2r + s + 1)\lambda_0 + s\lambda_1$$

of Γ. The highest weight of the G-module $\mathfrak{g}$ is $\gamma_1 = \gamma_{0,1}$.

The Lie algebra $\mathfrak{k}$ admits the decomposition

$$\mathfrak{k} = \mathfrak{k}_1 \oplus \mathfrak{z},$$

where $\mathfrak{z}$ is the one-dimensional center of $\mathfrak{k}$ and $\mathfrak{k}_1$ is a subalgebra of $\mathfrak{k}$ isomorphic to $\mathfrak{so}(n, \mathbb{C})$. In fact, a matrix

$$\begin{pmatrix} A & 0 \\ 0 & B \end{pmatrix}$$

of $\mathfrak{k}$, where $A \in \mathfrak{so}(2, \mathbb{C})$ and $B \in \mathfrak{so}(n, \mathbb{C})$, can be written as the sum of the two matrices

$$\begin{pmatrix} A & 0 \\ 0 & 0 \end{pmatrix}, \qquad \begin{pmatrix} 0 & 0 \\ 0 & B \end{pmatrix},$$

which belong to $\mathfrak{z}$ and $\mathfrak{k}_1$, respectively. The complexification $\mathfrak{p}$ of $\mathfrak{p}_0$ admits the decomposition

$$\mathfrak{p} = \mathfrak{p}_- \oplus \mathfrak{p}_+,$$

where $\mathfrak{p}_-$ and $\mathfrak{p}_+$ are the eigenspaces of the endomorphism $\operatorname{Ad} j$ of $\mathfrak{p}$ corresponding to the eigenvalues $+i$ and $-i$, respectively. Since j belongs to the center of K, this decomposition of $\mathfrak{p}$ is invariant under the action of K on $\mathfrak{p}$. We thus obtain the K-invariant decomposition

(5.104) $$\mathfrak{g} = \mathfrak{k}_1 \oplus \mathfrak{z} \oplus \mathfrak{p}_- \oplus \mathfrak{p}_+$$

of the Lie algebra $\mathfrak{g}$. The K-modules $\mathfrak{z}$, $\mathfrak{p}_-$ and $\mathfrak{p}_+$ are irreducible and their highest weights are equal to 0, $-\lambda_0 + \lambda_1$ and $\lambda_0 + \lambda_1$, respectively. If $n \neq 4$, the K-module $\mathfrak{k}_1$ is irreducible; its highest weight is equal to λ_1 when $n = 3$, and to $\lambda_1 + \lambda_2$ when $n \geq 5$. When $n = 4$, the Lie algebra admits the decomposition

(5.105) $$\mathfrak{k}_1 = \mathfrak{k}_1^+ \oplus \mathfrak{k}_1^-,$$

where $\mathfrak{k}_1^+$ and $\mathfrak{k}_1^-$ are simple subalgebras of $\mathfrak{k}_1$ isomorphic to $\mathfrak{so}(3, \mathbb{C})$; these factors $\mathfrak{k}_1^+$ and $\mathfrak{k}_1^-$ are irreducible K-modules whose highest weights are equal to $\lambda_1 + \lambda_2$ and $\lambda_1 - \lambda_2$, respectively.

We consider the subgroup

$$K' = G \cap (O(2) \times O(n))$$

of G, which we introduced in §3 and which contains the subgroup K. The decomposition (5.104) gives us the K'-invariant decomposition

(5.106) $$\mathfrak{g} = \mathfrak{k}_1 \oplus \mathfrak{z} \oplus \mathfrak{p}$$

of the Lie algebra $\mathfrak{g}$; in fact, the K'-modules $\mathfrak{z}$, $\mathfrak{k}_1$ and $\mathfrak{p}$ are irreducible.

We recall that the point x_0 of $\widetilde{G}_{2,n}^{\mathbb{R}}$ corresponding to the vector $e_0 \wedge e_1$ is identified with the point b of $X = Q_n$ defined in §2. We consider V_{x_0} as an $O(2)$-module and W_{x_0} as an $O(n)$-module. Let $\{\alpha_0, \alpha_1\}$ be a basis of $V_{x_0}^*$ dual to the basis $\{e_0, e_1\}$ of V_{x_0}. Since $\alpha_0 \wedge \alpha_1$ is a basis of the one-dimensional vector space $\bigwedge^2 V_{x_0}^*$, we see that it is a trivial $SO(2)$-module, but is not trivial as an $O(2)$-module. There are natural isomorphisms $\phi : \mathfrak{z} \to \bigwedge^2 V_{\mathbb{C}, x_0}^*$ of $O(2)$-modules and $\phi : \mathfrak{k}_1 \to \bigwedge^2 W_{\mathbb{C}, x_0}^*$ of $O(n)$-modules, which are both also isomorphisms of K'-modules.

We now suppose that $n = 4$. As we saw in §5, the orientation of the 2-plane V_{x_0} determines an orientation of the space W_{x_0} and a Hodge operator

$$* : \bigwedge^2 W_{x_0}^* \to \bigwedge^2 W_{x_0}^*;$$

if B is an element of $O(4)$, we easily verify that

$$(5.107) \qquad\qquad * B = (\det B) \cdot B *$$

as endomorphisms of $\bigwedge^2 W_{x_0}^*$. In fact, we have the decomposition

$$\bigwedge^2 W_{\mathbb{C},x_0}^* = \bigwedge^+ W_{\mathbb{C},x_0}^* \oplus \bigwedge^- W_{\mathbb{C},x_0}^*$$

of $\bigwedge^2 W_{\mathbb{C},x_0}^*$ into irreducible $SO(4)$-submodules, which are the eigenspaces of the involution $*$. The isomorphism $\phi : \mathfrak{k}_1 \to \bigwedge^2 W_{\mathbb{C},x_0}^*$ induces isomorphisms $\phi : \mathfrak{k}_1^+ \to \bigwedge^+ W_{\mathbb{C},x_0}^*$ and $\phi : \mathfrak{k}_1^- \to \bigwedge^- W_{\mathbb{C},x_0}^*$ of $SO(4)$-modules.

We no longer suppose that $n = 4$. Since $\bigwedge^2 V_{x_0}^*$ is a trivial K-module and since the $SO(n)$-modules $\mathfrak{k}_1$ and $\bigwedge^2 W_{\mathbb{C},x_0}^*$ are isomorphic, the K-modules $\mathfrak{k}_1$ and $(\bigwedge^2 V^* \otimes \bigwedge^2 W^*)_{\mathbb{C},x_0}$ are isomorphic. Hence from the last of the equalities (5.66), it follows that the fiber at the point b of the vector bundle $(S^2 T^*)_{\mathbb{C}}^{+-}$ is isomorphic to $\mathfrak{k}_1$ as a K-module.

The fibers at the point b of X of the vector bundles T' and T'' and of the vector bundles appearing in the decomposition (5.27) of $S^2 T_{\mathbb{C}}^*$ are irreducible K-modules, except for the fiber of $(S^2 T^*)_{\mathbb{C}}^{+-}$ which is irreducible only when $n \neq 4$. Using the description of these K-modules and of the action of K on T_b given in §2, we see that the highest weight of $(S^2 T^*)_{\mathbb{C},b}^{+-}$ is equal to λ_1 when $n = 3$, and to $\lambda_1 + \lambda_2$ when $n \geq 5$, and that the highest weights of the other such irreducible K-modules are given by the following table:

K-module	Highest weight	K-module	Highest weight
T_b'	$-\lambda_0 + \lambda_1$	T_b''	$\lambda_0 + \lambda_1$
L_b'	$2\lambda_0$	L_b''	$-2\lambda_0$
$(S^{2,0}T^*)_b^{\perp}$	$2\lambda_0 + 2\lambda_1$	$(S^{0,2}T^*)_b^{\perp}$	$-2\lambda_0 + 2\lambda_1$
$\{g\}_{\mathbb{C},a}$	0	$(S^2 T^*)_{0\mathbb{C},b}^{++}$	$2\lambda_1$

When $n = 4$, the fibers at the point b of X of the vector bundles $F_{\mathbb{C}}^+$ and $F_{\mathbb{C}}^-$ appearing in the decomposition (5.70) are irreducible K-modules whose highest weights are $\lambda_1 + \lambda_2$ and $\lambda_1 - \lambda_2$, respectively.

We recall that the equalities

$$E_1 = (S^2 T^*)_0^{++}, \qquad E_{2,\mathbb{C}} = L_{\mathbb{C}} = L' \oplus L'',$$

$$E_{3,\mathbb{C}} = (S^2 T^*)_{\mathbb{C}}^{-\perp} = (S^{2,0} T^*)^{\perp} \oplus (S^{0,2} T^*)^{\perp}$$

hold. Hence from the previous discussion, we obtain the following result:

LEMMA 5.15. *Let X be the complex quadric Q_n, with $n \geq 3$.*
(i) *We have*
$$\operatorname{Hom}_K(\mathfrak{g}, E_{j,\mathbb{C},b}) = \{0\},$$
for $j = 1, 2, 3$.
(ii) *If $n \neq 4$, we have*
$$\dim \operatorname{Hom}_K(\mathfrak{g}, (S^2 T^*)^{+-}_{\mathbb{C},b}) = 1.$$

(iii) *If $n = 4$, we have*
$$\dim \operatorname{Hom}_K(\mathfrak{g}, (S^2 T^*)^{+-}_{\mathbb{C},b}) = 2.$$

From Lemma 5.15 and the decomposition (5.26), we deduce that

(5.108) $$\dim \operatorname{Hom}_K(\mathfrak{g}, S_0^2 T^*_{\mathbb{C},b}) = 1$$

when $n \neq 4$, and that

(5.109) $$\dim \operatorname{Hom}_K(\mathfrak{g}, S_0^2 T^*_{\mathbb{C},b}) = 2$$

when $n = 4$.

According to the decomposition (5.106), the mapping

$$\phi : \mathfrak{g} \to (\textstyle\bigwedge^2 V^* \otimes \bigwedge^2 W^*)_{\mathbb{C},x_0}$$

is well-defined by

$$\phi(u) = (\alpha_0 \wedge \alpha_1) \otimes \phi(u_1),$$

where $u = u_1 + u_2$ is an element of $\mathfrak{g}$, with $u_1 \in \mathfrak{k}_1$ and $u_2 \in \mathfrak{z} \oplus \mathfrak{p}$; it is clearly a morphism of K-modules, but is not a morphism of K'-modules. When $n \neq 4$, by Lemma 5.15,(ii) this morphism ϕ is the generator of the space $\operatorname{Hom}_K(\mathfrak{g}, (\bigwedge^2 V^* \otimes \bigwedge^2 W^*)_{\mathbb{C},x_0})$, and so we have

$$\operatorname{Hom}_{K'}(\mathfrak{g}, (\textstyle\bigwedge^2 V^* \otimes \bigwedge^2 W^*)_{\mathbb{C},x_0}) = \{0\}.$$

On the other hand, when $n = 4$, by (5.107) we see that the mapping

$$\psi : \mathfrak{g} \to (\textstyle\bigwedge^2 V^* \otimes \bigwedge^2 W^*)_{\mathbb{C},x_0}$$

defined by

$$\psi(u) = (\alpha_0 \wedge \alpha_1) \otimes * \phi(u_1),$$

where $u = u_1 + u_2$ is an element of $\mathfrak{g}$, with $u_1 \in \mathfrak{k}_1$ and $u_2 \in \mathfrak{z} \oplus \mathfrak{p}$, is a morphism of K'-modules. When $n = 4$, by Lemma 5.15,(iii) we see that the space

$$\operatorname{Hom}_K(\mathfrak{g}, (\textstyle\bigwedge^2 V^* \otimes \bigwedge^2 W^*)_{\mathbb{C},x_0})$$

is generated by the morphisms ϕ and ψ; it follows that its subspace

$$\mathrm{Hom}_{K'}(\mathfrak{g}, (\textstyle\bigwedge^2 V^* \otimes \bigwedge^2 W^*)_{\mathbb{C},x_0})$$

is one-dimensional and is generated by the mapping ψ.

We consider the real Grassmannian $Y = G_{2,n}^{\mathbb{R}}$ and the natural projection $\varpi : X \to Y$. We recall that the isotropy group of the point $\varpi(b)$ is equal to K'; moreover the mapping $\varpi^* : S^2 T_{Y,\varpi(b)}^* \to S^2 T_b^*$ is an isomorphism of K-modules and the fiber at the point $\varpi(b)$ of the vector bundle $(S^2 T_Y^*)_{\mathbb{C}}^{+-}$ is isomorphic to $(\bigwedge^2 V^* \otimes \bigwedge^2 W^*)_{\mathbb{C},x_0}$ as a K'-module. By restriction, the isomorphism of G-modules $\varpi^* : C^\infty(Y, S^2 T_{Y,\mathbb{C}}^*) \to C^\infty(S^2 T_{\mathbb{C}}^*)^{\mathrm{ev}}$ induces isomorphisms of G-modules $C_\gamma^\infty(Y, S^2 T_{Y,\mathbb{C}}^*) \to C_\gamma^\infty(S^2 T_{\mathbb{C}}^*)^{\mathrm{ev}}$ and $C_\gamma^\infty(Y, (S^2 T_Y^*)_{\mathbb{C}}^{+-}) \to C_\gamma^\infty((S^2 T^*)_{\mathbb{C}}^{+-})^{\mathrm{ev}}$, for all $\gamma \in \Gamma$, given by (2.19). According to (2.21), since Y is an irreducible symmetric space which is not Hermitian, we have

$$\mathrm{Hom}_{K'}(\mathfrak{g}, \mathbb{C}) = \{0\}.$$

From these remarks, Lemma 5.15 and the decomposition (5.38), we obtain the following:

LEMMA 5.16. *Let Y be the real Grassmannian $G_{2,n}^{\mathbb{R}}$, with $n \geq 3$, and let y_0 be the point $\varpi(b)$ of Y.*

(i) *If $n \neq 4$, we have*

$$\mathrm{Hom}_{K'}(\mathfrak{g}, S^2 T_{Y,\mathbb{C},y_0}^*) = \{0\}.$$

(ii) *If $n = 4$, we have*

$$\dim \mathrm{Hom}_{K'}(\mathfrak{g}, S^2 T_{Y,\mathbb{C},y_0}^*) = \dim \mathrm{Hom}_{K'}(\mathfrak{g}, (S^2 T_Y^*)_{\mathbb{C},y_0}^{+-}) = 1.$$

According to the equalities (5.108) and (5.109) and Proposition 2.40, since the symmetric space X is irreducible and is not equal to a simple Lie group, we see that $E(X)$ vanishes when $n \neq 4$, and that $E(X)$ is isomorphic to the G-module $\mathfrak{g}_0$ when $n = 4$. By Lemma 5.15 and the Frobenius reciprocity theorem, we see that

$$C_{\gamma_{0,1}}^\infty(S_0^2 T_{\mathbb{C}}^*) = C_{\gamma_{0,1}}^\infty((S^2 T^*)_{\mathbb{C}}^{+-});$$

if $n \neq 4$, then Lemma 5.15 tells us that $C_{\gamma_{0,1}}^\infty(S_0^2 T_{\mathbb{C}}^*)$ is an irreducible G-module.

If $E(X)^{\mathrm{ev}}$ denotes the G-submodule $E(X) \cap C^\infty(S^2 T^*)^{\mathrm{ev}}$ of $E(X)$, the projection ϖ induces an isomorphism of G-modules $\varpi^* : E(Y) \to E(X)^{\mathrm{ev}}$ given by (2.7). Thus when $n \neq 4$, the vanishing of $E(X)$ implies that $E(Y) = \{0\}$. Since the symmetric space Y is irreducible, is not equal to a simple Lie group and is not Hermitian, from Lemma 5.16 and Proposition 2.40 we again obtain the vanishing of $E(Y)$ when $n \neq 4$; moreover when $n = 4$, we see that $E(Y)$ is isomorphic to the G-module $\mathfrak{g}_0$, and so we have the equality

$$E(X) = E(X)^{\mathrm{ev}}.$$

By Lemma 5.16 and the Frobenius reciprocity theorem, when $n \neq 4$ we see that $C^\infty_{\gamma_{0,1}}(Y, S^2 T^*_{Y,\mathbb{C}})$ and $C^\infty_{\gamma_{0,1}}(S^2 T^*_\mathbb{C})^{\mathrm{ev}}$ vanish; moreover when $n = 4$, the G-module $C^\infty_{\gamma_{0,1}}(Y, S^2 T^*_{Y,\mathbb{C}})$ is equal to $C^\infty_{\gamma_{0,1}}(Y, (S^2 T^*)^{+-}_{Y,\mathbb{C}})$ and is irreducible. When $n = 4$, from the relations (2.29) we therefore obtain the equality (5.112) of the next proposition; in turn, this equality implies that the relations (5.111) of the next proposition hold and that $C^\infty_{\gamma_{0,1}}(S^2 T^*_\mathbb{C})^{\mathrm{ev}}$ is an irreducible G-module. We have thus proved the following result:

PROPOSITION 5.17. *Let X be the complex quadric Q_n and Y be the real Grassmannian $G^\mathbb{R}_{2,n}$, with $n \geq 3$. If $n \neq 4$, we have*

$$E(X) = \{0\}, \qquad E(Y) = \{0\}.$$

If $n = 4$, the spaces $E(X)$ and $E(Y)$ are irreducible $SO(6)$-modules isomorphic to $\mathfrak{g}_0 = \mathfrak{so}(6)$; moreover, we have

$$(5.110) \qquad\qquad E(X) \subset C^\infty((S^2 T^*)^{+-})^{\mathrm{ev}},$$

$$(5.111) \qquad E(X)_\mathbb{C} = C^\infty_{\gamma_{0,1}}(S^2 T^*_\mathbb{C})^{\mathrm{ev}} = C^\infty_{\gamma_{0,1}}((S^2 T^*)^{+-}_\mathbb{C})^{\mathrm{ev}},$$

$$(5.112) \qquad\qquad E(Y)_\mathbb{C} = C^\infty_{\gamma_{0,1}}(Y, S^2 T^*_{Y,\mathbb{C}}).$$

When $n \neq 4$, the vanishing of the space $E(X)$ is also given by Theorem 1.22 (see Koiso [41] and [42]).

From the branching law for $G = SO(n + 2)$ and K described in Theorems 1.1 and 1.2 of [54], using the computation of the highest weights of the irreducible K-modules given above we obtain the following two propositions:

PROPOSITION 5.18. *Let X be the complex quadric Q_n, with $n \geq 3$. For $\gamma \in \Gamma$, the G-modules $C^\infty_\gamma(L')$ and $C^\infty_\gamma(L'')$ vanish unless $\gamma = \gamma_{r,s}$, with $r \geq 1$ and $s \geq 0$.*

PROPOSITION 5.19. *Let X be the complex quadric Q_n, with $n \geq 3$. For $r, s \geq 0$, the non-zero multiplicities of the G-modules $C^\infty_{\gamma_{r,s}}(F)$, where F is a homogeneous vector bundle over X equal either to T' or T'' or to one of the vector bundles appearing in the decomposition (5.27) of $S^2 T^*_\mathbb{C}$,* *are given by the following table:*

F	Conditions on r, s	Mult $C^{\infty}_{\gamma_{r,s}}(F)$
T'	$r + s \geq 1$	2 if $r, s \geq 1$
T''		1 otherwise
L'	$r \geq 1, s \geq 0$	1
L''		
$(S^{2,0}T^*)^{\perp}$	$r + s \geq 2$	2 if $r \geq 2, s = 1$
$(S^{2,0}T^*)^{\perp}$		or $r = 1, s \geq 2$
		3 if $r, s \geq 2$
		1 otherwise
$\{g\}_{\mathbb{C}}$	$r, s \geq 0$	1
$(S^2T^*)^{++}_{0\mathbb{C}}$	$r \geq 1$ or $s \geq 2$	2 if $r \geq 1, s = 1$
		3 if $r \geq 1, s \geq 2$
		1 otherwise
$(S^2T^*)^{+-}_{\mathbb{C}}$	$r \geq 0, s \geq 1$	1 if $n \neq 4$
		2 if $n = 4$

We note that, if F is one of the homogeneous sub-bundles of $S^2T^*_{\mathbb{C}}$ considered in the preceding proposition, the multiplicity of the G-module $C^{\infty}_{\gamma_{0,1}}(F)$ is also given by Lemma 5.15.

The Casimir element of $\mathfrak{g}_0$ operates by a scalar c_{γ} on an irreducible G-module which is a representative of $\gamma \in \Gamma$. We set $\lambda_{\gamma} = 4nc_{\gamma}$, for $\gamma \in \Gamma$. According to (5.5) and §7, Chapter II, we therefore see that

$$\Delta u = \lambda_{\gamma} u,$$

for all $u \in C^{\infty}_{\gamma}(S^pT^*_{\mathbb{C}})$. We shall write $\lambda_{r,s} = \lambda_{\gamma_{r,s}}$. Since $c_{\gamma_1} = 1$, we see that $\lambda_{0,1} = 4n$.

Let f be a complex-valued function on $\mathbb{C}^{n+2}$, whose restriction to the unit sphere S^{2n+3} of $\mathbb{C}^{n+2}$ is invariant under $U(1)$. As we saw in §4, Chapter III, the restriction of f to S^{2n+3} induces by passage to the quotient a function on $\mathbb{CP}^{n+1}$, which we also denote by f and whose restriction to X we denote by $\tilde{f}$. If ϕ is an element of $G = SO(n+2)$, the function $\phi^* f$ on $\mathbb{C}^{n+2}$ also gives rise to a function on $\mathbb{CP}^{n+1}$, whose restriction to X is equal to $\phi^* \tilde{f}$. We consider the standard complex coordinate system $\zeta = (\zeta_0, \zeta_1, \ldots, \zeta_{n+1})$ of $\mathbb{C}^{n+2}$, and the functions

$$f_{0,1}(\zeta) = (\zeta_0 + i\zeta_1)(\bar{\zeta}_2 + i\bar{\zeta}_3) - (\zeta_2 + i\zeta_3)(\bar{\zeta}_0 + i\bar{\zeta}_1),$$
$$f_{1,0}(\zeta) = (\zeta_0 + i\zeta_1)(\bar{\zeta}_0 + i\bar{\zeta}_1)$$

on $\mathbb{C}^{n+2}$. We also consider the function ρ on $\mathbb{C}^{n+2}$ given by

$$\rho(\zeta) = \zeta_0 \bar{\zeta}_0$$

and the space $\mathcal{H}$ of functions on $\mathbb{C}^{n+2}$ generated over $\mathbb{C}$ by the set

$$\{\,\zeta_j \bar{\zeta}_k - \zeta_k \bar{\zeta}_j \mid 0 \le j < k \le n+1\,\}.$$

If r, s are integers ≥ 0, the functions $f_{r,s} = f_{1,0}^r \cdot f_{0,1}^s$, the function ρ and the elements of $\mathcal{H}$ are all invariant under $U(1)$. The function $f_{0,1}$ and the function f', defined by

$$f'(\zeta) = (\zeta_0 + i\zeta_1)\bar{\zeta}_{n+1} - \zeta_{n+1}(\bar{\zeta}_0 + i\bar{\zeta}_1),$$

belong to $\mathcal{H}$. We shall consider the functions $\tilde{f}_{r,s}$ and $\tilde{f}'$ on X, and we set $\tilde{f}_{r,s} = 0$ when $r < 0$ or $s < 0$. The functions $f_{r,s}$ and the functions of $\mathcal{H}$ clearly satisfy the relations

$$(5.113) \qquad\qquad \tau^* \tilde{f}_{r,s} = (-1)^s \tilde{f}_{r,s}, \qquad \tau^* \tilde{f} = -\tilde{f},$$

for $f \in \mathcal{H}$. Clearly, the space $\tilde{\mathcal{H}} = \{\tilde{f} \mid f \in \mathcal{H}\}$ is a submodule of the G-module $C^\infty(X)^{\mathrm{odd}}$ and is isomorphic to the irreducible G-module $\mathfrak{g}$. Its subspace

$$\{\tilde{f} \mid f \in \mathcal{H}, \ f = \bar{f}\}$$

is isomorphic as a real G-module to the subalgebra $\mathfrak{g}_0$ of $\mathfrak{g}$.

Let ϕ be the element of G defined by

$$\phi(\zeta)_1 = \zeta_2, \quad \phi(\zeta)_2 = \zeta_3, \quad \phi(\zeta)_3 = \zeta_{n+1}, \quad \phi(\zeta)_{n+1} = \zeta_1$$

and $\phi(\zeta)_j = \zeta_j$ for $j = 0$ or $3 < j \le n$. For $r, s \ge 0$, we consider the function $f'_{r,s} = \phi^* f_{r,s}$ and $f'' = \phi^* f'$ on $\mathbb{C}^{n+2}$. We also consider the objects introduced in §4. Then on the open subset V of $\mathbb{CP}^{n+1}$, we see that

$$\rho(z) = \frac{1}{1 + |z|^2},$$

for $z \in V$, and that the functions $f'_{r,s}$ and f'' are determined by

$$f'_{1,0}(z) = \rho(z)(1 + iz_2)(1 + i\bar{z}_2),$$

$$f'_{0,1}(z) = \rho(z)((1 + iz_2)(\bar{z}_3 + i\bar{z}_{n+1}) - (z_3 + iz_{n+1})(1 + i\bar{z}_2)),$$

$$f''(z) = \rho(z)((1 + iz_2)\bar{z}_1 - z_1(1 + i\bar{z}_2)),$$

for $z \in V$. Then we have $\tilde{f}'_{r,s} = \phi^* \tilde{f}_{r,s}$, and we easily see that $\tilde{f}'_{r,s}(a) = 2^{-r}$ and $\tilde{f}''(a) = 0$. We also have

$$(5.114) \qquad d\tilde{f}'_{r,s} = \frac{1}{2^{r+1}} \left(i(2r+s)(dz_2 + d\bar{z}_2) - s(dz_3 - d\bar{z}_3) \right)$$

at the point a; thus by (5.50), we obtain the equality

$$(5.115) \qquad (d\tilde{f}'_{r,s})^{\sharp} = \frac{1}{2^{r-1}} \left(i(2r+s)(\eta_2 + \bar{\eta}_2) + s(\eta_3 - \bar{\eta}_3) \right)$$

at the point a. We set

$$c^j_{r,s} = 2(r+s), \qquad c^2_{r,s} = (2r+s)(2r+s+1), \qquad c^3_{r,s} = 2r + s^2 + s,$$

for $j = 1$ or $4 \le j \le n$. Using formulas (5.56) and (5.58), for $1 \le j, k \le n$ and all integers $r, s \ge 0$, we verify that the relations

$$(5.116) \qquad (\text{Hess}\,\tilde{f}'_{1,0})(\eta_j, \eta_k) = (\text{Hess}\,\tilde{f}'_{1,0})(\bar{\eta}_j, \bar{\eta}_k) = -\frac{1}{4}\,\delta_{jk},$$

$$(5.117) \qquad (\text{Hess}\,\tilde{f}'_{r,s})(\eta_1, \eta_1) = (\text{Hess}\,\tilde{f}'_{r,s})(\bar{\eta}_1, \bar{\eta}_1) = -\frac{r}{2^{r+1}},$$

$$(5.118) \qquad (\pi_{++}\text{Hess}\,\tilde{f}'_{r,s})(\eta_j, \bar{\eta}_k) = -\frac{1}{2^{r+2}}\,c^j_{r,s}\,\delta_{jk},$$

$$(5.119) \quad (\pi_{+-}\text{Hess}\,\tilde{f}'_{r,s})(\eta_j, \bar{\eta}_k) = \frac{i}{2^{r+2}}\,s(2r+s+1)(\delta_{j2}\delta_{k3} - \delta_{j3}\delta_{k2}),$$

$$(5.120) \qquad (\pi_{+-}\text{Hess}\,\tilde{f}'')(\eta_2, \bar{\eta}_1) = \frac{i}{2}$$

hold at the point a. From (5.115) and (5.120), we obtain

$$(5.121) \qquad \phi^* ((d\tilde{f}_{1,0})^{\sharp} \lrcorner \pi_{+-}\text{Hess}\,\tilde{f}')(\bar{\eta}_1) = -1$$

at the point a. Using the formulas (5.57), we see that

$$(5.122) \qquad \begin{aligned} (\nabla\text{Hess}\,\tilde{f}'_{1,0})(\bar{\eta}_1, \eta_1, \eta_j)(a) &= (\nabla\text{Hess}\,\tilde{f}'_{1,0})(\eta_1, \bar{\eta}_1, \bar{\eta}_j)(a) \\ &= (\nabla\text{Hess}\,\tilde{f}'_{1,0})(\eta_2, \eta_2, \eta_j)(a) = 0, \end{aligned}$$

for $j = 2, 3$.

When $r \ge 1$ or when $s \ge 2$, the section $\pi_{++}\text{Hess}\,\tilde{f}_{r,s}$ cannot be written in the form $c\tilde{f}_{r,s}g$, for some $c \in \mathbb{C}$. Indeed, if this were the case, the expression $(\pi_{++}\text{Hess}\,\tilde{f}'_{r,s})(\eta_j, \bar{\eta}_j)(a)$ would be independent of the index j, for $1 \le j \le 3$; our assertion is now a consequence of the formulas (5.118).

According to [50], for $r, s \geq 0$, the function $\tilde{f}_{r,s}$ on X is a highest weight vector of the irreducible G-module $\mathcal{H}_{r,s} = C^\infty_{\gamma_{r,s}}(X)$. According to §7, Chapter II, we know that $\mathcal{H}_{r,s}$ is the eigenspace of the Laplacian Δ acting on $C^\infty(X)$ with eigenvalue $\lambda_{r,s}$. In fact, since $f_{0,1}$ belongs to the space $\mathcal{H}$ and since the submodule $\tilde{\mathcal{H}}$ is isomorphic to the irreducible G-module $\mathfrak{g}$, the eigenspace $\mathcal{H}_{0,1}$ is equal to the space $\tilde{\mathcal{H}}$. It also follows that the space $\mathcal{H}_{r,s}$ is invariant under conjugation. Since $\mathcal{H}_{r,s}$ is an irreducible G-module, by (5.113) we see that

$$C^\infty_{\gamma_{r,2s}}(X) = C^\infty_{\gamma_{r,2s}}(X)^{\mathrm{ev}}, \qquad C^\infty_{\gamma_{r,2s+1}}(X) = C^\infty_{\gamma_{r,2s+1}}(X)^{\mathrm{odd}},$$

for $r, s \geq 0$. We also know that $C^\infty_\gamma(X) = 0$, whenever $\gamma \in \Gamma$ is not of the form $\gamma_{r,s}$, with $r, s \geq 0$.

Since the function $\tilde{f}'_{r,s} = \phi^* \tilde{f}_{r,s}$ also belongs to $\mathcal{H}_{r,s}$, using (5.50) and (5.118) we easily verify that the eigenvalue of the function $\tilde{f}'_{r,s}$ is equal to

$$(5.123) \qquad \lambda_{r,s} = 2\big(2(n-2)(r+s) + (2r+s)(2r+s+2) + s^2\big).$$

Let $r, s \geq 0$ be given integers and γ be an element of Γ; then $\gamma + \gamma_{r,s}$ also belongs to Γ. If $\gamma = \gamma_{r',s'}$, with $r', s' \geq 0$, then we have $\gamma + \gamma_{r,s} = \gamma_{r+r',s+s'}$. Let u be a highest weight vector of the G-module $C^\infty_\gamma(S^p T^*_\mathbb{C})$. According to §7, Chapter II, we know that u is a real-analytic section of $S^p T^*_\mathbb{C}$. Hence the section $\tilde{f}_{r,s} u$ of $S^p T^*_\mathbb{C}$ is non-zero; clearly, this element of $C^\infty_\gamma(S^p T^*_\mathbb{C})$ is of weight $\gamma + \gamma_{r,s}$. Since G is connected, the section $\tilde{f}_{r,s} u$ is a highest weight vector of $C^\infty_{\gamma + \gamma_{r,s}}(S^p T^*_\mathbb{C})$.

According to Proposition 5.19, since $\mathcal{H}_{r,s}$ is an irreducible G-module, we have

$$(5.124) \qquad\qquad C^\infty_{\gamma_{r,s}}(\{g\}_\mathbb{C}) = \mathcal{H}_{r,s} \cdot g,$$

and that the section $\tilde{f}_{r,s} g$ is a highest weight vector of this module; moreover, we know that

$$(5.125) \qquad C^\infty_{\gamma_{0,1}}(S^2 T^*_\mathbb{C}) = C^\infty_{\gamma_{0,1}}(\{g\}_\mathbb{C}) \oplus C^\infty_{\gamma_{0,1}}((S^2 T^*)^{+-}_\mathbb{C}).$$

Since $\mathrm{Hess} : C^\infty(X) \to C^\infty(S^2 T^*_\mathbb{C})$ is a homogeneous differential operator and $\mathcal{H}_{r,s}$ is an irreducible G-module, by (5.118) and (5.119) we know that the space $\pi_{+-}\mathrm{Hess}\,\mathcal{H}_{r,s}$ is an irreducible G-submodule of $C^\infty_{\gamma_{r,s}}((S^2 T^*)^{+-}_\mathbb{C})$ when $r \geq 0$ and $s \geq 1$, and that the space $\pi_{++}\mathrm{Hess}\,\mathcal{H}_{r,s}$ is an irreducible G-submodule of $C^\infty_{\gamma_{r,s}}((S^2 T^*)^{++}_\mathbb{C})$ when $r + s > 0$. Furthermore, when $r \geq 0$ and $s \geq 1$, the sections

$$\pi_{+-}\mathrm{Hess}\,\tilde{f}_{r,s}, \qquad \tilde{f}_{r,s-1}\pi_{+-}\mathrm{Hess}\,\tilde{f}_{0,1}$$

are highest weight vectors of the G-module $C^\infty_{\gamma_{r,s}}((S^2T^*)^{+-}_{\mathbb{C}})$; on the other hand, when $r+s > 0$, the section $\pi_{++}\mathrm{Hess}\,\tilde{f}_{r,s}$ is a highest weight vector of the G-module $C^\infty_{\gamma_{r,s}}((S^2T^*)^{++}_{\mathbb{C}})$. Using (5.118), we verified that the section

$$\pi^0_{++}\mathrm{Hess}\,\tilde{f}_{r,s} = \pi_{++}\mathrm{Hess}\,\tilde{f}_{r,s} + \frac{1}{2n}\lambda_{r,s}\tilde{f}_{r,s}g$$

of $(S^2T^*)^{++}_{0\mathbb{C}}$ is non-zero when $r \geq 1$ or when $s \geq 2$; therefore under these assumptions, this section is a highest weight vector of the irreducible G-submodule $\pi^0_{++}\mathrm{Hess}\,\mathcal{H}_{r,s}$ of $C^\infty_{\gamma_{r,s}}((S^2T^*)^{++}_{0\mathbb{C}})$. By (5.113) and (5.35), we see that all these sections of $(S^2T^*)^+_{\mathbb{C}}$ are even (resp. odd) when s is an even (resp. odd) integer. According to Proposition 5.19 we have

$$(5.126) \qquad\qquad C^\infty_{\gamma_{r,0}}((S^2T^*)^{+-}_{\mathbb{C}}) = \{0\},$$

for $r \geq 0$; moreover when $n \neq 4$, the G-module $C^\infty_{\gamma_{r,s}}((S^2T^*_{\mathbb{C}})^{+-})$ is irreducible for $r \geq 0$ and $s \geq 1$. Therefore when $n \neq 4$, for $r \geq 0$ and $s \geq 1$, we have the equality

$$(5.127) \qquad\qquad C^\infty_{\gamma_{r,s}}((S^2T^*_{\mathbb{C}})^{+-}) = \pi_{+-}\mathrm{Hess}\,\mathcal{H}_{r,s}$$

of irreducible G-modules, and there is a non-zero constant $c_{r,s} \in \mathbb{C}$ such that

$$(5.128) \qquad\qquad \pi_{+-}\mathrm{Hess}\,\tilde{f}_{r,s} = c_{r,s}\cdot\tilde{f}_{r,s-1}\pi_{+-}\mathrm{Hess}\,\tilde{f}_{0,1}.$$

Furthermore, when $n \neq 4$, we have the equalities

$$(5.129) \qquad \begin{aligned} C^\infty_{\gamma_{r,2s}}((S^2T^*)^{+-}_{\mathbb{C}}) &= C^\infty_{\gamma_{r,2s}}((S^2T^*)^{+-}_{\mathbb{C}})^{\mathrm{ev}}, \\ C^\infty_{\gamma_{r,2s+1}}((S^2T^*)^{+-}_{\mathbb{C}}) &= C^\infty_{\gamma_{r,2s+1}}((S^2T^*)^{+-}_{\mathbb{C}})^{\mathrm{odd}}, \end{aligned}$$

for $r, s \geq 0$; by (5.124), (5.125) and (5.129), we obtain the relation

$$(5.130) \qquad\qquad C^\infty_{\gamma_{0,1}}(S^2T^*_{\mathbb{C}}) = C^\infty_{\gamma_{0,1}}(S^2T^*_{\mathbb{C}})^{\mathrm{odd}}.$$

Some of the above facts concerning G-submodules of $C^\infty_{\gamma_{0,1}}(S^2T^*_{\mathbb{C}})$, and in particular the equalities (5.125) and (5.130), were previously derived from Lemmas 5.15 and 5.16.

Now suppose that $n = 4$ and let $r \geq 0$ and $s \geq 1$ be given integers. Since the involutive morphism of vector bundles $* : (S^2T^*)^{+-} \to (S^2T^*)^{+-}$ is G-equivariant, the sections

$$\tilde{f}_{r,s-1}*\pi_{+-}\mathrm{Hess}\,\tilde{f}_{0,1}, \qquad *\pi_{+-}\mathrm{Hess}\,\tilde{f}_{r,s}$$

are highest weight vectors of the G-module $C^\infty_{\gamma_{r,s}}((S^2 T^*_{\mathbb{C}})^{+-})$; by (5.113), (5.35) and (5.71), we see that these sections are even (resp. odd) when s is an odd (resp. even) integer. Thus when s is even (resp. odd) integer, the highest weight vector $\tilde{f}_{r,s-1}\mathrm{Hess}\,\tilde{f}_{0,1}$ is even (resp. odd), while the highest weight vector $\tilde{f}_{r,s-1} * \pi_{+-}\mathrm{Hess}\,\tilde{f}_{0,1}$ is odd (resp. even). Since the multiplicity of the G-module $C^\infty_{\gamma_{r,s}}((S^2 T^*_{\mathbb{C}})^{+-})$, which is given by Proposition 5.19, is equal to 2, we see that the G-modules $C^\infty_{\gamma_{r,s}}((S^2 T^*_{\mathbb{C}})^{+-})^{\mathrm{ev}}$ and $C^\infty_{\gamma_{r,s}}((S^2 T^*_{\mathbb{C}})^{+-})^{\mathrm{odd}}$ are irreducible, that a highest weight vector h of the G-module $C^\infty_{\gamma_{r,s}}((S^2 T^*_{\mathbb{C}})^{+-})$ can be written in the form

$$h = b_1 \tilde{f}_{r,s-1}\mathrm{Hess}\,\tilde{f}_{0,1} + b_2 \tilde{f}_{r,s-1} * \pi_{+-}\mathrm{Hess}\,\tilde{f}_{0,1},$$

where b_1, b_2 are complex numbers which do not both vanish, and that there is a non-zero constant $c_{r,s} \in \mathbb{C}$ such that the relation (5.128) holds. From these remarks, we obtain the following equalities among irreducible G-modules

(5.131)
$$C^\infty_{\gamma_{r,s}}((S^2 T^*_{\mathbb{C}})^{+-})^{\mathrm{ev}} = \begin{cases} \pi_{+-}\mathrm{Hess}\,\mathcal{H}_{r,s} & \text{if } s \text{ is even,} \\ * \pi_{+-}\mathrm{Hess}\,\mathcal{H}_{r,s} & \text{if } s \text{ is odd,} \end{cases}$$

$$C^\infty_{\gamma_{r,s}}((S^2 T^*_{\mathbb{C}})^{+-})^{\mathrm{odd}} = \begin{cases} * \pi_{+-}\mathrm{Hess}\,\mathcal{H}_{r,s} & \text{if } s \text{ is even,} \\ \pi_{+-}\mathrm{Hess}\,\mathcal{H}_{r,s} & \text{if } s \text{ is odd.} \end{cases}$$

By (5.124), (5.125) and (5.131), we see that the relations

$$C^\infty_{\gamma_{0,1}}(S^2 T^*_{\mathbb{C}})^{\mathrm{ev}} = C^\infty_{\gamma_{0,1}}((S^2 T^*_{\mathbb{C}})^{+-})^{\mathrm{ev}} = * \pi_{+-}\mathrm{Hess}\,\mathcal{H}_{0,1},$$

(5.132)
$$C^\infty_{\gamma_{0,1}}(S^2 T^*_{\mathbb{C}})^{\mathrm{odd}} = C^\infty_{\gamma_{0,1}}(\{g\}_{\mathbb{C}}) \oplus C^\infty_{\gamma_{0,1}}((S^2 T^*_{\mathbb{C}})^{+-})^{\mathrm{odd}}$$
$$= \mathcal{H}_{0,1} \cdot g \oplus \pi_{+-}\mathrm{Hess}\,\mathcal{H}_{0,1}$$

hold. The first equality of (5.132) is also given by the relation (5.111) of Proposition 5.17. From the first equalities of (5.132) and Proposition 5.17, we obtain:

PROPOSITION 5.20. *Let X be the complex quadric Q_4. Then the irreducible $SO(6)$-module $E(X)$ is equal to the $SO(6)$-submodule*

$$\{ * \pi_{+-}\mathrm{Hess}\, f \mid f \in \mathcal{H}_{0,1}, \ f = \bar{f} \}$$

of $C^\infty((S^2 T^*)^{+-})$.

We no longer assume that $n = 4$ and return to the situation where n is an arbitrary integer ≥ 3. Since $\mathrm{Hess} : C^\infty(X) \to C^\infty(S^2 T^*_{\mathbb{C}})$ is a homogeneous differential operator and $\mathcal{H}_{1,0}$ is an irreducible G-module, by (5.116) we know that the spaces $\pi'\mathrm{Hess}\,\mathcal{H}_{1,0}$ and $\pi''\mathrm{Hess}\,\mathcal{H}_{1,0}$ are irreducible G-submodules of $C^\infty_{\gamma_{1,0}}(S^{2,0}T^*)$ and $C^\infty_{\gamma_{1,0}}(S^{0,2}T^*)$, respectively. Moreover, the sections $\pi'\mathrm{Hess}\,\tilde{f}_{1,0}$ and $\pi''\mathrm{Hess}\,\tilde{f}_{1,0}$ are highest weight vectors of these modules. When $r \geq 1$, we therefore see that the sections $\tilde{f}_{r-1,s}\pi'\mathrm{Hess}\,\tilde{f}_{1,0}$ and $\tilde{f}_{r-1,s}\pi''\mathrm{Hess}\,\tilde{f}_{1,0}$ of $(S^2T^*)^-_{\mathbb{C}}$ are highest weight vectors of the G-modules $C^\infty_{\gamma_{r,s}}(S^{2,0}T^*)$ and $C^\infty_{\gamma_{r,s}}(S^{0,2}T^*)$, respectively.

According to Proposition 5.19, for $r \geq 1$ and $s \geq 0$, the G-modules $C^\infty_{\gamma_{r,s}}(L')$ and $C^\infty_{\gamma_{r,s}}(L'')$ are irreducible, and we have the equalities

$$C^\infty_{\gamma_{1,0}}(L') = C^\infty_{\gamma_{1,0}}(S^{2,0}T^*), \qquad C^\infty_{\gamma_{1,0}}(L'') = C^\infty_{\gamma_{1,0}}(S^{0,2}T^*).$$

From the preceding remarks, it follows that

$$C^\infty_{\gamma_{1,0}}(L') = \pi'\mathrm{Hess}\,\mathcal{H}_{1,0}, \qquad C^\infty_{\gamma_{1,0}}(L'') = \pi''\mathrm{Hess}\,\mathcal{H}_{1,0};$$

thus $\pi'\mathrm{Hess}\,\tilde{f}_{1,0}$ and $\pi''\mathrm{Hess}\,\tilde{f}_{1,0}$ are sections of L' and L'' and are highest weight vectors of the irreducible G-modules $C^\infty_{\gamma_{1,0}}(L')$ and $C^\infty_{\gamma_{1,0}}(L'')$, respectively. If r, s are integers ≥ 0, we see that the sections $\tilde{f}_{r,s}\pi'\mathrm{Hess}\,\tilde{f}_{1,0}$ and $\tilde{f}_{r,s}\pi''\mathrm{Hess}\,\tilde{f}_{1,0}$ are highest weight vectors of the irreducible G-modules $C^\infty_{\gamma_{r+1,s}}(L')$ and $C^\infty_{\gamma_{r+1,s}}(L'')$, respectively. Thus a highest weight vector h of the G-module $C^\infty_{\gamma_{r+1,s}}(L_{\mathbb{C}})$ can be written in the form

$$h = \tilde{f}_{r,s}(b'\pi' + b''\pi'')\,\mathrm{Hess}\,\tilde{f}_{1,0},$$

where b', b'' are complex numbers which do not both vanish. According to (5.113) and (5.36), the sections

$$(\pi' + \pi'')\mathrm{Hess}\,\tilde{f}_{1,0}, \qquad (\pi' - \pi'')\mathrm{Hess}\,\tilde{f}_{1,0}$$

are highest weight vectors of the irreducible G-modules $C^\infty_{\gamma_{1,0}}(L_{\mathbb{C}})^{\mathrm{ev}}$ and $C^\infty_{\gamma_{1,0}}(L_{\mathbb{C}})^{\mathrm{odd}}$, respectively. Moreover by (1.4) and the second equality of (1.75), we know that the highest weight vector $(\pi' - \pi'')\mathrm{Hess}\,\tilde{f}_{1,0}$ of $C^\infty_{\gamma_{1,0}}(L_{\mathbb{C}})^{\mathrm{odd}}$ belongs to $D_0C^\infty(T_{\mathbb{C}})$, and so we have

(5.133) $$C^\infty_{\gamma_{1,0}}(L_{\mathbb{C}})^{\mathrm{odd}} = (\pi' - \pi'')\mathrm{Hess}\,\mathcal{H}_{1,0} \subset D_0C^\infty(T_{\mathbb{C}}).$$

We define integers $d_{r,s}$ by

$$d_{r,s} = \begin{cases} 0 & \text{if } r = s = 0, \\ 2 & \text{if } r, s \geq 1, \\ 1 & \text{otherwise.} \end{cases}$$

The following lemma is a consequence of Proposition 5.19 and of the proof of Lemma 9.1 of [23].

LEMMA 5.21. *The multiplicity of the G-module $C^\infty_{\gamma_{r,s}}(T_{\mathbb{C}})^{\mathrm{ev}}$ or of the G-module $C^\infty_{\gamma_{r,s}}(T_{\mathbb{C}})^{\mathrm{odd}}$ is equal to $d_{r,s}$.*

As in §7, Chapter II, we identify the complexification $\mathcal{K}_{\mathbb{C}}$ of the space $\mathcal{K}$ of all Killing vector fields on X with a G-submodule of $C^\infty_{\gamma_{0,1}}(T_{\mathbb{C}})$. From the relations (2.28), (5.37) and (5.113), we obtain the equalities

(5.134)
$$C^\infty_{\gamma_{0,1}}(T_{\mathbb{C}})^{\mathrm{ev}} = \mathcal{K}_{\mathbb{C}},$$

$$C^\infty_{\gamma_{0,1}}(T^*_{\mathbb{C}})^{\mathrm{ev}} = (\partial - \bar\partial)\mathcal{H}_{0,1}, \qquad C^\infty_{\gamma_{0,1}}(T^*_{\mathbb{C}})^{\mathrm{odd}} = d\mathcal{H}_{0,1},$$

of irreducible G-modules.

The first relation of (5.134), Lemma 5.21 and Proposition 2.42,(i), with $X = Q_n$, $\Sigma = \{\tau\}$ and $\varepsilon = +1$, give us the following result:

PROPOSITION 5.22. *For $\gamma \in \Gamma$, the equality*

$$\mathcal{N}_{2,\mathbb{C}} \cap C^\infty_\gamma(S^2 T^*_{\mathbb{C}})^{\mathrm{ev}} = D_0 C^\infty_\gamma(T_{\mathbb{C}})^{\mathrm{ev}}$$

is equivalent to

$$\mathrm{Mult}\,(\mathcal{N}_{2,\mathbb{C}} \cap C^\infty_\gamma(S^2 T^*_{\mathbb{C}})^{\mathrm{ev}}) \leq \mathrm{Mult}\,C^\infty_\gamma(T_{\mathbb{C}})^{\mathrm{ev}}$$

when $\gamma \neq \gamma_{0,1}$, or to

$$\mathcal{N}_{2,\mathbb{C}} \cap C^\infty_{\gamma_{0,1}}(S^2 T^*_{\mathbb{C}})^{\mathrm{ev}} = \{0\}$$

when $\gamma = \gamma_{0,1}$.

The first relation of (5.134), Lemma 5.21 and Proposition 2.42,(ii), with $X = Q_n$, $\Sigma = \{\tau\}$ and $\varepsilon = -1$, give us the following result:

PROPOSITION 5.23. *For $\gamma \in \Gamma$, the equality*

$$\mathcal{Z}_{2,\mathbb{C}} \cap C^\infty_\gamma(S^2 T^*_{\mathbb{C}})^{\mathrm{odd}} = D_0 C^\infty_\gamma(T_{\mathbb{C}})^{\mathrm{odd}}$$

is equivalent to

$$\mathrm{Mult}\,(\mathcal{Z}_{2,\mathbb{C}} \cap C^\infty_\gamma(S^2 T^*_{\mathbb{C}})^{\mathrm{odd}}) \leq \mathrm{Mult}\,C^\infty_\gamma(T_{\mathbb{C}})^{\mathrm{odd}}.$$

Proposition 2.27, with $X = Q_n$, $p = 2$, $F = (S^2 T^*)^{+-}_{\mathbb{C}}$, $\Sigma = \{\tau\}$ and $\varepsilon = +1$, gives us the following result:

PROPOSITION 5.24. *Let X be the complex quadric Q_n, with $n \geq 3$. The submodule*

$$\bigoplus_{\gamma \in \Gamma} (\mathcal{N}_{2,\mathbb{C}} \cap C^\infty_\gamma((S^2 T^*)^{+-}_{\mathbb{C}})^{\mathrm{ev}})$$

is a dense subspace of $\mathcal{N}_{2,\mathbb{C}} \cap C^\infty((S^2 T^*)^{+-}_{\mathbb{C}})^{\mathrm{ev}}$.

Proposition 2.28, with $X = Q_n$, $p = 2$, $F = (S^2 T^*)^{+-}_{\mathbb{C}}$, $\Sigma = \{\tau\}$ and $\varepsilon = -1$, gives us the following result:

PROPOSITION 5.25. *Let X be the complex quadric Q_n, with $n \geq 3$. The submodule*

$$\bigoplus_{\gamma \in \Gamma} \left(\mathcal{Z}_{2,\mathbb{C}} \cap C_\gamma^\infty ((S^2 T^*)_{\mathbb{C}}^{+-})^{\text{odd}} \right)$$

is a dense subspace of $\mathcal{Z}_{2,\mathbb{C}} \cap C^\infty((S^2 T^*)_{\mathbb{C}}^{+-})^{\text{odd}}$.

§8. Vanishing results for symmetric forms

This section is mainly devoted to results concerning the sections of the vector bundle L and to the proofs of the following two results:

PROPOSITION 5.26. *Let X be the complex quadric Q_n, with $n \geq 3$. A section h of L over X, which satisfies the relation $\operatorname{div} h = 0$, vanishes identically.*

THEOREM 5.27. *Let X be the complex quadric Q_n, with $n \geq 3$. An even section of L over X, which belongs to the space $D_0 C^\infty(T)$, vanishes identically. Moreover, we have the equality*

$$D_0 C^\infty(T_\mathbb{C}) \cap C^\infty(L_\mathbb{C}) = C_{\gamma_{1,0}}^\infty(L_\mathbb{C})^{\text{odd}} = (\pi' - \pi'') \operatorname{Hess} \mathcal{H}_{1,0}.$$

Theorem 5.27 may be restated as follows:

THEOREM 5.28. *Let X be the complex quadric Q_n, with $n \geq 3$. A section h of L over X is a Lie derivative of the metric if and only if there is a real-valued function f on X satisfying*

$$h = \pi' \operatorname{Hess} f - \pi'' \operatorname{Hess} f$$

and $\Delta f = \lambda_{1,0} f$.

Since $\operatorname{Hess} : C^\infty(X) \to C^\infty(S^2 T_\mathbb{C}^*)$ is a homogeneous differential operator, the Hessian of an element of $\mathcal{H}_{0,1}$ belongs to $C_{\gamma_0,1}^\infty(S^2 T_\mathbb{C}^*)$. Hence by (5.125), we see that

$$\operatorname{Hess} f = \pi_+ \operatorname{Hess} f = \pi_{+-} \operatorname{Hess} f - \frac{1}{2n}(\Delta f) \cdot g,$$

for all $f \in \mathcal{H}_{0,1}$. Since $\lambda_{0,1} = 4n$, from the previous equalities we deduce that

$$\pi_{+-} \operatorname{Hess} f = \pi_+ \operatorname{Hess} f + 2fg,$$

for $f \in \mathcal{H}_{0,1}$. Hence from formulas (1.76) and (1.8), we obtain the relation

(5.135) $\operatorname{div} \pi_{+-} \operatorname{Hess} f = 2(n-1) df,$

for all $f \in \mathcal{H}_{0,1}$.

When $n \neq 4$, from the relations (2.25), (5.125), (5.127) and (5.135) we obtain the vanishing of the space $E(X)$, given by Proposition 5.17, without having recourse to Proposition 2.40.

By (5.123), we know that $\lambda_{1,0} = 4(n + 2)$; hence according to (5.5) and (1.77), we see that

$$\text{(5.136)} \qquad \text{div } \pi'\text{Hess } f = 4\partial f, \qquad \text{div } \pi''\text{Hess } f = 4\bar{\partial}f,$$

for all $f \in \mathcal{H}_{1,0}$.

We consider the element ϕ of $G = SO(n + 2)$ of §7, and, for $r, s \geq 0$, the function $\tilde{f}'_{r,s} = \phi^* \tilde{f}_{r,s}$ on X, which belongs to $\mathcal{H}_{r,s}$; we write

$$h_{r,s} = \text{Hess } \tilde{f}'_{r,s}.$$

By (1.8), (5.114)–(5.116) and (5.136), for $r \geq 1$ and $s \geq 0$ we see that the equalities

$$\text{div}\left(\tilde{f}'_{r-1,s}\pi'h_{1,0}\right) = \frac{1}{2^r}\left(i(2r + s + 2)dz_2 + sdz_3\right),$$

(5.137)

$$\text{div}\left(\tilde{f}'_{r-1,s}\pi''h_{1,0}\right) = \frac{1}{2^r}\left(i(2r + s + 2)d\bar{z}_2 - sd\bar{z}_3\right)$$

hold at the point a. By (1.8), (5.114), (5.115), (5.119) and (5.135), for $r \geq 0$ and $s \geq 1$ we see that the equality

$$\text{div}\left(\tilde{f}'_{r,s-1}\pi_{+-}h_{0,1}\right) = \frac{1}{2^r} i(s + n - 2)(dz_2 + d\bar{z}_2)$$

(5.138)

$$- \frac{1}{2^r}(2r + s + n - 2)(dz_3 - d\bar{z}_3)$$

holds at the point a.

LEMMA 5.29. *Let $r, s \geq 0$, with $r + s \geq 1$, and $b, b', b'' \in \mathbb{C}$. Suppose that the element*

$$h = b'\tilde{f}_{r-1,s}\pi'\text{Hess } \tilde{f}_{1,0} + b''\tilde{f}_{r-1,s}\pi''\text{Hess } \tilde{f}_{1,0}$$

(5.139)

$$+ b\tilde{f}_{r,s-1}\pi_{+-}\text{Hess } \tilde{f}_{0,1}$$

of $C^\infty_{\gamma_{r,s}}\left(L_{\mathbb{C}} \oplus (S^2T^*)^{+-}_{\mathbb{C}}\right)$ satisfies $\text{div } h = 0$. Then h vanishes identically.

PROOF: By (5.137) and (5.138), when $r, s \geq 1$, we see that the relation $\text{div } \phi^* h = 0$ implies that the equalities

$$0 = 2^r \cdot (\text{div } \phi^* h)(\eta_2) = i((2r + s + 2)b' + (s + n - 2)b),$$

$$0 = 2^r \cdot (\text{div } \phi^* h)(\eta_3) = sb' - (2r + s + n - 2)b,$$

$$0 = 2^r \cdot (\text{div } \phi^* h)(\bar{\eta}_2) = i((2r + s + 2)b'' + (s + n - 2)b),$$

$$0 = 2^r \cdot (\text{div } \phi^* h)(\bar{\eta}_3) = -sb'' + (2r + s + n - 2)b$$

hold at a. Since the determinant of the matrix

$$\begin{pmatrix} 2r+s+2 & s+n-2 \\ -s & 2r+s+n-2 \end{pmatrix}$$

is positive, when $r, s \geq 1$ the coefficients b', b'' and b vanish, and so in this case h vanishes. Since

$$\text{div}\,(S^2 T^{2,0}) \subset T^{1,0}, \qquad \text{div}\,(S^2 T^{0,2}) \subset T^{0,1},$$

when either $r = 0$ or $s = 0$, by (5.138) and (5.139) we see that the relation $\text{div}\,\phi^* h = 0$ implies that h vanishes.

Let $r, s \geq 0$ be given integers, with $r + s \geq 1$. According to §7, when $r \geq 1$ and $s \geq 0$, a highest weight vector h of the G-module $C_{\gamma_{r,s}}^\infty(L_\mathbb{C})$ can be written in the form (5.139), with $b = 0$ and $b', b'' \in \mathbb{C}$; moreover when $n \neq 4$, a highest weight vector h of the G-module $C_{\gamma_{r,s}}^\infty(L_\mathbb{C} \oplus (S^2 T^*)_\mathbb{C}^{+-})$ can be written in the form (5.139), with $b, b', b'' \in \mathbb{C}$. Hence since the differential operator $\text{div} : S^2 T_\mathbb{C}^* \to T_\mathbb{C}^*$ is homogeneous, from Lemma 5.29 and the relation (2.1) we deduce the following:

PROPOSITION 5.30. *Let X be the complex quadric Q_n, with $n \geq 3$. Let $r, s \geq 0$ be given integers, with $r + s \geq 1$.*
  (i) *When $r \geq 1$, an element h of $C_{\gamma_{r,s}}^\infty(L_\mathbb{C})$ satisfying $\text{div}\, h = 0$ vanishes identically.*
  (ii) *When $n \neq 4$, an element h of $C_{\gamma_{r,s}}^\infty(L_\mathbb{C} \oplus (S^2 T^*)_\mathbb{C}^{+-})$ satisfying $\text{div}\, h = 0$ vanishes identically.*

Since $\text{div} : S^2 T_\mathbb{C}^* \to T_\mathbb{C}^*$ is a homogeneous differential operator, by Propositions 2.3 (with $Q_1 = 0$), 5.18 and 5.30,(i) we see that the operator

$$\text{div} : C^\infty(L_\mathbb{C}) \to C^\infty(T_\mathbb{C}^*)$$

is injective. This result implies the assertion of Proposition 5.26.

By (5.116)–(5.119) and (5.50), we easily verify that the relations

$$(5.140) \quad (\pi'' h_{1,0})((\eta_1 \lrcorner\, h_{r,s})^\sharp, \bar{\eta}_1) = (\pi' h_{1,0})((\bar{\eta}_1 \lrcorner\, h_{r,s})^\sharp, \eta_1) = \frac{r}{2^{r+1}},$$

$$(5.141) \quad (\pi' h_{1,0})((\eta_2 \lrcorner\, h_{r,s})^\sharp, \eta_2) = \frac{1}{2^{r+2}}\,(2r+s)(2r+s+1)$$

hold at the point a.

PROPOSITION 5.31. *Let $r, s \geq 0$ be given integers and $b', b'' \in \mathbb{C}$. Suppose that the element*

$$h = \tilde{f}_{r,s}'(b'\pi' + b''\pi'')\,\text{Hess}\,\tilde{f}_{1,0}'$$

of $C^\infty(S^2 T_{\mathbb{C}}^*)$ satisfies the relation

$$(D^1 \operatorname{div} h)(\eta_1, \bar{\eta}_1)(a) = 0.$$

Then we have $b' + b'' = 0$.

PROOF: We write

$$h_1 = (b'\pi' + b''\pi'') h_{1,0}.$$

According to (5.137), we have

$$\operatorname{div} h_1 = 4(b'\partial \tilde{f}'_{1,0} + b''\bar{\partial} \tilde{f}'_{1,0});$$

hence by the first equality of (1.75), we see that

$$\pi_+ D^1 \operatorname{div} h_1 = 2(b' + b'')\pi_+ h_{1,0}.$$

By (5.114), (5.115), (5.117), (5.122) and (5.140), we therefore obtain the equalities

$$(D^1 \operatorname{div} h_1)(\eta_1, \bar{\eta}_1) = -\frac{1}{2}(b' + b''), \qquad (d\tilde{f}'_{r,s} \cdot \operatorname{div} h_1)(\eta_1, \bar{\eta}_1) = 0,$$

$$(\nabla h_1)(\eta_1, (d\tilde{f}'_{r,s})^\sharp, \bar{\eta}_1) = (\nabla h_1)(\bar{\eta}_1, (d\tilde{f}'_{r,s})^\sharp, \eta_1) = 0,$$

$$h_1((\eta_1 \lrcorner h_{r,s})^\sharp, \bar{\eta}_1) + h_1((\bar{\eta}_1 \lrcorner h_{r,s})^\sharp, \eta_1) = \frac{r}{2^{r+1}}(b' + b'')$$

at the point a. According to formulas (1.9) and (1.10) and the preceding relations, we see that

$$(D^1 \operatorname{div}(\tilde{f}'_{r,s} h_1))(\eta_1, \bar{\eta}_1)(a) = -\frac{1}{2^{r+2}}(r + 2)(b' + b'').$$

Thus our hypotheses imply that $b' + b'' = 0$.

PROPOSITION 5.32. *Let $r, s \geq 0$ be given integers. Suppose that the element*

$$\tilde{f}'_{r,s}(\pi' - \pi'') \operatorname{Hess} \tilde{f}'_{1,0}$$

of $C^\infty(S^2 T_{\mathbb{C}}^*)$ belongs to the space $D_0 C^\infty(T_{\mathbb{C}})$. Then the integers r and s vanish.

PROOF: We consider the sections $h_1 = (\pi' - \pi'') \operatorname{Hess} \tilde{f}_{1,0}$ and $h'_1 = \phi^* h_1$ of $S_0^2 T_{\mathbb{C}}^*$. According to (5.136), we have

$$\operatorname{div} h'_1 = 4(\partial \tilde{f}'_{1,0} - \bar{\partial} \tilde{f}'_{1,0}).$$

Thus by (5.114), (5.115), (5.122) and (5.141), the equalities

$$(d\tilde{f}'_{r,s} \cdot \operatorname{div} h'_1)(\eta_2, \eta_2) = -\frac{1}{2^{r-1}}(2r+s), \qquad (\nabla h'_1)(\eta_2, (d\tilde{f}'_{r,s})^\sharp, \eta_2) = 0,$$

$$h'_1((\eta_2 \lrcorner h_{r,s})^\sharp, \eta_2) = \frac{1}{2^{r+2}}(2r+s)(2r+s+1)$$

hold at the point a. Let h_2 denote the section of $S^2 T^*_{\mathbb{C}}$ equal to the left-hand side of the relation (2.24), with $f = \tilde{f}'_{r,s}$, $h = h'_1$, $\gamma = \gamma_{1,0}$ and $\gamma' = \gamma_{r+1,s}$. Then by formula (1.10), the previous equalities, and the relations (5.116) and (5.123), we see that

$$h_2(\eta_2, \eta_2)(a) = \frac{1}{2^{r+2}}(s^2 + (2n-3)s + 2(n-1)r).$$

According to (5.133) and observations made in §7, we know that the section h_1 of $S^2 T^*_{\mathbb{C}}$ belongs to the spaces $D_0 C^\infty(T_{\mathbb{C}})$ and is a highest weight vector of the G-module $C^\infty_{\gamma_{1,0}}(S^2_0 T^*_{\mathbb{C}})$, and that the section $\tilde{f}_{r,s} h_1$ is a highest weight vector of the G-module $C^\infty_{\gamma_{r+1,s}}(S^2 T^*_{\mathbb{C}})$. Therefore the section $h'_1 = \phi^* h_1$ belongs to the space $D_0 C^\infty(T_{\mathbb{C}})$ and the section $\tilde{f}'_{r,s} h'_1 = \phi^*(\tilde{f}_{r,s} h_1)$ belongs to the G-module $C^\infty_{\gamma_{r+1,s}}(S^2 T^*_{\mathbb{C}})$. According to Lemma 2.38, with $\gamma = \gamma_{1,0}$ and $\gamma' = \gamma_{r+1,s}$, and our hypothesis, we see that h_2 vanishes. Therefore we obtain the relation

$$s^2 + (2n-3)s + 2(n-1)r = 0,$$

and so we have $r = s = 0$.

PROPOSITION 5.33. *Let $r, s \geq 0$ be given integers and $b', b'' \in \mathbb{C}$. Suppose that the element*

$$h = \tilde{f}_{r,s}(b'\pi' + b''\pi'') \operatorname{Hess} \tilde{f}_{1,0}$$

of $C^\infty(S^2 T^*_{\mathbb{C}})$ belongs to the space $D_0 C^\infty(T_{\mathbb{C}})$. Then we have $b' + b'' = 0$; moreover if h is non-zero, the integers r and s vanish.

PROOF: Since h is a section of the sub-bundle $(S^2 T^*)^-_{\mathbb{C}}$, so is the section Δh. Clearly, we have $\operatorname{Tr} h = 0$; hence by formula (1.39), we see that $\pi_+ D^1 \operatorname{div} h = 0$. Thus the section

$$h' = \phi^* h = \tilde{f}'_{r,s}(b'\pi' + b''\pi'') \operatorname{Hess} \tilde{f}'_{1,0}$$

satisfies $\pi_+ D^1 \operatorname{div} h' = 0$. From Proposition 5.31, we therefore obtain the relation $b' + b'' = 0$. According to our hypothesis, h' belongs to the space $D_0 C^\infty(T_\mathbb{C})$, and so the other assertion of the proposition is a consequence of Proposition 5.32.

Let $r, s \geq 0$ be given integers. According to Proposition 5.33 and the description of the highest weight vectors of the G-module $C^\infty_{\gamma_{r+1,s}}(L_\mathbb{C})$ given above, we see that the space

$$C^\infty_{\gamma_{r+1,s}}(L_\mathbb{C}) \cap D_0 C^\infty(T_\mathbb{C})$$

vanishes unless $r = s = 0$ and that

$$C^\infty_{\gamma_{1,0}}(L_\mathbb{C})^{\mathrm{ev}} \cap D_0 C^\infty(T_\mathbb{C}) = \{0\}.$$

Since D_0 is a homogeneous differential operator, by Proposition 5.18 and the relations (2.1) and (5.133), we see that Theorem 5.27 is a consequence of these results.

§9. The complex quadric of dimension two

This section is devoted to the proof of Proposition 4.3. We once again consider the natural projection $\pi : \mathbb{C}^{m+1} - \{0\} \to \mathbb{CP}^m$ and the Fubini-Study metric $\tilde{g}$ on $\mathbb{CP}^m$ of constant holomorphic curvature 4. As in §2, we view the complex quadric Q_2 as a hypersurface of $\mathbb{CP}^3$ endowed with the Kähler metric g induced by the metric $\tilde{g}$ of $\mathbb{CP}^3$. We recall that the complex conjugation of $\mathbb{C}^4$ induces the involutive isometry τ of Q_2.

We endow the manifold $\mathbb{CP}^1 \times \mathbb{CP}^1$ with the Kähler metric which is the product of the metrics $\tilde{g}$ on each factor. It is well-known that the Segre imbedding

$$\sigma : \mathbb{CP}^1 \times \mathbb{CP}^1 \to \mathbb{CP}^3,$$

which sends the point $(\pi(u), \pi(v))$, where $u = (u_0, u_1)$ and $v = (v_0, v_1)$ are non-zero vectors of $\mathbb{C}^2$, into the point $\pi(u_0 v_0, u_0 v_1, u_1 v_0, u_1 v_1)$ of $\mathbb{CP}^3$, is an isometry. The element A of $SU(4)$ determined by

$$A(\zeta) = \frac{1}{\sqrt{2}} (\zeta_0 + \zeta_3, \zeta_1 - \zeta_2, i(\zeta_3 - \zeta_0), i(\zeta_1 + \zeta_2)),$$

for $\zeta = (\zeta_0, \zeta_1, \zeta_2, \zeta_3) \in \mathbb{C}^4$, induces an isometry of $\mathbb{CP}^3$ which we also denote by A. We now easily verify that the image of the mapping $\alpha = A \circ \sigma$ from $\mathbb{CP}^1 \times \mathbb{CP}^1$ to $\mathbb{CP}^3$ is equal to the complex quadric Q_2; thus the mapping α induces an isometry

$$\alpha : \mathbb{CP}^1 \times \mathbb{CP}^1 \to Q_2.$$

We consider the involutive isometry Ψ of $\mathbb{CP}^1$ defined in §4, Chapter III; according to the commutativity of diagram (3.26), it sends the point $\pi(u)$, where $u = (u_0, u_1)$ is a non-zero vector of $\mathbb{C}^2$, into the point $\pi(v)$, where $v = (-\bar{u}_1, \bar{u}_0)$ is a non-zero vector of $\mathbb{C}^2$ orthogonal to u. We easily verify that the diagram

$$
\begin{array}{ccc}
\mathbb{CP}^1 \times \mathbb{CP}^1 & \xrightarrow{\;\alpha\;} & Q_2 \\
\downarrow{\scriptstyle \Psi \times \Psi} & & \downarrow{\scriptstyle \tau} \\
\mathbb{CP}^1 \times \mathbb{CP}^1 & \xrightarrow{\;\alpha\;} & Q_2
\end{array}
$$

(5.142)

is commutative.

Now we consider the diffeomorphism $\Theta : \widetilde{G}_{2,2}^{\mathbb{R}} \to Q_2$ defined in §5 and the involutive isometry Ψ of $\widetilde{G}_{2,2}^{\mathbb{R}}$ defined in §1, Chapter IV, which sends an oriented 2-plane of $\mathbb{R}^4$ into its orthogonal complement endowed with the appropriate orientation. If $\psi : Q_2 \to Q_2$ is the involutive isometry equal to the composition $\Theta^{-1} \circ \Psi \circ \Theta$, the diagram

$$
\begin{array}{ccc}
\mathbb{CP}^1 \times \mathbb{CP}^1 & \xrightarrow{\;\alpha\;} & Q_2 \\
\downarrow{\scriptstyle \mathrm{id} \times \Psi} & & \downarrow{\scriptstyle \psi} \\
\mathbb{CP}^1 \times \mathbb{CP}^1 & \xrightarrow{\;\alpha\;} & Q_2
\end{array}
$$

(5.143)

is also commutative. In fact, let $u = (u_0, u_1)$ and $v = (v_0, v_1)$ be non-zero vectors of $\mathbb{C}^2$. Then we see that $\alpha(\pi(u), \pi(v))$ is equal to $\pi(\zeta)$, where ζ is the non-zero vector of $\mathbb{C}^4$ given by

$$\zeta = (u_0 v_0 + u_1 v_1, u_0 v_1 - u_1 v_0, i(u_1 v_1 - u_0 v_0), i(u_0 v_1 + u_1 v_0)).$$

If h is the complex bilinear form on $\mathbb{C}^4$ defined in §2, an elementary computation shows that the point $\alpha(\pi(u), \Psi(\pi(v)))$ can be written in the form $\pi(\zeta')$, where ζ' is a non-zero vector of $\mathbb{C}^4$ satisfying

$$\langle \zeta, \zeta' \rangle = h(\zeta, \zeta') = 0.$$

This observation, together with the fact that the points $\pi(\zeta)$ and $\pi(\zeta')$ belong to the quadric Q_2, allows us to see that $\{\operatorname{Re}\zeta, \operatorname{Im}\zeta, \operatorname{Re}\zeta', \operatorname{Im}\zeta'\}$ is a positively oriented orthogonal basis of $\mathbb{R}^4$. If $x, y \in \widetilde{G}_{2,2}^{\mathbb{R}}$ are the oriented 2-planes of $\mathbb{R}^4$ corresponding to the vectors $\operatorname{Re}\zeta \wedge \operatorname{Im}\zeta$ and $\operatorname{Re}\zeta' \wedge \operatorname{Im}\zeta'$ of $\bigwedge^2 \mathbb{R}^4$, respectively, we have

$$\Theta(x) = \alpha(\pi(u), \pi(v)), \qquad \Theta(y) = \alpha(\pi(u), \Psi(\pi(v))).$$

The commutativity of the diagram (5.143) is now a consequence of the relations $\Psi(x) = y$.

We endow the manifolds $S^2 \times S^2$ and $\mathbb{RP}^2 \times \mathbb{RP}^2$ with the Riemannian metrics which are the product of the metrics g_0 of constant curvature 1 on each factor. We recall that the diffeomorphism $\varphi : S^2 \to \mathbb{CP}^1$ defined in §4, Chapter III is an isometry of (S^2, g_0) onto $(\mathbb{CP}^1, 4\tilde{g})$. We denote by g' the Riemannian metrics on the Grassmannians $\widetilde{G}^{\mathbb{R}}_{2,2}$ and $G^{\mathbb{R}}_{2,2}$ defined in §1, Chapter IV. When we endow the space $\widetilde{G}^{\mathbb{R}}_{2,2}$ with the metric $2g'$, by (5.59) we see that the mapping

$$\beta = \Theta^{-1} \circ \alpha \circ (\varphi \times \varphi) : S^2 \times S^2 \to \widetilde{G}^{\mathbb{R}}_{2,2}$$

is an isometry.

We consider the anti-podal involution τ_1 of S^2 which is denoted by τ in §3, Chapter III. The group Λ_1 (resp. Λ_2) of isometries of the Riemannian manifold $S^2 \times S^2$ generated by the mapping $\tau_1 \times \tau_1$ (resp. the mappings $\mathrm{id} \times \tau_1$ and $\tau_1 \times \mathrm{id}$), which is of order 2 (resp. order 4), acts freely on $S^2 \times S^2$. Clearly, Λ_1 is a subgroup of Λ_2. We consider the Riemannian manifolds $(S^2 \times S^2)/\Lambda_1$ and $(S^2 \times S^2)/\Lambda_2$ endowed with the Riemannian metrics induced by the metric of $S^2 \times S^2$. The relation (5.60) and the commutativity of the diagrams (3.25) and (5.142) now tell us that β induces an isometry from the quotient $(S^2 \times S^2)/\Lambda_1$ onto the Grassmannian $G^{\mathbb{R}}_{2,2}$ endowed with the metric $2g'$.

On the other hand, we observe that the quotient $(S^2 \times S^2)/\Lambda_2$ is diffeomorphic to $\mathbb{RP}^2 \times \mathbb{RP}^2$. If $g_{\bar{Y}}$ is the Riemannian metric on $\bar{Y} = \bar{G}^{\mathbb{R}}_{2,2}$ defined in §1, Chapter IV, from the commutativity of the diagrams (3.25), (5.142) and (5.143), we infer that β induces an isometry from $\mathbb{RP}^2 \times \mathbb{RP}^2$ to the space $\bar{G}^{\mathbb{R}}_{2,2}$ endowed with the metric $2g_{\bar{Y}}$. This completes the proof of Proposition 4.3.

CHAPTER VI

THE RIGIDITY OF THE COMPLEX QUADRIC

§1. Outline

In §2, we describe an explicit totally geodesic flat torus of the complex quadric Q_n, with $n \geq 3$, viewed as a complex hypersurface of projective space. In §3, we introduce certain symmetric 2-forms on the quadric; later, in §7 we shall see that they provide us with explicit bases for the highest weight subspaces of the isotypic components of the $SO(5)$-module of complex symmetric 2-forms on the three-dimensional quadric Q_3. In §§4 and 5, we compute integrals over closed geodesics in order to prove that linear combinations of the symmetric 2-forms of §3 satisfying the zero-energy condition must verify certain relations. As the space of complex symmetric 2-forms on Q_3 satisfying the Guillemin condition (resp. the zero-energy condition) is invariant under the group $SO(5)$, the rigidity results (Theorems 6.35 and 6.36) for the quadric Q_3 are obtained by establishing appropriate bounds for the dimensions of certain spaces of symmetric 2-forms consisting of vectors of highest weight satisfying either the Guillemin condition or the zero-energy condition. These bounds were obtained in the case of the Guillemin condition in [23] and are recalled in §3. In §6, we establish the corresponding bounds for odd forms on Q_3 satisfying the zero-energy condition. In §8, we prove the rigidity theorems for the quadric of dimension ≥ 3. In particular, we show that the quadric Q_3 is infinitesimally rigid; this result is the last remaining one needed to complete our study of the rigidity of the quadric. Since the quadric Q_n is a two-fold covering of the real Grassmannian $G^{\mathbb{R}}_{2,n}$ of 2-planes in $\mathbb{R}^{n+2}$, we deduce the rigidity in the sense of Guillemin of this Grassmannian from these results of §8 when $n \geq 3$; we note that the proof presented here does not rely, as does the one given in [23], on the infinitesimal rigidity of the quadrics of dimension ≥ 4. In §§9 and 10, we present detailed outlines of the various other proofs of the infinitesimal rigidity of the quadric of dimension ≥ 4 following [18] and [22]. Also in §9, we give a new proof of the infinitesimal rigidity of the quadric of dimension ≥ 5, which follows some of the lines of the proof of the infinitesimal rigidity of the quadric Q_4 of [22]. One of the main ingredients of our new proof for the quadric of dimension ≥ 5 is a vanishing result for sections of the vector bundle L proved in §8, Chapter V; in fact, it is quite different from the one found in [18], which totally avoids the use of harmonic analysis. In §11, we present the results of [20] and [23] which concern 1-forms; in particular, we show that a 1-form on Q_n satisfying the zero-energy condition or an even 1-form on Q_n satisfying the Guillemin condition is exact.

§2. Totally geodesic flat tori of the complex quadric

Throughout this chapter, we suppose that X is the complex quadric Q_n, with $n \geq 3$, endowed with the Kähler metric g introduced in §2, Chapter V. We shall consider the objects and use the notations established in Chapter V.

If Z is a flat totally geodesic 2-torus of X, we denote by ∇^Z the Levi-Civita connection of the Riemannian manifold Z endowed with the metric induced by g; if ζ is a vector field on Z, we consider the complex vector fields on X defined along Z by

$$\zeta' = \tfrac{1}{2}(\zeta - iJ\zeta), \qquad \zeta'' = \tfrac{1}{2}(\zeta + iJ\zeta) = \overline{\zeta'},$$

which are of type $(1,0)$ and $(0,1)$, respectively.

We now study an explicit maximal flat totally geodesic torus of X. The image Z_0 of the imbedding

$$\iota : \widetilde{G}_1^{\mathbb{R}}(V_1) \times \widetilde{G}_1^{\mathbb{R}}(V_2) \to \widetilde{G}_{2,n}^{\mathbb{R}}$$

of §2, Chapter IV, corresponding to the orthogonal 2-planes $V_1 = \mathbb{R}e_0 \oplus \mathbb{R}e_1$ and $V_2 = \mathbb{R}e_n \oplus \mathbb{R}e_{n+1}$ of $\mathbb{R}^{n+2}$, is a totally geodesic flat 2-torus of $\widetilde{G}_{2,n}^{\mathbb{R}}$. When we identify the Grassmannian $\widetilde{G}_{2,n}^{\mathbb{R}}$ with the complex quadric X as in §5, Chapter V, the torus Z_0 can be viewed as follows. As in Chapter V, we consider the point a of X which is the image of the point $(1, 0, \ldots, 0, i)$ of $\mathbb{C}^{n+2}$ under the natural projection

$$\pi : \mathbb{C}^{n+2} - \{0\} \to \mathbb{C}\mathbb{P}^{n+1}.$$

If K_0 is the subgroup of $G = SO(n + 2)$ consisting of all matrices

$$\begin{pmatrix} A & 0 & 0 \\ 0 & I_{n-2} & 0 \\ 0 & 0 & B \end{pmatrix},$$

where $A, B \in SO(2)$ and I_{n-2} is the $(n - 2) \times (n - 2)$ identity matrix, the submanifold Z_0 of X is equal to $K_0 \cdot a$ and is the image of the mapping $\sigma : \mathbb{R}^2 \to X$ defined by $\sigma(\theta, \varphi) = \pi\tilde{\sigma}(\theta, \varphi)$, where

$$\tilde{\sigma}(\theta, \varphi) = (\cos\theta, \sin\theta, 0, \ldots, 0, -i\sin\varphi, i\cos\varphi) \in \mathbb{C}^{n+2},$$

for $(\theta, \varphi) \in \mathbb{R}^2$. This mapping σ satisfies $\sigma(0, 0) = a$ and

$$\sigma(\theta, \varphi) = \sigma(\theta + 2k\pi, \varphi + 2l\pi) = \sigma(\theta + k\pi, \varphi + k\pi),$$

for all $k, l \in \mathbb{Z}$ and $(\theta, \varphi) \in \mathbb{R}^2$. We consider the group of translations Γ of $\mathbb{R}^2$ generated by the vectors $(2\pi, 0)$ and (π, π) and the flat torus $\mathbb{R}^2/\Gamma$, which is the quotient of $\mathbb{R}^2$ by the group Γ. According to the preceding relations, we see that σ induces an imbedding

$$\bar{\sigma} : \mathbb{R}^2/\Gamma \to X.$$

Let (θ, φ) be the standard coordinate system of $\mathbb{R}^2$. It is easily verified that

$$(6.1) \qquad \sigma^* g = \tfrac{1}{2}(d\theta \otimes d\theta + d\varphi \otimes d\varphi).$$

Therefore, if the quotient $\mathbb{R}^2/\Gamma$ is endowed with the flat metric induced by the metric $\tfrac{1}{2}(d\theta \otimes d\theta + d\varphi \otimes d\varphi)$ on $\mathbb{R}^2$, the mapping $\bar{\sigma}$ is a totally geodesic isometric imbedding. Throughout this chapter, we shall often identify a function f on $\mathbb{R}^2$ satisfying

$$f(\theta, \varphi) = f(\theta + 2k\pi, \varphi + 2l\pi) = f(\theta + k\pi, \varphi + k\pi),$$

for all $k, l \in \mathbb{Z}$ and $(\theta, \varphi) \in \mathbb{R}^2$, with the unique function $\hat{f}$ on the torus Z_0 satisfying the equality $\sigma^* \hat{f} = f$ on $\mathbb{R}^2$. The restriction of the mapping σ to the subset $\tilde{Z}_0 = [0, 2\pi] \times [0, \pi]$ of $\mathbb{R}^2$ is a diffeomorphism from $\tilde{Z}_0$ to Z_0. Therefore if f is a function on Z_0, then we see that

$$(6.2) \qquad \begin{aligned} \int_{Z_0} f\, dZ_0 &= \frac{1}{2} \int_{\tilde{Z}_0} (\sigma^* f)(\theta, \varphi)\, d\theta\, d\varphi = \frac{1}{2} \int_0^\pi \int_0^{2\pi} (\sigma^* f)(\theta, \varphi)\, d\theta\, d\varphi \\ &= \frac{1}{2} \int_0^\pi \int_0^\pi \left((\sigma^* f)(\theta, \varphi) + (\sigma^* f)(\theta + \pi, \varphi) \right) d\theta\, d\varphi. \end{aligned}$$

We now consider the objects introduced in §4, Chapter V. If (θ, φ) is an element of $\mathbb{R}^2$ satisfying $\cos \theta \neq 0$, the point $\sigma(\theta, \varphi)$ belongs to the open subset V of $\mathbb{CP}^{n+1}$ defined in §4, Chapter V and we have

$$z_1(\sigma(\theta, \varphi)) = \tan \theta, \qquad\qquad z_j(\sigma(\theta, \varphi)) = 0,$$

$$z_n(\sigma(\theta, \varphi)) = -i\,\frac{\sin \varphi}{\cos \theta}, \qquad z_{n+1}(\sigma(\theta, \varphi)) = i\,\frac{\cos \varphi}{\cos \theta},$$

for $2 \leq j \leq n-1$. The vector fields $\partial/\partial\theta$ and $\partial/\partial\varphi$ on $\mathbb{R}^2$ are σ-projectable; in other words, there exist well-defined parallel vector fields ξ_0 and η_0 on Z_0 such that

$$\xi_0(\sigma(\theta, \varphi)) = \sigma_* \left(\left(\frac{\partial}{\partial\theta} \right)(\theta, \varphi) \right), \qquad \eta_0(\sigma(\theta, \varphi)) = \sigma_* \left(\left(\frac{\partial}{\partial\varphi} \right)(\theta, \varphi) \right),$$

for $(\theta, \varphi) \in \mathbb{R}^2$. In fact, $\{\xi_0, \eta_0\}$ is a basis for the space of parallel vector fields on Z_0; according to (6.1), we have

$$g(\xi_0, \xi_0) = g(\eta_0, \eta_0) = \frac{1}{2}, \qquad g(\xi_0, \eta_0) = 0.$$

We see that the equalities

(6.3)

$$\xi_0' = \frac{1}{\cos^2 \theta} \left(\frac{\partial}{\partial z_1} - i \sin \theta \sin \varphi \, \frac{\partial}{\partial z_n} + i \sin \theta \cos \varphi \, \frac{\partial}{\partial z_{n+1}} \right),$$

$$\eta_0' = -\frac{i}{\cos \theta} \left(\cos \varphi \, \frac{\partial}{\partial z_n} + \sin \varphi \, \frac{\partial}{\partial z_{n+1}} \right)$$

hold at the point $\sigma(\theta, \varphi)$ of Z_0, for $(\theta, \varphi) \in \mathbb{R}^2$ satisfying $\cos \theta \neq 0$.

When $\cos \theta \cos \varphi \neq 0$, the point $\sigma(\theta, \varphi)$ belongs to the open subset $U_0 = Z_0 \cap U$ of Z_0. The restrictions of the complex vector fields η_k to U_0 are determined by the equalities

(6.4) $\eta_1 = \cos^2 \theta \cdot \xi_0', \qquad \eta_j = \partial/\partial z_j, \qquad \eta_n = i \cos \theta \cos \varphi \cdot \eta_0',$

with $2 \leq j \leq n-1$, which hold at the point $\sigma(\theta, \varphi)$ whenever $(\theta, \varphi) \in \mathbb{R}^2$ satisfies $\cos \theta \cos \varphi \neq 0$. We consider the section ν of the bundle S over U given by (5.40) and the involutive endomorphism K_ν of $T_{|U}$ defined by $K_\nu(x) = K_{\nu(x)}$, for $x \in U$; from the relation (5.51), we obtain the equalities

(6.5) $K_\nu \xi_0 = \xi_0, \qquad K_\nu \eta_0 = -\eta_0$

on the open subset U_0 of Z_0. It follows that the restriction of this involution K_ν to T_{Z_0} satisfies

(6.6) $K_\nu(T_{Z_0|U_0}) \subset T_{Z_0|U_0}.$

As in Chapter II, we denote by Ξ the space of all maximal flat totally geodesic tori of X. Since the point a belongs to Z_0, according to the description of Ξ given in §6, Chapter V, we see that

$$Z_0 = \mathrm{Exp}_a(\mathbb{R}\xi_0(a) \oplus \mathbb{R}\eta_0(a)).$$

LEMMA 6.1. *Let Z be a totally geodesic flat 2-torus of X and let $x_0 \in Z$. Then there exist an open neighborhood U' of x_0 in Z, an involution κ of $T_{|U'}$ which preserves the tangent bundle of Z and a section μ of S over U' such that $\kappa(x) = K_{\mu(x)}$, for all $x \in U'$. Moreover, the restriction of* *this involution* κ *to* T_Z *is an endomorphism of* $T_{Z|U'}$ *which is parallel with respect to the connection* ∇^Z.

PROOF: Since the group G acts transitively both on Ξ and on the torus $Z \in \Xi$, without loss of generality by (5.14) we may assume that Z is the torus Z_0 described above and that x_0 is the point a of the subset U_0 of Z_0. Then if ν is the section of S over U given by (5.40), according to (6.6) we know that the involution K_ν of $T_{|U}$ preserves the tangent bundle of Z_0. Let ξ, η be tangent vectors to Z_0 at $x \in U_0$; if κ is the restriction of the involution K_ν to T_{Z_0}, we know that

$$(\nabla^Z_\xi \kappa)\eta = (\nabla_\xi K_\nu)\eta.$$

According to (5.22), the right-hand side of this equality belongs to $JT_{Z_0,x}$; since Z_0 is a totally real submanifold of X, it vanishes. Thus we have $\nabla^Z \kappa = 0$.

LEMMA 6.2. *Let* Z *be a totally geodesic flat 2-torus of* X. *Then there exists a unique (up to a sign) involution* κ *of* $T_{|Z}$ *which preserves the tangent bundle of* Z *and which at every point* x *of* Z *is equal to a real structure* K_μ *of* X, *where* $\mu \in S_x$. *Moreover, the restriction of this involution* κ *to* T_Z *is an endomorphism of* T_Z *which is parallel with respect to the connection* ∇^Z.

PROOF: Let x be a point of Z. According to §6, Chapter V, we may write

$$Z = \mathrm{Exp}_x(\mathbb{R}\xi \oplus \mathbb{R}J\eta),$$

where μ is an appropriately chosen element of S_x and $\{\xi, \eta\}$ is an orthonormal set of elements of T_μ^+. Clearly, K_μ preserves the tangent space to Z at x. According to (5.9), a real structure κ' of the quadric X associated with another unit normal of S_x can be written in the form

$$\kappa' = \cos\theta \cdot K_\mu + \sin\theta \cdot JK_\mu,$$

where $\theta \in \mathbb{R}$. We see that κ' preserves the tangent space to Z at x if and only if $\sin\theta = 0$, that is, if $\kappa' = \pm K_\mu$. From this observation and the orientability of Z, by Lemma 6.1 we obtain the desired endomorphism κ of $T_{|Z}$; clearly, it is unique up to a sign and is parallel with respect to the connection ∇^Z.

The involution κ, which Lemma 6.2 associates with a totally geodesic flat 2-torus Z contained in X, is called a real structure of the torus Z; it is uniquely determined up to a sign.

According to Lemma 6.2 and its proof, there exists a unique real structure κ_0 of the torus Z_0 such that

$$\kappa_0(a) = K_{\nu(a)},$$

and we know that the vector fields $\kappa_0\xi_0$ and $\kappa_0\eta_0$ on Z_0 are parallel. Hence by (6.5), we see that

$$(6.7) \qquad\qquad \kappa_0\xi_0 = \xi_0, \qquad \kappa_0\eta_0 = -\eta_0$$

and that the restriction of κ_0 to U_0 is equal to the involutive endomorphism K_ν of $T_{|U_0}$.

Let Z be a totally geodesic flat 2-torus contained in X; we choose a real structure κ of Z. Since the restriction of κ to T_Z is parallel, the tangent bundle T_Z admits an orthogonal decomposition

$$T_Z = T_Z^+ \oplus T_Z^-$$

invariant under ∇^Z, where T_Z^+ and T_Z^- are the eigenbundles of the restriction of κ to T_Z, corresponding to the eigenvalues $+1$ and -1. Clearly, this decomposition of T_Z is independent of the choice of κ. It is easily seen that there exist unitary parallel sections ξ of T_Z^+ and η of T_Z^-; these two vector fields are unique up to a sign and $\{\xi,\eta\}$ is a basis for the space of parallel vector fields on Z.

Since the group G acts transitively on Ξ, we see that an element h of $C^\infty(S^2T^*)$ satisfies the Guillemin condition if and only if

$$\int_{Z_0} (\phi^*h)(\xi_0,\xi_0)\,dZ_0 = \int_{Z_0} (\phi^*h)(\eta_0,\eta_0)\,dZ_0 = \int_{Z_0} (\phi^*h)(\xi_0,\eta_0)\,dZ_0 = 0,$$

for all $\phi \in G$.

Now let ξ, η be arbitrary vector fields on Z satisfying $\kappa\xi = \xi$ and $\kappa\eta = -\eta$. If h is a section of $(S^2T^*)^+$ over X, we have

$$(\pi_{++}h)(\xi,\xi) = h(\xi,\xi), \qquad (\pi_{++}h)(\eta,\eta) = h(\eta,\eta),$$

$$(6.8) \qquad (\pi_{++}h)(\xi,\eta) = (\pi_{+-}h)(\xi,\xi) = (\pi_{+-}h)(\eta,\eta) = 0,$$

$$(\pi_{+-}h)(\xi,\eta) = h(\xi,\eta).$$

Let f be an element of $C^\infty(X)$. If ζ_1,ζ_2 are vector fields on Z, according to Lemma 1.25, we have

$$(\pi_+\mathrm{Hess}\, f)(\zeta_1',\zeta_2'') = (\partial\bar\partial f)(\zeta_1',\zeta_2'').$$

Thus by (6.8), if ζ is a vector field on Z equal either to ξ or to η, we obtain the relations

$$(6.9) \quad (\pi_{++}\mathrm{Hess}\, f)(\zeta,\zeta) = 2(\pi_+\mathrm{Hess}\, f)(\zeta',\zeta'') = 2(\partial\bar\partial f)(\zeta',\zeta''),$$

$$(6.10) \quad \begin{aligned}(\pi_{+-}\mathrm{Hess}\, f)(\xi,\eta) &= (\pi_+\mathrm{Hess}\, f)(\xi',\eta'') + (\pi_+\mathrm{Hess}\, f)(\eta',\xi'') \\ &= (\partial\bar\partial f)(\xi',\eta'') + (\partial\bar\partial f)(\eta',\xi'').\end{aligned}$$

From the above discussion concerning the parallel vector fields on the totally geodesic flat 2-torus Z and from the relations (6.8), we deduce the following result:

PROPOSITION 6.3. *Let h be a section of $(S^2T^*)^+$ over $X = Q_n$, with $n \geq 3$. The symmetric 2-form h satisfies the Guillemin condition if and only if the two symmetric 2-forms $\pi_{++}h$ and $\pi_{+-}h$ satisfy the Guillemin condition.*

Let N be an odd integer ≥ 1; we consider the closed geodesic

$$\delta_N : [0, \pi] \to Z_0$$

of Z_0, defined by $\delta_N(t) = \sigma(t, Nt)$, for $0 \leq t \leq \pi$. The tangent vector field $\dot{\delta}_N$ along the geodesic δ_N is determined by

$$\dot{\delta}_N(t) = (\xi_0 + N\eta_0)(\delta_N(t)),$$

for $0 \leq t \leq \pi$; we see that $\dot{\delta}_N$ has constant length equal to $\sqrt{(N^2+1)/2}$. Thus if h is a symmetric 2-form on X, we have

$$
\begin{aligned}
\int_{\delta_N} h &= \frac{2}{N^2+1} \int_0^\pi h(\dot{\delta}_N(t), \dot{\delta}_N(t))\, dt \\
&= \frac{2}{N^2+1} \int_0^\pi h(\xi_0 + N\eta_0, \xi_0 + N\eta_0)(\delta_N(t))\, dt.
\end{aligned}
$$

(6.11)

Hence by the formulas (6.7) and (6.8), if h is a section of $(S^2T^*)^{++}$ over X, we have

$$(6.12) \qquad \int_{\delta_N} h = c_N \left(\int_0^\pi h(\xi_0, \xi_0)(\delta_N(t))\, dt + N^2 \int_0^\pi h(\eta_0, \eta_0)(\delta_N(t))\, dt \right),$$

where $c_N = 2/(N^2+1)$, while if h is a section of $(S^2T^*)^{+-}$ over X, we have

$$(6.13) \qquad \int_{\delta_N} h = \frac{2N}{N^2+1} \int_0^\pi h(\xi_0, \eta_0)(\delta_N(t))\, dt.$$

We shall use the formulas (6.12) and (6.13) with $N = 1$ or 3, and we write $\delta = \delta_1$.

§3. Symmetric forms on the complex quadric

We now introduce certain symmetric 2-forms on X which are defined in [23, §4]; we shall also recall some of their properties which are established in [23].

We remark that the Hermitian symmetric 2-forms

$$u_1 = ((\zeta_0 + i\zeta_1)d(\zeta_2 + i\zeta_3) - (\zeta_2 + i\zeta_3)d(\zeta_0 + i\zeta_1))$$
$$\cdot ((\bar\zeta_0 + i\bar\zeta_1)d\bar\zeta_{n+1} - \bar\zeta_{n+1}d(\bar\zeta_0 + i\bar\zeta_1)),$$

$$u_2 = ((\zeta_0 + i\zeta_1)d\zeta_{n+1} - \zeta_{n+1}d(\zeta_0 + i\zeta_1))$$
$$\cdot ((\bar\zeta_0 + i\bar\zeta_1)d(\bar\zeta_2 + i\bar\zeta_3) - (\bar\zeta_2 + i\bar\zeta_3)d(\bar\zeta_0 + i\bar\zeta_1))$$

on $\mathbb{C}^{n+2}$ are $U(1)$-invariant. If ζ^0 is the complex vector field on $\mathbb{C}^{n+2}$ introduced in §4, Chapter III, we easily verify that

$$\zeta^0 \lrcorner u_j = \bar\zeta^0 \lrcorner u_j = 0,$$

for $j = 1, 2$. Therefore, as we saw in §4, Chapter III, the symmetric 2-forms u_1 and u_2 induce Hermitian symmetric 2-forms on $\mathbb{CP}^{n+1}$, which we also denote by u_1 and u_2 and whose restrictions to X we denote by $\tilde u_1$ and $\tilde u_2$, respectively. Clearly, if τ is the involutive isometry of $\mathbb{CP}^{n+1}$ induced by the complex conjugation of $\mathbb{C}^{n+2}$, we have the equality $\tau^* u_1 = u_2$ on $\mathbb{CP}^{n+1}$, and thus the symmetric 2-form

$$k = \tilde u_1 - \tilde u_2$$

on X is an odd form, and so is an element of $C^\infty((S^2T^*)_{\mathbb{C}}^+)^{\mathrm{odd}}$. By (5.35), the sections

$$k^+ = \pi_{++}k, \qquad k^- = \pi_{+-}k$$

of $(S^2T^*)_{\mathbb{C}}^+$ are elements of $C^\infty((S^2T^*)_{\mathbb{C}}^{++})^{\mathrm{odd}}$ and $C^\infty((S^2T^*)_{\mathbb{C}}^{+-})^{\mathrm{odd}}$, respectively.

We consider the sections

$$h_1' = \pi'\mathrm{Hess}\,\tilde f_{1,0}, \qquad h_1'' = \pi''\mathrm{Hess}\,\tilde f_{1,0},$$
$$h_2' = \partial\tilde f_{1,0} \cdot \partial\tilde f_{1,0}, \qquad h_2'' = \bar\partial\tilde f_{1,0} \cdot \bar\partial\tilde f_{1,0},$$
$$h_3' = \partial\tilde f_{1,0} \cdot \partial\tilde f_{0,1}, \qquad h_3'' = \bar\partial\tilde f_{1,0} \cdot \bar\partial\tilde f_{0,1},$$
$$h_4' = \partial\tilde f_{0,1} \cdot \partial\tilde f_{0,1}, \qquad h_4'' = \bar\partial\tilde f_{0,1} \cdot \bar\partial\tilde f_{0,1},$$
$$k_1' = \partial\tilde f_{1,0} \cdot (\tilde f'\partial\tilde f_{0,1} - \tilde f_{0,1}\partial\tilde f'), \qquad k_1'' = \bar\partial\tilde f_{1,0} \cdot (\tilde f'\bar\partial\tilde f_{0,1} - \tilde f_{0,1}\bar\partial\tilde f'),$$
$$k_2' = \partial\tilde f_{0,1} \cdot (\tilde f'\partial\tilde f_{0,1} - \tilde f_{0,1}\partial\tilde f'), \qquad k_2'' = \bar\partial\tilde f_{0,1} \cdot (\tilde f'\bar\partial\tilde f_{0,1} - \tilde f_{0,1}\bar\partial\tilde f')$$

of $(S^2T^*)_{\mathbb{C}}^-$, the sections

$$w_1 = \pi_{++}(\partial\tilde f_{1,0} \cdot \bar\partial\tilde f_{0,1} - \partial\tilde f_{0,1} \cdot \bar\partial\tilde f_{1,0}),$$
$$w_2 = \pi_{++}(\tilde f_{0,1}\,\mathrm{Hess}\,(\tilde f_{0,1}\tilde f') - \tilde f'\,\mathrm{Hess}\,\tilde f_{0,2})$$

of $(S^2 T^*)_{\mathbb{C}}^{++}$ and the section

$$w_3 = \pi_{+-}\left(\tilde{f}_{0,1} \operatorname{Hess} \tilde{f}' - \tilde{f}' \operatorname{Hess} \tilde{f}_{0,1}\right)$$

of $(S^2 T^*)_{\mathbb{C}}^{+-}$.

We fix integers $r, s \geq 0$. We consider the subspace $V_{r,s}$ of $C^\infty(S^2 T_{\mathbb{C}}^*)$ generated (over $\mathbb{C}$) by the sections

$$h_1 = \tilde{f}_{r,s} g, \quad h_2 = \tilde{f}_{r-1,s}\pi_{++}\operatorname{Hess} \tilde{f}_{1,0}, \quad h_3 = \tilde{f}_{r,s-2}\pi_{++}(\partial \tilde{f}_{0,1} \cdot \bar{\partial} \tilde{f}_{0,1})$$

of $(S^2 T^*)_{\mathbb{C}}^{++}$, the sections

$$h_4 = \tilde{f}_{r-1,s}(h_1' + h_1''), \qquad h_5 = \tilde{f}_{r-2,s}(h_2' + h_2''),$$
$$h_6 = \tilde{f}_{r-1,s-1}(h_3' + h_3''), \qquad h_7 = \tilde{f}_{r,s-2}(h_4' + h_4'')$$

of $(S^2 T^*)_{\mathbb{C}}^-$ and the section $h_8 = \tilde{f}_{r,s-1}\pi_{+-}\operatorname{Hess} \tilde{f}_{0,1}$ of $(S^2 T^*)_{\mathbb{C}}^{+-}$. By the relations (5.35)–(5.37) and (5.113), we see that, when s is an even (resp. odd) integer, all the sections h_j, with $1 \leq j \leq 8$, are even (resp. odd). Thus we have the inclusions

$$(6.14) \qquad V_{r,2p} \subset C^\infty(S^2 T_{\mathbb{C}}^*)^{\mathrm{ev}}, \qquad V_{r,2p+1} \subset C^\infty(S^2 T_{\mathbb{C}}^*)^{\mathrm{odd}},$$

for all $p \geq 0$. Lemmas 7.2, 7.5 and 7.6,(i) of [23] give us the following:

LEMMA 6.4. *Let $r, s \geq 0$ be given integers. The non-zero elements of the set $\{h_j\}_{1 \leq j \leq 8}$ of generators of the space $V_{r,s}$ form a basis of $V_{r,s}$. More precisely, the dimension and a basis of $V_{r,s}$ are given by the following table:*

	dim $V_{r,s}$	Basis of $V_{r,s}$
$r = s = 0$	1	h_1
$r = 1, s = 0$	3	h_1, h_2, h_4
$r \geq 2, s = 0$	4	h_1, h_2, h_4, h_5
$r = 0, s = 1$	2	h_1, h_8
$r = 1, s = 1$	5	h_1, h_2, h_4, h_6, h_8
$r \geq 2, s = 1$	6	$h_1, h_2, h_4, h_5, h_6, h_8$
$r = 0, s \geq 2$	4	h_1, h_3, h_7, h_8
$r = 1, s \geq 2$	7	$h_1, h_2, h_3, h_4, h_6, h_7, h_8$
$r, s \geq 2$	8	h_j, with $1 \leq j \leq 8$

We also consider the subspace $W_{r,s}$ of $C^\infty(S^2T_{\mathbb{C}}^*)$ generated (over $\mathbb{C}$) by the section $\tilde{h}_1 = \tilde{f}_{r-1,s-1}w_1$ of $(S^2T^*)_{\mathbb{C}}^{++}$ and the sections

$$\tilde{h}_2 = \tilde{f}_{r-1,s}(h_1' - h_1''), \qquad \tilde{h}_3 = \tilde{f}_{r-2,s}(h_2' - h_2''),$$
$$\tilde{h}_4 = \tilde{f}_{r-1,s-1}(h_3' - h_3''), \qquad \tilde{h}_5 = \tilde{f}_{r,s-2}(h_4' - h_4'')$$

of $(S^2T^*)_{\mathbb{C}}^-$. By the relations (5.35)–(5.37) and (5.113), we see that, when s is an even (resp. odd) integer, all these generators of $W_{r,s}$ are odd (resp. even). Thus we have the inclusions

$$(6.15) \qquad W_{r,2p+1} \subset C^\infty(S^2T_{\mathbb{C}}^*)^{\mathrm{ev}}, \qquad W_{r,2p} \subset C^\infty(S^2T_{\mathbb{C}}^*)^{\mathrm{odd}},$$

for all $p \geq 0$. Lemma 7.7 of [23] asserts the following:

LEMMA 6.5. *Let $r, s \geq 0$ be given integers. The non-zero elements of the set of generators of the space $W_{r,s}$ form a basis of $W_{r,s}$. More precisely, the dimensions and bases of the non-zero spaces $W_{r,s}$ are given by the following table:*

	dim $W_{r,s}$	Basis of $W_{r,s}$
$r = 1,\, s = 0$	1	$\tilde{h}_2$
$r \geq 2,\, s = 0$	2	$\tilde{h}_2, \tilde{h}_3$
$r = 1,\, s = 1$	3	$\tilde{h}_1, \tilde{h}_2, \tilde{h}_4$
$r \geq 2,\, s = 1$	4	$\tilde{h}_1, \tilde{h}_2, \tilde{h}_3, \tilde{h}_4$
$r = 0,\, s \geq 2$	1	$\tilde{h}_5$
$r = 1,\, s \geq 2$	4	$\tilde{h}_1, \tilde{h}_2, \tilde{h}_4, \tilde{h}_5$
$r, s \geq 2$	5	$\tilde{h}_j$, with $1 \leq j \leq 5$

If $s \geq 1$, we consider the subspace $V_{r,s}'$ of $C^\infty(S^2T_{\mathbb{C}}^*)$ generated (over $\mathbb{C}$) by the sections $\vartheta_1 = \tilde{f}_{r,s-1}k^-$ and $\vartheta_2 = \tilde{f}_{r,s-1}k^+$ of $(S^2T^*)_{\mathbb{C}}^+$ and the sections

$$\vartheta_3 = \tilde{f}_{r-1,s-1}(k_1' - k_1''), \qquad \vartheta_4 = \tilde{f}_{r,s-2}(k_2' - k_2'')$$

of $(S^2T^*)_{\mathbb{C}}^-$. Let $W_{r,s}'$ be the subspace of $C^\infty(S^2T_{\mathbb{C}}^*)$ generated (over $\mathbb{C}$) by the sections $\tilde{\vartheta}_1 = \tilde{f}_{r,s-2}w_2$ and $\tilde{\vartheta}_2 = \tilde{f}_{r,s-1}w_3$ of $(S^2T^*)_{\mathbb{C}}^+$ and the sections

$$\tilde{\vartheta}_3 = \tilde{f}_{r-1,s-1}(k_1' + k_1''), \qquad \tilde{\vartheta}_4 = \tilde{f}_{r,s-2}(k_2' + k_2'')$$

of $(S^2T^*)_{\mathbb{C}}^{-}$. By (5.35), (5.37) and (5.113), we see that, when s is an even (resp. odd) integer, these generators of $V'_{r,s}$ are even (resp. odd) and these generators of $W'_{r,s}$ are odd (resp. even). Thus we have the inclusions

$$\text{(6.16)} \qquad \begin{aligned} V'_{r,2p+2} &\subset C^\infty(S^2T_{\mathbb{C}}^*)^{\mathrm{ev}}, & V'_{r,2p+1} &\subset C^\infty(S^2T_{\mathbb{C}}^*)^{\mathrm{odd}}, \\ W'_{r,2p+1} &\subset C^\infty(S^2T_{\mathbb{C}}^*)^{\mathrm{ev}}, & W'_{r,2p} &\subset C^\infty(S^2T_{\mathbb{C}}^*)^{\mathrm{odd}}, \end{aligned}$$

for all $p \geq 0$.

Lemma 7.9 of [23] asserts the following:

LEMMA 6.6. *If $r \geq 0$, $s \geq 1$, the non-zero elements of the set of generators of the spaces $V'_{r,s}$ and $W'_{r,s}$ form bases of these spaces. More precisely, the dimensions and bases of the non-zero spaces $V'_{r,s}$ and $W'_{r,s}$ are given by the following table:*

	dim $V'_{r,s}$	Basis of $V'_{r,s}$	dim $W'_{r,s}$	Basis of $W'_{r,s}$
$r = 0, s = 1$	2	ϑ_1, ϑ_2	1	$\tilde\vartheta_2$
$r = 0, s \geq 2$	3	$\vartheta_1, \vartheta_2, \vartheta_4$	3	$\tilde\vartheta_1, \tilde\vartheta_2, \tilde\vartheta_4$
$r \geq 1, s = 1$	3	$\vartheta_1, \vartheta_2, \vartheta_3$	2	$\tilde\vartheta_2, \tilde\vartheta_3$
$r \geq 1, s \geq 2$	4	$\vartheta_1, \vartheta_2, \vartheta_3, \vartheta_4$	4	$\tilde\vartheta_1, \tilde\vartheta_2, \tilde\vartheta_3, \tilde\vartheta_4$

We remark that

$$h_8 = \tilde{f}_{r,s-1}\mathrm{Hess}\,\tilde{f}_{0,1}, \qquad \tilde\vartheta_2 = \tilde{f}_{r,s-1}w_3$$

are sections of the bundle $(S^2T^*)_{\mathbb{C}}^{+-}$, while all the other generators of the spaces $V_{r,s}$, $W_{r,s}$, with $s \geq 0$, and of the spaces $V'_{r,s}$, $W'_{r,s}$, with $s \geq 1$, are sections of the bundle $(S^2T^*)_{\mathbb{C}}^{-} \oplus (S^2T^*)_{\mathbb{C}}^{++}$, which is the orthogonal complement of $(S^2T^*)_{\mathbb{C}}^{+-}$ in $S^2T_{\mathbb{C}}^*$.

The following two lemmas are given by Lemmas 7.6, 7.8 and 7.10 of [23].

LEMMA 6.7. *Let $r, s \geq 0$ be given integers.*

(i) *If s is even, we have*

$$\dim (\mathcal{N}_{2,\mathbb{C}} \cap V_{r,s}) \leq d_{r,s}.$$

(ii) *If s is odd, we have*

$$\dim (\mathcal{N}_{2,\mathbb{C}} \cap W_{r,s}) \leq d_{r,s}.$$

LEMMA 6.8. *Let $r \geq 0$, $s \geq 1$ be given integers.*

(i) *If s is even, we have*

$$\dim\left(\mathcal{N}_{2,\mathbb{C}} \cap V'_{r,s}\right) \le 1.$$

(ii) *If s is odd, we have*

$$\dim\left(\mathcal{N}_{2,\mathbb{C}} \cap W'_{r,s}\right) \le 1.$$

§4. Computing integrals of symmetric forms

In this section and the next one, we shall compute integrals of symmetric 2-forms over specific closed geodesics of X. Each of these geodesics is contained in one of the families of flat 2-tori of X considered in [23, §4]. Thus many of the computations appearing in these two sections of this chapter are the same or similar to those of [23, §4].

We consider the torus Z_0 introduced in §2 and we shall use the objects associated there with this torus. In particular, we shall identify a function f on $\mathbb{R}^2$ satisfying

$$f(\theta, \varphi) = f(\theta + 2k\pi, \varphi + 2l\pi) = f(\theta + k\pi, \varphi + k\pi),$$

for all $k, l \in \mathbb{Z}$ and $(\theta, \varphi) \in \mathbb{R}^2$, with the unique function $\hat{f}$ on the torus Z_0 satisfying the equality $\sigma^*\hat{f} = f$ on $\mathbb{R}^2$.

The restriction $\tilde{\rho}$ to X of the function ρ on $\mathbb{CP}^{n+1}$ satisfies

$$(6.17) \qquad \tilde{\rho}(\sigma(\theta, \varphi)) = \frac{\cos^2\theta}{2},$$

for $(\theta, \varphi) \in \mathbb{R}^2$. Using formulas (6.3) and (6.17), we easily verify that the equalities

$$\langle \xi'_0, \partial\tilde{\rho} \rangle = -\frac{\sin\theta\cos\theta}{2}, \qquad \langle \eta'_0, \partial\tilde{\rho} \rangle = 0,$$

$$(6.18) \qquad (\partial\bar{\partial}\tilde{\rho})(\xi'_0, \xi''_0) = \frac{1}{4}\left(3\sin^2\theta - 1\right), \qquad (\partial\bar{\partial}\tilde{\rho})(\xi'_0, \eta''_0) = 0,$$

$$(\partial\bar{\partial}\tilde{\rho})(\eta'_0, \eta''_0) = -\frac{\cos^2\theta}{4}$$

hold at all points $\sigma(\theta, \varphi)$ of Z_0, with $(\theta, \varphi) \in \mathbb{R}^2$, for which $\cos\theta \ne 0$, and hence at all points $\sigma(\theta, \varphi)$ of Z_0.

For $\alpha \in \mathbb{R}$, let ψ_α be the element of $G = SO(n+2)$ defined by

$$\psi_\alpha(\zeta)_0 = \sin\alpha \cdot \zeta_{n+1} + \cos\alpha \cdot \zeta_2, \qquad \psi_\alpha(\zeta)_{n+1} = \cos\alpha \cdot \zeta_{n+1} - \sin\alpha \cdot \zeta_2,$$

$$\psi_\alpha(\zeta)_2 = \zeta_n, \qquad \psi_\alpha(\zeta)_3 = \zeta_0, \qquad \psi_\alpha(\zeta)_n = \delta_3^n \zeta_0 + (1 - \delta_3^n)\zeta_3,$$

and $\psi_\alpha(\zeta)_j = \zeta_j$ for $j = 0$ or $3 < j < n$, where $\zeta \in \mathbb{C}^{n+2}$. We set $\mu = \cos\alpha$ and $\lambda = \sin\alpha$. Then for $r, s \geq 0$, at a point z of the complex hypersurface $\tilde{V}$ of the open subset V of $\mathbb{CP}^{n+1}$ defined by the equation $z_2 = 0$, the functions $\psi_\alpha^* f_{r,s}$ are determined by

$$(\psi_\alpha^* f_{1,0})(z) = \rho(z)(\lambda z_{n+1} + iz_1)(\lambda \bar{z}_{n+1} + i\bar{z}_1),$$

$$(\psi_\alpha^* f_{0,1})(z) = \rho(z)((\lambda z_{n+1} + iz_1)(\bar{z}_n + i) - (z_n + i)(\lambda \bar{z}_{n+1} + i\bar{z}_1)).$$

Using (6.3), (6.17) and (6.18), we easily verify that the equalities

$$\psi_\alpha^* \tilde{f}_{1,0} = \tfrac{1}{2}(\lambda^2 \cos^2\varphi - \sin^2\theta),$$

$$\psi_\alpha^* \tilde{f}_{0,1} = -(\lambda \cos\theta \cos\varphi + \sin\theta \sin\varphi),$$

(6.19)

$$(\psi_\alpha^* \partial\bar{\partial}\tilde{f}_{0,1})(\xi_0', \eta_0'') = (\psi_\alpha^* \partial\bar{\partial}\tilde{f}_{0,1})(\eta_0', \xi_0'')$$

$$= -\tfrac{1}{2}(\cos\theta \cos\varphi + \lambda \sin\theta \sin\varphi),$$

hold at the point $\sigma(\theta, \varphi)$ of Z_0, with $(\theta, \varphi) \in \mathbb{R}^2$. For $r, s \geq 0$, by (6.7) and (6.10) we now obtain

(6.20)
$$\psi_0^*(\tilde{f}_{r,s}\pi_{+-}\text{Hess }\tilde{f}_{0,1})(\xi_0, \eta_0)$$

$$= \frac{(-1)^{r+s+1}}{2^r} \sin^{2r+s}\theta \cdot \cos\theta \cdot \sin^s\varphi \cdot \cos\varphi$$

at this point $\sigma(\theta, \varphi)$ of Z_0.

LEMMA 6.9. For $r, s \geq 0$, the integral

$$\int_\delta \psi_0^*(\tilde{f}_{r,s}\pi_{+-}\text{Hess }\tilde{f}_{0,1})$$

does not vanish.

PROOF: According to formula (6.20), we have

$$\psi_0^*(\tilde{f}_{r,s}\pi_{+-}\text{Hess }\tilde{f}_{0,1})(\xi_0, \eta_0)(\delta(t)) = \frac{(-1)^{r+s+1}}{2^r} \cos^2 t \cdot \sin^{2(r+s)} t,$$

for $0 \leq t \leq \pi$. Since $\cos^2 t \cdot \sin^{2(r+s)} t \geq 0$, the lemma is a consequence of the equality (6.13) and the above relation.

Let α be a given real number; we set $\mu = \cos \alpha$ and $\lambda = \sin \alpha$. If $j : \tilde{V} \to V$ is the inclusion mapping of the complex hypersurface $\tilde{V}$ into V, the symmetric 2-forms $j^*\psi_\alpha^* u_1$ and $j^*\psi_\alpha^* u_2$ on $\tilde{V}$ satisfy the relations

$$j^*\psi_\alpha^* u_1 = i\mu\rho^2((\lambda z_{n+1} + iz_1)dz_n - (z_n + i)(\lambda dz_{n+1} + idz_1))$$
$$\cdot (\bar{z}_1 d\bar{z}_{n+1} - \bar{z}_{n+1} d\bar{z}_1),$$
$$j^*\psi_\alpha^* u_2 = i\mu\rho^2(z_1 dz_{n+1} - z_{n+1}dz_1)$$
$$\cdot ((\lambda \bar{z}_{n+1} + i\bar{z}_1)d\bar{z}_n - (\bar{z}_n + i)(\lambda d\bar{z}_{n+1} + id\bar{z}_1))$$

at $z \in \tilde{V}$. We again consider the symmetric 2-forms $k = \tilde{u}_1 - \tilde{u}_2$ and $k^- = \pi_{+-} k$ on X introduced in §3. Using formulas (6.3) and (6.17) and the preceding relations, we verify that the equalities

$$(\psi_\alpha^* k)(\xi_0', \eta_0'') = -\frac{\mu}{4}(\lambda \cos \varphi(\cos \theta + \sin \varphi) + \sin \theta(\sin \varphi - \cos \theta)),$$
$$(\psi_\alpha^* k)(\eta_0', \xi_0'') = -\frac{\mu}{4}(\lambda \cos \varphi(\cos \theta - \sin \varphi) + \sin \theta(\sin \varphi + \cos \theta))$$

hold at the point $\sigma(\theta, \varphi)$ of Z_0, with $(\theta, \varphi) \in \mathbb{R}^2$; by (6.7) and (6.8), we see that

$$(6.21) \quad \begin{aligned} (\psi_\alpha^* k^-)(\xi_0, \eta_0) &= (\psi_\alpha^* k)(\xi_0, \eta_0) = (\psi_\alpha^* k)(\xi_0', \eta_0'') + (\psi_\alpha^* k)(\eta_0', \xi_0'') \\ &= -\frac{\mu}{2}(\lambda \cos \theta \cos \varphi + \sin \theta \sin \varphi) \end{aligned}$$

at this point of Z_0.

Let $r, s \geq 0$ be given integers. For $u, t \in \mathbb{R}$, we set

$$q_{r,s}(u, t) = (u^2 \cos^2 t - \sin^2 t)^r \cdot (u \cos^2 t + \sin^2 t)^{s+1},$$

and we consider the polynomial

$$Q_{r,s}(u) = \int_0^\pi q_{r,s}(u, t) \, dt$$

in u. From formulas (6.19) and (6.21), we obtain

$$(6.22) \qquad \psi_\alpha^*(\tilde{f}_{r,s} k^-)(\xi_0, \eta_0)(\delta(t)) = \frac{(-1)^{s+1}}{2^{r+1}} \cos \alpha \cdot q_{r,s}(\sin \alpha, t),$$

for $0 \leq t \leq \pi$.

LEMMA 6.10. *Let $r, s \geq 0$ be given integers. Then there exists $\alpha_0 \in \mathbb{R}$ such that the integral*

$$\int_\delta \psi_{\alpha_0}^* (\tilde{f}_{r,s} k^-)$$

does not vanish.

PROOF: The coefficient of u^{2r+s+1} of the polynomial $Q_{r,s}(u)$ is equal to the integral

$$\int_0^\pi \cos^{2(r+s+1)} t \, dt,$$

which is positive, and so the polynomial $Q_{r,s}$ is non-zero. Hence there exists a real number α_0 such that the expressions $\cos \alpha_0$ and $Q_{r,s}(\sin \alpha_0)$ do not vanish. Therefore by (6.12) and (6.22), we infer that the integral of the lemma corresponding to this element $\alpha_0 \in \mathbb{R}$ does not vanish.

Let ψ' be the element of G defined by

$$\psi'(\zeta)_1 = \zeta_n, \qquad \psi'(\zeta)_3 = \zeta_{n+1},$$

$$\psi'(\zeta)_{n+1} = \zeta_1, \qquad \psi'(\zeta)_n = \delta_3^n \zeta_{n+1} + (\delta_3^n - 1)\zeta_3,$$

and $\psi'(\zeta)_j = \zeta_j$, for $j = 0, 2$, or $3 < j < n$, where $\zeta \in \mathbb{C}^{n+2}$. Then for $r, s \geq 0$, the restrictions of the functions $\psi'^* f_{r,s}$ and $\psi'^* f'$ to $\tilde{V}$ are determined by

$$(\psi'^* f_{1,0})(z) = \rho(z)(1 + iz_n)(1 + i\bar{z}_n),$$
$$(\psi'^* f_{0,1})(z) = i\rho(z)((1 + iz_n)\bar{z}_{n+1} - z_{n+1}(1 + i\bar{z}_n)),$$
$$(\psi'^* f')(z) = \rho(z)((1 + iz_n)\bar{z}_1 - z_1(1 + i\bar{z}_n)),$$

for $z \in \tilde{V}$. Using (6.3), (6.17) and (6.18), we easily verify that the equalities

$$\psi'^* \tilde{f}_{1,0} = \tfrac{1}{2}(\cos^2 \theta - \sin^2 \varphi), \qquad \psi'^* \tilde{f}_{0,1} = \cos \theta \cos \varphi,$$

$$\psi'^* \tilde{f}' = \sin \theta \sin \varphi,$$

$$(\psi'^* \partial \bar{\partial} \tilde{f}_{0,1})(\xi_0', \eta_0'') = (\psi'^* \partial \bar{\partial} \tilde{f}_{0,1})(\eta_0', \xi_0'') = \tfrac{1}{2} \sin \theta \sin \varphi,$$

$$(\psi'^* \partial \bar{\partial} \tilde{f}')(\xi_0', \eta_0'') = (\psi'^* \partial \bar{\partial} \tilde{f}')(\eta_0', \xi_0'') = \tfrac{1}{2} \cos \theta \cos \varphi$$

hold at the point $\sigma(\theta, \varphi)$ of Z_0, with $(\theta, \varphi) \in \mathbb{R}^2$. By (6.7) and (6.10), we now obtain the equality

$$(6.23) \qquad \psi'^* (\tilde{f}_{r,s} w_3)(\xi_0, \eta_0) = \frac{1}{2^r} (\cos^2 \theta - \sin^2 \varphi)^{r+1}(\cos \theta \cos \varphi)^s$$

at this point of Z_0.

For $k, l \geq 0$, we consider the integral

$$I_{k,l} = \int_0^\pi \cos^k 2t \cdot \cos^l t \, dt.$$

If k and l are even integers, we clearly have

(6.24) $$I_{k,l} > 0$$

and we easily see that

(6.25) $$I_{k,l} - I_{k,l+2} > 0.$$

When k is odd, we have

(6.26) $$I_{k,0} = \frac{1}{2} \int_0^{2\pi} \cos^k u \, du = 0.$$

We use integration by parts to obtain the formula

(6.27) $$(l+1) I_{k,l+1} = 4k(I_{k-1,l+1} - I_{k-1,l+3}) + l I_{k,l-1}.$$

When k is an odd integer and $s \geq 1$ is an arbitrary integer, from the relations (6.25)–(6.27) we deduce by induction on s that

(6.28) $$I_{k,2s} > 0.$$

LEMMA 6.11. *Let $r \geq 0$ and $s \geq 1$ be given integers. Then the integral*

$$\int_\delta \psi'^*(\tilde{f}_{r,s} w_3)$$

does not vanish.

PROOF: For $r, s \geq 0$, according to formula (6.23), we have

$$\psi'^*(\tilde{f}_{r,s} w_3)(\xi_0, \eta_0)(\delta(t)) = \frac{1}{2^r} \cos^{r+1} 2t \cdot \cos^{2s} t,$$

with $0 \leq t \leq \pi$. If $s \geq 1$, from this equality and the relations (6.12), (6.24) and (6.28) we infer that the integral of the lemma does not vanish.

Since the elements ψ_α and ψ' of G induce isometries of X, by Lemmas 6.9, 6.10 and 6.11, for $r, s \geq 0$ we see that the symmetric 2-forms $\tilde{f}_{r,s} \pi_+ - \mathrm{Hess}\, \tilde{f}_{0,1}$, $\tilde{f}_{r,s} k^-$ and $\tilde{f}_{r,s+1} w_3$ do not satisfy the zero-energy condition.

§5. Computing integrals of odd symmetric forms

This section is a continuation of the preceding one. Its results will only be used in the proofs of the lemmas of §6, which we require for Propositions 6.28 and 6.34 and for Theorem 6.36. This last theorem is needed in §8 to establish the infinitesimal rigidity of the quadric Q_3 of dimension 3.

If $k, l \geq 0$ are given integers, we set

$$J_{k,l} = \int_0^\pi \sin^{2k} t \cdot \cos^{l+1} 3t \cdot \cos t \, dt,$$

$$\tilde{J}_{k,l} = \int_0^\pi \sin^{2k} t \cdot \cos^l 3t \, dt.$$

Clearly, we have $J_{0,0} = 0$. Using elementary trigonometric relations and integration by parts, we verify that

$$J_{0,l} = \frac{1}{4}\left(2 + \frac{3l-2}{3l+2} + \frac{3l-4}{3l+4}\right) J_{0,l-1},$$

for $l \geq 1$; it follows that $J_{0,l} = 0$, for all $l \geq 0$. We easily see that

(6.29) $$4J_{k+1,l} = J_{k,l} - \tilde{J}_{k,l+2},$$

for $k, l \geq 0$. Let $l \geq 0$ be a given even integer. Then we know that $\tilde{J}_{k,l} > 0$, for $k \geq 0$. Since $J_{0,l} = 0$, from (6.29) we infer by induction on k that $J_{k,l} < 0$, for all $k \geq 1$. Thus we have proved the following result:

LEMMA 6.12. *If $k \geq 1$ and $l \geq 0$ are given integers, with l even, then we have $J_{k,l} < 0$.*

In this section, we again consider the torus Z_0 and the objects associated with Z_0, and we shall use the conventions and notations of §4.

For $\alpha, \beta \in \mathbb{R}$, let $\psi_{\alpha,\beta}$ be the element of G defined by

$$\psi_{\alpha,\beta}(\zeta)_0 = \cos\alpha \cdot \zeta_1 + \sin\alpha \cdot \zeta_2, \quad \psi_{\alpha,\beta}(\zeta)_1 = \zeta_{n+1},$$

$$\psi_{\alpha,\beta}(\zeta)_2 = \sin\alpha \cdot \zeta_1 - \cos\alpha \cdot \zeta_2, \quad \psi_{\alpha,\beta}(\zeta)_3 = \cos\beta \cdot \zeta_0 + \sin\beta \cdot \zeta_n,$$

$$\psi_{\alpha,\beta}(\zeta)_n = \delta_3^n (\cos\beta \cdot \zeta_0 + \sin\beta \cdot \zeta_n) + (\delta_3^n - 1)\zeta_3,$$

$$\psi_{\alpha,\beta}(\zeta)_{n+1} = \cos\beta \cdot \zeta_n - \sin\beta \cdot \zeta_0,$$

and $\psi_{\alpha,\beta}(\zeta)_j = \zeta_j$ for $3 < j < n$, where $\zeta \in \mathbb{C}^{n+2}$. We write $\check{\psi}_\alpha = \psi_{\alpha,0}$ and we set $\mu = \cos\alpha$ and $\lambda = \sin\alpha$. Then for $r, s \geq 0$, the restrictions of the functions $\check{\psi}_\alpha^* f_{r,s}$ to $\tilde{V}$ are determined by

$$(\check{\psi}_\alpha^* f_{1,0})(z) = \rho(z)(\mu z_1 + i z_{n+1})(\mu \bar{z}_1 + i \bar{z}_{n+1}),$$

$$(\check{\psi}_\alpha^* f_{0,1})(z) = \rho(z)((\mu z_1 + i z_{n+1})(\lambda \bar{z}_1 + i) - (\lambda z_1 + i)(\mu \bar{z}_1 + i \bar{z}_{n+1})),$$

for $z \in \widetilde{V}$. By (6.17), we see that the equalities

(6.30)
$$\check{\psi}_\alpha^* \tilde{f}_{1,0} = -\tfrac{1}{2}((\lambda^2 - 1)\sin^2 \theta + \cos^2 \varphi),$$
$$\check{\psi}_\alpha^* \tilde{f}_{0,1} = -\cos \varphi\,(\lambda \sin \theta + i \cos \theta)$$

hold at the point $\sigma(\theta, \varphi)$ of Z_0, with $(\theta, \varphi) \in \mathbb{R}^2$.

We denote by f the function $\check{\psi}_\alpha^* \tilde{f}_{1,0}$ on $\mathbb{C}^{n+2}$ and we consider the symmetric 2-forms $h' = \pi' \mathrm{Hess}\, \tilde{f}$ and $h'' = \pi'' \mathrm{Hess}\, \tilde{f}$ on X. Then we have $h' = \check{\psi}_\alpha^* h_1'$ and $h'' = \check{\psi}_\alpha^* h_1''$. Using the formulas (5.55), we verify that the equalities

$$h'(\eta_1, \eta_1) = -\tfrac{1}{4}\cos^4 \theta\,(\mu \sin \theta + \cos \varphi)^2,$$
$$h'(\eta_1, \eta_n) = 0,$$
$$h'(\eta_n, \eta_n) = -\tfrac{1}{4}\cos^2 \theta \cos^2 \varphi\,(\mu \sin \theta + \cos \varphi)^2,$$
$$h''(\bar{\eta}_1, \bar{\eta}_1) = -\tfrac{1}{4}\cos^4 \theta\,(\mu \sin \theta - \cos \varphi)^2,$$
$$h''(\bar{\eta}_1, \bar{\eta}_n) = 0,$$
$$h''(\bar{\eta}_n, \bar{\eta}_n) = -\tfrac{1}{4}\cos^2 \theta \cos^2 \varphi\,(\mu \sin \theta - \cos \varphi)^2$$

hold at the point $\sigma(\theta, \varphi)$ of Z_0, with $(\theta, \varphi) \in \mathbb{R}^2$. By means of (6.4) and the preceding formulas, we see that

(6.31)
$$h'(\eta_0, \eta_0) = -h'(\xi_0, \xi_0) = \tfrac{1}{4}(\mu \sin \theta + \cos \varphi)^2,$$
$$h''(\eta_0, \eta_0) = -h''(\xi_0, \xi_0) = \tfrac{1}{4}(\mu \sin \theta - \cos \varphi)^2,$$
$$h'(\xi_0, \eta_0) = h''(\xi_0, \eta_0) = 0$$

at this point of Z_0.

We consider the vector field

$$\zeta_0 = \xi_0 + 3\eta_0$$

on Z_0. If h is a symmetric 2-form on X, by (6.11) we have

(6.32)
$$\int_{\delta_3} h = \frac{1}{5} \int_0^\pi h(\zeta_0, \zeta_0)(\delta_3(t))\,dt,$$

where $\delta_3(t) = \sigma(t, 3t)$, for $0 \le t \le \pi$.

Let $r, s \ge 0$ be given integers. We define

$$p_{r,s}(u, \theta, \varphi) = (u^2 \sin^2 \theta + \cos^2 \varphi - \sin^2 \theta)^r \cdot (u \sin \theta \cos \varphi + i \cos \theta \cos \varphi)^s,$$
$$p_{r,s}^1(u, \theta, \varphi) = ((1 - u^2)\sin^2 \theta + \cos^2 \varphi) \cdot p_{r,s}(u, \theta, \varphi),$$
$$p_{r,s}^2(u, \theta, \varphi) = \sin \theta \cos \varphi \cdot p_{r,s}(u, \theta, \varphi),$$

for $u, \theta, \varphi \in \mathbb{R}$. For $j = 1, 2$, we consider the polynomial

$$P_{r,s}^j(u) = \int_0^\pi p_{r,s}^j(u, t, 3t)\, dt$$

in u. We write

$$c_{r,s} = \frac{(-1)^{r+s}}{5 \cdot 2^{r-3}}.$$

By (6.30) and (6.31), we now obtain

(6.33) $\check{\psi}_\alpha^*(\tilde{f}_{r,s}(h_1' + h_1''))(\zeta_0, \zeta_0) = 2c_{r,s}\, p_{r,s}^1(\sin \alpha, \theta, \varphi),$

(6.34) $\check{\psi}_\alpha^*(\tilde{f}_{r,s}(h_1' - h_1''))(\zeta_0, \zeta_0) = c_{r,s}\cos \alpha \cdot p_{r,s}^2(\sin \alpha, \theta, \varphi).$

LEMMA 6.13. *Let $r \geq 0$ and $s \geq 1$ be given integers. Assume that s is odd and that $2r + s > 1$. Then there exists $\alpha_0 \in \mathbb{R}$ such that the integral*

$$\int_{\delta_3} \check{\psi}_{\alpha_0}^*\left(\tilde{f}_{r,s}(h_1' + h_1'')\right)$$

does not vanish.

PROOF: We write $s = 2l + 1$, with $l \geq 0$. The coefficient of u^{2r+s-1} of the polynomial $P_{r,s}^1(u)$ is equal to $-is J_{r+l,s-1}$. Since $r + l > 0$, by Lemma 6.12 this expression is non-zero. Thus the polynomial $P_{r,s}^1$ is non-zero, and so there exists a real number α_0 such that $P_{r,s}^1(\sin \alpha_0)$ does not vanish. From the relations (6.32) and (6.33), we infer that the integral of the lemma corresponding to this element $\alpha_0 \in \mathbb{R}$ does not vanish.

LEMMA 6.14. *Let $r \geq 0$ and $s \geq 2$ be given integers, with s even. Then there exists $\alpha_0 \in \mathbb{R}$ such that the integral*

$$\int_{\delta_3} \check{\psi}_{\alpha_0}^*\left(\tilde{f}_{r,s}(h_1' - h_1'')\right)$$

does not vanish.

PROOF: We write $s = 2l$, with $l \geq 0$. The coefficient of u^{2r+s-1} of the polynomial $P_{r,s}^2(u)$ is equal to $is J_{r+l,s}$. By Lemma 6.12, this expression is non-zero. Thus the polynomial $P_{r,s}^2$ is non-zero, and so there exists a real number α_0 such that the expressions $\cos \alpha_0$ and $P_{r,s}^2(\sin \alpha_0)$ do not vanish. From the relations (6.32) and (6.34), we infer that the integral of the lemma corresponding to this element $\alpha_0 \in \mathbb{R}$ does not vanish.

Let $r \geq 0$ and $s \geq 1$ be given integers. Since the elements $\check{\psi}_\alpha$ of G induce isometries of X, according to Lemmas 6.13 and 6.14, the symmetric

2-form $\tilde{f}_{r,s}(h'_1+h''_1)$, when $2r+s > 1$ and s is odd, and the symmetric 2-form $\tilde{f}_{r,s}(h'_1 - h''_1)$, when s is even, do not satisfy the zero-energy condition.

Let A, B, C be the functions on $\mathbb{R}^2$ defined by

$$A = (\lambda \sin\theta + i\cos\theta)\sin\varphi, \qquad B = (\lambda \sin\theta + i\cos\theta)\cos\varphi,$$

$$C = (i\sin\theta - \lambda\cos\theta)\cos\varphi,$$

for $(\theta, \varphi) \in \mathbb{R}^2$. Using (6.4), (6.17) and (6.18) and the expressions for the functions $\check{\psi}^*_\alpha \tilde{f}_{1,0}$ and $\check{\psi}^*_\alpha \tilde{f}_{0,1}$ on $\tilde{V}$, we verify that the equalities

$$\langle \xi'_0, \check{\psi}^*_\alpha \partial \tilde{f}_{1,0}\rangle = \tfrac{1}{2}\mu\cos\theta\,(\mu\sin\theta + \cos\varphi),$$

$$\langle \xi''_0, \check{\psi}^*_\alpha \bar\partial \tilde{f}_{1,0}\rangle = \tfrac{1}{2}\mu\cos\theta\,(\mu\sin\theta - \cos\varphi),$$

$$\langle \eta'_0, \check{\psi}^*_\alpha \partial \tilde{f}_{1,0}\rangle = \tfrac{1}{2}\sin\varphi\,(\cos\varphi + \mu\sin\theta),$$

$$\langle \eta''_0, \check{\psi}^*_\alpha \bar\partial \tilde{f}_{0,1}\rangle = \tfrac{1}{2}\sin\varphi\,(\cos\varphi - \mu\sin\theta),$$

$$\langle \xi'_0, \check{\psi}^*_\alpha \partial \tilde{f}_{0,1}\rangle = \tfrac{1}{2}(C + i\mu), \qquad \langle \xi''_0, \check{\psi}^*_\alpha \bar\partial \tilde{f}_{0,1}\rangle = \tfrac{1}{2}(C - i\mu),$$

$$\langle \eta'_0, \check{\psi}^*_\alpha \partial \tilde{f}_{0,1}\rangle = \tfrac{1}{2}A = \langle \eta''_0, \check{\psi}^*_\alpha \bar\partial \tilde{f}_{0,1}\rangle,$$

$$(\check{\psi}^*_\alpha \partial\bar\partial \tilde{f}_{1,0})(\xi'_0, \xi''_0) = \tfrac{1}{4}(\mu^2(2 - 3\sin^2\theta) + \cos^2\varphi),$$

$$(\check{\psi}^*_\alpha \partial\bar\partial \tilde{f}_{1,0})(\eta'_0, \eta''_0) = \tfrac{1}{4}(\cos^2\varphi - 2\sin^2\varphi - \mu^2\sin^2\theta),$$

$$(\check{\psi}^*_\alpha \partial\bar\partial \tilde{f}_{0,1})(\xi'_0, \eta''_0) = \tfrac{1}{2}\sin\varphi\,(\lambda\cos\theta - i\sin\theta) = (\check{\psi}^*_\alpha \partial\bar\partial \tilde{f}_{0,1})(\eta'_0, \xi''_0)$$

hold at the point $\sigma(\theta, \varphi)$ of Z_0, with $(\theta, \varphi) \in \mathbb{R}^2$.

Let $r, s \geq 0$ be fixed integers; we now consider the sections h_j, with $1 \leq j \leq 8$, defined in §3. We first suppose that $r = s = 1$. Using (6.5), (6.8), (6.9), (6.10) and the preceding formulas, we see that

$$(\check{\psi}^*_\alpha h_1)(\zeta_0, \zeta_0) = \frac{5}{2}B((\lambda^2 - 1)\sin^2\theta + \cos^2\varphi),$$

$$(\check{\psi}^*_\alpha h_2)(\zeta_0, \zeta_0) = -B((\lambda^2 - 1)(6\sin^2\theta - 1) + 5\cos^2\varphi - 9\sin^2\varphi),$$

(6.35)

$$(\check{\psi}^*_\alpha h_6)(\zeta_0, \zeta_0) = (C + 3A)((1 - \lambda^2)\sin\theta\cos\theta + 3\sin\varphi\cos\varphi)$$

$$+ i(1 - \lambda^2)(\cos\theta\cos\varphi + 3\sin\theta\sin\varphi),$$

$$(\check{\psi}^*_\alpha h_8)(\zeta_0, \zeta_0) = 3((1 - \lambda^2)\sin^2\theta - \cos^2\varphi)(\lambda\cos\theta - i\sin\theta)\sin\varphi.$$

LEMMA 6.15. *Suppose that $r = s = 1$ and let $a_1, a_2, a_4, a_6, a_8 \in \mathbb{C}$. If the symmetric 2-form $h = a_1 h_1 + a_2 h_2 + a_4 h_4 + a_6 h_6 + a_8 h_8$ satisfies the zero-energy condition, then we have the relation*

(6.36) $$5a_1 - 12a_2 + 4a_6 + 6a_8 = 0.$$

PROOF: By formulas (6.32) and (6.35), there exists an explicit polynomial $P(u)$ of degree ≤ 3 in u such that

$$\int_{\delta_3} \check{\psi}_\alpha^* h = P(\sin \alpha),$$

for all $\alpha \in \mathbb{R}$; moreover, we verify that the coefficient of u^2 of $P(u)$ is equal to

$$\frac{i}{10} \left(5a_1 - 12a_2 + 4a_6 + 6a_8\right) J_{1,0}.$$

If the symmetric 2-form h satisfies the zero-energy condition, the polynomial P vanishes and so, by Lemma 6.12, we obtain the relation (6.36).

When $r = 0$ and $s = 3$, by the methods used to verify the relations (6.35) we obtain the equalities

(6.37)
$$(\check{\psi}_\alpha^* h_1)(\zeta_0, \zeta_0) = -5B^3,$$
$$(\check{\psi}_\alpha^* h_3)(\zeta_0, \zeta_0) = -\frac{B}{2}\left(1 - \lambda^2 + C^2 + 9A^2\right),$$
$$(\check{\psi}_\alpha^* h_7)(\zeta_0, \zeta_0) = B\left(1 - \lambda^2 - (C + 3A)^2\right),$$
$$(\check{\psi}_\alpha^* h_8)(\zeta_0, \zeta_0) = 6B^2 \left(\lambda \cos \theta - i \sin \theta\right) \sin \varphi.$$

LEMMA 6.16. *Suppose that $r = 0$ and $s = 3$ and let $a_1, a_3, a_7, a_8 \in \mathbb{C}$. If the symmetric 2-form $h = a_1 h_1 + a_3 h_3 + a_7 h_7 + a_8 h_8$ satisfies the zero-energy condition, then the relation*

(6.38)
$$15a_1 + 4a_3 + 12a_7 + 6a_8 = 0$$

holds.

PROOF: By formulas (6.32) and (6.37), there exists an explicit polynomial $Q(u)$ of degree ≤ 3 in u such that

$$\int_{\delta_3} \check{\psi}_\alpha^* h = Q(\sin \alpha),$$

for all $\alpha \in \mathbb{R}$; moreover, we verify that the coefficient of u^2 of $Q(u)$ is equal to

$$-\frac{i}{5} \left(15a_1 + 4a_3 + 12a_7 + 6a_8\right) J_{1,1}.$$

If the symmetric 2-form h satisfies the zero-energy condition, the polynomial Q vanishes and so, by Lemma 6.12, we obtain the relation (6.38).

LEMMA 6.17. *Let $r \geq 0$ and $s \geq 1$ be given integers, with s odd. Let $a_1, \ldots, a_8$ be given complex numbers; suppose that $a_j = 0$ when $h_j = 0$, for $1 \leq j \leq 8$. Suppose that the section*

$$h = \sum_{j=1}^{8} a_j h_j$$

of $S^2 T^*_{\mathbb{C}}$ satisfies the zero-energy condition. Then we have relations

$$(6.39) \quad \begin{aligned}(2r + 2s - 3)&\big((2r + 2s - 1)a_1 - 4((r + s - 1)a_2 - a_5 - a_6) + 2a_8\big) \\ &- (2r^2 + 4rs - 4r + 2s^2 - 4s + 3)a_3 \\ &+ 4(r^2 + 2rs + r + s^2 + s - 3)a_7 = 0,\end{aligned}$$

$$(6.40) \quad \begin{aligned}s(2r + 2s - 5)&\big((2r + 2s - 3)a_1 - 4(r + s - 3)a_2 + 12a_5\big) \\ &- (2s^3 + 4rs^2 + 2r^2 s + 4r^2 - 4s^2 - s - 16r)a_3 \\ &- 4(2r + 2s - 5)(2r^2 + 4rs + 2s^2 - 3s)a_6 - 4ca_7 \\ &+ 2(2r + 2s - 5)(4r^2 + 8rs + 4s^2 - 5s - 8r)a_8 = 0,\end{aligned}$$

$$(6.41) \quad \begin{aligned}s^2(2r + s - 2)a_1 &- 4s^2(r - 2)a_2 + 4s^2(s + 2)a_5 \\ &- (2r + s - 2)(s^2 - 2r - 2s)(a_3 + 2a_8) \\ &- 4(rs^2 + s^2 + 2r^2 + 5rs - 2s - 2r)a_6 \\ &+ 4s(s^2 + 2rs + 2r^2 - 4r - 2s)a_7 = 0,\end{aligned}$$

where

$$c = 3s^3 + 4r^3 - 4r^2 + 10rs^2 + 11r^2 s - 9s^2 - 13rs + 3s - 12r.$$

When r is odd, we have the relations

$$(6.42) \quad (r + s - 1)(a_1 - 4a_2 - a_3 - 2a_8) + 4(r + s + 1)(a_5 + a_6 + a_7) = 0,$$

$$(6.43) \quad \begin{aligned}r(2r + s + 2)a_1 &- 4(2r^2 + rs - 2)a_2 - r(2r + s)(a_3 - 2a_8) \\ &+ 4(2r^2 + rs + 2r + 2s)a_5 - 4ra_6 = 0.\end{aligned}$$

When r is even, we have the relations

$$(6.44) \quad (r - 1)(a_1 - 4a_2 - a_3) + 4(r + 1)a_5 + 2(r - 1)a_8 = 0,$$

$$(6.45) \quad \begin{aligned}(2r + s)(r + s - 2)a_1 &- (2r + s - 2)(r + s - 2)(4a_2 + a_3 + 2a_8) \\ &+ 4(r + s)(2r + s - 4)a_5 + 4(r + s)(2r + s - 3)a_6 \\ &+ 4(r + s)(2r + s - 2)a_7 = 0.\end{aligned}$$

PROOF: By (6.30) and the formulas involving $\check{\psi}_\alpha$ which appear after Lemma 6.15, we obtain expressions for the functions $(\check{\psi}_\alpha^* h_j)(\xi_0 + \eta_0, \xi_0 + \eta_0)$ on Z_0, with $1 \leq j \leq 8$. By the relations (6.11)–(6.13), with $N = 1$, there exists an explicit polynomial $P(u)$ of degree $\leq 2r + s$ in u such that

$$\int_\delta \check{\psi}_\alpha^* h = P(\sin \alpha),$$

for all $\alpha \in \mathbb{R}$. Since the symmetric 2-form h satisfies the zero-energy condition, this polynomial P vanishes. The vanishing of the coefficient of u^{2r+s-1} of $P(u)$ gives us the relation (6.41). Next, we give equalities analogous to those appearing after Lemma 6.14, with $\check{\psi}_\alpha$ replaced by the element ψ_α of G defined in §4. Then by (6.19) and these formulas, we obtain expressions for the functions $(\psi_\alpha^* h_j)(\xi_0 + \eta_0, \xi_0 + \eta_0)$ on Z_0, with $1 \leq j \leq 8$. Using relations (6.11)–(6.13), with $N = 1$, we compute an explicit polynomial $Q(u)$ of degree $\leq 2r + s$ in u such that

$$\int_\delta \psi_\alpha^* h = Q(\sin \alpha),$$

for all $\alpha \in \mathbb{R}$. Our hypotheses imply that this polynomial Q vanishes. The vanishing of the coefficient of u^{2r+s} (resp. of u^{2r+s-1}) of $Q(u)$ gives us the relation (6.39) (resp. the relation (6.40)). Moreover, when r is odd, the equality $Q(-1) = 0$ is equivalent to (6.42), while the equality $Q'(1) = 0$ is equivalent to (6.43). On the other hand, when r is even, the equality $Q(1) = 0$ is equivalent to (6.44), while the equality $Q'(-1) = 0$ is equivalent to (6.45).

Let $r, s \geq 0$ be fixed integers; we now consider the sections $\tilde{h}_j$, with $1 \leq j \leq 5$, defined in §3.

LEMMA 6.18. Let $r \geq 1$ and $s \geq 2$ be given integers, with s even. Let a_1, a_2, a_3, a_4, a_5 be given complex numbers; suppose that $a_3 = 0$ when $r = 1$. If the section

$$h = \sum_{j=1}^{5} a_j \tilde{h}_j$$

of $S^2 T_{\mathbb{C}}^*$ satisfies the zero-energy condition, then we have relations

(6.46) $$(s - 1)(a_1 - a_4) + 2ra_5 = 0,$$

(6.47)
$$(s^2 - 2r - 3s + 3)a_1 + 4s(s - 1)a_3 - (s - 1)(s^2 + 2r - 3s - 1)a_4$$
$$+ 2r(s^2 - 3s + 1)a_5 = 0.$$

PROOF: We first derive equalities analogous to (6.30) and to those appearing after Lemma 6.14, with $\check{\psi}_\alpha$ replaced by $\psi_{\alpha,\beta}$, with $\alpha, \beta \in \mathbb{R}$.

Since we have $\check{\psi}_\alpha = \psi_{\alpha,0}$, these new formulas generalize those given above; we also remark that $\psi^*_{\alpha,\beta} f_{1,0} = \check{\psi}^*_\alpha f_{1,0}$. We then obtain expressions for the functions $(\check{\psi}^*_\alpha \tilde{h}_j)(\xi_0 + \eta_0, \xi_0 + \eta_0)$ and

$$\frac{\partial}{\partial \beta}(\psi^*_{\alpha,\beta}\tilde{h}_j)(\xi_0 + \eta_0, \xi_0 + \eta_0)_{|\beta=0}$$

on Z_0, with $1 \leq j \leq 5$. Using the relations (6.11)–(6.13), with $N = 1$, we find explicit polynomials $P_1(u)$ of degree $\leq 2r + s - 1$ and $P_2(u)$ of degree $\leq 2r + s$ in u such that

$$\int_\delta \check{\psi}^*_\alpha h = \cos \alpha \cdot P_1(\sin \alpha), \qquad \frac{\partial}{\partial \beta}\left(\int_\delta \psi^*_{\alpha,\beta} h\right)_{|\beta=0} = P_2(\sin \alpha),$$

for all $\alpha \in \mathbb{R}$. If the symmetric 2-form h satisfies the zero-energy condition, the polynomials P_1 and P_2 vanish; the vanishing of the coefficients of u^{2r+s-1} in the polynomials $P_1(u)$ and $P_2(u)$ give us the relations (6.46) and (6.47), respectively.

We consider the sections

$$k_1 = \tfrac{1}{2}(k'_2 + k''_2 + w_2), \qquad k_2 = \tfrac{1}{2}(k'_2 + k''_2 - w_2)$$

of $S^2 T^*_\mathbb{C}$.

LEMMA 6.19. *Let $r, s \geq 0$ be given integers, with s even, and let a_1, a_2, a_3, a_4 be given complex numbers. Suppose that $a_3 = 0$ when $r = 0$. If the section*

$$h = a_1 \tilde{f}_{r,s} k_1 + a_2 \tilde{f}_{r,s} k_2 + a_3 \tilde{f}_{r-1,s+1}(k'_1 + k''_1) + a_4 \tilde{f}_{r,s+1} w_3$$

of $S^2 T^*_\mathbb{C}$ satisfies the zero-energy condition, then we have the relations

(6.48) $(4r + 3s + 7)a_1 + (4r + s + 5)a_2 - 4(s + 1)(a_3 - a_4) = 0,$

(6.49) $c_1 a_1 + c_2 a_2 - 4(s + 1)(c_3 a_3 - c_4 a_4) = 0,$

(6.50) $(s + 1)c'_1 a_1 + c'_2 a_2 - 4(s + 1)(c'_3 a_3 - c'_4 a_4) = 0,$

where

$$c_1 = 28 + 56r + 45s + 44r^2 + 25s^2 + 48rs + 12r^2 s + 12rs^2 + 4rs^3$$
$$+ 16r^3 + 11s^3 + 3s^4,$$
$$c_2 = 20 + 40r + 39s + 20r^2 + 27s^2 + 64rs + 4r^2 s + 28rs^2 + 4rs^3$$
$$+ 16r^3 + 9s^3 + s^4,$$

$$c_3 = 4r^2 + 4rs + 8r + s^3 + 12s^2 + 39s + 36,$$

$$c_4 = 4r^2 - 4rs - 8r + s^3 + 4s^2 + 7s + 4,$$

$$c_1' = 3s^3 + 10rs^2 + 8r^2s + 2s^2 + 4rs - 4r^2 - 16r - 12s - 7,$$

$$c_2' = s^4 + 8r^2s^2 + 26rs^2 + 6rs^3 + 20r^2s + 7s^3 + 12s^2$$
$$+ 16rs - 4r^2 + 3s - 20r - 3,$$

$$c_3' = s^3 + 2rs^2 + 4r^2 + 12rs + 6s^2 + 12r + 8s - 1,$$

$$c_4' = (2r + s + 1)(s^2 + s - 2r - 1).$$

PROOF: By the methods used in proving Lemma 6.18, we compute explicit polynomials $Q_1(u)$ of degree $\leq 2r + s + 2$ and $Q_2(u)$ of degree $\leq 2r + s + 3$ in u such that

$$\int_\delta \check{\psi}_\alpha^* h = \cos\alpha \cdot Q_1(\sin\alpha), \qquad \frac{\partial}{\partial\beta}\left(\int_\delta \psi_{\alpha,\beta}^* h\right)_{|\beta=0} = Q_2(\sin\alpha),$$

for all $\alpha \in \mathbb{R}$. If the symmetric 2-form h satisfies the zero-energy condition, the polynomials Q_1 and Q_2 vanish; the vanishing of the coefficients of u^{2r+s+2} and u^{2r+s} in the polynomial $Q_1(u)$ give us the relations (6.48) and (6.49), respectively, while the vanishing of the coefficient of u^{2r+s+2} in the polynomial $Q_2(u)$ give us the relation (6.50).

LEMMA 6.20. Let $r, s \geq 0$ be given integers, with s even, and let a_1, a_2, a_3, a_4 be given complex numbers. Suppose that $a_3 = 0$ when $r = 0$, and that $a_4 = 0$ when $s = 0$. If the section

$$h = a_1 \tilde{f}_{r,s} k^- + a_2 \tilde{f}_{r,s} k^+ + a_3 \tilde{f}_{r-1,s}(k_1' - k_1'') + a_4 \tilde{f}_{r,s-1}(k_2' - k_2'')$$

of $S^2 T_{\mathbb{C}}^*$ satisfies the zero-energy condition, then we have the relations

$$(s+1)(2r+2s-1)a_1 + (4r^2 + 8rs + 4s^2 + 3s - 1)a_2$$
(6.51)
$$- 2(4r^2 + 8rs + 4s^2 + 4r + s - 3)a_3$$
$$- 2(2r^2 + 3rs + s^2 + 7r + 3s + 2)a_4 = 0,$$

(6.52) $\qquad (2r + 2s + 1)a_1 + a_2 + 2a_3 + 2(r + s + 2)a_4 = 0,$

(6.53) $\qquad\qquad a_1 + a_2 - 2a_3 = 0.$

PROOF: By the methods used to prove Lemma 6.18, we compute an explicit polynomial $P(u)$ of degree $\leq 2r + s + 1$ in u such that

$$\int_\delta \check{\psi}_\alpha^* h = \cos\alpha \cdot P(\sin\alpha),$$

for all $\alpha \in \mathbb{R}$. Here we require formulas (6.19) and (6.21) and certain formulas used in proving Lemma 6.17. The equality $P(0) = 0$ is equivalent to (6.52), while the equality $P'(0) = 0$ is equivalent to (6.51). On the other hand, when r is even, the equality $P(1) = 0$ is equivalent to (6.53). When r is odd, the equality $P(-1) = 0$ is equivalent to the relation

$$(6.54) \qquad (r + s)(a_1 - a_2) + 2(r + s + 2)(a_3 + a_4) = 0.$$

We easily see that the relations (6.52) and (6.54) imply that (6.53) holds. If the symmetric 2-form h satisfies the zero-energy condition, the polynomial P vanishes, and we obtain the desired equalities.

Lemmas 6.17–6.20 are due to Tela Nlenvo; details of the proofs of these lemmas can be found in [52].

§6. Bounds for the dimensions of spaces of symmetric forms

In this section, we use the results of §§4 and 5 to give bounds for the dimension of certain spaces which we shall need in §7.

LEMMA 6.21. *Let $r, s \geq 0$ be given integers, with s even. Then we have*

$$\dim (\mathcal{Z}_{2,\mathbb{C}} \cap V'_{r,s+1}) \leq 1.$$

PROOF: For $r \geq 1$, $s \geq 2$, the determinant of the 3×3 matrix, whose entries are the coefficients of a_1, a_2 and a_3 in the relations (6.51), (6.52) and (6.53), is equal to

$$16s(r + s + 1).$$

If $r = 0$ and $s \geq 2$ (resp. if $r \geq 1$ and $s = 0$), the relations (6.51) and (6.52), with $a_3 = 0$ (resp. with $a_4 = 0$), are clearly linearly independent. Finally, if $r = s = 0$, the relations (6.51), (6.52) and (6.53) reduce to $a_1 + a_2 = 0$. We remark that the symmetric 2-form h of Lemma 6.20 is an element of $V'_{r,s+1}$. From Lemmas 6.6 and 6.20, we then deduce the desired inequality.

LEMMA 6.22. *Let $r, s \geq 0$ be given integers, with s even. Then we have*

$$\dim (\mathcal{Z}_{2,\mathbb{C}} \cap W'_{r,s+2}) \leq 1.$$

PROOF: If $r \geq 1$, the determinant of the 3×3 matrix, whose entries are the coefficients of a_2, a_3 and a_4 in the relations (6.48), (6.49) and (6.50), is equal to

$$2^8 (s + 2)(s + 1)^3 (r + s + 2)(2r + s + 1).$$

If $r = 0$, the relations (6.48) and (6.50), with $a_3 = 0$, are linearly independent; in fact, the determinant of the 2×2 matrix, whose entries are the coefficients of a_2 and a_4 in these two relations, is equal to $-8(s + 1)^2$. We remark that the symmetric 2-form h of Lemma 6.19 is an element of $W'_{r,s+2}$. From Lemmas 6.6 and 6.19, we then obtain the desired inequality.

LEMMA 6.23. *Let $r, s \geq 0$ be given integers, with s even. Then we have*

$$\dim \left(\mathcal{Z}_{2,\mathbb{C}} \cap W_{r,s} \right) \leq d_{r,s}.$$

PROOF: If $r, s \geq 2$, the determinant of the 2×2 matrix, whose entries are the coefficients of a_3 and a_4 in the relations (6.46) and (6.47), is equal to the non-zero expression $4s(s-1)^2$. If $r = 1$ and $s \geq 2$, or if $r \geq 2$ and $s = 0$, the relation (6.46) is non-trivial. We note that the relations (6.46) and (6.47) do not involve the coefficient a_2; when $r \geq 1$ and $s \geq 2$, according to Lemma 6.14, the symmetric 2-form $\tilde{h}_2$, which belongs to $W_{r,s}$, does not satisfy the zero-energy condition. We remark that the symmetric 2-form h of Lemma 6.18 is an element of $W_{r,s}$. These observations, together with Lemmas 6.5 and 6.18, give us the desired inequality.

LEMMA 6.24. *Let $r \geq 0$, $s \geq 1$ be given integers, with s odd. Then we have*

$$\dim \left(\mathcal{Z}_{2,\mathbb{C}} \cap V_{r,s} \right) \leq d_{r,s}.$$

PROOF: We note that the relations (6.39)–(6.45) do not involve the coefficient a_4; when $r \geq 1$ and $2r + s - 3 > 0$, according to Lemma 6.13, the symmetric 2-form h_4, which belongs to $V_{r,s}$, does not satisfy the zero-energy condition. Also the symmetric 2-form h of Lemma 6.17 is an element of $V_{r,s}$.

(i) We first consider the case when $r, s \geq 3$ and r is odd. In view of the above observations and Lemmas 6.4 and 6.17, it suffices to show that the 5×7 matrix, corresponding to the linear system consisting of the equations (6.39)–(6.43) in the scalars a_j, with $1 \leq j \leq 8$ and $j \neq 4$, is of maximal rank. The determinant of the 5×5 matrix, whose entries are the coefficients of the a_j, with $j = 1$ and $5 \leq j \leq 8$, in these relations, is equal to

$$\Delta = 2^{10}(s-1)(s-2)(r+s-1)(2r+s)(r+s)^3 \Delta',$$

where

$$\Delta' = 4r^3 - 10r^2 + 12r^2 s - 20rs + 9rs^2 + 4r + 8 + 5s - 8s^2 + 2s^3.$$

We verify that the expression Δ' is > 0; indeed, since $r, s \geq 3$, we have

$$9rs^2 + 12r^2 s = 3rs^2 + 6rs^2 + 10r^2 s + 2r^2 s$$

$$\geq 9s^2 + 18rs + 10r^2 + 2rs$$

$$> 8s^2 + 20rs + 10r^2.$$

Therefore the determinant Δ is > 0 and our 5×7 matrix is of maximal rank.

(ii) We next consider the case when $r \geq 2$ is even and $s \geq 3$. In view of the observations which precede the case (i), and Lemmas 6.4 and 6.17, it suffices to show that the 5×7 matrix, corresponding to the linear system consisting of the equations (6.39)–(6.41), and (6.44) and (6.45) in the scalars a_j, with $1 \leq j \leq 8$ and $j \neq 4$, is of maximal rank. The determinant of the 5×5 matrix, whose entries are the coefficients of the a_j, with $j = 1$ and $5 \leq j \leq 8$, in these relations, is equal to

$$\Delta = -2^{10}(s-1)(s-2)(2r+s)(r+s-1)^2(r+s)^2 A(r,s),$$

where

$$A(r,s) = 4r^3 - 18r^2 + 12r^2 s - 38rs + 9rs^2 + 28r + 31s - 15s^2 + 2s^3 - 12.$$

We now show that the expression $A(r,s)$ is positive. In fact, when $r = 2$, we have

$$A(2,s) = 2s^3 + 3s^2 + 3s + 4 > 0.$$

On the other hand, when $r \geq 4$, we have

$$2s^3 + 9rs^2 + 12r^2 s = 2s^3 + 3rs^2 + 6rs^2 + 5r^2 s + 7r^2 s$$

$$\geq 6s^2 + 12s^2 + 18rs + 20rs + 21r^2$$

$$> 15s^2 + 38rs + 18r^2;$$

since $31s - 12 > 0$, we see that $A(r,s) > 0$ in this case. Therefore the determinant Δ is always < 0 and our 5×7 matrix is of maximal rank.

(iii) We now consider the case when $r = 1$ and $s \geq 3$. We set $a_5 = 0$; then the relations (6.39) and (6.41)–(6.43) are equivalent to the system of equations

$$(6.55) \quad \begin{aligned} (4s^2 - 1)a_1 - 2(2s - 1)(2sa_2 - 2a_6 - a_8) - (2s^2 + 1)a_3 \\ + 4(s^2 + 3s - 1)a_7 = 0, \end{aligned}$$

$$(6.56) \quad \begin{aligned} s^2 a_1 + 4sa_2 - (s^2 - 2s - 2)(a_3 + 2a_8) - 4(2s + 3)a_6 \\ + 4(s^2 - 2)a_7 = 0, \end{aligned}$$

$$(6.57) \quad s(a_1 - 4a_2 - a_3 - 2a_8) + 4(s + 2)(a_6 + a_7) = 0,$$

$$(6.58) \quad (s + 4)a_1 - 4sa_2 - (s + 2)(a_3 - 2a_8) - 4a_6 = 0.$$

Since $2r + s - 3 = s - 1 > 0$, in view of the observations which precede the case (i), and Lemmas 6.4 and 6.17, it suffices to show that the 4×6 matrix, corresponding to the linear system consisting of equations (6.55)–(6.58) in the scalars a_j, with $1 \le j \le 8$ and $j \ne 4, 5$, is of maximal rank. The determinant of the 4×4 matrix, whose entries are the coefficients of the scalars a_1, a_2, a_3, a_6 in the equations (6.55)–(6.58), is equal to

$$32s(s-1)(s-2)(s+1)^3,$$

and so our 4×6 matrix is of maximal rank.

(iv) We now consider the case when $r \ge 2$ and $s = 1$. We set $a_3 = a_7 = 0$. Then the relations (6.39) and (6.40) are equivalent to the system of equations

(6.59) $$(2r+1)a_1 - 4ra_2 + 4a_5 + 4a_6 + 2a_8 = 0,$$

(6.60) $$\begin{aligned}(2r-1)a_1 - 4(r-2)a_2 + 12a_5 - 4(2r^2 + 4r - 1)a_6 \\ + 2(4r^2 - 1)a_8 = 0.\end{aligned}$$

On the other hand, the equation (6.42) is equivalent to

(6.61) $$r(a_1 - 4a_2 - 2a_8) + 4(r+2)(a_5 + a_6) = 0,$$

while the equation (6.44) is equivalent to

(6.62) $$(r-1)(a_1 - 4a_2) + 4(r+1)a_5 + 2(r-1)a_8 = 0.$$

Since $2r + s - 3 = 2r - 2 > 0$, in view of the observations which precede the case (i), and Lemmas 6.4 and 6.17, when r is odd (resp. even) it suffices to know that the 3×5 matrix corresponding to the linear system consisting of equations (6.59)–(6.61) (resp. of equations (6.59), (6.60) and (6.62)) in the scalars a_1, a_2, a_5, a_6, a_8 is of maximal rank. Since the determinant of the matrix

$$\begin{pmatrix} 4 & 4 & 2 \\ 12 & -4(2r^2 + 4r - 1) & 2(4r^2 - 1) \\ 4(r+2) & 4(r+2) & -2r \end{pmatrix}$$

is equal to $128(r+1)^3$ (resp. of the matrix

$$\begin{pmatrix} 4 & 4 & 2 \\ 12 & -4(2r^2 + 4r - 1) & 2(4r^2 - 1) \\ 4(r+1) & 0 & 2(r-1) \end{pmatrix}$$

is equal to $128r(r+1)^2$), our 3×5 matrix is of maximal rank when r is odd (resp. even).

(v) We now consider the case when $r = s = 1$. We set

$$a_3 = a_5 = a_7 = 0;$$

then the relations (6.40) and (6.42) are equivalent to the system of equations

(6.63) $\qquad a_1 + 4a_2 - 20a_6 + 6a_8 = 0, \qquad a_1 - 4a_2 + 12a_6 - 2a_8 = 0.$

We also consider the relation (6.36) of Lemma 6.15. Since the determinant of the matrix

$$\begin{pmatrix} 1 & 4 & 6 \\ 1 & -4 & -2 \\ 5 & -12 & 6 \end{pmatrix}$$

is equal to -64, we see that the 3×4 matrix of the linear system consisting of equations (6.63) and (6.36) in the scalars a_1, a_2, a_6, a_8 is of maximal rank. From Lemmas 6.4, 6.15 and 6.17, we obtain the desired inequality in this case.

(vi) We now consider the case when $r = 0$ and $s \geq 3$. We set

$$a_2 = a_4 = a_5 = a_6 = 0;$$

then the relations (6.40), (6.44) and (6.45) are equivalent to the system of equations

(6.64) $\qquad (2s - 3)(2s - 5)a_1 - (2s^2 - 4s - 1)a_3 - 12(s^2 - 3s + 1)a_7$
$$+ 2(2s - 5)(4s - 5)a_8 = 0,$$

(6.65) $\qquad\qquad a_1 - a_3 + 2a_8 = 0,$

(6.66) $\qquad\qquad sa_1 + (2 - s)a_3 + 4sa_7 - (2s - 4)a_8 = 0.$

Since the determinant of the matrix

$$\begin{pmatrix} 2s^2 - 4s - 1 & 12(s^2 - 3s + 1) & 2(2s - 5)(4s - 5) \\ 1 & 0 & 2 \\ s - 2 & -4s & -2s + 4 \end{pmatrix}$$

is equal to $-32(s - 1)(s - 3)$, we see that the 3×4 matrix of the linear system consisting of equations (6.64)–(6.66) in the scalars a_1, a_3, a_7, a_8 is of maximal rank when $s \geq 5$. If $s = 3$, the relation (6.64) gives us the equation

(6.67) $\qquad\qquad 3a_1 - a_3 + 12a_7 - 2a_8 = 0;$

in this case, we also consider the relation (6.38) of Lemma 6.16. Since the determinant of the matrix

$$\begin{pmatrix} 1 & -1 & 2 \\ 3 & -1 & -2 \\ 15 & 4 & 6 \end{pmatrix}$$

is equal to 104, the 3×4 matrix of the linear system consisting of equations (6.65), (6.67) and (6.38) in the scalars a_1, a_3, a_7, a_8 is of maximal rank. For all $s \geq 3$, we then obtain the desired inequality from Lemmas 6.4, 6.16 and 6.17.

(vii) We finally consider the case when $r = 0$ and $s = 1$. In this case, we set

$$a_2 = a_3 = a_4 = a_5 = a_6 = a_7 = 0.$$

The equation (6.39) is equivalent to

$$a_1 + 2a_8 = 0,$$

and then Lemmas 6.4 and 6.17 imply the desired result.

§7. The complex quadric of dimension three

In this section, we suppose that $n = 3$ and that X is the quadric Q_3 of dimension 3, which is a homogeneous space of the group $G = SO(5)$. Let Γ be the dual of the group G.

The Casimir element of the Lie algebra $\mathfrak{g}_0$ of G operates by a scalar c_γ on an irreducible G-module which is a representative of $\gamma \in \Gamma$. We know that, for $\gamma \in \Gamma$, the G-module $C_\gamma^\infty(S^p T_{\mathbb{C}}^*)$ is an eigenspace of the Lichnerowicz Laplacian Δ with eigenvalue $\lambda_\gamma = 12 c_\gamma$. If W is a G-submodule of $C_\gamma^\infty(S^p T_{\mathbb{C}}^*)$, with $\gamma \in \Gamma$, we denote by $\mathcal{C}(W)$ the weight subspace of W corresponding to its highest weight γ; we recall that the multiplicity of the G-module W is equal to the dimension of the space $\mathcal{C}(W)$.

If we set

$$c(t, s) = \frac{(t + s)(t + s + 3) + s(s + 1)}{6},$$

by Freudenthal's formula we have

$$c_{\gamma_{r,s}} = c(2r, s), \qquad c_{\gamma'_{r,s}} = c(2r + 1, s);$$

in particular, we see that $c_{\gamma'_{0,0}} = 2/3$. In fact, we easily verify that the expression for $\lambda_{r,s}$ given by (5.123) is equal to $\lambda_{\gamma_{r,s}} = 12 c_{\gamma_{r,s}}$.

From the branching law for $SO(5)$ and its subgroup K described in Theorem 1.2 of [54], using the computation of the highest weights of the irreducible K-modules given in §7, Chapter V, we obtain the following result given by Proposition 9.1 of [23] (see also [54, §4]):

PROPOSITION 6.25. *For $\gamma \in \Gamma$, the $SO(5)$-module $C_\gamma^\infty(F)$, where F is a homogeneous vector bundle over $X = Q_3$, equal either to T' or T'' or to one of the vector bundles appearing in the decomposition (5.27) of $S^2 T_{\mathbb{C}}^*$, vanishes unless γ is equal to $\gamma_{r,s}$ or to $\gamma'_{r,s}$, for some $r, s \geq 0$. For $r, s \geq 0$, the non-zero multiplicities of the $SO(5)$-modules $C_{\gamma_{r,s}}^\infty(F)$, where F is one of these homogeneous vector bundles, are given by the following table:*

F	Conditions on r, s	Mult $C_{\gamma_{r,s}}^\infty(F)$
T'	$r \geq 0, \ s \geq 1$	1
T''		
$(S^{2,0}T^*)^\perp$	$r + s \geq 2$ and $s \geq 1$	2 if $r \geq 1, \ s \geq 2$
$(S^{0,2}T^*)^\perp$		1 otherwise
$(S^2T^*)_{0\mathbb{C}}^{++}$	$r \geq 0, \ s \geq 1$	2 if $r \geq 0, \ s \geq 2$
		1 otherwise
$(S^2T^*)_{\mathbb{C}}^{+-}$	$r, s \geq 0$	2 if $r \geq 0, \ s \geq 1$
		1 otherwise

REMARK. According to Proposition 6.25, we see that the irreducible $SO(5)$-module $C_{\gamma'_{0,0}}^\infty(S^2T^*)_{\mathbb{C}}^{+-}$ is a non-zero eigenspace of the Lichnerowicz Laplacian Δ with eigenvalue $\lambda_{\gamma'_{0,0}} = 12 c_{\gamma'_{0,0}} = 8$. Clearly the numbers $c_{\gamma_{r,s}}$ and $c_{\gamma'_{r,s}}$ are $> c_{\gamma'_{0,0}}$ when $r + s > 0$; therefore the first eigenvalue of the Lichnerowicz Laplacian Δ acting on $C^\infty(S^2T_{\mathbb{C}}^*)$ is equal to 8 and is strictly less than the first eigenvalue $\lambda_{0,1} = 12$ of the Laplacian Δ acting on the space of functions $C^\infty(X)$.

In [23], we verified that all the non-zero vectors of the spaces $V_{r,s}$ and $W_{r,s}$ are highest weight vectors of the $SO(5)$-module $C_{\gamma_{r,s}}^\infty(S^2T_{\mathbb{C}}^*)$. From Proposition 5.19, Lemmas 6.4 and 6.5, and the inclusions (6.14) and (6.15), we infer the following result given by Lemma 9.2 of [23]:

LEMMA 6.26. *For $r, s \geq 0$, we have*

$$
\begin{aligned}
&\mathcal{C}(C_{\gamma_{r,2s}}^\infty(S^2T_{\mathbb{C}}^*)^{\mathrm{ev}}) = V_{r,2s}, \quad \mathcal{C}(C_{\gamma_{r,2s+1}}^\infty(S^2T_{\mathbb{C}}^*)^{\mathrm{odd}}) = V_{r,2s+1}, \\
(6.68) \quad &\mathcal{C}(C_{\gamma_{r,2s}}^\infty(S^2T_{\mathbb{C}}^*)^{\mathrm{odd}}) = W_{r,2s}, \quad \mathcal{C}(C_{\gamma_{r,2s+1}}^\infty(S^2T_{\mathbb{C}}^*)^{\mathrm{ev}}) = W_{r,2s+1}.
\end{aligned}
$$

In §7, Chapter V, we saw that, for $r \geq 0$ and $s \geq 1$, the section $\tilde{f}_{r,s-1}\pi_{+-}\mathrm{Hess}\,\hat{f}_{0,1}$ is a highest weight vector of the irreducible $SO(5)$-module $C_{\gamma_{r,s}}^\infty((S^2T_{\mathbb{C}}^*)^{+-})$ and we derived the relations (5.126), (5.129) and (5.130); in fact, this section is even (resp. is odd) when s is an even (resp. odd) integer. These results can also be obtained using Proposition 5.19 and Lemma 6.26.

From Proposition 5.22 and from Lemmas 5.21, 6.7 and 6.26, we obtain the following result:

PROPOSITION 6.27. *For $r, s \geq 0$, we have*

$$\mathcal{N}_{2,\mathbb{C}} \cap C^{\infty}_{\gamma_{r,s}}(S^2 T^*_{\mathbb{C}})^{\text{ev}} = D_0 C^{\infty}_{\gamma_{r,s}}(T_{\mathbb{C}})^{\text{ev}}.$$

From Proposition 5.23 and from Lemmas 5.21, 6.23, 6.24 and 6.26, we then obtain the following result:

PROPOSITION 6.28. *For $r, s \geq 0$, we have*

$$\mathcal{Z}_{2,\mathbb{C}} \cap C^{\infty}_{\gamma_{r,s}}(S^2 T^*_{\mathbb{C}})^{\text{odd}} = D_0 C^{\infty}_{\gamma_{r,s}}(T_{\mathbb{C}})^{\text{odd}}.$$

The following lemma is a consequence of Proposition 6.25 and of the proof of Lemma 9.3 of [23].

LEMMA 6.29. *For $r \geq 0, s \geq 1$, the $SO(5)$-modules $C^{\infty}_{\gamma_{r,s}}(T_{\mathbb{C}})^{\text{ev}}$ and $C^{\infty}_{\gamma_{r,s}}(T_{\mathbb{C}})^{\text{odd}}$ are irreducible. For $r \geq 0$, we have*

$$C^{\infty}_{\gamma_{r,0}}(T_{\mathbb{C}}) = \{0\}.$$

In [23], we verified that all the non-zero vectors of the spaces $V'_{r,s}$, with $s \geq 1$, and $W'_{r,s}$, with $s \geq 0$, are highest weight vectors of the $SO(5)$-module $C^{\infty}_{\gamma_{r,s}}(S^2 T^*_{\mathbb{C}})$.

The next lemma is given by Lemma 9.4 of [23].

LEMMA 6.30. *The $SO(5)$-module $C^{\infty}_{\gamma_{0,0}}((S^2 T^*)^{+-}_{\mathbb{C}})^{\text{ev}}$ is irreducible, and it possesses a highest weight vector h_0 satisfying*

$$k^- = \tilde{f}_{0,1} \cdot h_0.$$

PROOF: If ϕ is the element of $SO(5)$ defined in §7, Chapter V, we have $(\phi^* \tilde{f}_{0,1})(a) = 1$ and $(\phi^* \tilde{f}')(a) = 0$. By (5.121), we obtain

$$(6.69) \qquad \phi^*((d\tilde{f}_{1,0})^{\sharp} \lrcorner w_3)(\bar{\eta}_1) = \phi^*((d\tilde{f}_{1,0})^{\sharp} \lrcorner \pi_{+-} \operatorname{Hess} \tilde{f}')(\bar{\eta}_1) = -1$$

at the point a. By Proposition 6.25, the $SO(5)$-module $C^{\infty}_{\gamma_{0,0}}((S^2 T^*)^{+-}_{\mathbb{C}})$ is irreducible; let h'_0 be a highest weight vector of this module. Clearly the section h'_0 is either even or odd. For $r, s \geq 0$, from the equality $\gamma'_{r,s} = \gamma'_{0,0} + \gamma_{r,s}$ we infer that $\tilde{f}_{r,s}h'_0$ is a highest weight vector of the $SO(5)$-module $C^{\infty}_{\gamma_{r,s}}((S^2 T^*)^{+-}_{\mathbb{C}})$. Since $\operatorname{div} : S^2 T^*_{\mathbb{C}} \to T^*_{\mathbb{C}}$ is a homogeneous differential operator, by Lemma 6.29 we see that

$$(6.70) \qquad \operatorname{div}(\tilde{f}_{r,0} h'_0) = 0,$$

for $r \geq 0$. The relation (6.70), with $r = 0, 1$, and formula (1.8) imply that

$$(d\tilde{f}_{1,0})^{\#} \lrcorner\, h_0' = 0.$$

Hence the highest weight vector $h = \tilde{f}_{0,1} h_0'$ of $C^{\infty}_{\gamma'_{0,1}}((S^2 T^*)^{+-}_{\mathbb{C}})$ satisfies the equation

(6.71) $$(d\tilde{f}_{1,0})^{\#} \lrcorner\, h = 0.$$

According to Proposition 6.25, we know that the multiplicity of the $SO(5)$-module $C^{\infty}_{\gamma'_{0,1}}((S^2 T^*)^{+-}_{\mathbb{C}})$ is equal to 2. By the remark preceding this lemma and the inclusions (6.16), the sections w_3 and k^- are highest weight vectors of the modules $C^{\infty}_{\gamma'_{0,1}}((S^2 T^*)^{+-}_{\mathbb{C}})^{\mathrm{ev}}$ and $C^{\infty}_{\gamma'_{0,1}}((S^2 T^*)^{+-}_{\mathbb{C}})^{\mathrm{odd}}$, respectively. Therefore these vectors are linearly independent and there exist unique scalars $c_1, c_2 \in \mathbb{C}$ such that

$$h = c_1 w_3 + c_2 k^-.$$

Moreover since the section h is either even or odd, we know that one (and only one) of the coefficients c_1, c_2 must vanish. If $c_2 = 0$, then from (6.71) we obtain the relation

$$(d\tilde{f}_{1,0})^{\#} \lrcorner\, w_3 = 0,$$

which contradicts (6.69). Therefore we must have $c_1 = 0$, and the vector h is a non-zero multiple of k^- and so is an odd section of $(S^2 T^*)^{+-}_{\mathbb{C}}$. Hence the vector $h_0 = (1/c_2) h_0'$ satisfies the conclusion of the lemma.

For $r \geq 0$, the highest weight vector $\tilde{f}_{r,0} h_0$ of the $SO(5)$-module $C^{\infty}((S^2 T^*)^{+-}_{\mathbb{C}})^{\mathrm{ev}}$ generates (over $\mathbb{C}$) a subspace $V'_{r,0}$ of this module. By Lemma 6.30 we have

$$\tilde{f}_{r,s} k^- = \tilde{f}_{r,s+1} h_0,$$

for $r, s \geq 0$, and so the symmetric 2-form $\tilde{f}_{r,s} h_0$ belongs to the subspace $V'_{r,s}$, for all $r, s \geq 0$.

From Proposition 6.25, Lemma 6.6, the inclusions (6.16) and the remark preceding Lemma 6.30, we deduce the following result given by Lemma 9.5 of [23]:

LEMMA 6.31. *For $r, s \geq 0$, we have*

(6.72)
$$\mathcal{C}(C^{\infty}_{\gamma_{r,2s}}(S^2 T^*_{\mathbb{C}})^{\mathrm{ev}}) = V'_{r,2s}, \quad \mathcal{C}(C^{\infty}_{\gamma_{r,2s+1}}(S^2 T^*_{\mathbb{C}})^{\mathrm{odd}}) = V'_{r,2s+1},$$
$$\mathcal{C}(C^{\infty}_{\gamma_{r,2s}}(S^2 T^*_{\mathbb{C}})^{\mathrm{odd}}) = W'_{r,2s}, \quad \mathcal{C}(C^{\infty}_{\gamma_{r,2s+1}}(S^2 T^*_{\mathbb{C}})^{\mathrm{ev}}) = W'_{r,2s+1}.$$

Let $r, s \geq 0$ be given integers. By Lemma 6.31, the $SO(5)$-modules $C^\infty_{\gamma'_{r,s}}((S^2T^*)^{+-}_{\mathbb{C}})^{\mathrm{ev}}$ and $C^\infty_{\gamma'_{r,s+1}}((S^2T^*)^{+-}_{\mathbb{C}})^{\mathrm{odd}}$ are irreducible. Moreover, we have

$$C^\infty_{\gamma'_{r,0}}((S^2T^*)^{+-}_{\mathbb{C}}) = C^\infty_{\gamma'_{r,0}}((S^2T^*)^{+-}_{\mathbb{C}})^{\mathrm{ev}}.$$

In fact, when s is even (resp. odd), the section $\tilde{f}_{r,s}h_0$ (resp. $\tilde{f}_{r,s-1}w_3$) of $(S^2T^*)^{+-}_{\mathbb{C}}$ is a highest weight vector of the irreducible $SO(5)$-module $C^\infty_{\gamma'_{r,s}}((S^2T^*)^{+-}_{\mathbb{C}})^{\mathrm{ev}}$. When s is ≥ 1 and even (resp. odd), the section $\tilde{f}_{r,s-1}w_3$ (resp. $\tilde{f}_{r,s}h_0 = \tilde{f}_{r,s-1}k^-$) is a highest weight vector of the irreducible $SO(5)$-module $C^\infty_{\gamma'_{r,s}}((S^2T^*)^{+-}_{\mathbb{C}})^{\mathrm{odd}}$.

The following result is proved in [23, §9].

LEMMA 6.32. For $r, s \geq 0$, with s even, the symmetric 2-form $\tilde{f}_{r,s}h_0$ on X does not satisfy the Guillemin condition.

For $r \geq 0$, we have $W'_{r,0} = \{0\}$. From Proposition 5.22 and from Lemmas 6.8, 6.29, 6.31 and 6.32, we then obtain the following result:

PROPOSITION 6.33. For $r, s \geq 0$, we have

$$\mathcal{N}_{2,\mathbb{C}} \cap C^\infty_{\gamma'_{r,s}}(S^2T^*_{\mathbb{C}})^{\mathrm{ev}} = D_0 C^\infty_{\gamma'_{r,s}}(T_{\mathbb{C}})^{\mathrm{ev}}.$$

From Proposition 5.23 and from Lemmas 6.21, 6.22, 6.29 and 6.31, we obtain the following result:

PROPOSITION 6.34. For $r, s \geq 0$, we have

$$\mathcal{Z}_{2,\mathbb{C}} \cap C^\infty_{\gamma'_{r,s}}(S^2T^*_{\mathbb{C}})^{\mathrm{odd}} = D_0 C^\infty_{\gamma'_{r,s}}(T_{\mathbb{C}})^{\mathrm{odd}}.$$

We now complete the proof of the following result of [23].

THEOREM 6.35. An even symmetric 2-form on the quadric $X = Q_3$ satisfies the Guillemin condition if and only if it is a Lie derivative of the metric.

PROOF: From Proposition 2.30,(i), with $X = Q_3$, $\Sigma = \{\tau\}$ and $\varepsilon = +1$, and Propositions 6.25, 6.27 and 6.33, we obtain the equality

$$\mathcal{N}_{2,\mathbb{C}} \cap C^\infty(S^2T^*_{\mathbb{C}})^{\mathrm{ev}} = D_0 C^\infty(T_{\mathbb{C}})^{\mathrm{ev}},$$

which implies the desired result.

The following is a consequence of joint work with Tela Nlenvo (see [52] and §5).

THEOREM 6.36. *An odd symmetric 2-form on the quadric* $X = Q_3$ *satisfies the zero-energy condition if and only if it is a Lie derivative of the metric.*

PROOF: From Proposition 2.30,(ii), with $X = Q_3$, $\Sigma = \{\tau\}$ and $\varepsilon = -1$, and Propositions 6.25, 6.28 and 6.34, we obtain the equality

$$\mathcal{Z}_{2,\mathbb{C}} \cap C^\infty(S^2 T_\mathbb{C}^*)^{\text{odd}} = D_0 C^\infty(T_\mathbb{C})^{\text{odd}},$$

which implies the desired result.

Let $r, s \geq 0$ be given integers. By Lemma 4.5 of [23], when $s \geq 2$ is even, the highest weight vector

$$\tilde{f}_{r,s-1}\pi_{+-}\text{Hess } \tilde{f}_{0,1}$$

of the irreducible $SO(5)$-module $C^\infty_{\gamma_{r,s}}((S^2 T^*)^{+-}_\mathbb{C})^{\text{ev}}$ does not satisfy the Guillemin condition. When s is even (resp. odd), by Lemma 6.32 (resp. Lemma 4.10 of [23]), the highest weight vector $\tilde{f}_{r,s}h_0$ (resp. $\tilde{f}_{r,s-1}w_3$) of the irreducible $SO(5)$-module $C^\infty_{\gamma'_{r,s}}((S^2 T^*)^{+-}_\mathbb{C})^{\text{ev}}$ does not satisfy the Guillemin condition. Thus from the remarks following Lemma 6.31 and the equalities (5.126) and (5.129), we obtain the following:

PROPOSITION 6.37. *Let* X *be the complex quadric* Q_3. *For* $r, s \geq 0$, *we have*

$$\mathcal{N}_{2,\mathbb{C}} \cap C^\infty_{\gamma_{r,s}}((S^2 T^*)^{+-}_\mathbb{C})^{\text{ev}} = \mathcal{N}_{2,\mathbb{C}} \cap C^\infty_{\gamma'_{r,s}}((S^2 T^*)^{+-}_\mathbb{C})^{\text{ev}} = \{0\}.$$

The following theorem is a direct consequence of Propositions 5.24, 6.25 and 6.37.

THEOREM 6.38. *An even section of* $(S^2 T^*)^{+-}$ *over the quadric* Q_3, *which satisfies the Guillemin condition, vanishes identically.*

Let $r \geq 0$, $s \geq 1$ be given integers. By Lemma 6.9, when s is an odd integer, the highest weight vector

$$\tilde{f}_{r,s-1}\pi_{+-}\text{Hess } \tilde{f}_{0,1}$$

of the irreducible $SO(5)$-module $C^\infty_{\gamma_{r,s}}((S^2 T^*)^{+-}_\mathbb{C})^{\text{odd}}$ does not satisfy the zero-energy condition. When s is even (resp. odd), by Lemma 6.11 (resp. Lemma 6.10), the highest weight vector $\tilde{f}_{r,s-1}w_3$ (resp. $\tilde{f}_{r,s}h_0 = \tilde{f}_{r,s-1}k^-$) of the irreducible $SO(5)$-module $C^\infty_{\gamma'_{r,s}}((S^2 T^*)^{+-}_\mathbb{C})^{\text{odd}}$ does not satisfy the zero-energy condition. Thus from the remarks following Lemma 6.31 and the equalities (5.126) and (5.129), we obtain the following:

PROPOSITION 6.39. *Let X be the complex quadric Q_3. For $r, s \geq 0$, we have*

$$\mathcal{Z}_{2,\mathbb{C}} \cap C^\infty_{\gamma_{r,s}}((S^2 T^*)^{+-}_\mathbb{C})^{\mathrm{odd}} = \mathcal{Z}_{2,\mathbb{C}} \cap C^\infty_{\gamma'_{r,s}}((S^2 T^*)^{+-}_\mathbb{C})^{\mathrm{odd}} = \{0\}.$$

The following theorem is a direct consequence of Propositions 5.25, 6.25 and 6.39.

THEOREM 6.40. *An odd section of $(S^2 T^*)^{+-}$ over the quadric Q_3, which satisfies the zero-energy condition, vanishes identically.*

§8. The rigidity of the complex quadric

In this section, we assume that X is the complex quadric Q_n, with $n \geq 3$, and we extend the main results of §7 to the quadric Q_n; in particular, we shall prove the four following theorems.

THEOREM 6.41. *An even symmetric 2-form on the quadric $X = Q_n$, with $n \geq 3$, satisfies the Guillemin condition if and only if it is a Lie derivative of the metric.*

THEOREM 6.42. *An odd symmetric 2-form on the quadric $X = Q_n$, with $n \geq 3$, satisfies the zero-energy condition if and only if it is a Lie derivative of the metric.*

THEOREM 6.43. *An even section of $(S^2 T^*)^{+-}$ over the quadric Q_n, with $n \geq 3$, which satisfies the Guillemin condition, vanishes identically.*

THEOREM 6.44. *An odd section of $(S^2 T^*)^{+-}$ over the quadric Q_n, with $n \geq 3$, which satisfies the zero-energy condition, vanishes identically.*

From Theorems 6.41 and 6.42, we shall deduce the following:

THEOREM 6.45. *The complex quadric $X = Q_n$, with $n \geq 3$, is infinitesimally rigid.*

From Theorems 6.43 and 6.44, we shall deduce the following:

THEOREM 6.46. *A section of $(S^2 T^*)^{+-}$ over the complex quadric Q_n, with $n \geq 3$, which satisfies the zero-energy condition, vanishes identically.*

We now prove these last two theorems simultaneously. Let h be a symmetric 2-form on X satisfying the zero-energy condition. We write $h = h^+ + h^-$, where

$$h^+ = \tfrac{1}{2}(h + \tau^* h), \qquad h^- = \tfrac{1}{2}(h - \tau^* h)$$

are the even and odd parts of h, respectively. Clearly, since τ is an isometry, both these 2-forms h^+ and h^- satisfy the zero-energy condition. If h is a section of $(S^2T^*)^{+-}$, so are the forms h^+ and h^-. By Lemma 2.11, the even form h^+ satisfies the Guillemin condition. First, by Theorems 6.41 and 6.42 we know that the even form h^+ and the odd form h^- are Lie derivatives of the metric. Finally, if h is a section of $(S^2T^*)^{+-}$, by Theorems 6.43 and 6.44 we know that the forms h^+ and h^- vanish.

From Theorems 6.41 and 6.43, and from Propositions 2.18 and 2.21, with $\Lambda = \{\mathrm{id}, \tau\}$ and $F = E = (S^2T^*)^{+-}$, we deduce the following two results:

THEOREM 6.47. *The real Grassmannian* $Y = G^{\mathbb{R}}_{2,n}$, *with* $n \geq 3$, *is rigid in the sense of Guillemin.*

THEOREM 6.48. *A section of the vector bundle* E_Y *over the real Grassmannian* $Y = G^{\mathbb{R}}_{2,n}$, *with* $n \geq 3$, *which satisfies the Guillemin condition, vanishes identically.*

We now proceed to prove Theorems 6.41–6.44. If x is a point of X, let $\mathcal{F}'_x$ be the family of all closed connected totally geodesic submanifolds of X passing through $x \in X$ which can be written as $\mathrm{Exp}_x V_x \otimes W_1$, where W_1 is a three-dimensional subspace of W_x. In §6, Chapter V, we considered the family

$$\mathcal{F}' = \bigcup_{x \in X} \mathcal{F}'_x$$

of submanifolds of X and we saw that a submanifold of X belonging to $\mathcal{F}'$ is isometric to the complex quadric Q_3 of dimension 3.

From Lemma 4.8, with $p = 2$ and $q = 3$, we obtain:

LEMMA 6.49. *Let* X *be the complex quadric* Q_n, *with* $n \geq 3$. *Let* $x \in X$ *and* h *be an element of* $S^2T^*_x$. *If the restriction of* h *to an arbitrary submanifold of the family* $\mathcal{F}'_x$ *vanishes, then* h *vanishes.*

PROPOSITION 6.50. *Let* h *be a symmetric 2-form on the quadric* $X = Q_n$, *with* $n \geq 3$.

(i) *If* h *is an even form satisfying the Guillemin condition, then* h *belongs to* $\mathcal{L}(\mathcal{F}')$.

(ii) *If* h *is an odd form satisfying the zero-energy condition, then* h *belongs to* $\mathcal{L}(\mathcal{F}')$.

(iii) *If* h *is a section of* $(S^2T^*)^{+-}$ *and satisfies the hypotheses of* (i) *or* (ii), *then* h *vanishes.*

PROOF: We consider the complex quadric $Z = Q_3$ of dimension 3. Let X' be a submanifold of X belonging to the family $\mathcal{F}'$. According to Lemma 4.6 and the equality (5.64), there is a totally geodesic imbedding $i : Z \to X$ whose image is equal to X' and which possesses the following properties:

(a) if h is a section of $(S^2T^*)^{+-}$, then the symmetric 2-form i^*h on Z is a section of the sub-bundle $(S^2T_Z^*)^{+-}$ of $S^2T_Z^*$;

(b) if h is an even (resp. odd) form on X, then the symmetric 2-form i^*h on Z is even (resp. odd).

Since the rank of the symmetric space X' is equal to that of X, if h satisfies the Guillemin condition, then the form i^*h on Z also satisfies the Guillemin condition. First, if h satisfies the hypotheses of (i) (resp. of (ii)), by Theorem 6.35 (resp. Theorem 6.36) the form i^*h is a Lie derivative of the metric of Z, and hence the restriction of h to X' is a Lie derivative of the metric of X'. Next, if h is a section of $(S^2T^*)^{+-}$ and satisfies the hypotheses of (i) (resp. of (ii)), by Theorem 6.38 (resp. Theorem 6.40), we infer that i^*h and the restriction of h to X' vanish. Now suppose that h satisfies the hypotheses of (i) or (ii). Then we have shown that h belongs to $\mathcal{L}(\mathcal{F}')$; moreover, if h is a section of $(S^2T^*)^{+-}$, by Lemma 6.49 we see that h vanishes.

We note that Proposition 6.50,(iii) gives us Theorems 6.43 and 6.44. We therefore know that Theorem 6.46 also holds.

We consider the G-invariant family $\mathcal{F} = \mathcal{F}_1$ of closed connected totally geodesic surfaces of X defined in §6, Chapter V; there we saw that each surface of $\mathcal{F}$ is contained in a totally geodesic submanifold belonging to the family $\mathcal{F}'$, and that all the totally geodesic flat 2-tori of X belong to the family $\mathcal{F}$. Thus any closed geodesic of X is contained in a totally geodesic surface of X belonging to the family $\mathcal{F}$; it follows that

$$(6.73) \qquad\qquad \mathcal{L}(\mathcal{F}') \subset \mathcal{Z}_2.$$

According to the inclusion (6.73) and Theorem 6.46, we know that the equality

$$(6.74) \qquad\qquad C^\infty((S^2T^*)^{+-}) \cap \mathcal{L}(\mathcal{F}') = \{0\}$$

holds. By Lemma 5.8, relation (6.74) and Proposition 5.17, we see that the families $\mathcal{F}$ and $\mathcal{F}'$ and the vector bundle $E = (S^2T^*)^{+-}$ satisfy the hypotheses of Theorem 2.48,(iii). Hence from this theorem, we deduce the following result:

THEOREM 6.51. Let h be a symmetric 2-form on the quadric $X = Q_n$, with $n \geq 3$. If h belongs to $\mathcal{L}(\mathcal{F}')$, then h is a Lie derivative of the metric of X.

This theorem together with the first two parts of Proposition 6.50 implies Theorems 6.41 and 6.42. According to the proofs of Theorem 2.45,(iii) and Proposition 6.50, we see that the only results of §7 which we require for the proof of Theorem 6.41 (resp. Theorem 6.42) are Theorems 6.35 and 6.38 (resp. Theorem 6.36 and 6.40).

None of our results concerning forms on the quadric satisfying the zero-energy condition enter into our proof of Theorem 6.41 given above. Previously, in [23] we deduced Theorem 6.41 for the quadric Q_n, with $n \geq 4$, from Theorem 6.35 by means of the infinitesimal rigidity of this quadric. In fact, if h is an even symmetric 2-form on $X = Q_n$, with $n \geq 4$, satisfying the Guillemin condition, by Proposition 6.50,(i) and the inclusion (6.73) we know that h satisfies the zero-energy condition; the infinitesimal rigidity of Q_n implies that h is a Lie derivative of the metric.

From Theorems 5.27 and 6.41, we obtain the following:

THEOREM 6.52. *An even section of L over the quadric Q_n, with $n \geq 3$, which satisfies the Guillemin condition, vanishes identically.*

From Theorem 6.41 and the decomposition (1.11), we obtain the relation

$$(6.75) \qquad \mathcal{N}_2 \cap \{\, h \in C^\infty(S^2 T^*)^{\mathrm{ev}} \mid \operatorname{div} h = 0 \,\} = \{0\}.$$

§9. Other proofs of the infinitesimal rigidity of the quadric

In this section, we suppose that X is the complex quadric Q_n, with $n \geq 4$. This section and the next one are devoted to other proofs of the infinitesimal rigidity of the quadric $X = Q_n$, with $n \geq 4$. Some of the methods used here were introduced in [18] and [22].

The essential aspects of the proof of the following proposition were first given by Dieng in [10].

PROPOSITION 6.53. *The infinitesimal rigidity of the quadric Q_3 implies that all the quadrics Q_n, with $n \geq 3$, are infinitesimally rigid.*

PROOF: We consider the G-invariant family $\mathcal{F}_3$ of closed connected totally geodesic surfaces of X introduced in §6, Chapter V and the family $\mathcal{F}'$ of closed connected totally geodesic submanifolds of X isometric to the quadric Q_3 introduced in §6, Chapter V and in §8. According to a remark made in §6, Chapter V, we know that each surface belonging to the family $\mathcal{F}_3$ is contained in a totally geodesic submanifold of X belonging to the family $\mathcal{F}'$. Assume that we know that the quadric Q_3 is infinitesimally rigid; then the family $\mathcal{F}'$ possesses property (III) of §8, Chapter II; moreover, by Propositions 5.13 and 5.14, the families $\mathcal{F} = \mathcal{F}_3$ and $\mathcal{F}'$ satisfy the hypotheses of Theorem 2.47,(iii). From this last theorem, we deduce the infinitesimal rigidity of X.

We now consider the G-invariant family $\mathcal{F} = \mathcal{F}_2$ of closed connected totally geodesic surfaces of X. The sub-bundle $N_2 = N_{\mathcal{F}}$ of B consisting of those elements of B, which vanish when restricted to the closed totally geodesic submanifolds of $\mathcal{F}$, was introduced in §8, Chapter II and was considered in §6, Chapter V. We also consider the differential operator

$$D_{1,\mathcal{F}} : S^2 T^* \to \mathcal{B}/\mathcal{N}_2$$

of §8, Chapter II.

We consider the families $\tilde{\mathcal{F}}^1$, $\tilde{\mathcal{F}}^6$ and $\tilde{\mathcal{F}}^7$ of closed connected totally geodesic submanifolds of X introduced in §6, Chapter V and we set

$$\mathcal{F}' = \tilde{\mathcal{F}}^1 \cup \tilde{\mathcal{F}}^6 \cup \tilde{\mathcal{F}}^7.$$

A submanifold of X belonging to $\tilde{\mathcal{F}}^1$ (resp. to $\tilde{\mathcal{F}}^6$) is a surface isometric to the flat 2-torus (resp. to the real projective plane $\mathbb{RP}^2$), while a submanifold of X belonging to $\tilde{\mathcal{F}}^7$ is isometric to the complex projective space $\mathbb{CP}^2$. In §6, Chapter V, we saw that each surface belonging to $\tilde{\mathcal{F}}^2$ is contained in a submanifold of X belonging to the family $\tilde{\mathcal{F}}^7$; therefore each surface of X belonging to $\mathcal{F}$ is contained in a submanifold of X belonging to the family $\mathcal{F}'$. According to Proposition 3.19 and Theorems 3.7 and 3.39, we see that the family $\mathcal{F}'$ possesses property (III) of §8, Chapter II. Hence a symmetric 2-form h on X satisfying the zero-energy condition belongs to $\mathcal{L}(\mathcal{F}')$, and, by Proposition 2.44, verifies the relation

$$D_{1,\mathcal{F}}h = 0.$$

PROPOSITION 6.54. *Let h be a symmetric 2-form on quadric $X = Q_n$, with $n \geq 4$, satisfying the zero-energy condition and the relation $\operatorname{div} h = 0$. Then when $n \geq 5$, the symmetric form h is a section of the vector bundle L; when $n = 4$, it is a section of the vector bundle $L \oplus (S^2 T^*)^{+-}$.*

PROOF: We know that h belongs to $\mathcal{L}(\mathcal{F}')$. We suppose that $n \geq 5$ (resp. that $n = 4$). According to Proposition 5.10 (resp. Proposition 5.11), we see that the hypotheses of Theorem 2.48,(i) hold, with $E = L$ (resp. with $E = L \oplus (S^2 T^*)^{+-}$). By Proposition 5.17, we know that $E(X) = \{0\}$ (resp. that $E(X) \subset C^\infty((S^2 T^*)^{+-})$). Then Theorem 2.48,(i) tells us that h is a section of L (resp. of $L \oplus (S^2 T^*)^{+-}$).

In §10, we shall prove the following result:

PROPOSITION 6.55. *Let X be the quadric Q_4. A section h of the vector bundle $L \oplus (S^2 T^*)^{+-}$ satisfying the relations*

$$\operatorname{div} h = 0, \qquad D_{1,\mathcal{F}}h = 0$$

vanishes identically.

We now give an alternate proof of Theorem 6.45, with $n \geq 4$, using Propositions 6.54, 6.55 and 5.26. In the case $n = 4$, this proof appears in [22]. Let h be a symmetric 2-form on the quadric $X = Q_n$, with $n \geq 4$, satisfying the zero-energy condition and the relation $\operatorname{div} h = 0$. When $n \geq 5$, Proposition 6.54 tells us that h is a section of L; by Proposition 5.26, we see that h vanishes identically. When $n = 4$, Proposition 6.54 tells us that h is a section of $L \oplus (S^2 T^*)^{+-}$, and, as we saw above, Proposition 2.44 gives us the relation $D_{1,\mathcal{F}} h = 0$; by Proposition 6.55, we see that h vanishes. Then Proposition 2.13 gives us the infinitesimal rigidity of X.

Finally, we present an outline of the proof of the infinitesimal rigidity of Q_n, with $n \geq 5$, given in [18]. This proof completely avoids the use of harmonic analysis on the quadric; it requires the description of the explicit complement of $\tilde{B}$ in the vector bundle $N_2 = N_{\mathcal{F}}$ given by Proposition 5.12.

We consider the natural projection $\alpha : B \to B/\tilde{B}$ and the differential operator $D_1' : \mathcal{B}/\tilde{\mathcal{B}} \to \mathcal{B}_1$ defined in §3, Chapter I. We also consider the morphism of vector bundles

$$\psi : \bigwedge\nolimits^2 T^* \to B, \qquad \hat{\tau}_B : S^2 T^* \to B$$

of Chapter I. In [18], using the equalities (5.73), (5.19) and (5.22), by purely algebraic computations and elementary operations involving differential forms we were able to prove the following result:

PROPOSITION 6.56. *Let β be a section of $(\bigwedge^2 T^*)^-$ and v be a section of L over an open subset of X satisfying*

$$(6.76) \qquad\qquad D_1' \alpha(\psi(\beta) + \hat{\tau}_B(v)) = 0.$$

Then we have $v = 0$ and $\nabla \beta = 0$.

We suppose that $n \geq 5$. Let h be a symmetric 2-form on X satisfying the zero-energy condition. As we have seen above, we have $D_{1,\mathcal{F}} h = 0$. By Proposition 5.12, there exist elements β of $C^\infty((\bigwedge^2 T^*)^-)$ and $v \in C^\infty(L)$ such that

$$D_1 h = \alpha(\psi(\beta) + \hat{\tau}_B(v)).$$

By Lemma 1.17, we obtain the relation (6.76). Then Proposition 6.56 tells us that $v = 0$ and $\nabla \beta = 0$. Since a harmonic differential 2-form on X is a constant multiple of the Kähler form of X, we immediately deduce that $\beta = 0$. Thus we know that $D_1 h = 0$. We know that the sequence (1.24) is exact (see §6, Chapter V); therefore h is a Lie derivative of the metric. We have thus shown that the infinitesimal rigidity of the quadric Q_n, with $n \geq 5$, is a consequence of Propositions 5.12, 5.14 and 6.56.

§10. The complex quadric of dimension four

In this section, we suppose that X is the complex quadric Q_4 of dimension 4, which is a homogeneous space of the group $G = SO(6)$, and we present an outline of the proof of Proposition 6.55.

For $r, s \geq 0$, we consider the elements

$$\gamma_{r,s}^1 = (2r+s+2)\lambda_0 + (s+1)\lambda_1 + \lambda_2, \quad \gamma_{r,s}^2 = (2r+s+2)\lambda_0 + (s+1)\lambda_1 - \lambda_2$$

of Γ.

We recall that the $SO(6)$-equivariant involution $*$ of $(S^2T^*)^{+-}$ defined in §5, Chapter V gives rise to the decomposition (5.70) of the subbundle $(S^2T^*)_{\mathbb{C}}^{+-}$. In §7, Chapter V, we saw that the highest weight of the irreducible K-submodules $F_{\mathbb{C},b}^+$ and $F_{\mathbb{C},b}^-$ of $(S^2T^*)_{\mathbb{C},b}^{+-}$ are equal to $\lambda_1 + \lambda_2$ and $\lambda_1 - \lambda_2$, respectively. Hence from the branching law for $SO(6)$ and K described in Theorem 1.1 of [54], we obtain the following result given by Proposition 4.2 of [21] (see also [54, §4]):

PROPOSITION 6.57. *For* $\gamma \in \Gamma$, *the* $SO(6)$-*module* $C_\gamma^\infty((S^2T^*)_{\mathbb{C}}^{+-})$ *vanishes unless* γ *is equal to* $\gamma_{r,s+1}$, *to* $\gamma_{r,s}^1$ *or to* $\gamma_{r,s}^2$, *for some* $r, s \geq 0$. *For* $r, s \geq 0$, *the multiplicities of the* $SO(6)$-*modules* $C_{\gamma_{r,s}^1}^\infty((S^2T^*)_{\mathbb{C}}^{+-})$ *and* $C_{\gamma_{r,s}^2}^\infty((S^2T^*)_{\mathbb{C}}^{+-})$ *are equal to* 2.

The function f_1 on $\mathbb{C}^{n+2}$ defined by

$$f_1(\zeta) = (\zeta_0 + i\zeta_1)(\bar{\zeta}_n + i\bar{\zeta}_{n+1}) - (\zeta_n + \zeta_{n+1})(\bar{\zeta}_0 + i\bar{\zeta}_1)$$

belongs to $\mathcal{H}$. We consider the element $\tilde{f}_1$ of $\tilde{\mathcal{H}}$ induced by f_1, and also the section

$$w_4 = \pi_{+-}\left(\tilde{f}_{0,1} \operatorname{Hess} \tilde{f}_1 - \tilde{f}_1 \operatorname{Hess} \tilde{f}_{0,1}\right)$$

of $(S^2T^*)_{\mathbb{C}}^{+-}$.

The following lemma is given by Lemma 3.9 of [22].

LEMMA 6.58. *Let* $r, s \geq 0$. *The elements*

$$\operatorname{div}(\tilde{f}_{r,s}w_4), \qquad \operatorname{div}(\tilde{f}_{r,s} * w_4)$$

of $C^\infty(T_{\mathbb{C}}^*)$ *are linearly independent.*

According to Lemma 6.58, we know that w_4 is a non-zero section of $(S^2T^*)_{\mathbb{C}}^{+-}$; in [22, §3], we saw that $\tilde{f}_{r,s}w_4$ is a highest weight vector of the $SO(6)$-module $C_{\gamma_{r,s}^1}^\infty((S^2T^*)_{\mathbb{C}}^{+-})$, for $r, s \geq 0$. Since the morphism $*$ is an $SO(6)$-equivariant involution of $(S^2T^*)^{+-}$, it follows that $\tilde{f}_{r,s} * w_4$ is also a highest weight vector of the $SO(6)$-module $C_{\gamma_{r,s}^1}^\infty((S^2T^*)_{\mathbb{C}}^{+-})$. From Lemma 6.58 and Proposition 6.57, we infer that the space $\mathcal{C}(C_{\gamma_{r,s}^1}^\infty((S^2T^*)_{\mathbb{C}}^{+-}))$ is generated by the sections $\tilde{f}_{r,s}w_4$ and $\tilde{f}_{r,s} * w_4$. Since $\operatorname{div} : S^2T_{\mathbb{C}}^* \to T_{\mathbb{C}}^*$ is a homogeneous differential operator, Lemma 6.58 tells us that the mapping

(6.77) $$\operatorname{div} : C_{\gamma_{r,s}^j}^\infty((S^2T^*)_{\mathbb{C}}^{+-}) \to C_{\gamma_{r,s}^j}^\infty(T_{\mathbb{C}}^*)$$

is injective for $j = 1$.

In [22, §3], we showed that the $SO(6)$-module $C^\infty_{\gamma^2_{r,s}}((S^2T^*)^{+-}_\mathbb{C})$ is equal to the image of $C^\infty_{\gamma^1_{r,s}}((S^2T^*)^{+-}_\mathbb{C})$ under the conjugation mapping sending the section h of $(S^2T^*)^{+-}_\mathbb{C}$ into $\bar{h}$. Therefore the mapping (6.77), with $j = 2$, is also injective.

We again consider the family $\mathcal{F} = \mathcal{F}_2$ of closed connected surfaces of X, the sub-bundle N_2 of B and the differential operator $D_{1,\mathcal{F}}$ of §9. The vector bundle $T^*_\mathbb{C} \oplus (B/N_2)_\mathbb{C}$ endowed with the Hermitian scalar product induced by the metric g is homogeneous and unitary; moreover the differential operator

$$Q = \mathrm{div} \oplus D_{1,\mathcal{F}} : S^2T^*_\mathbb{C} \to T^*_\mathbb{C} \oplus (B/N_2)_\mathbb{C}$$

is homogeneous.

The following result is given by Lemma 3.7 of [22]:

LEMMA 6.59. *If $r \geq 0$, $s \geq 1$, we have*

$$D_{1,\mathcal{F}} * \pi_{+-}\mathrm{Hess}\,\tilde{f}_{r,s} \neq 0.$$

According to the results of §7, Chapter V, concerning the highest weight vectors of the $SO(6)$-module $C^\infty_{\gamma_{r,s}}((S^2T^*)^{+-}_\mathbb{C})$, with $r \geq 0$ and $s \geq 1$, and by the relation (5.128), a highest weight vector h of this $SO(6)$-module can be written in the form

$$h = b_1\tilde{f}_{r,s-1}\pi_{+-}\mathrm{Hess}\,\tilde{f}_{0,1} + b_2 * \pi_{+-}\mathrm{Hess}\,\tilde{f}_{r,s},$$

where $b_1, b_2 \in \mathbb{C}$.

The following result is a consequence of Lemma 3.8 of [22].

LEMMA 6.60. *Let $r, s \geq 0$, with $r + s \geq 1$, and $b, b', b'' \in \mathbb{C}$. Suppose that the element*

(6.78)
$$\begin{aligned}h &= b'\tilde{f}_{r-1,s}\pi'\mathrm{Hess}\,\tilde{f}_{1,0} + b''\tilde{f}_{r-1,s}\pi''\mathrm{Hess}\,\tilde{f}_{1,0} \\ &\quad + b_1\tilde{f}_{r,s-1}\pi_{+-}\mathrm{Hess}\,\tilde{f}_{0,1} + b_2 * \pi_{+-}\mathrm{Hess}\,\tilde{f}_{r,s}\end{aligned}$$

of $C^\infty_{\gamma_{r,s}}(L_\mathbb{C} \oplus (S^2T^*)^{+-}_\mathbb{C})$ satisfies $Qh = 0$. Then h vanishes identically.

PROOF: According to Lemmas 5.2 and 5.29, the relation $\mathrm{div}\,h = 0$ gives us the vanishing of the coefficients b' and b'' when $r \geq 1$, and of the coefficient b_1 when $s \geq 1$. Therefore when $s \geq 1$, we see that

$$h = b_2 * \pi_{+-}\mathrm{Hess}\,\tilde{f}_{r,s};$$

then Lemma 6.59 tells us that b_2 also vanishes.

If $r, s \geq 0$ are integers satisfying $r + s \geq 1$, according to §7, Chapter V and the remarks appearing above, a highest weight vector h of the $SO(6)$-module

$$C^\infty_{\gamma_{r,s}}(L_{\mathbb{C}} \oplus (S^2 T^*)^{+-}_{\mathbb{C}})$$

can be written in the form (6.78), with $b_1, b_2, b', b'' \in \mathbb{C}$. Since the differential operator Q is homogeneous, from Lemma 6.60 we deduce the following:

PROPOSITION 6.61. *Let $r, s \geq 0$ be given integers, with $r + s \geq 1$. An element h of $C^\infty_{\gamma_{r,s}}(L_{\mathbb{C}} \oplus (S^2 T^*)^{+-}_{\mathbb{C}})$ satisfying $Qh = 0$ vanishes identically.*

Propositions 5.18, 5.19, 6.57 and 6.61, together with the injectivity of the mappings (6.77), with $j = 1, 2$, imply the following:

PROPOSITION 6.62. *Let γ be an element of Γ. Then an element h of $C^\infty_\gamma(L_{\mathbb{C}} \oplus (S^2 T^*)^{+-}_{\mathbb{C}})$ satisfying $Qh = 0$ vanishes identically.*

Now Proposition 6.55 is a direct consequence of Propositions 6.62 and 2.3 (with $Q_1 = 0$).

According to Proposition 5.17 and (5.132), we know that the relations

$$E(X) \subset C^\infty(S^2 T^*)^{\text{ev}}, \qquad E(X)_{\mathbb{C}} = C^\infty_{\gamma_{0,1}}(S^2 T^*_{\mathbb{C}})^{\text{ev}} = * \pi_{+-} \text{Hess} \, \mathcal{H}_{0,1}$$

hold. The last equalities imply that the result of Lemma 5.2 holds for all $f \in \mathcal{H}_{0,1}$. From the preceding relations and (6.75), we obtain the following:

PROPOSITION 6.63. *Let X be the quadric Q_4. Then we have*

$$(6.79) \qquad \mathcal{N}_2 \cap E(X) = \mathcal{N}_2 \cap C^\infty_{\gamma_{0,1}}(S^2 T^*_{\mathbb{C}})^{\text{ev}} = \{0\}.$$

§11. Forms of degree one

We now return to the study of the complex quadric $X = Q_n$ of dimension $n \geq 3$, viewed as a homogeneous space of the group $G = SO(n+2)$. Most of the results and proofs of this section can be found in [20] and in [23, §11].

We say that a differential form θ of degree p on X is even (resp. odd) if $\tau^* \theta = \varepsilon \theta$, where $\varepsilon = 1$ (resp. $\varepsilon = -1$). Clearly, if θ is an even (resp. odd) differential p-form on X, so is the $(p+1)$-form $d\theta$. In particular, we have

$$(6.80) \qquad dC^\infty(X)^{\text{ev}} \subset C^\infty(T^*_{\mathbb{C}})^{\text{ev}}, \qquad dC^\infty(X)^{\text{odd}} \subset C^\infty(T^*_{\mathbb{C}})^{\text{odd}}.$$

We consider the G-invariant families $\tilde{\mathcal{F}}^1$, $\tilde{\mathcal{F}}^2$ and $\tilde{\mathcal{F}}^4$ of closed connected totally geodesic surfaces of X introduced in §6, Chapter V. We know that $\tilde{\mathcal{F}}^1$ consists of all the totally geodesic flat 2-tori of X. We easily see that an element of T^*, which vanishes when restricted to the surfaces belonging to the family $\tilde{\mathcal{F}}^1$, vanishes. The set C_1 (resp. C_2) consisting of those elements of $\bigwedge^2 T^*$, which vanish when restricted to the surfaces belonging to the family $\tilde{\mathcal{F}}^1$ (resp. the family $\tilde{\mathcal{F}}^4$), is a sub-bundle of $\bigwedge^2 T^*$. The set C_3 consisting of those elements of $\bigwedge^2 T^*$, which vanish when restricted to the surfaces belonging to the G-invariant family $\tilde{\mathcal{F}}^1 \cup \tilde{\mathcal{F}}^2$, is clearly a sub-bundle of C_1.

Let $x \in X$ and let β_1 be an element of $C_{1,x}$ and β_2 an element of $C_{2,x}$. Let ν be an element of S_x, and let $\{\xi, \eta\}$ be an orthonormal set of elements of T_ν^+. An element θ of T_x^*, which vanishes when restricted to the surfaces belonging to the family $\tilde{\mathcal{F}}^1$, satisfies

$$\theta(\xi) = 0, \qquad \theta(J\eta) = 0,$$

and therefore vanishes. According to the definition of the families $\tilde{\mathcal{F}}^1$ and $\tilde{\mathcal{F}}^4$, we see that β_1 and β_2 satisfy the relations

(6.81) $$\beta_1(\xi, J\eta) = 0, \qquad \beta_2(\xi, \eta) = 0.$$

Since ν is an arbitrary element of S_x, from relation (5.12) we infer that β_1 and β_2 also satisfy

(6.82)
$$\beta_1(\xi + tJ\xi, J\eta - t\eta) = 0,$$
$$\beta_2(J\xi, J\eta) = 0, \qquad \beta_2(\xi + tJ\xi, \eta + tJ\eta) = 0,$$

for all $t \in \mathbb{R}$. From the first and last identities of (6.82), we infer that

(6.83) $$\beta_1(J\xi, J\eta) = \beta_1(\xi, \eta), \qquad \beta_2(\xi, J\eta) + \beta_2(J\xi, \eta) = 0.$$

Hence, if β_2 belongs to $(T_\mathbb{R}^{1,1})^-$, we see that

$$\beta_2(\xi, J\eta) = 0;$$

therefore by the second relations of (6.81) and (6.82), we obtain the equality

(6.84) $$C_2 \cap (\bigwedge^2 T^*)^- = \{0\}.$$

Since the vectors $\xi + \eta$ and $\xi - \eta$ of T_ν^+ are orthogonal and have the same length, the first equality of (6.81) tells us that

$$\beta_1(\xi - \eta, J\xi + J\eta) = 0;$$

the preceding relation, together with the first equality of (6.81), implies that the element β_1 of C_1 satisfies

(6.85) $$\beta_1(\xi, J\xi) = \beta_1(\eta, J\eta).$$

By (6.81) and (6.83), we see that β_1 belongs to $(T_{\mathbb{R}}^{1,1})$. Clearly, the 2-form ω is a section of C_1. If we set $a = \beta_1(\xi, J\xi)$, then according to (6.85) and (6.81), the element $\beta' = \beta - a\omega$ satisfies

$$\beta'(\zeta, J\zeta) = 0,$$

for all elements ζ of T_ν^+. By (5.24), we therefore know that β' belongs to $(T_{\mathbb{R}}^{1,1})^+$. We have thus proved that C_1 is a sub-bundle of $(T_{\mathbb{R}}^{1,1})^+ \oplus \{\omega\}$. Clearly, by (5.24) we also see that $(T_{\mathbb{R}}^{1,1})^+ \subset C_1$. We have thus verified the first relation given by the following result:

LEMMA 6.64. *We have*

(6.86) $$C_1 = (T_{\mathbb{R}}^{1,1})^+ \oplus \{\omega\}, \qquad C_2 = (T_{\mathbb{R}}^{1,1})^-, \qquad C_3 = \{0\}.$$

PROOF: It remains to verify the last two equalities of (6.86). Let $x \in X$ and let β be an element of $\bigwedge^2 T_x^*$. According to (6.81) and (6.82), we know that β belongs to C_2 if and only if the relations

$$\beta(\xi, \eta) = \beta(J\xi, J\eta) = 0$$

hold for all elements $\nu \in S_x$ and all orthonormal sets $\{\xi, \eta\}$ of elements of $T_{\nu,x}^+$. It follows that an element of $\bigwedge^2 T_x^*$ belongs to C_2 if and only if both its components in $(\bigwedge^2 T^*)^-$ and $T_{\mathbb{R}}^{1,1}$ are elements of C_2. Hence from the equality (6.84), we see that C_2 is a sub-bundle of $T_{\mathbb{R}}^{1,1}$; more precisely, by (5.24) we obtain the second relation of (6.86). Finally, suppose that β belongs to C_3; then we may write

$$\beta = \beta_1 + c\omega,$$

where β_1 is an element of $(T_{\mathbb{R}}^{1,1})^+$ and $c \in \mathbb{R}$. Let ν be an element of S_x and $\{\xi, \eta\}$ be an orthonormal set of elements of $T_{\nu,x}^+$. According to the definition of the family $\tilde{\mathcal{F}}^2$, if the element β of C_1 belongs to C_3, it satisfies

$$\beta(\xi + J\eta, J\xi - \eta) = 0.$$

Since β_1 is a Hermitian form, the left-hand side of the preceding relation is equal to

$$-2\beta_1(\xi, \eta) + 2c,$$

and so we obtain

$$\beta_1(\xi, \eta) = \beta_1(\eta, \xi) = 2c.$$

From these last equalities, we deduce that $\beta_1(\xi, \eta) = 0$ and $c = 0$. Since β_1 also satisfies the first relation of (6.81), we see that $\beta = 0$ and so we obtain the last equality of (6.86).

According to the decompositions (1.66) and (5.28), the sub-bundles

$$F_1 = (T_{\mathbb{R}}^{1,1})_0^- \oplus (\wedge^2 T^*)^-, \qquad F_2 = (T_{\mathbb{R}}^{1,1})^+ \oplus (\wedge^2 T^*)^-$$

are the orthogonal complements of C_1 and of C_2, respectively, in $\wedge^2 T^*$; for $j = 1, 2$, we consider the orthogonal projection $\pi_j : \wedge^2 T^* \to F_j$. Then $F_{1,\mathbb{C}}$ and $F_{2,\mathbb{C}}$ are homogeneous complex sub-bundles of $\wedge^2 T_{\mathbb{C}}^*$. We consider the first-order homogeneous differential operator $\tilde{d}_j = \pi_j d : T^* \to F_j$ and the kernel M_j of the induced mapping $\tilde{d}_j : C^\infty(T_{\mathbb{C}}^*) \to C^\infty(F_{j,\mathbb{C}})$, for $j = 1, 2$. For $\gamma \in \Gamma$ and $j = 1, 2$, we consider the G-submodules

$$M_{j,\gamma} = M_j \cap C_\gamma^\infty(T_{\mathbb{C}}^*), \quad M_{2,\gamma}^{\mathrm{ev}} = M_2 \cap C_\gamma^\infty(T_{\mathbb{C}}^*)^{\mathrm{ev}}$$

of $C_\gamma^\infty(T_{\mathbb{C}}^*)$. For $j = 1, 2$, we consider the complex

$$(6.87) \qquad C^\infty(X) \xrightarrow{d} C^\infty(T^*) \xrightarrow{\tilde{d}_j} C^\infty(F_j).$$

LEMMA 6.65. *Let $\gamma \in \Gamma$ and let θ be an element of $C_\gamma^\infty(T_{\mathbb{C}}^*)$. Suppose that $d\theta$ is a form of type $(1, 1)$.*

(i) *If γ is not of the form $\gamma_{r,s}$, with $r + s > 0$, then we have $d\theta = 0$.*

(ii) *If γ is equal to $\gamma_{r,s}$, with $r + s > 0$, and θ is a highest weight vector of $C_\gamma^\infty(T_{\mathbb{C}}^*)$, then there exists a scalar $c \in \mathbb{C}$ such that*

$$(6.88) \qquad d\theta = c\partial\bar{\partial}\tilde{f}_{r,s}.$$

PROOF: Since $d\theta$ is a form of type $(1, 1)$ and X is a simply-connected Kähler manifold, there exists a unique function $f \in C^\infty(X)$ satisfying $d\theta = \partial\bar{\partial}f$ and

$$\int_X f\, dX = 0.$$

Since the differential operator $\partial\bar{\partial}$ is homogeneous, this function f belongs to $C_\gamma^\infty(X)$. Hence if γ is not of the form $\gamma_{r,s}$, with $r + s > 0$, this function f vanishes and $d\theta = 0$. On the other hand, if $\gamma = \gamma_{r,s}$, with $r + s > 0$, and θ is a highest weight vector of $C_\gamma^\infty(T_{\mathbb{C}}^*)$, then f is a multiple of the highest weight vector $\tilde{f}_{r,s}$ of $C_\gamma^\infty(X)$; thus there exists a complex number $c \in \mathbb{C}$ such that the relation (6.88) holds.

LEMMA 6.66. *Let γ be an element of Γ.*

(i) *If γ is not equal to $\gamma_1 = \gamma_{0,1}$, we have*

$$M_1 \cap C_\gamma^\infty(T_{\mathbb{C}}^*) = dC_\gamma^\infty(X).$$

(ii) *If γ is not of the form $\gamma_{r,0}$, with $r \geq 1$, we have*

$$M_2 \cap C_\gamma^\infty(T_{\mathbb{C}}^*) = dC_\gamma^\infty(X).$$

(iii) *If $\gamma = \gamma_{r,0}$, with $r \geq 1$, we have*

$$M_2 \cap C_\gamma^\infty(T_{\mathbb{C}}^*)^{\mathrm{ev}} = dC_\gamma^\infty(X)^{\mathrm{ev}}.$$

PROOF: We first remark that, since X is simply-connected and d is a homogeneous differential operator, a closed 1-form of $C_\gamma^\infty(T_{\mathbb{C}}^*)$ belongs to $dC_\gamma^\infty(X)$. Therefore, to prove the desired results, it suffices to show that $dM_{1,\gamma} = 0$ (resp. $dM_{2,\gamma} = 0$) when γ satisfies the hypothesis of (i) (resp. of (ii)), and that $dM_{2,\gamma}^{\mathrm{ev}} = 0$ when $\gamma = \gamma_{r,0}$, with $r \geq 1$. Let j be an integer equal to 1 or 2. Assume that $M_{j,\gamma} \neq 0$ and let θ be a highest weight vector of $M_{j,\gamma}$. The relation $\tilde{d}_j\theta = 0$ implies that $d\theta$ is a form of type $(1,1)$. First, if γ is not of the form $\gamma_{r,s}$, with $r + s > 0$, according to Lemma 6.65,(i), we have $d\theta = 0$ and so $dM_{j,\gamma} = \{0\}$. Now suppose that γ is equal to $\gamma_{r,s}$, with $r + s > 0$; by Lemma 6.65,(ii), we know that there exists a constant $c \in \mathbb{C}$ such that the equality (6.88) holds. Since the morphism (1.67) induces the isomorphisms (5.25), by Lemma 1.25, when $j = 1$ (resp. $j = 2$), we see that the relation $\tilde{d}_j\theta = 0$ is equivalent to the fact that the symmetric 2-form $c\pi_+\mathrm{Hess}\,\tilde{f}_{r,s}$ is a section of the sub-bundle $\{g\}_{\mathbb{C}} \oplus (S^2T^*)_{\mathbb{C}}^{+-}$ (resp. of the sub-bundle $(S^2T^*)_{\mathbb{C}}^{++}$), or to the equality $c\pi_{++}^0\mathrm{Hess}\,\tilde{f}_{r,s} = 0$ (resp. to the equality $c\pi_{+-}\mathrm{Hess}\,\tilde{f}_{r,s} = 0$). According to §7, Chapter V, when $r \geq 1$ or $s \geq 2$ (resp. when $s \geq 1$), the section $\pi_{++}^0\mathrm{Hess}\,\tilde{f}_{r,s}$ (resp. the section $\pi_{+-}\mathrm{Hess}\,\tilde{f}_{r,s}$) is non-zero; therefore for $j = 1$ (resp. $j = 2$), under this hypothesis, the constant c vanishes and we see that $d\theta = 0$ and $dM_{j,\gamma} = \{0\}$. We have thus completed the proof of assertions (i) and (ii). For $r \geq 1$, by (5.37) and (5.113) the 2-form $\partial\bar{\partial}\tilde{f}_{r,0}$ is odd. If θ is an even section of $T_{\mathbb{C}}^*$, by (6.80) we know that the 2-form $d\theta$ is even. Thus if $j = 2$ and $\gamma = \gamma_{r,0}$, with $r \geq 1$, and if θ is an even section of $T_{\mathbb{C}}^*$, the relation (6.88) implies that $d\theta$ vanishes. Thus when $\gamma = \gamma_{r,0}$, with $r \geq 1$, we have shown that $dM_{2,\gamma}^{\mathrm{ev}} = \{0\}$, and so assertion (iii) holds.

PROPOSITION 6.67. *An even 1-form θ on X satisfying the condition $\tilde{d}_2\theta = 0$ is exact.*

PROOF: Let θ be an even 1-form on X satisfying $\tilde{d}_2\theta = 0$, and let γ be an arbitrary element of Γ. Since $\tilde{d}_2$ is a homogeneous differential operator, by (2.1) we know that $P_\gamma\theta$ belongs to $M_{2,\gamma}^{\mathrm{ev}}$. Lemma 6.66 then tells us that $P_\gamma\theta$ is an element of $dC_\gamma^\infty(X)$. Since d is an elliptic operator, from Proposition 2.2,(iii) it follows that θ belongs to $dC^\infty(X)$.

PROPOSITION 6.68. *The space $g^\flat(\mathcal{K})$ belongs to the kernel of $\tilde{d}_1$ and is isomorphic to the cohomology of the complex (6.87), with $j = 1$.*

PROOF: By Proposition 3.10,(i), we know that $g^\flat(\mathcal{K})$ belongs to the kernel of $\tilde{d}_1$. Let θ be a 1-form on X satisfying $\tilde{d}_1\theta = 0$. We consider the element $\theta' = P_{\gamma_1}\theta$ of $C^\infty_{\gamma_1}(T^*_\mathbb{C})$; from the equality (2.28), we obtain the existence of elements $\xi \in \mathcal{K}_\mathbb{C}$ and $f \in C^\infty(X)$ such that

$$\theta' = df + g^\flat(\xi).$$

The element $\theta'' = \theta - \theta'$ of $C^\infty(T^*_\mathbb{C})$ satisfies $P_{\gamma_1}\theta'' = 0$. Since $\tilde{d}_1$ is a homogeneous differential operator, by (2.1) we know that $P_\gamma\theta'' = P_\gamma\theta$ belongs to $M_{1,\gamma}$, for all $\gamma \in \Gamma$, with $\gamma \neq \gamma_1$. By Lemma 6.66,(i), we see that $P_\gamma\theta''$ is an element of $dC^\infty_\gamma(X)$, for all $\gamma \in \Gamma$. Since d is an elliptic operator, by Proposition 2.2,(iii), it follows that θ'' belongs to $dC^\infty(X)$. Therefore we may write

$$\theta = df' + g^\flat(\xi),$$

where $f' \in C^\infty(X)$. Since the relation $d^*g^\flat(\eta) = 0$ is satisfied by all Killing vector fields η on X, we now obtain the desired result.

We consider the real Grassmannian $Y = G^\mathbb{R}_{2,n}$ and the natural projection $\varpi : X \to Y$.

PROPOSITION 6.69. *An even 1-form θ on X satisfying the Guillemin condition verifies the relation $\tilde{d}_2\theta = 0$.*

PROOF: Let θ be an even 1-form on X satisfying the Guillemin condition and let $\hat\theta$ be the unique 1-form on $Y = G^\mathbb{R}_{2,n}$ satisfying $\theta = \varpi^*\hat\theta$. Let Z be a closed totally geodesic submanifold of X belonging to the family $\tilde{\mathcal{F}}^4$; in §6, Chapter V, we saw that Z is isometric to a 2-sphere of constant curvature 2 and that its projection Z' in Y is a closed totally geodesic submanifold of Y isometric to a real projective plane; in fact, by Lemma 4.6, we know that Z' can be written in the form $\mathrm{Exp}_y v \otimes W'$, where y is a point of Y, and v is a unit vector of $V_{Y,y}$, and W' is a two-dimensional subspace of $W_{Y,y}$. Thus, according to Lemma 4.7 and Proposition 4.15, with $m = r = 2$, the restriction of $\hat\theta$ to Z' is exact. Hence the restriction of $d\theta$ to Z vanishes. By Lemma 6.64, it follows that $d\theta$ is a section of $(T^{1,1}_\mathbb{R})^-$, and so $\tilde{d}_2\theta$ vanishes.

From Propositions 2.20, 6.67 and 6.69, we deduce the following two results given by Theorems 11.1 and 11.2 of [23]:

THEOREM 6.70. *An even 1-form on the quadric Q_n, with $n \geq 3$, satisfies the Guillemin condition if and only if it is exact.*

THEOREM 6.71. *A 1-form on the Grassmannian $G^\mathbb{R}_{2,n}$, with $n \geq 3$, satisfies the Guillemin condition if and only if it is exact.*

The following result is given by Theorem 2 of [20].

THEOREM 6.72. *A 1-form on the quadric* $X = Q_n$, *with* $n \geq 3$, *satisfies the zero-energy condition if and only if it is exact.*

PROOF: Let θ be a 1-form on X satisfying the zero-energy condition. From Theorem 3.8, it follows that the restriction of θ to a flat torus of X belonging to the family $\tilde{\mathcal{F}}^1$ is exact. Therefore we have $\tilde{d}_1\theta = 0$. According to Proposition 6.68, we may write

$$\theta = df + g^\flat(\xi),$$

where f is a real-valued function and ξ is a Killing vector field on X. Let Z be a surface of X belonging to the family $\tilde{\mathcal{F}}^1$ and let $i : Z \to X$ be the natural imbedding. Since the 1-form $g^\flat(\xi) = \theta - df$ on X satisfies the zero-energy condition and Z is a flat torus, Proposition 3.10,(ii) tells us that $i^* g^\flat(\xi) = 0$. Thus the restriction of the 1-form $g^\flat(\xi)$ to an arbitrary surface of X belonging to the family $\tilde{\mathcal{F}}^1$ vanishes. As we have seen above, this implies that $g^\flat(\xi)$ vanishes; we thus obtain the equality $\theta = df$.

When $n \geq 4$, we are able to give a proof of the preceding theorem which avoids the use of Proposition 6.68. In §6, Chapter V, we saw that each surface belonging to the family $\tilde{\mathcal{F}}^2$ is contained in a submanifold belonging to the family $\tilde{\mathcal{F}}^7$ of closed totally geodesic submanifolds of X introduced there. Since a submanifold belonging to the family $\tilde{\mathcal{F}}^7$ is isometric to the complex projective space $\mathbb{CP}^2$, according to Theorems 3.8 and 3.40 we know that the family $\tilde{\mathcal{F}}^1 \cup \tilde{\mathcal{F}}^7$ possesses property (VI) of §8, Chapter II. When $n \geq 4$, Theorem 6.72 is thus a consequence of the last equality of (6.86) and Theorem 2.51,(ii), with $\mathcal{F} = \tilde{\mathcal{F}}^3$ and $\mathcal{F}' = \tilde{\mathcal{F}}^1 \cup \tilde{\mathcal{F}}^7$.

CHAPTER VII

THE RIGIDITY OF THE REAL GRASSMANNIANS

§1. The rigidity of the real Grassmannians

Let $m \geq 2$ and $n \geq 3$ be given integers. We consider the real Grassmannians $X = \widetilde{G}_{m,n}^{\mathbb{R}}$ and $Y = G_{m,n}^{\mathbb{R}}$, endowed with the Riemannian metrics g and g_Y defined in §1, Chapter IV, and the natural Riemannian submersion $\varpi : X \to Y$. As in §1, Chapter IV, we view these Grassmannians as irreducible symmetric spaces and as homogeneous spaces of the group $G = SO(m + n)$. We identify the tangent bundle T of X with the vector bundle $V \otimes W$. We shall also consider the Kähler metric $\tilde{g}$ on the complex quadric Q_n defined in §2, Chapter V and denoted there by g.

Let x be a point of X. Let $\mathcal{F}_x$ be the family of all closed connected totally geodesic surfaces of X passing through x of the form $\mathrm{Exp}_x F$, where F is a subspace of the tangent space T_x satisfying one of the following three conditions:

(i) F is generated by the vectors $\{v_1 \otimes w_1, v_2 \otimes w_2\}$, where $\{v_1, v_2\}$ is an orthonormal set of elements of V_x and $\{w_1, w_2\}$ is an orthonormal set of elements of W_x;

(ii) $F = V_1 \otimes w$, where V_1 is a two-dimensional subspace of V_x and w is a unit vector of W_x;

(iii) $F = v \otimes W_1$, where v is a unit vector of V_x and W_1 is a two-dimensional subspace of W_x.

According to the formula for the curvature of the Grassmannian $\widetilde{G}_{m,n}^{\mathbb{R}}$ given in §1, Chapter IV, we know that a surface of $\mathcal{F}_x$ corresponding to a subspace F of T_x of type (i) is a totally geodesic flat 2-torus; on the other hand, by Lemma 4.6, a surface of $\mathcal{F}_x$ corresponding to a subspace F of T_x of type (ii) or of type (iii) is isometric to a 2-sphere of constant curvature 1.

Let $\mathcal{F}_x'$ be the family of all closed connected totally geodesic submanifolds of X passing through x which can be written as $\mathrm{Exp}_x V_1 \otimes W_x$, where V_1 is a two-dimensional subspace of V_x. Clearly, each surface of $\mathcal{F}_x$ is contained in a totally geodesic submanifold of X belonging to the family $\mathcal{F}_x'$. According to the relation (5.59) and Lemma 4.6, we know that a submanifold of X belonging to the family $\mathcal{F}_x'$ is isometric to the complex quadric Q_n of dimension n endowed with the Riemannian metric $2\tilde{g}$.

From Lemma 4.8, with $p = 2$ and $q = n$, we obtain:

LEMMA 7.1. *Let X be the real Grassmannian $\widetilde{G}_{m,n}^{\mathbb{R}}$, with $m \geq 2$, $n \geq 3$. Let u be an element of $\bigotimes^2 T_x^*$, with $x \in X$. If the restriction of u to an arbitrary submanifold of the family $\mathcal{F}_x$ vanishes, then u vanishes.*

We consider the G-invariant families

$$\mathcal{F} = \bigcup_{x \in X} \mathcal{F}_x, \qquad \mathcal{F}' = \bigcup_{x \in X} \mathcal{F}'_x$$

of closed connected totally geodesic submanifolds of X. We consider the sub-bundle $N = N_{\mathcal{F}}$ of B introduced in §8, Chapter II consisting of those elements of B, which vanish when restricted to the closed totally geodesic submanifolds of $\mathcal{F}$.

LEMMA 7.2. *Let X be the real Grassmannian $\widetilde{G}^{\mathbb{R}}_{m,n}$, with $m \geq 2$, $n \geq 3$. Then we have*

$$\operatorname{Tr} N \subset E.$$

PROOF: Let x be a point of X. Let v be an arbitrary unit vector of V_x and w be an arbitrary unit vector of W_x. Let $\{v_1, \ldots, v_m\}$ be an orthonormal basis of V_x and $\{w_1, \ldots, w_n\}$ be an orthonormal basis of W_x, with $v_1 = v$ and $w_1 = w$. If $1 \leq i \leq m$ and $1 \leq j \leq n$ are given integers, the two vectors $v \otimes w$ and $v_i \otimes w_j$ are tangent to a surface belonging to the family $\mathcal{F}_x$. Thus if u is an element of N_x, we see that

$$(\operatorname{Tr} u)(v \otimes w, v \otimes w) = \sum_{\substack{1 \leq i \leq m \\ 1 \leq j \leq n}} u(v \otimes w, v_i \otimes w_j, v \otimes w, v_i \otimes w_j) = 0.$$

Hence $\operatorname{Tr} N_x$ is a subspace of E_x.

PROPOSITION 7.3. *Let h be a section of E over the real Grassmannian $X = \widetilde{G}^{\mathbb{R}}_{m,n}$, with $m \geq 2$ and $n \geq 3$. If the restriction of h to an arbitrary submanifold X' of X belonging to the family $\mathcal{F}'$ is a Lie derivative of the metric of X', then h vanishes.*

PROOF: We consider the complex quadric $Z = Q_n$ endowed with the metric $2\tilde{g}$. Let X' be a submanifold of X belonging to the family $\mathcal{F}'$. According to Lemma 4.6 and the equality (5.64), there is a totally geodesic isometric imbedding $i : Z \to X$ whose image is equal to X' such that i^*h is a section of the sub-bundle $(S^2 T^*_Z)^{+-}$ of $S^2 T^*_Z$. If the restriction of h to X' is a Lie derivative of the metric of X', by Lemma 2.6 the symmetric 2-form i^*h satisfies the zero-energy condition; by Theorem 6.46, we infer that i^*h and the restriction of h to X' vanish. The desired result is now a consequence of Lemma 7.1.

The infinitesimal rigidity of the complex quadric Q_n, given by Theorem 6.45, implies that the family $\mathcal{F}'$ possesses property (III) of §8, Chapter II; therefore we have

(7.1) $$\mathcal{Z}_2 \subset \mathcal{L}(\mathcal{F}').$$

By relation (4.8), we know that E is a G-invariant sub-bundle of $S_0^2 T^*$. According to Lemma 7.2 and Proposition 7.3, the families $\mathcal{F}$ and $\mathcal{F}'$ and the vector bundle E satisfy the relations (2.33) and (2.37). Now assume that $m \geq 3$. Then Proposition 4.2 tells us that $E(X)$ is a subspace of $C^\infty(E)$; hence the relations (2.37) and (7.1) give us the equality (2.36). Thus the families $\mathcal{F}$ and $\mathcal{F}'$ and the vector bundle E satisfy the hypotheses of Theorem 2.49,(ii) and 2.48,(iii). Hence from these two theorems, we deduce the infinitesimal rigidity of the space $\widetilde{G}_{m,n}^{\mathbb{R}}$ and the equality $\mathcal{L}(\mathcal{F}') = D_0 C^\infty(T)$ when $m, n \geq 3$. As the space $\widetilde{G}_{2,n}^{\mathbb{R}}$ is isometric to the complex quadric Q_n endowed with the metric $2\tilde{g}$, the Grassmannian $\widetilde{G}_{2,n}^{\mathbb{R}}$ is infinitesimally rigid. Since the Grassmannian $\widetilde{G}_{n,2}^{\mathbb{R}}$ is isometric to $\widetilde{G}_{2,n}^{\mathbb{R}}$, we have therefore proved the following two theorems:

THEOREM 7.4. *The real Grassmannian $X = \widetilde{G}_{m,n}^{\mathbb{R}}$, with $m, n \geq 2$ and $m + n \geq 5$, is infinitesimally rigid.*

THEOREM 7.5. *Let h be a symmetric 2-form on the real Grassmannian $X = \widetilde{G}_{m,n}^{\mathbb{R}}$, with $m, n \geq 3$. If h belongs to $\mathcal{L}(\mathcal{F}')$, then h is a Lie derivative of the metric of X.*

In Chapter X, we shall show that the Grassmannian $\widetilde{G}_{2,2}^{\mathbb{R}}$ is not infinitesimally rigid; on the other hand, Theorem 10.20 tells us that the Grassmannian $G_{2,2}^{\mathbb{R}}$ is infinitesimally rigid

Let y be a point of Y; choose a point $x \in X$ satisfying $\varpi(x) = y$. The family $\mathcal{F}_{Y,y}$ of all closed connected totally geodesic submanifolds of Y of the form $\mathrm{Exp}_y F$, where F is a subspace of the tangent space T_x satisfying conditions (i), (ii) or (iii), does not depend on the choice of the point x and so is well-defined. Similarly the family $\mathcal{F}'_{Y,y}$ consisting of all closed connected totally geodesic submanifolds of Y which can be written as $\mathrm{Exp}_x V_1 \otimes W_x$, where V_1 is a two-dimensional subspace of V_x, is well-defined. Clearly, each surface of $\mathcal{F}_{Y,y}$ is contained in a totally geodesic submanifold of Y belonging to the family $\mathcal{F}'_{Y,y}$. According to Lemma 4.6, we know that a submanifold of Y belonging to the family $\mathcal{F}'_{Y,y}$ is isometric to the real Grassmannian $G_{2,n}^{\mathbb{R}}$.

In the next section, we shall require the following result; its proof is similar to the proof of Lemma 4.8 and shall be omitted.

LEMMA 7.6. *Let Y be the real Grassmannian $G_{m,n}^{\mathbb{R}}$, with $m, n \geq 2$. Let u be an element of $\bigotimes^2 T_{Y,y}^*$, with $y \in Y$. If the restriction of u to an arbitrary submanifold of the family $\mathcal{F}_{Y,y}$ vanishes, then u vanishes.*

We consider the G-invariant families

$$\mathcal{F}_Y = \bigcup_{y \in Y} \mathcal{F}_{Y,y}, \qquad \mathcal{F}'_Y = \bigcup_{y \in Y} \mathcal{F}'_{Y,y}$$

of closed connected totally geodesic submanifolds of X. If Y' is a surface of Y belonging to the family $\mathcal{F}_Y$, there is a subgroup of G which acts transitively on Y'; thus we see that an element u of $\bigotimes^q T_x^*$, with $x \in X$, vanishes when restricted to an arbitrary surface belonging to the family $\mathcal{F}_Y$ if and only if it vanishes when restricted to an arbitrary surface belonging to the family $\mathcal{F}_{Y,y}$. We consider the sub-bundle $N_{\mathcal{F}_Y}$ of B consisting of those elements of B, which vanish when restricted to the closed totally geodesic surfaces of $\mathcal{F}_Y$.

If Y' is a submanifold of Y belonging to $\mathcal{F}_{Y,y}$ (resp. to $\mathcal{F}'_{Y,y}$) which can be written as $\mathrm{Exp}_y F$, where F is a subspace of T_x, then $\varpi^{-1}Y'$ is a submanifold of X equal to the submanifold $\mathrm{Exp}_x F$ belonging to the family $\mathcal{F}_x$ (resp. to the family $\mathcal{F}'_x$); moreover $\varpi : X' \to Y'$ is a two-fold covering. Thus if h is a symmetric 2-form on Y belonging to $\mathcal{L}(\mathcal{F}'_Y)$, the even 2-form $\varpi^* h$ on X belongs to $\mathcal{L}(\mathcal{F}')$. Also from Lemma 7.2, we obtain the relation

$$(7.2) \qquad\qquad \mathrm{Tr}\, N_{\mathcal{F}_Y} \subset E_Y.$$

PROPOSITION 7.7. *Let h be a section of E_Y over the real Grassmannian $Y = G_{m,n}^{\mathbb{R}}$, with $m \geq 2$ and $n \geq 3$. If the restriction of h to an arbitrary submanifold Y' of Y belonging to the family $\mathcal{F}'_Y$ is a Lie derivative of the metric of Y', then h vanishes.*

PROOF: The symmetric 2-form $\varpi^* h$ on $X = \widetilde{G}_{m,n}^{\mathbb{R}}$ is even and is a section of the vector bundle E over X. If h is an element of $\mathcal{L}(\mathcal{F}'_Y)$, then the form $\varpi^* h$ belongs to $\mathcal{L}(\mathcal{F}')$. The desired result is a consequence of Proposition 7.3.

We remark that the preceding proposition actually requires only Theorem 6.43 rather than Theorem 6.46, whose proof relies upon both Theorems 6.43 and 6.44 and which is needed for the case of an arbitrary symmetric 2-form considered in Proposition 7.3.

We now suppose that $m < n$. According to Theorem 6.47, the real Grassmannian $Y = G_{2,n}^{\mathbb{R}}$ is rigid in the sense of Guillemin, and so the family $\mathcal{F}'_Y$ possesses property (II) of §8, Chapter II. By Proposition 4.12, with $q = 2$ and $r = n$, the family $\mathcal{F}'_Y$ also possesses property (I) of §8, Chapter II. By relation (4.8), we know that E_Y is a G-invariant sub-bundle of $S_0^2 T_Y^*$. According to (7.2) and Proposition 7.7, the families $\mathcal{F}_Y$ and $\mathcal{F}'_Y$ and the vector bundle E_Y satisfy the relations (2.33) and (2.37) of Theorem 2.48. When $m \geq 3$, Proposition 4.2,(i) tells us the equality (2.35) holds; thus the families $\mathcal{F}_Y$ and $\mathcal{F}'_Y$ and the vector bundle E_Y satisfy the hypotheses of Theorem 2.49,(i). Hence from this theorem, we deduce the Guillemin rigidity of the space $G_{m,n}^{\mathbb{R}}$ when $m < n$ and $m \geq 3$. Since the Grassmannian $G_{p,q}^{\mathbb{R}}$ is isometric to $G_{q,p}^{\mathbb{R}}$, we have therefore proved the following:

THEOREM 7.8. *The real Grassmannian* $X = G_{m,n}^{\mathbb{R}}$, *with* $m, n \geq 2$ *and* $m \neq n$, *is rigid in the sense of Guillemin.*

Since the space $\widetilde{G}_{2,n}^{\mathbb{R}}$ is isometric to the complex quadric Q_n endowed with the metric $2\tilde{g}$, we know that the sequence (1.24) is exact for $\widetilde{G}_{2,n}^{\mathbb{R}}$, with $n \geq 3$ (see §6, Chapter V). We now proceed to show that the sequence (1.24) is also exact for the Grassmannians $\widetilde{G}_{m,n}^{\mathbb{R}}$, with $m, n \geq 3$; we begin by verifying the following:

PROPOSITION 7.9. *Let* h *be a symmetric 2-form on the real Grassmannian* $X = \widetilde{G}_{m,n}^{\mathbb{R}}$, *with* $m, n \geq 3$, *satisfying* $D_1 h = 0$. *Then* h *belongs to* $\mathcal{L}(\mathcal{F}')$.

PROOF: Since a submanifold Z of X belonging to the family $\mathcal{F}'$ is isometric to the complex quadric of dimension n endowed with the metric $2\tilde{g}$, we know that the sequence (1.24) for Z is exact. The desired result is given by Proposition 2.46,(ii).

From Proposition 7.9 and Theorem 7.5, we obtain the exactness of the sequence (1.24) for the real Grassmannian $\widetilde{G}_{m,n}^{\mathbb{R}}$, with $m, n \geq 3$. Since the real Grassmannian $G_{m,n}^{\mathbb{R}}$ is a quotient of $\widetilde{G}_{m,n}^{\mathbb{R}}$ by the group of isometries Λ of order 2 considered in §1, Chapter IV, the sequence (1.24) is also exact for the real Grassmannian $G_{m,n}^{\mathbb{R}}$, with $m, n \geq 2$ and $m + n \geq 5$. Thus we have demonstrated the following:

PROPOSITION 7.10. *Let* $m, n \geq 3$ *be given integers. If* X *is the real Grassmannian* $\widetilde{G}_{m,n}^{\mathbb{R}}$ *or* $G_{m,n}^{\mathbb{R}}$, *then the sequence* (1.24) *is exact.*

The result of the preceding proposition is given by Theorem 1.23 for all the real Grassmannians $X = \widetilde{G}_{m,n}^{\mathbb{R}}$, with $m, n \geq 3$, other than the space $\widetilde{G}_{3,3}^{\mathbb{R}}$. In fact, when $m, n \geq 3$ and $m + n \geq 7$, according to Proposition 4.2,(i), we know that the space $E(X)$ vanishes; Lemma 1.12 then tells us that the sequence (1.24) is exact.

It is easily seen that the sub-bundle $C_{\mathcal{F}'}$ of $\bigwedge^2 T^*$, consisting of those elements of $\bigwedge^2 T^*$ which vanish when restricted to the closed totally geodesic submanifolds of $\mathcal{F}'$, vanishes. It follows that the sub-bundle $C_{\mathcal{F}'_Y}$ of $\bigwedge^2 T_Y^*$, consisting of those elements of $\bigwedge^2 T_Y^*$ which vanish when restricted to the closed totally geodesic submanifolds of $\mathcal{F}'_Y$, also vanishes.

By Theorem 6.72, we know that $\mathcal{F}'$ possesses property (VI) of §8, Chapter II. When $m < n$, Proposition 4.12, with $q = 2$ and $r = n$, tells us that the family $\mathcal{F}'_Y$ satisfies property (IV) of §8, Chapter II; moreover by Theorem 6.71, we know that $\mathcal{F}'_Y$ also possesses property (V) of §8, Chapter II. From these observations and Theorem 2.51, we obtain the following two theorems:

THEOREM 7.11. *Let $m, n \geq 2$ be given integers, with $m \neq n$. A 1-form on the real Grassmannian $Y = G^{\mathbb{R}}_{m,n}$ satisfies the Guillemin condition if and only if it is exact.*

THEOREM 7.12. *Let $m, n \geq 2$ be given integers, with $m + n \geq 5$. A 1-form on the real Grassmannian $X = \tilde{G}^{\mathbb{R}}_{m,n}$ satisfies the zero-energy condition if and only if it is exact.*

Theorem 7.12 is given by Theorem 3 of [20]. In Chapter X, we shall show that a 1-form on the Grassmannian $G^{\mathbb{R}}_{2,2}$ satisfies the zero-energy condition if and only if it is exact (see Theorem 10.21).

§2. The real Grassmannians $\bar{G}^{\mathbb{R}}_{n,n}$

Let F be a real vector space of dimension $m + n$, where $m, n \geq 1$, endowed with a positive definite scalar product. We consider the real Grassmannians $G^{\mathbb{R}}_m(F)$ and $G^{\mathbb{R}}_n(F)$ endowed with the Riemannian metrics induced by the scalar product of F, which are defined in §1, Chapter IV. There we also saw that the natural mapping

$$\Psi : G^{\mathbb{R}}_m(F) \to G^{\mathbb{R}}_n(F),$$

sending an m-plane of F into its orthogonal complement, is an isometry. When $m = n$, then $\Psi = \Psi_X$ is an involution of $X = G^{\mathbb{R}}_n(F)$; as in §1, Chapter IV, we say that a symmetric p-form u on X is even (resp. odd) if $\Psi^*_X u = \varepsilon u$, where $\varepsilon = 1$ (resp. $\varepsilon = -1$).

Let $n \geq 2$ be a given integer. We now suppose that the dimension of F is equal to $2n + 2$ and we consider the Grassmannian $X = G^{\mathbb{R}}_{n+1}(F)$. We identify the tangent bundle T of X with the vector bundle $V \otimes W$. Let U be a subspace of F of dimension $\geq n$. If the orthogonal complement $U^\perp$ of U in F is non-zero, according to §2, Chapter IV, there is totally geodesic imbedding

(7.3) $$\iota : G^{\mathbb{R}}_1(U^\perp) \times G^{\mathbb{R}}_n(U) \to G^{\mathbb{R}}_{n+1}(F),$$

sending the pair (z, y), where $z \in G^{\mathbb{R}}_1(U^\perp)$ and $y \in G^{\mathbb{R}}_n(U)$, into the point of $G^{\mathbb{R}}_{n+1}(F)$ corresponding to the $(n+1)$-plane of F generated by the subspaces of $U^\perp$ and U corresponding to the points z and y, respectively.

We henceforth suppose that the subspace U is of dimension $2n$ and we write $Y = G^{\mathbb{R}}_n(U)$. Since $\dim U^\perp = 2$, the manifold $Z = G^{\mathbb{R}}_n(U^\perp)$ is isometric to the circle S^1. We consider the totally geodesic imbedding $\iota : Z \times Y \to X$ given by (7.3). For $z \in Z$, let $\iota_z : Y \to X$ be the mapping sending $y \in Y$ into $\iota(z, y)$; for $y \in Y$, let $\rho_y : Z \to X$ be the mapping sending $z \in Z$ into $\iota(z, y)$. A symmetric p-form θ on X determines a symmetric p-form θ_U on Y as follows. If $\xi_1, \ldots, \xi_p \in T_Y$, we consider the real-valued function f on Z defined by

$$f(z) = (\iota_z^* \theta)(\xi_1, \ldots, \xi_p),$$

for $z \in Z$, and we define θ_U by setting

$$\theta_U(\xi_1, \ldots, \xi_p) = \int_Z f \, dZ.$$

In particular, if θ is a function on X, we have

$$\theta_U(y) = \int_Z \rho_y^* \theta \, dZ,$$

for $y \in Y$.

If $z \in Z$ and $y \in Y$, the mapping $\iota_z : T_{Y,y} \to T_x$, where $x = \iota(z, y)$, is equal to the natural inclusion of $(V_Y \otimes W_Y)_y$ into $(V \otimes W)_x$. Thus if h is a section of the sub-bundle E of $S^2 T^*$ over X, then the symmetric 2-form $\iota_z^* h$ is a section of the sub-bundle E_Y of $S^2 T_Y^*$; therefore the symmetric 2-form h_U on Y is also a section of E_Y.

For $z \in Z$ and $y \in Y$, we easily see that

$$\Psi_X \iota(z, y) = \iota(\Psi_Z(z), \Psi_Y(y)),$$

and so we have

(7.4) $$\Psi_X \circ \iota_z = \iota_{\Psi_Z(z)} \circ \Psi_Y,$$

as mappings from Y to X. If θ is a symmetric p-form on X and $\xi_1, \ldots, \xi_p$ are vectors of T_Y, we consider the symmetric p-form $\theta' = \Psi_X^* \theta$ on X and the real-valued functions f_1 and f_2 on Z defined by

$$f_1(z) = (\iota_z^* \theta')(\xi_1, \ldots, \xi_p), \qquad f_2(z) = (\iota_z^* \theta)(\Psi_{Y*}\xi_1, \ldots, \Psi_{Y*}\xi_p),$$

for $z \in Z$. From (7.4), it follows that

$$f_1 = \Psi_Z^* f_2.$$

Since Ψ_Z is an isometry of Z, this relation implies that

$$\theta_U'(\xi_1, \ldots, \xi_p) = \int_Z f_1 \, dZ = \int_Z f_2 \, dZ = \theta_U(\Psi_{Y*}\xi_1, \ldots, \Psi_{Y*}\xi_p).$$

The following lemma is a direct consequence of the preceding equalities.

LEMMA 7.13. *Let $n \geq 2$ be a given integer. Let F be a real vector space of dimension $2n + 2$ endowed with a positive definite scalar product and let U be a subspace of F of dimension $2n$. If θ is an even symmetric p-form on $G_{n+1}^{\mathbb{R}}(F)$, then the symmetric p-form θ_U on $G_n^{\mathbb{R}}(U)$ is even.*

Let Y' be a maximal flat totally geodesic torus of Y; then we easily see that $X' = \iota(Z \times Y')$ is a maximal flat totally geodesic torus of X. Let $\xi_1, \ldots, \xi_p$ be arbitrary parallel vector fields on Y'; they induce parallel vector fields $\xi_1', \ldots, \xi_p'$ on X' determined by

$$\xi_j'(\iota(z, y)) = \iota_{z*}\xi_j(y),$$

for $z \in Z$, $y \in Y'$ and $1 \leq j \leq p$. Then from the definition of θ_U and Fubini's theorem, we infer that

$$\int_{Y'} \theta_U(\xi_1, \ldots, \xi_p) \, dY' = \int_{X'} \theta(\xi_1', \ldots, \xi_p') \, dX'.$$

From this equality, we deduce the following result:

LEMMA 7.14. *Let $n \geq 2$ be a given integer. Let F be a real vector space of dimension $2n + 2$ endowed with a positive definite scalar product and let U be a subspace of F of dimension $2n$. If θ is a symmetric p-form on $G_{n+1}^{\mathbb{R}}(F)$ satisfying the Guillemin condition, then the symmetric p-form θ_U on $G_n^{\mathbb{R}}(U)$ satisfies the Guillemin condition.*

We now suppose that the vector space F is equal to the space $\mathbb{R}^{2n+2}$ endowed with the standard Euclidean scalar product. Then X is the Grassmannian $G_{n+1,n+1}^{\mathbb{R}}$ endowed with the Riemannian metric g defined in §1, Chapter IV; we view X as an irreducible symmetric space and as a homogeneous space of the group $G = SO(2n + 2)$. We consider the G-invariant family $\mathcal{F}_X$ of closed connected totally geodesic surfaces of X which was introduced in §1; in fact, for the Grassmannian $G_{n,n}^{\mathbb{R}}$, this family was denoted there by $\mathcal{F}_Y$. We also consider the sub-bundle $N = N_{\mathcal{F}_X}$ of B consisting of those elements of B, which vanish when restricted to the closed totally geodesic submanifolds of $\mathcal{F}_X$, and the differential operator

$$D_{1,\mathcal{F}_X} : S^2 T^* \to B/N$$

of §8, Chapter II. According to Lemma 7.2, we have the relation

(7.5) $\operatorname{Tr} N \subset E_X,$

which is also given by (7.2).

PROPOSITION 7.15. *Let f be a real-valued function on the real Grassmannian $X = G_{n+1,n+1}^{\mathbb{R}}$, with $n \geq 2$. Suppose that, for any subspace U of $\mathbb{R}^{2n+2}$ of dimension $2n$, the function f_U on $G_n^{\mathbb{R}}(U)$ vanishes. Then the function f vanishes.*

PROOF: Let x be a point of X; we choose a subspace V_1 of V_x of codimension one and we consider the orthogonal complement U' of V_1 in $F = \mathbb{R}^{2n+2}$, whose dimension is equal to $n+2$. We also consider the totally geodesic imbedding

$$i : G_1^{\mathbb{R}}(U') \to G_{n+1}^{\mathbb{R}}(F),$$

sending $z \in G_1^{\mathbb{R}}(U')$ into the $(n+1)$-plane of F generated by the subspace V_1 and the line corresponding to the point z. There is a unique point z_0 of $G_1^{\mathbb{R}}(U')$ such that $x = i(z_0)$. Let γ be a closed geodesic of the projective space $G_1^{\mathbb{R}}(U')$. Then there is a subspace U'' of U' of dimension 2 such that the image of γ is equal to the submanifold $G_1^{\mathbb{R}}(U'')$ of $G_1^{\mathbb{R}}(U')$. The orthogonal complement U of U'' in F is of dimension $2n$ and contains V_1; thus the subspace V_1 of U corresponds to a point y_0 of $G_n^{\mathbb{R}}(U)$. We consider the totally geodesic imbedding

$$\iota : G_1^{\mathbb{R}}(U'') \times G_n^{\mathbb{R}}(U) \to G_{n+1}^{\mathbb{R}}(F)$$

given by (7.3). Then we have $i(z) = \iota(z, y_0)$, for $z \in G_1^{\mathbb{R}}(U'')$, and so the equality

$$\int_\gamma i^* f = f_U(y_0)$$

holds. Our hypothesis tells us that f_U vanishes, and hence the function $i^* f$ on $G_1^{\mathbb{R}}(U')$ satisfies the zero-energy condition. Since the dimension of $G_1^{\mathbb{R}}(U')$ is ≥ 3, the injectivity of the Radon transform on this real projective space, given by Theorem 2.23,(ii), implies that the function $i^* f$ vanishes. Since we have $x = i(z_0)$, we obtain the vanishing of the function f at the point x.

PROPOSITION 7.16. *Let h be a symmetric 2-form and θ be a 1-form on the real Grassmannian $X = G_{n+1,n+1}^{\mathbb{R}}$, with $n \geq 3$.*

(i) Suppose that, for any subspace U of $\mathbb{R}^{2n+2}$ of dimension $2n$, the symmetric 2-form h_U on $G_n^{\mathbb{R}}(U)$ vanishes. Then the symmetric form h vanishes.

(ii) Suppose that, for any subspace U of $\mathbb{R}^{2n+2}$ of dimension $2n$, the symmetric 2-form h_U on $G_n^{\mathbb{R}}(U)$ is a Lie derivative of the metric of $G_n^{\mathbb{R}}(U)$. Then we have the relation

$$D_{1,\mathcal{F}_X} h = 0.$$

(iii) *Suppose that, for any subspace U of $\mathbb{R}^{2n+2}$ of dimension $2n$, the 1-form θ_U on $G^{\mathbb{R}}_n(U)$ is closed. Then the restriction of $d\theta$ to an arbitrary submanifold of X belonging to the family $\mathcal{F}_X$ vanishes.*

PROOF: Let x be a point of X, and let $\xi_1, \xi_2 \in T_x$ be vectors tangent to a given totally geodesic surface X' of X belonging to the family $\mathcal{F}_X$ and containing x. By Lemma 7.6, we see that, in order to prove assertion (i) (resp. assertion (iii)) it suffices to show that $h(\xi_1, \xi_2) = 0$ (resp. that $(d\theta)(\xi_1, \xi_2) = 0$) under the hypothesis of (i) (resp. of (iii)). According to Proposition 1.14,(ii), in order to prove assertion (ii) it suffices to show that the relation

$$(7.6) \qquad\qquad (D_g h)(\xi_1, \xi_2, \xi_1, \xi_2) = 0$$

holds under the hypothesis of (ii). According to the definition of $\mathcal{F}_X$, there are subspaces V_1 of V_x of codimension one and W_1 of W_x of dimension 2 such that the vectors ξ_1, ξ_2 belong to the subspace $V_1 \otimes W_1$ of $T_x = (V \otimes W)_x$. We set $U' = V_1 \oplus W_1$; the subspace V_1 of U' corresponds to a point y_0 of $Y' = G^{\mathbb{R}}_n(U')$. We consider the orthogonal complements V'_1 of V_1 in V_x and U'' of U' in $F = \mathbb{R}^{2n+2}$. Then V'_1 is a one-dimensional subspace of U'' corresponding to a point z_0 of the real projective space $G^{\mathbb{R}}_1(U'')$ of dimension $n-1$. The totally geodesic imbedding

$$\iota' : G^{\mathbb{R}}_1(U'') \times G^{\mathbb{R}}_n(U') \to G^{\mathbb{R}}_{n+1}(F),$$

given by (7.3), sends the point (z_0, y_0) into the point x of X. If z is a point of $G^{\mathbb{R}}_1(U'')$, we consider the mapping $\iota'_z : Y' \to X$ sending $y \in Y'$ into $\iota'(z, y)$. The mapping $\iota'_{z_0*} : T_{Y', y_0} \to T_{x_0}$ is equal to the natural inclusion of $V_1 \otimes W_1$ into $(V \otimes W)_x$. Thus there are unique tangent vectors $\eta_1, \eta_2 \in T_{Y', y_0}$ such that $\iota'_{z_0*}\eta_1 = \xi_1$ and $\iota'_{z_0*}\eta_2 = \xi_2$; moreover these vectors are tangent to a closed totally geodesic surface of Y'. If $g_{Y'}$ is the Riemannian metric of the Grassmannian Y', by formula (1.57) we see that

$$(7.7) \qquad (D_g h)(\xi_1, \xi_2, \xi_1, \xi_2) = (D_{g_{Y'}} \iota'^*_{z_0} h)(\eta_1, \eta_2, \eta_1, \eta_2);$$

also we have

$$(d\theta)(\xi_1, \xi_2) = (d\iota'^*_{z_0}\theta)(\eta_1, \eta_2).$$

We consider the real-valued functions f_1, f_2 and f_3 on $G^{\mathbb{R}}_1(U'')$ defined by

$$f_1(z) = (\iota'^*_z h)(\eta_1, \eta_2), \qquad f_2(z) = (D_{g_{Y'}} \iota'^*_z h)(\eta_1, \eta_2, \eta_1, \eta_2),$$

$$f_3(z) = (d\iota'^*_z \theta)(\eta_1, \eta_2),$$

for $z \in G^{\mathbb{R}}_1(U'')$. Let γ be a closed geodesic of the projective space $G^{\mathbb{R}}_1(U'')$. Then there is a subspace U''_1 of U'' of dimension 2 such that the image of γ is equal to the submanifold $G_1^{\mathbb{R}}(U_1'')$ of $G_1^{\mathbb{R}}(U'')$. The orthogonal complement U of U_1'' in F is of dimension $2n$ and contains U'. Thus $G_n^{\mathbb{R}}(U')$ is a submanifold of $Y = G_n^{\mathbb{R}}(U)$ and η_1, η_2 may be considered as tangent vectors to Y. We denote by g_Y the Riemannian metric of the Grassmannian Y. We consider the totally geodesic imbedding

$$\iota : G_1^{\mathbb{R}}(U_1'') \times G_n^{\mathbb{R}}(U) \to G_{n+1}^{\mathbb{R}}(F)$$

given by (7.3) and the induced mapping $\iota_z : Y \to X$ corresponding to the point z of $G_1^{\mathbb{R}}(U_1'')$. Then by (1.57), we have

$$f_1(z) = (\iota_z^* h)(\eta_1, \eta_2), \qquad f_2(z) = (D_{g_Y} \iota_z^* h)(\eta_1, \eta_2, \eta_1, \eta_2),$$
$$f_3(z) = (d\iota_z^* \theta)(\eta_1, \eta_2),$$

for $z \in G_1^{\mathbb{R}}(U_1'')$. Using the preceding equalities, we easily verify that the relations

(7.8)
$$\int_\gamma f_1 = h_U(\eta_1, \eta_2), \qquad \int_\gamma f_2 = (D_{g_Y} h_U)(\eta_1, \eta_2, \eta_1, \eta_2),$$
$$\int_\gamma f_3 = (d\theta_U)(\eta_1, \eta_2)$$

hold. If h_U is a Lie derivative of the metric g_Y, since η_1, η_2 are tangent to a totally geodesic surface of $G_n^{\mathbb{R}}(U)$ of constant curvature, by Lemma 1.1 and the relations (1.49) and (1.57) the expression $(D_{g_Y} h_U)(\eta_1, \eta_2, \eta_1, \eta_2)$ vanishes. We now assume that the hypothesis of the j-th assertion of the proposition holds, where $j = 1, 2$ or 3. Then by (7.8) the integrals of f_j over the closed geodesics of the real projective space $G_1^{\mathbb{R}}(U'')$ vanish. Since the dimension of $G_1^{\mathbb{R}}(U'')$ is ≥ 2, the injectivity of the Radon transform on this real projective space, given by Theorem 2.23,(ii), implies that the function f_j vanishes on $G_1^{\mathbb{R}}(U'')$. From the vanishing of this function at the point z_0 of $G_1^{\mathbb{R}}(U'')$ and (7.7), we obtain the relation $h(\xi_1, \xi_2) = 0$ when $j = 1$, the relation (7.6) when $j = 2$, or the relation $(d\theta)(\xi_1, \xi_2) = 0$ when $j = 3$.

Let n be an integer ≥ 2, and let U be a real vector space of dimension $2n$ endowed with a positive definite scalar product. According to an observation made in §1, Chapter IV, if all even functions on $G_{n,n}^{\mathbb{R}}$ satisfying the Guillemin condition vanish, then the analogous result is also true for the Grassmannian $G_n^{\mathbb{R}}(U)$; moreover if all even symmetric 2-forms (resp. 1-forms) on $G_{n,n}^{\mathbb{R}}$ satisfying the Guillemin condition are Lie derivatives of the metric (resp. are exact), then the analogous result is also true for the Grassmannian $G_n^{\mathbb{R}}(U)$. We shall use these remarks in the course of the proofs of the next three propositions.

PROPOSITION 7.17. *For $n \geq 2$, the maximal flat Radon transform for functions on the symmetric space $\bar{G}^{\mathbb{R}}_{n,n}$ is injective.*

PROOF: We proceed by induction on $n \geq 2$. Proposition 4.4 tells us that the desired result is true for $n = 2$. Next, let $n \geq 2$ be a given integer and suppose that the maximal flat Radon transform for functions on the symmetric space $\bar{G}^{\mathbb{R}}_{n,n}$ is injective. Let f be an even real-valued function on $X = G^{\mathbb{R}}_{n+1,n+1}$ satisfying the Guillemin condition. Let U be an arbitrary subspace of $\mathbb{R}^{2n+2}$ of dimension $2n$. According to Lemmas 7.13 and 7.14, the function f_U on $G^{\mathbb{R}}_n(U)$ is even and satisfies the Guillemin condition. From Lemma 4.5 and our induction hypothesis, we infer that the function f_U vanishes. Then by Proposition 7.15, we know that f vanishes. According to Lemma 4.5, this argument gives us the desired result for the space X.

The preceding proposition is also given by Theorem 2.24.

PROPOSITION 7.18. *Let n be an integer ≥ 3 and suppose that all even symmetric 2-forms on $G^{\mathbb{R}}_{n,n}$ satisfying the Guillemin condition are Lie derivatives of the metric. Then an even symmetric 2-form on $X = G^{\mathbb{R}}_{n+1,n+1}$ satisfying the Guillemin condition is a Lie derivative of the metric.*

PROOF: Let k be an even symmetric 2-form on X satisfying the Guillemin condition. According to the decomposition (1.11), we may decompose k as

$$k = h + D_0\xi,$$

where h is an even symmetric 2-form on X satisfying $\operatorname{div} h = 0$, which is uniquely determined by k, and where ξ is a vector field on X. Then by Lemma 2.10, h also satisfies the Guillemin condition. Let U be an arbitrary subspace of $\mathbb{R}^{2n+2}$ of dimension $2n$ and consider the Grassmannian $Y = G^{\mathbb{R}}_n(U)$. According to Lemmas 7.13 and 7.14 and our hypothesis, we see that the symmetric 2-form h_U on Y is a Lie derivative of the metric of Y. Therefore by Proposition 7.15,(ii), we know that $D_{1,\mathcal{F}_X}h = 0$. According to the relations (4.8) and (7.5), the vector bundle E and the symmetric 2-form h satisfy the hypotheses of Theorem 2.48,(i), with $\mathcal{F}' = \mathcal{F} = \mathcal{F}_X$. By Proposition 4.2,(i), we know that $E(X) = \{0\}$. Then Theorem 2.48,(i) tells us that h is a section of E. Therefore h_U is a section of E_Y over Y. From Proposition 7.7, we now infer that the form h_U vanishes. Proposition 7.15,(i) tells us that h vanishes; thus the symmetric 2-form k is a Lie derivative of the metric of X.

PROPOSITION 7.19. *Let n be an integer ≥ 3 and suppose that all even 1-forms on $G^{\mathbb{R}}_{n,n}$ satisfying the Guillemin condition are exact. Then an even 1-form on $X = G^{\mathbb{R}}_{n+1,n+1}$ satisfying the Guillemin condition is exact.*

PROOF: Let θ be an even 1-form on X satisfying the Guillemin condition. Let U be an arbitrary subspace of $\mathbb{R}^{2n+2}$ of dimension $2n$ and consider the Grassmannian $Y = G_n^{\mathbb{R}}(U)$. According to Lemmas 7.13 and 7.14 and our hypothesis, the 1-form θ_U on Y is closed. Therefore by Proposition 7.15,(iii), we know that the restriction of $d\theta$ to an arbitrary submanifold of X belonging to the family $\mathcal{F}_X$ vanishes. Then Lemma 7.6 tells us that $d\theta = 0$. Since the cohomology group $H^1(X, \mathbb{R})$ vanishes, the form θ is exact.

The following theorem is a direct consequence of Propositions 2.18 and 7.18.

THEOREM 7.20. *If the symmetric space $\bar{G}_{n,n}^{\mathbb{R}}$, with $n \geq 3$, is rigid in the sense of Guillemin, then so is the space $\bar{G}_{n+1,n+1}^{\mathbb{R}}$.*

The following proposition is a direct consequence of Propositions 2.20 and 7.19.

PROPOSITION 7.21. *Let n be an integer ≥ 3 and suppose that all 1-forms on $\bar{G}_{n,n}^{\mathbb{R}}$ satisfying the Guillemin condition are exact. Then a 1-form on $X = \bar{G}_{n+1,n+1}^{\mathbb{R}}$ satisfying the Guillemin condition is exact.*

According to Proposition 4.3, the symmetric space $\bar{G}_{2,2}^{\mathbb{R}}$ is isometric to the product $\mathbb{R}P^2 \times \mathbb{R}P^2$. In Chapter X, we shall show that this space is not rigid in the sense of Guillemin (Theorem 10.5) and that there exist 1-forms on this space which satisfy the Guillemin condition and which are not exact (Theorem 10.6).

CHAPTER VIII

THE COMPLEX GRASSMANNIANS

§1. Outline

This chapter is devoted to the geometry of the complex Grassmannians. In §2, we study the complex Grassmannian $G_{m,n}^{\mathbb{C}}$ of complex m-planes in $\mathbb{C}^{m+n}$, with $m, n \geq 2$, and show that it is a Hermitian symmetric space and a homogeneous space of the group $SU(m+n)$; we also consider the Grassmannian $\bar{G}_{n,n}^{\mathbb{C}}$, which is the adjoint space of $G_{n,n}^{\mathbb{C}}$. We introduce certain vector bundles over $G_{m,n}^{\mathbb{C}}$ and use them to decompose the bundle of symmetric 2-forms on $G_{m,n}^{\mathbb{C}}$ into irreducible $SU(m+n)$-invariant sub-bundles. We then determine the highest weights of the fibers of these vector bundles in §3. We define certain complex-valued functions on $G_{m,n}^{\mathbb{C}}$ by means of the corresponding Stiefel manifold; then these functions and specific symmetric 2-forms on $G_{m,n}^{\mathbb{C}}$, arising from the complexification $\mathfrak{g}$ of the Lie algebra of $SU(m+n)$, allow us to describe explicitly and study the $SU(m+n)$-modules of functions and complex symmetric 2-forms on $G_{m,n}^{\mathbb{C}}$ isomorphic to $\mathfrak{g}$. In particular, we examine the case when $m = n$ and determine explicitly the space of infinitesimal Einstein deformations of $G_{n,n}^{\mathbb{C}}$. In §5, we define the natural isometry between the Grassmannian $G_{2,2}^{\mathbb{C}}$ and the complex quadric Q_4 and use it to relate their geometries; from the results of Chapter VI, it follows that this complex Grassmannian is infinitesimally rigid and that its quotient $\bar{G}_{2,2}^{\mathbb{C}}$ is rigid in the sense of Guillemin. In the next section, we show that the Guillemin condition for forms on $G_{m,n}^{\mathbb{C}}$, with $m \neq n$, is hereditary with respect to certain totally geodesic submanifolds. The remainder of this chapter is mainly devoted to the proof of the following result, which plays an essential role in our study of the rigidity of the complex Grassmannians presented in Chapter IX: an infinitesimal Einstein deformation of $G_{m,n}^{\mathbb{C}}$ satisfying the Guillemin condition vanishes. In order to prove this result in the case when $m \neq n$, we compute the integrals of some of the symmetric 2-forms considered in §4 over explicit closed geodesics. By means of these computations, in §8 we also establish relations among the symmetric 2-forms of §4. Finally, in §9 we study forms on the Grassmannian $G_{n,n}^{\mathbb{C}}$ and we introduce an averaging process which assigns to a p-form u on $G_{n+1,n+1}^{\mathbb{C}}$ a class of p-forms on $G_{n,n}^{\mathbb{C}}$ that are obtained by integrating u over closed geodesics. We then consider a certain explicit complex symmetric 2-form $h_1^{(n)}$ on $G_{n,n}^{\mathbb{C}}$ and show that one of the averages of the 2-form $h_1^{(n+1)}$ on $G_{n+1,n+1}^{\mathbb{C}}$ is equal to the form $h_1^{(n)}$. From this last result, we deduce by induction on n that an Einstein deformation of the space $G_{n,n}^{\mathbb{C}}$ which satisfies the Guillemin condition vanishes.

§2. The complex Grassmannians

Let X be a manifold and let E be a real vector bundle over X endowed with a complex structure J. The complexification $E_{\mathbb{C}}$ of E admits the decomposition

$$E_{\mathbb{C}} = E' \oplus E'',$$

where E' and E'' are the eigenbundles corresponding to the eigenvalues $+i$ and $-i$, respectively, of the endomorphism J of $E_{\mathbb{C}}$. Let $(S^2 E)^+$ and $(S^2 E)^-$ (resp. $(\wedge^2 E)^+$ and $(\wedge^2 E)^-$) be the eigenbundles of $S^2 E$ (resp. of $\wedge^2 E$) corresponding to the eigenvalues $+1$ and -1, respectively, of the involution of $S^2 E$ (resp. of $\wedge^2 E$) induced by J. We then have the decompositions

$$S^2 E = (S^2 E)^+ \oplus (S^2 E)^-, \qquad \wedge^2 E = (\wedge^2 E)^+ \oplus (\wedge^2 E)^-.$$

The complex structure J induces a complex structure on the vector bundle E^* dual to E, which we also denote by J. We identify $S^2 E^*$ with the bundle of symmetric 2-forms on E in such a way that

$$(\alpha \cdot \alpha)(\xi_1, \xi_2) = 2 \langle \xi_1, \alpha \rangle \langle \xi_2, \alpha \rangle,$$

for $\xi_1, \xi_2 \in E$ and $\alpha \in E^*$. We consider the sub-bundle $S^{2,0} E^*$ of $(S^2 E^*)_{\mathbb{C}}^-$ (resp. the sub-bundle $\wedge^{2,0} E^*$ of $(\wedge^2 E^*)_{\mathbb{C}}^-$) consisting of all elements u of $S^2 E_{\mathbb{C}}^*$ (resp. of $\wedge^2 E_{\mathbb{C}}^*$) which satisfy $u(\xi, \eta) = 0$, for $\xi \in E''$ and $\eta \in E_{\mathbb{C}}$, and the sub-bundle $S^{0,2} E^*$ of $(S^2 E^*)_{\mathbb{C}}^-$ (resp. the sub-bundle $\wedge^{0,2} E^*$ of $(\wedge^2 E^*)_{\mathbb{C}}^-$) consisting of all elements u of $S^2 E_{\mathbb{C}}^*$ (resp. of $\wedge^2 E_{\mathbb{C}}^*$) which satisfy $u(\xi, \eta) = 0$, for $\xi \in E'$ and $\eta \in E_{\mathbb{C}}$. We then have the decompositions

$$(S^2 E^*)_{\mathbb{C}}^- = S^{2,0} E^* \oplus S^{0,2} E^*, \qquad (\wedge^2 E^*)_{\mathbb{C}}^- = \wedge^{2,0} E^* \oplus \wedge^{0,2} E^*.$$

Now suppose that E is a complex vector bundle over X whose complex structure we denote by J. A sesquilinear form h on E satisfies the relation

$$h(u, v) = \operatorname{Re} h(u, v) + i \operatorname{Re} h(u, Jv),$$

for all $u, v \in E$. If we consider E as a real vector bundle endowed with the complex structure J and if h is a Hermitian form on E, then $\operatorname{Re} h$ is a section of $(S^2 E^*)^+$.

We suppose that X is a complex manifold. Then the vector bundles T', T'', $(S^2 T^*)^+$, $(S^2 T^*)^-$ and $(\wedge^2 T^*)^-$ of §4, Chapter I coincide with the bundles associated above with the vector bundles T and T^*; moreover, the bundle $T_{\mathbb{R}}^{1,1}$ is equal to $(\wedge^2 T^*)^+$. If (p, q) is equal to $(2, 0)$ or to $(0, 2)$, the bundles $S^{p,q}T^*$ and $\bigwedge^{p,q}T^*$ coincide with the bundles associated above with the vector bundle $E = T$.

Let $m, n \geq 0$ be given integers and let F be a complex vector space of dimension $m+n$ endowed with a positive definite Hermitian scalar product. We now suppose that X is the complex Grassmannian $G_m^{\mathbb{C}}(F)$ of all complex m-planes in F; then X is a complex manifold whose complex structure we denote by J. If either $m = 0$ or $n = 0$, the manifold $G_m^{\mathbb{C}}(F)$ is a point. When $m = 1$, the manifold $G_1^{\mathbb{C}}(F)$ is the complex projective space of all complex lines of F.

Let $V = V_X$ be the canonical complex vector bundle of rank m over X whose fiber at $x \in X$ is equal to the subspace x of F. We denote by $W = W_X$ the complex vector bundle of rank n over X whose fiber over $x \in X$ is the orthogonal complement W_x of V_x in F. We consider the complex vector bundles V^* and W^* dual to V and W. We identify the complex vector bundles $\text{Hom}_{\mathbb{C}}(V, W)$ and $V^* \otimes_{\mathbb{C}} W$, and we also identify V with the dual of the vector bundle V^*. We denote by J the complex structure of any one of these complex vector bundles. We denote by $\alpha \otimes w$ the tensor product of $\alpha \in V^*$ and $w \in W$ in $V^* \otimes_{\mathbb{C}} W$; we have

$$J(\alpha \otimes w) = J\alpha \otimes w = \alpha \otimes Jw.$$

We have a natural isomorphism of vector bundles

(8.1) $$V^* \otimes_{\mathbb{C}} W \to T$$

over X, which sends an element $\theta \in (V^* \otimes_{\mathbb{C}} W)_x$ into the tangent vector $dx_t/dt|_{t=0}$ to X at x, where x_t is the point of X corresponding to the m-plane

$$\{\, v + t\theta(v) \mid v \in V_x \,\},$$

for $t \in \mathbb{R}$. The isomorphism (8.1) allows us to identify these two vector bundles $V^* \otimes_{\mathbb{C}} W$ and T together with their complex structures.

Throughout the remainder of this section, we suppose that $m, n \geq 1$. It is easily verified that the involution θ of $(S^2T^*)^-$ determined by

$$(\theta h)(\alpha_1 \otimes w_1, \alpha_2 \otimes w_2) = h(\alpha_2 \otimes w_1, \alpha_1 \otimes w_2),$$

for $h \in S^2T^*$ and $\alpha_1, \alpha_2 \in V^*$ and $w_1, w_2 \in W$, is well-defined. If $(S^2T^*)^{-+}$ and $(S^2T^*)^{--}$ are the eigenbundles of $(S^2T^*)^-$ corresponding to the eigenvalues $+1$ and -1, respectively, of this involution θ, then we have the decomposition

$$(S^2T^*)^- = (S^2T^*)^{-+} \oplus (S^2T^*)^{--}.$$

If either $m = 1$ or $n = 1$, we see that the bundle $(S^2T^*)^{--}$ vanishes. If (p, q) is equal to $(2,0)$ or to $(0,2)$, we consider the complex vector bundles

$$(S^{p,q}T^*)^+ = S^{p,q}T^* \cap (S^2T^*)_{\mathbb{C}}^{-+}, \qquad (S^{p,q}T^*)^- = S^{p,q}T^* \cap (S^2T^*)_{\mathbb{C}}^{--};$$

then we have the decompositions

$$(8.2) \quad \begin{aligned} S^{2,0}T^* &= (S^{2,0}T^*)^+ \oplus (S^{2,0}T^*)^-, \\ S^{0,2}T^* &= (S^{0,2}T^*)^+ \oplus (S^{0,2}T^*)^-. \end{aligned}$$

It is easily verified that the morphisms

$$\tau_1 : (S^2V)^+ \otimes (S^2W^*)^+ \to (S^2T^*)^+,$$
$$\tau_2 : (S^2V)^- \otimes (S^2W^*)^- \to (S^2T^*)^{-+},$$
$$\tau_3 : (\wedge^2V)^- \otimes (\wedge^2W^*)^- \to (S^2T^*)^{--},$$

which send the element $k_1 \otimes k_2$ into the symmetric 2-form $k_1 \circ k_2$ of S^2T^* determined by

$$(k_1 \circ k_2)(\alpha_1 \otimes w_1, \alpha_2 \otimes w_2) = k_1(\alpha_1, \alpha_2)k_2(w_1, w_2) - k_1(\alpha_1, J\alpha_2)k_2(w_1, Jw_2),$$

for all $\alpha_1, \alpha_2 \in V^*$ and $w_1, w_2 \in W$, are well-defined. Clearly if k_1 is an element of $(S^2V)^+$ and k_2 is an element of $(S^2W^*)^+$, the element $k_1 \circ k_2$ vanishes if and only if one of the two elements k_1 and k_2 vanishes. The morphisms τ_2 and τ_3 induce morphisms of vector bundles

$$(8.3) \quad S^{2,0}V \otimes_{\mathbb{C}} S^{2,0}W^* \to (S^{2,0}T^*)^+, \quad S^{0,2}V \otimes_{\mathbb{C}} S^{0,2}W^* \to (S^{0,2}T^*)^+,$$

$$(8.4) \quad \wedge^{2,0}V \otimes_{\mathbb{C}} \wedge^{2,0}W^* \to (S^{2,0}T^*)^-, \quad \wedge^{0,2}V \otimes_{\mathbb{C}} \wedge^{0,2}W^* \to (S^{0,2}T^*)^-;$$

it is easily seen that the morphisms (8.3) are non-zero and that, when $m, n \geq 2$, the morphisms (8.4) are also non-zero at each point of X. In fact, we shall later verify that the morphisms (8.3) and (8.4) are isomorphisms and that the equalities

$$(8.5) \quad \begin{aligned} (S^{2,0}T^*)^+ &= S^{2,0}V \circ S^{2,0}W^*, \quad (S^{0,2}T^*)^+ = S^{0,2}V \circ S^{0,2}W^*, \\ (S^{2,0}T^*)^- &= \wedge^{2,0}V \circ \wedge^{2,0}W^*, \quad (S^{0,2}T^*)^- = \wedge^{0,2}V \circ \wedge^{0,2}W^* \end{aligned}$$

hold.

Since the vector bundles V and W are complex sub-bundles of the trivial complex vector bundle over X whose fiber is F, a sesquilinear form on F induces by restriction sesquilinear forms on the vector bundles V and W. In particular, the Hermitian scalar product on F induces by restriction positive definite Hermitian scalar products g_1 and g_2 on the vector bundles V and W, respectively. We consider the mappings

$$V^* \to V, \qquad W \to W^*$$

sending $\alpha \in V^*$ into the element $\alpha^\sharp \in V$ and w into the element $w^\flat \in W^*$ determined by

$$g_1(v, \alpha^\sharp) = \langle v, \alpha \rangle, \qquad \langle u, w^\flat \rangle = g_2(u, w),$$

for all $v \in V$ and $u \in W$. We remark that

$$(c\alpha)^\sharp = \bar{c}\alpha^\sharp, \qquad (cw)^\flat = \bar{c}w^\flat,$$

for all $c \in \mathbb{C}$, $\alpha \in V^*$ and $w \in W$.

A sesquilinear form q on the complex vector bundle V induces a sesquilinear form on the vector bundle V^*, which we also denote by q and which is determined by

$$q(\alpha_1, \alpha_2) = q(\alpha_2^\sharp, \alpha_1^\sharp),$$

for all $\alpha_1, \alpha_2 \in V^*$. In particular, the sesquilinear form g_1 on V^* induced by the Hermitian scalar product g_1 on V is also a positive definite Hermitian scalar product.

Two sesquilinear forms q_1 on V and q_2 on W determine a section $q_1 \cdot q_2$ of $(S^2 T^*)^+_{\mathbb{C}}$ over X, which is well-defined by

$$(q_1 \cdot q_2)(\alpha_1 \otimes w_1, \alpha_2 \otimes w_2) = q_1(\alpha_1, \alpha_2) \cdot q_2(w_1, w_2) + q_1(\alpha_2, \alpha_1) \cdot q_2(w_2, w_1),$$

for all $\alpha_1, \alpha_2 \in V^*$ and $w_1, w_2 \in W$.

As we have seen above, the real positive definite scalar products h_1 and h_2 on the vector bundles V^* and W determined by

$$h_1(\alpha_1, \alpha_2) = \operatorname{Re} g_1(\alpha_1, \alpha_2), \qquad h_2(w_1, w_2) = \operatorname{Re} g_2(w_1, w_2),$$

for all $\alpha_1, \alpha_2 \in V^*$ and $w_1, w_2 \in W$, are sections of $(S^2 V)^+$ and $(S^2 W^*)^+$, respectively. From an observation made above, we infer that the subbundles

$$\tilde{E}_1 = h_1 \circ (S^2 W^*)^+, \qquad \tilde{E}_2 = (S^2 V)^+ \circ h_2$$

of $(S^2 T^*)^+$ are isomorphic to $(S^2 W^*)^+$ and $(S^2 V)^+$, respectively.

If q_2 is a sesquilinear form on W, then we easily verify that

$$g_1 \cdot q_2 = h_1 \circ q_2' + i h_1 \circ q_2'',$$

where q_2' and q_2'' are the sections of $(S^2 W^*)^+$ defined by

$$q_2'(w_1, w_2) = \operatorname{Re}(q_2(w_1, w_2) + q_2(w_2, w_1)),$$
$$q_2''(w_1, w_2) = \operatorname{Re}(q_2(w_1, Jw_2) + q_2(w_2, Jw_1)),$$

for all $w_1, w_2 \in W$; therefore $g_1 \cdot q_2$ is a section of the sub-bundle $\tilde{E}_{1,\mathbb{C}}$ of $(S^2 T^*)_{\mathbb{C}}^+$. If q_1 is a sesquilinear form on V, then we easily verify that

$$q_1 \cdot g_2 = q_1' \circ h_2 + i q_1'' \circ h_2,$$

where q_1' and q_1'' are the sections of $(S^2 V)^+$ defined by

$$q_1'(\alpha_1, \alpha_2) = \operatorname{Re}(q_1(\alpha_1, \alpha_2) + q_1(\alpha_2, \alpha_1)),$$
$$q_1''(\alpha_1, \alpha_2) = \operatorname{Re}(q_1(\alpha_1, J\alpha_2) + q_1(\alpha_2, J\alpha_1)),$$

for all $\alpha_1, \alpha_2 \in W$; therefore $q_1 \cdot g_2$ is a section of the sub-bundle $\tilde{E}_{2,\mathbb{C}}$ of $(S^2 T^*)_{\mathbb{C}}^+$.

The Riemannian metric g on X determined by the section $\tilde{g} = \frac{1}{2} g_1 \cdot g_2$ of $(S^2 T^*)_{\mathbb{C}}^+$ is Hermitian and is related to $\tilde{g}$ by the formula

$$\tilde{g}(\xi, \eta) = g(\xi, \eta) + i g(\xi, J\eta),$$

for $\xi, \eta \in T$. In fact, according to the above formulas the metric g is equal to $h_1 \circ h_2$

The curvature R of the Riemannian manifold (X, g) can be computed in terms of the scalar products g_1 and g_2; in fact, if $\alpha_j \in V^*$, $w_j \in W$, with $1 \le j \le 4$, the expression

$$R(\alpha_1 \otimes w_1, \alpha_2 \otimes w_2, \alpha_3 \otimes w_3, \alpha_4 \otimes w_4)$$

is equal to the real part of the sum

$$g_1(\alpha_1, \alpha_4) g_1(\alpha_3, \alpha_2) g_2(w_1, w_2) g_2(w_3, w_4)$$
$$- g_1(\alpha_1, \alpha_3) g_1(\alpha_4, \alpha_2) g_2(w_1, w_2) g_2(w_4, w_3)$$
$$+ g_2(w_1, w_4) g_2(w_3, w_2) g_1(\alpha_1, \alpha_2) g_1(\alpha_3, \alpha_4)$$
$$- g_2(w_1, w_3) g_2(w_4, w_2) g_1(\alpha_1, \alpha_2) g_1(\alpha_4, \alpha_3).$$

It follows that g is an Einstein metric; in fact, its Ricci tensor is given by

(8.6) $$\operatorname{Ric} = 2(m+n)\, g.$$

We consider the trace mappings

$$\operatorname{Tr} : (S^2 V)^+ \to \mathbb{R}, \qquad \operatorname{Tr} : (S^2 W^*)^+ \to \mathbb{R}$$

defined by

$$\operatorname{Tr} k_1 = \sum_{j=1}^{2m} k_1(\beta_j, \beta_j), \qquad \operatorname{Tr} k_2 = \sum_{l=1}^{2n} k_2(t_l, t_l),$$

for $k_1 \in S^2V_x$ and $k_2 \in S^2W_x^*$, where $x \in X$ and $\{\beta_1, \ldots, \beta_{2m}\}$ is an orthonormal basis of V_x^* and $\{t_1, \ldots, t_{2n}\}$ is an orthonormal basis of W_x with respect to the scalar products h_1 and h_2, respectively. We denote by $(S^2V)_0^+$ and $(S^2W^*)_0^+$ the sub-bundles of $(S^2V)^+$ and $(S^2W^*)^+$ equal to the kernels of these trace mappings. The sub-bundles

$$E_1 = h_1 \circ (S^2W^*)_0^+, \qquad E_2 = (S^2V)_0^+ \circ h_2$$

of $\tilde{E}_1$ and $\tilde{E}_2$ are isomorphic to $(S^2W^*)_0^+$ and $(S^2V)_0^+$, respectively. Clearly if E_0 is the line bundle $\{g\}$ generated by the section g of $(S^2T^*)^+$, we have

$$\tilde{E}_1 = E_0 \oplus E_1, \qquad \tilde{E}_2 = E_0 \oplus E_2.$$

We consider the trace mappings

$$\text{Tr}_1 : (S^2T^*)^+ \to (S^2W^*)^+, \qquad \text{Tr}_2 : (S^2T^*)^+ \to (S^2V)^+$$

determined by

$$(\text{Tr}_1 h)(w_1, w_2) = \sum_{j=1}^{2m} h(\beta_j \otimes w_1, \beta_j \otimes w_2),$$

$$(\text{Tr}_2 h)(\alpha_1, \alpha_2) = \sum_{l=1}^{2n} h(\alpha_1 \otimes t_l, \alpha_2 \otimes t_l),$$

for $h \in S^2T_x^*$, $k_1 \in S^2V_x$, $k_2 \in S^2W_x^*$, $\alpha_1, \alpha_2 \in V_x^*$ and $w_1, w_2 \in W_x$, where $x \in X$ and $\{\beta_1, \ldots, \beta_{2m}\}$ is an orthonormal basis of V_x^* and $\{t_1, \ldots, t_{2n}\}$ is an orthonormal basis of W_x (over $\mathbb{R}$) with respect to the scalar products h_1 and h_2, respectively. We have

$$\text{Tr}\,\text{Tr}_1 h = \text{Tr}\,\text{Tr}_2 h = 2\,\text{Tr}\, h,$$

for $h \in (S^2T^*)^+$. For $k_1 \in (S^2V)^+$, $k_2 \in (S^2W^*)^+$, we see that the relations

$$\text{Tr}_1(k_1 \circ k_2) = (\text{Tr}\, k_1) \cdot k_2, \qquad \text{Tr}_2(k_1 \circ k_2) = (\text{Tr}\, k_2) \cdot k_1$$

hold. Clearly the kernel E_3 of the morphism

$$\text{Tr}_1 \oplus \text{Tr}_2 : (S^2T^*)^+ \to (S^2W^*)^+ \oplus (S^2V)^+$$

contains the sub-bundle $(S^2V)_0^+ \circ (S^2W^*)_0^+$ of $(S^2T^*)^+$. Using the above relations involving the trace mappings, we obtain the orthogonal decomposition

(8.7)
$$(S^2T^*)^+ = \bigoplus_{j=0}^{3} E_j,$$

and we see that the orthogonal projections $\pi_j : (S^2T^*)^+ \to E_j$ are given by

$$\pi_1 h = \frac{1}{2m} h_1 \circ \left(\mathrm{Tr}_1 h - \frac{1}{n} (\mathrm{Tr}\, h) \cdot h_2 \right),$$

$$(8.8) \qquad \pi_2 h = \frac{1}{2n} \left(\mathrm{Tr}_2 h - \frac{1}{m} (\mathrm{Tr}\, h) \cdot h_1 \right) \circ h_2,$$

$$\pi_3 h = h - \frac{1}{2m} h_1 \circ (\mathrm{Tr}_1 h) - \frac{1}{2n} (\mathrm{Tr}_2 h) \circ h_2 + \frac{1}{2mn} (\mathrm{Tr}\, h) \cdot g,$$

for $h \in (S^2T^*)^+$. If $\alpha \in V^*$ and $w \in W$ are unit vectors and if h is an element of $(S^2T^*)^+$, we easily verify that

$$(8.9) \qquad \begin{aligned} (\pi_1 h)(\alpha \otimes w, \alpha \otimes w) &= \frac{1}{2m} \left((\mathrm{Tr}_1 h)(w, w) - \frac{1}{n} \mathrm{Tr}\, h \right), \\ (\pi_2 h)(\alpha \otimes w, \alpha \otimes w) &= \frac{1}{2n} \left((\mathrm{Tr}_2 h)(\alpha, \alpha) - \frac{1}{m} \mathrm{Tr}\, h \right). \end{aligned}$$

We shall later verify that the equality

$$(8.10) \qquad E_3 = (S^2V)_0^+ \circ (S^2W^*)_0^+$$

holds. The relations (8.7) and (8.10) imply that

$$(S^2T^*)^+ = (S^2V)^+ \circ (S^2W^*)^+$$

and that the mapping τ_1 is an isomorphism.

From the decompositions (1.69), (8.2) and (8.7), we obtain the decomposition

$$(8.11) \quad S^2T_{\mathbb{C}}^* = \bigoplus_{j=0}^3 E_{j,\mathbb{C}} \oplus (S^{2,0}T^*)^+ \oplus (S^{2,0}T^*)^- \oplus (S^{0,2}T^*)^+ \oplus (S^{0,2}T^*)^-.$$

We consider the sub-bundle $E = E_X$ of S^2T^* consisting of all elements h of S^2T^* which satisfy

$$h(\xi, \xi) = 0,$$

for all elements ξ of $V^* \otimes_{\mathbb{C}} W$ of rank one. Clearly, we have

$$(8.12) \qquad \mathrm{Tr}\, E = \{0\}.$$

Moreover, if $m = 1$, the vector bundle E vanishes. It is easily seen that $(S^2T^*)^{--}$ is a sub-bundle of E; in fact, when $m, n \geq 2$, in §3 we shall verify that these two vector bundles are equal.

We also consider the Grassmannian $X' = G_n^{\mathbb{C}}(F)$. Let V' be the canonical complex vector bundle of rank n over X' whose fiber at $a \in X'$ is the n-plane a, and let W' the complex vector bundle of rank m over X' whose fiber over $a \in X'$ is the orthogonal complement W_a' of V_a' in F. As above, we identify the tangent bundle of X' with the bundle $V'^* \otimes W'$, and the standard Hermitian scalar product on F induces a Riemannian metric on X'. There is a natural diffeomorphism

$$\Psi : G_m^{\mathbb{C}}(F) \to G_n^{\mathbb{C}}(F),$$

sending an m-plane of F into its orthogonal complement; in fact, Ψ sends $x \in G_m^{\mathbb{C}}(F)$ into the n-plane W_x. For $x \in X$, we have $V_{\Psi(x)}' = W_x$ and $W_{\Psi(x)}' = V_x$. If θ is an element of $\operatorname{Hom}_{\mathbb{C}}(V_x, W_x)$, we consider the adjoint ${}^t\theta \in \operatorname{Hom}_{\mathbb{C}}(W_x, V_x)$ of θ defined in terms of the Hermitian scalar product on F; if $\alpha \in V_x^*$ and $w \in W_x$, the element $\alpha \otimes w$ of $\operatorname{Hom}_{\mathbb{C}}(V_x, W_x)$ satisfies

$$ {}^t(\alpha \otimes w) = w^\flat \otimes \alpha^\sharp. $$

It is easily verified that the induced mapping

$$\Psi_* : (V^* \otimes_{\mathbb{C}} W)_x \to (V'^* \otimes_{\mathbb{C}} W')_{\Psi(x)}$$

sends $\theta \in \operatorname{Hom}_{\mathbb{C}}(V_x, W_x)$ into $-{}^t\theta$; therefore Ψ is an isometry and we have

(8.13) $$\Psi_*(\alpha \otimes w) = -w^\flat \otimes \alpha^\sharp,$$

for all $\alpha \in V_x^*$ and $w \in W_x$. Thus we see that $\Psi^* E_{X',\Psi(x)} = E_{X,x}$, for all $x \in X$.

When $m = n$, the mapping $\Psi = \Psi_X$ is an involutive isometry of $X = G_n^{\mathbb{C}}(F)$ which preserves the bundle $E = E_X$; in this case, we say that a symmetric p-form u on X is even (resp. odd) if $\Psi^* u = \varepsilon u$, where $\varepsilon = 1$ (resp. $\varepsilon = -1$).

Let $\{e_1, \ldots, e_{m+n}\}$ be the standard basis of $\mathbb{C}^{m+n}$. We henceforth suppose that F is the vector space $\mathbb{C}^{m+n}$ endowed with the standard Hermitian scalar product. We now consider the complex Grassmannian

$$X = G_{m,n}^{\mathbb{C}} = G_m^{\mathbb{C}}(\mathbb{C}^{m+n}),$$

endowed with the Riemannian metric g induced by the standard Hermitian scalar product on $\mathbb{C}^{m+n}$.

The action of the group $G = SU(m+n)$ on $\mathbb{C}^{m+n}$ gives rise to an action of G on X. In fact, the group G acts transitively on the Riemannian manifold (X, g) by holomorphic isometries. The isotropy group of the point x_0 of X equal to the m-plane V_{x_0} of $\mathbb{C}^{m+n}$ spanned by the vectors

$\{e_1, \ldots, e_m\}$ is the subgroup $K = S(U(m) \times U(n))$ of G consisting of the matrices

(8.14)
$$\phi = \begin{pmatrix} A & 0 \\ 0 & B \end{pmatrix},$$

where $A \in U(m)$ and $B \in U(n)$, with determinant 1. The diffeomorphism

$$\Phi : G/K \to X,$$

which sends the class $\phi \cdot K$, where $\phi \in G$, into the m-plane of $\mathbb{C}^{m+n}$ spanned by the vectors $\{\phi(e_1), \ldots, \phi(e_m)\}$, is compatible with the actions of G on G/K and X.

The element

$$j = \begin{pmatrix} I_m & 0 \\ 0 & i \cdot I_n \end{pmatrix}$$

of $U(m+n)$ belongs to the centralizer of K. The element $s = j^2$ of $U(m+n)$ determines an involution σ of G which sends $\phi \in G$ into $s\phi s^{-1}$. Then K is equal to the set of fixed points of σ, and (G, K) is a Riemannian symmetric pair. The center S of G consists of all matrices λI_n, where λ is an $(m+n)$-th root of unity; it is invariant under the involution σ. Then we see that

$$K_S = \{\phi \in G \mid \phi^{-1}\sigma(\phi) \in S\}$$

is a subgroup of G containing K and S. The Cartan decomposition of the Lie algebra $\mathfrak{g}_0$ of G corresponding to σ is

$$\mathfrak{g}_0 = \mathfrak{k}_0 \oplus \mathfrak{p}_0,$$

where $\mathfrak{k}_0$ is the Lie algebra of K and $\mathfrak{p}_0$ is the space of all matrices

(8.15)
$$\begin{pmatrix} 0 & -{}^t\bar{Z} \\ Z & 0 \end{pmatrix}$$

of $\mathfrak{g}_0$, where Z is a complex $n \times m$ matrix and ${}^t\bar{Z}$ is its conjugate transpose. We identify $\mathfrak{p}_0$ with the vector space $M_{n,m}^{\mathbb{C}}$ of all complex $n \times m$ matrices and, in particular, the element (8.15) of $\mathfrak{p}_0$ with the matrix $Z \in M_{n,m}^{\mathbb{C}}$. The adjoint action of K on $\mathfrak{p}_0$ is expressed by

$$\text{Ad}\,\phi \cdot Z = B \cdot Z \cdot A^{-1},$$

where ϕ is the element (8.14) of K and $Z \in M_{n,m}^{\mathbb{C}}$.

We identify $\mathfrak{p}_0$ with the tangent space of G/K at the coset of the identity element of G; the diffeomorphism Φ sends this coset into the point x_0

of X. Since V_{x_0} is the subspace of $\mathbb{C}^{m+n}$ generated by $\{e_1, \ldots, e_m\}$, clearly W_{x_0} is the subspace generated by $\{e_{m+1}, \ldots, e_{m+n}\}$. If $\{e_1^*, \ldots, e_m^*\}$ is the basis of $V_{x_0}^*$ dual to the basis $\{e_1, \ldots, e_m\}$ of V_{x_0}, then it is easily verified that the isomorphism $\Phi_* : \mathfrak{p}_0 \to (V^* \otimes W)_{x_0}$ sends the element (8.15) of $\mathfrak{p}_0$ corresponding to the matrix $Z = (z_{jk})$ of $M_{n,m}^{\mathbb{C}}$, with $1 \le j \le n$ and $1 \le k \le m$, into the vector

$$\sum_{\substack{1 \le j \le n \\ 1 \le k \le m}} z_{jk} e_k^* \otimes e_{j+m}$$

of $(V^* \otimes W)_{x_0}$.

The restriction of $\operatorname{Ad} j$ to $\mathfrak{p}_0$ is a complex structure on $\mathfrak{p}_0$, and so gives rise to a G-invariant complex structure on G/K. If B is the Killing form of $\mathfrak{g}_0$, the restriction to $\mathfrak{p}_0$ of the scalar product $-B$ is invariant under the adjoint action of K and therefore induces a G-invariant metric g_0 on the homogeneous space G/K. Endowed with this complex structure and the metric g_0, the manifold G/K is an irreducible Hermitian symmetric space of compact type of rank $\min(m, n)$ (see Proposition 4.2 in Chapter VIII of [36]). When $m \ne n$, we easily see that the group K_S is equal to K; then according to §9 in Chapter VII of [36], it follows that G/K is equal to its adjoint space.

The group K acts on T_{x_0} and, for $\phi \in K$, we have the equality

$$\Phi_* \circ \operatorname{Ad} \phi = \phi \cdot \Phi_*$$

as mappings from $\mathfrak{p}_0$ to T_{x_0}. We also see that $\Phi_* \circ \operatorname{Ad} j = J \circ \Phi_*$ and that

$$(8.16) \qquad\qquad g_0 = 4(m + n)\Phi^* g.$$

Thus Φ is a holomorphic isometry from the symmetric space G/K, endowed with the metric $(1/4(m+n)) \cdot g_0$, to X; henceforth, we shall identify these Hermitian manifolds by means of this G-equivariant isometry. Therefore the metric g is Kähler and, from Lemma 1.21, we again obtain the equality (8.6).

The Grassmannian $G_{n,m}^{\mathbb{C}}$ is also a homogeneous space of the group G. From (8.13), it follows that the mapping $\Psi : G_{m,n}^{\mathbb{C}} \to G_{n,m}^{\mathbb{C}}$ is an anti-holomorphic isometry. It is easily verified that the isometry Ψ satisfies

$$\Psi \circ \phi = \phi \circ \Psi,$$

for all $\phi \in G$. Thus $G_{m,n}^{\mathbb{C}}$ and $G_{n,m}^{\mathbb{C}}$ are isometric as Riemannian symmetric spaces, but not as complex manifolds.

The vector bundles V and W are homogeneous G-sub-bundles of the trivial complex vector bundle over X whose fiber is $\mathbb{C}^{m+n}$. Therefore the tensor product $V^* \otimes_{\mathbb{C}} W$ is a homogeneous G-bundle and it is easily seen that (8.1) is an isomorphism of homogeneous G-bundles over X. All the vector bundles appearing in the decomposition (8.11) and the bundle E are homogeneous sub-bundles of $S^2 T_{\mathbb{C}}^*$; hence the fibers at x_0 of these vector bundles are K-submodules of $S^2 T_{\mathbb{C},x_0}^*$. Moreover under the action of the group K on $\mathbb{C}^{m+n}$, the subspaces V_{x_0} and W_{x_0} of $\mathbb{C}^{m+n}$ are preserved; in fact, the fiber V_{x_0} is a $U(m)$-module, while the fiber W_{x_0} is a $U(n)$-module. The fibers at $x_0 \in X$ of the vector bundles

$$(8.17) \qquad (S^2 V)_{0,\mathbb{C}}^+, \quad S^{2,0}V, \quad S^{0,2}V, \quad {\textstyle\bigwedge}^{2,0}V, \quad {\textstyle\bigwedge}^{0,2}V$$

are $U(m)$-modules, while the fibers at $x_0 \in X$ of the vector bundles

$$(8.18) \qquad (S^2 W^*)_{0,\mathbb{C}}^+, \quad S^{2,0}W^*, \quad S^{0,2}W^*, \quad {\textstyle\bigwedge}^{2,0}W^*, \quad {\textstyle\bigwedge}^{0,2}W^*$$

are $U(n)$-modules. Each of these modules is either irreducible or vanishes; in fact, they are all irreducible when $m, n \geq 2$. The tensor products

$$(8.19) \qquad \begin{gathered} ((S^2 V)_0^+ \otimes (S^2 W^*)_0^+)_{x_0}, \\ (S^{2,0}V \otimes_{\mathbb{C}} S^{2,0}W^*)_{x_0}, \quad (S^{0,2}V \otimes_{\mathbb{C}} S^{0,2}W^*)_{x_0}, \\ ({\textstyle\bigwedge}^{2,0}V \otimes_{\mathbb{C}} {\textstyle\bigwedge}^{2,0}W^*)_{x_0}, \quad ({\textstyle\bigwedge}^{0,2}V \otimes_{\mathbb{C}} {\textstyle\bigwedge}^{0,2}W^*)_{x_0} \end{gathered}$$

possess natural structures of K-modules. Therefore when $m, n \geq 2$, the K-modules E_{1,x_0} and E_{2,x_0} and the K-modules (8.19) are irreducible. The morphism of vector bundles

$$(8.20) \qquad \tau_1 : (S^2 V)_0^+ \otimes (S^2 W^*)_0^+ \to (S^2 V)_0^+ \circ (S^2 W^*)_0^+$$

and the morphisms of vector bundles (8.3) and (8.4) are G-equivariant, and hence the restrictions of these morphisms of vector bundles to the fibers at x_0 are morphisms of K-modules. When $m, n \geq 2$, we know that these morphisms of K-modules are non-zero. From these remarks, we infer that the morphism (8.20) is an isomorphism of vector bundles and that the morphisms of vector bundles (8.3) and (8.4) are injective. Since the rank of the vector bundles $(S^2 V)_0^+$ and $(S^2 W^*)_0^+$ are equal to $m^2 - 1$ and $n^2 - 1$, respectively, from the decomposition (8.7) we now see that the vector bundle E_3 and its sub-bundle $(S^2 V)_0^+ \circ (S^2 W^*)_0^+$ have the same rank; we thus obtain the equality (8.10). On the other hand, by a dimension-counting argument, from the decomposition (8.2) we now obtain the equalities (8.5) and see that the morphisms (8.3) and (8.4) are isomorphisms. In fact, we have

$$\text{(8.21)} \qquad \text{rank}\,(S^2T^*)^{--} = 2\binom{m}{2}\binom{n}{2}.$$

When $m = 1$, we easily see that $G_{1,n}^{\mathbb{C}}$ is isometric to the complex projective space $\mathbb{CP}^n$ endowed with its Fubini-Study metric of constant holomorphic curvature 4, and we know that $E = \{0\}$.

If F is an arbitrary complex vector space of dimension $m + n$ endowed with a positive definite Hermitian scalar product, then an isometry $\varphi : \mathbb{C}^{m+n} \to F$ induces a holomorphic isometry $\varphi : G_{m,n}^{\mathbb{C}} \to G_m^{\mathbb{C}}(F)$. Thus the Riemannian manifold $G_m^{\mathbb{C}}(F)$ is a Hermitian symmetric space. From (8.13), it follows that the mapping $\Psi : G_m^{\mathbb{C}}(F) \to G_n^{\mathbb{C}}(F)$ is an anti-holomorphic isometry. If we write $X = G_{m,n}^{\mathbb{C}}$ and $X' = G_m^{\mathbb{C}}(F)$, for $x \in X$ the isomorphism $\varphi : \mathbb{C}^{m+n} \to F$ induces by restriction isomorphisms

$$\varphi : V_{X,x} \to V_{X',\varphi(x)}, \qquad \varphi : W_{X,x} \to W_{X',\varphi(x)};$$

hence the isomorphism $\varphi_* : T_{X,x} \to T_{X',\varphi(x)}$ is equal to the natural mapping

$$\varphi^{-1*} \otimes \varphi : V_{X,x}^* \otimes W_{X,x} \to V_{X',\varphi(x)}^* \otimes W_{X',\varphi(x)}.$$

It follows that

$$\varphi^* E_{X',\varphi(x)} = E_{X,x},$$

for all $x \in X$. When $m = 1$, the Hermitian metric of the complex projective space $G_1^{\mathbb{C}}(F)$ induced by the Hermitian scalar product of F has constant holomorphic curvature 4. When $m = n$, since $\Psi \circ \varphi$ is equal to $\varphi \circ \Psi$ as mappings from $G_{n,n}^{\mathbb{C}}$ to $G_n^{\mathbb{C}}(F)$, we see that, if u is an even (resp. odd) symmetric form on $G_n^{\mathbb{C}}(F)$, then $\varphi^* u$ is an even (resp. odd) symmetric form on $G_{n,n}^{\mathbb{C}}$.

For the remainder of this section, we suppose that $m = n \geq 1$. We consider the involutive isometry Ψ of $X = G_{n,n}^{\mathbb{C}}$. From formulas (8.9) and (8.13), we easily infer that

$$\text{(8.22)} \qquad\qquad \Psi^* \pi_1 h = \pi_2 \Psi^* h,$$

for all $h \in (S^2T^*)^+$. We saw above that the isometry Ψ preserves the sub-bundle E of S^2T^*. The group Λ of isometries of X generated by Ψ, which is of order 2, acts freely on X and we may consider the Riemannian manifold $\bar{X} = \bar{G}_{n,n}^{\mathbb{C}}$ equal to the quotient X/Λ endowed with the Riemannian metric $g_{\bar{X}}$ induced by g. The natural projection $\varpi : X \to \bar{X}$ is a two-fold covering. The action of the group $SU(2n)$ on X passes to the quotient $\bar{X}$. In fact,

$SU(2n)$ acts transitively on $\bar{X}$, and it is easily verified that the isotropy group of the point $\varpi(x_0)$ is equal to the subgroup of $SU(2n)$ generated by K and the matrix

$$\begin{pmatrix} 0 & -I_n \\ I_n & 0 \end{pmatrix}$$

of $SU(2n)$. This isotropy group is precisely the group K_S; thus according to §9 in Chapter VII of [36], we see that $\bar{X}$ is a symmetric space of compact type which is equal to the adjoint space of X. Moreover, the space $\bar{X}$ is irreducible and has rank n.

The notion of even or odd tensor on X (with respect to the involutive isometry Ψ) defined here coincides with the one considered in §4, Chapter II. In fact, a section u of $S^p T^*$ over X is even if and only if we can write $u = \varpi^* u'$, where u' is a symmetric p-form on $\bar{X}$. Lemma 2.17 gives us the following result:

LEMMA 8.1. *A symmetric p-form u on $\bar{G}^{\mathbb{C}}_{n,n}$ satisfies the Guillemin condition if and only if the even symmetric p-form $\varpi^* u$ on $G^{\mathbb{C}}_{n,n}$ satisfies the Guillemin condition.*

From Proposition 2.18, we obtain the following:

PROPOSITION 8.2. *The symmetric space $\bar{G}^{\mathbb{C}}_{n,n}$ is rigid in the sense of Guillemin if and only if every even symmetric 2-form on $G^{\mathbb{C}}_{n,n}$ satisfying the Guillemin condition is a Lie derivative of the metric.*

The notion of even (resp. odd) symmetric p-form on $\mathbb{CP}^1$ defined in §4, Chapter III, coincides with the one introduced here on $G^{\mathbb{C}}_{1,1}$. Hence from Lemma 8.1 and Propositions 2.20 and 3.29, we obtain the following result:

PROPOSITION 8.3. *Let X be the symmetric space $\bar{G}^{\mathbb{C}}_{1,1}$.*
 (i) *The X-ray transform for functions on X is injective.*
 (ii) *A differential form of degree 1 on X satisfies the zero-energy condition if and only if it is exact.*

Since the space $\bar{G}^{\mathbb{C}}_{1,1}$ has rank one, the first assertion of this proposition is also given by Theorem 2.24.

§3. Highest weights of irreducible modules associated with the complex Grassmannians

Let $m, n \geq 2$ be given integers. We pursue our study of the complex Grassmannian $X = G^{\mathbb{C}}_{m,n}$. We consider the Lie algebras $\mathfrak{g}_0$ and $\mathfrak{k}_0$ of the compact Lie group $G = SU(m+n)$ and its subgroup K. The complexification $\mathfrak{g}$ of $\mathfrak{g}_0$ is equal to $\mathfrak{sl}(m+n, \mathbb{C})$, and the complexification $\mathfrak{k}$ of the Lie algebra $\mathfrak{k}_0$ admits the decomposition

$$\mathfrak{k} = \mathfrak{k}_1 \oplus \mathfrak{k}_2 \oplus \mathfrak{z},$$

where $\mathfrak{z}$ is the center of $\mathfrak{k}$, which is one-dimensional, and where $\mathfrak{k}_1$ and $\mathfrak{k}_2$ are simple subalgebras of $\mathfrak{k}$ isomorphic to $\mathfrak{sl}(m, \mathbb{C})$ and $\mathfrak{sl}(n, \mathbb{C})$, respectively. In fact, a matrix

$$\begin{pmatrix} A & 0 \\ 0 & B \end{pmatrix}$$

of $\mathfrak{k}$, where $A \in \mathfrak{gl}(m, \mathbb{C})$, $B \in \mathfrak{gl}(n, \mathbb{C})$ satisfy $\operatorname{Tr} A + \operatorname{Tr} B = 0$, can be written as the sum of the three matrices

$$\begin{pmatrix} A - \frac{1}{m}(\operatorname{Tr} A) \cdot I_m & 0 \\ 0 & 0 \end{pmatrix}, \quad \begin{pmatrix} 0 & 0 \\ 0 & B - \frac{1}{n}(\operatorname{Tr} B) \cdot I_n \end{pmatrix},$$

$$\begin{pmatrix} \frac{1}{m}(\operatorname{Tr} A) \cdot I_m & 0 \\ 0 & \frac{1}{n}(\operatorname{Tr} B) \cdot I_n \end{pmatrix},$$

which belong to $\mathfrak{k}_1$, $\mathfrak{k}_2$ and $\mathfrak{z}$, respectively. The complexification $\mathfrak{p}$ of the subspace $\mathfrak{p}_0$ of $\mathfrak{g}_0$ can be written as

$$\mathfrak{p} = \mathfrak{p}_- \oplus \mathfrak{p}_+,$$

where $\mathfrak{p}_-$ and $\mathfrak{p}_+$ are the eigenspaces of the endomorphism $\operatorname{Ad} j$ of $\mathfrak{p}$ corresponding to the eigenvalues $+i$ and $-i$, respectively. Since j belongs to the center of $U(m + n)$, this decomposition of $\mathfrak{p}$ is invariant under the action of K on $\mathfrak{p}$. We thus obtain the decomposition

$$(8.23) \qquad \mathfrak{g} = \mathfrak{k}_1 \oplus \mathfrak{k}_2 \oplus \mathfrak{z} \oplus \mathfrak{p}_- \oplus \mathfrak{p}_+$$

of the Lie algebra $\mathfrak{g}$ into irreducible K-modules. If $E_{ij} = (c_{kl})$ is the matrix of $\mathfrak{gl}(m + n, \mathbb{C})$ which is determined by $c_{ij} = 1$ and $c_{kl} = 0$ whenever $(k, l) \neq (i, j)$, the subspace $\mathfrak{p}_+$ of $\mathfrak{p}$ is generated (over $\mathbb{C}$) by the matrices

$$\{ E_{ij} \mid 1 \leq i \leq m \text{ and } m + 1 \leq j \leq m + n \},$$

while the subspace $\mathfrak{p}_-$ of $\mathfrak{p}$ is generated (over $\mathbb{C}$) by the matrices

$$\{ E_{ij} \mid m + 1 \leq i \leq m + n \text{ and } 1 \leq j \leq m \}.$$

The group of all diagonal matrices of G is a maximal torus of G and of K. The complexification $\mathfrak{t}$ of the Lie algebra $\mathfrak{t}_0$ of this torus is a Cartan subalgebra of the semi-simple Lie algebra $\mathfrak{g}$ and also of the reductive Lie algebra $\mathfrak{k}$. For $1 \leq j \leq m + n$, the linear form $\lambda_j : \mathfrak{t} \to \mathbb{C}$, sending the diagonal matrix with $a_1, \dots, a_{m+n} \in \mathbb{C}$ as its diagonal entries into a_j, is purely imaginary on $\mathfrak{t}_0$. Then

$$\Delta = \{ \lambda_i - \lambda_j \mid 1 \leq i, j \leq m + n \text{ and } i \neq j \}$$

is the system of roots of $\mathfrak{g}$ with respect to $\mathfrak{t}$; if we set

$$\Delta_1 = \{\lambda_i - \lambda_j \mid 1 \leq i, j \leq m \text{ and } i \neq j\},$$
$$\Delta_2 = \{\lambda_i - \lambda_j \mid m+1 \leq i, j \leq m+n \text{ and } i \neq j\},$$

then $\Delta' = \Delta_1 \cup \Delta_2$ is the system of roots of $\mathfrak{k}$ with respect to $\mathfrak{t}$. We fix the positive system

$$\Delta^+ = \{\lambda_i - \lambda_j \mid 1 \leq i < j \leq m+n\}$$

for the roots of $\mathfrak{g}$, and the positive system $\Delta'^+ = \Delta' \cap \Delta^+$ for the roots of $\mathfrak{k}$.

If α is the root $\lambda_i - \lambda_j$ of Δ, with $1 \leq i, j \leq m+n$ and $i \neq j$, we consider the subspace $\mathfrak{g}_\alpha$ of $\mathfrak{g}$ generated by E_{ij}. If $\Delta_{\mathfrak{z}}$ is the set of roots of Δ which do not vanish identically on the center $\mathfrak{z}$, we write $Q_+ = \Delta^+ \cap \Delta_{\mathfrak{z}}$; then we see that

$$\mathfrak{p}_+ = \bigoplus_{\alpha \in Q_+} \mathfrak{g}_\alpha, \qquad \mathfrak{p}_- = \bigoplus_{\alpha \in Q_+} \mathfrak{g}_{-\alpha}.$$

It is easily verified that the highest weights of the K-modules $\mathfrak{k}_1$, $\mathfrak{k}_2$, $\mathfrak{z}$, $\mathfrak{p}_-$ and $\mathfrak{p}_+$ are equal to $\lambda_1 - \lambda_m$, $\lambda_{m+1} - \lambda_{m+n}$, 0, $-\lambda_m + \lambda_{m+1}$ and $\lambda_1 - \lambda_{m+n}$, respectively.

We consider the vector bundles

$$V_0^{1,1} = (S^2 V)_{0,\mathbb{C}}^+, \qquad W_0^{1,1} = (S^2 W^*)_{0,\mathbb{C}}^+.$$

As we saw in §2, the fibers at $x_0 \in X$ of the vector bundles (8.17) and (8.18) are irreducible K-modules. It is easily verified (see [39, pp. 222–223]) that the highest weights of these irreducible K-modules are given by the following table:

K-module	Highest weight	K-module	Highest weight
$V_{0,x_0}^{1,1}$	$\lambda_1 - \lambda_m$	$W_{0,x_0}^{1,1}$	$\lambda_{m+1} - \lambda_{m+n}$
$(S^{2,0}V)_{x_0}$	$2\lambda_1$	$(S^{2,0}W^*)_{x_0}$	$-2\lambda_{m+n}$
$(S^{0,2}V)_{x_0}$	$-2\lambda_m$	$(S^{0,2}W^*)_{x_0}$	$2\lambda_{m+1}$
$(\bigwedge^{2,0}V)_{x_0}$	$\lambda_1 + \lambda_2$	$(\bigwedge^{2,0}W^*)_{x_0}$	$-\lambda_{m+n-1} - \lambda_{m+n}$
$(\bigwedge^{0,2}V)_{x_0}$	$-\lambda_{m-1} - \lambda_m$	$(\bigwedge^{0,2}W^*)_{x_0}$	$\lambda_{m+1} + \lambda_{m+2}$

In §2, we saw that the K-modules E_{1,x_0} and E_{2,x_0} and the K-modules (8.19) are irreducible. We know that the mappings (8.3), (8.4) and (8.20) are G-equivariant isomorphisms of vector bundles. Therefore the fibers at x_0 of the homogeneous sub-bundles of $S^2 T_{\mathbb{C}}^*$ which appear in the right-hand side of (8.11) are irreducible K-modules, and the morphism

$$\tau_1 : ((S^2 V)_0^+ \otimes (S^2 W^*)_0^+)_{x_0} \to E_{3,x_0}$$

is an isomorphism of irreducible K-modules. Using these facts, from the above table we deduce that the highest weights of these irreducible K-modules are given by the following table:

K-module	Highest weight
$E_{0,\mathbb{C},x_0}$	0
$E_{1,\mathbb{C},x_0}$	$\lambda_{m+1} - \lambda_{m+n}$
$E_{2,\mathbb{C},x_0}$	$\lambda_1 - \lambda_m$
$E_{3,\mathbb{C},x_0}$	$\lambda_1 - \lambda_m + \lambda_{m+1} - \lambda_{m+n}$
$(S^{2,0}T^*)_{x_0}^+$	$2\lambda_1 - 2\lambda_{m+n}$
$(S^{2,0}T^*)_{x_0}^-$	$\lambda_1 + \lambda_2 - \lambda_{m+n-1} - \lambda_{m+n}$
$(S^{0,2}T^*)_{x_0}^+$	$-2\lambda_m + 2\lambda_{m+1}$
$(S^{0,2}T^*)_{x_0}^-$	$-\lambda_{m-1} - \lambda_m + \lambda_{m+1} + \lambda_{m+2}$

The fact that the K-modules $(S^{2,0}T^*)_{x_0}^+$ and $(S^{2,0}T^*)_{x_0}^-$ are irreducible is asserted in §2 of [9, Chapter 3].

Since the irreducible K-modules appearing in the above table are pairwise non-isomorphic, from the decomposition (8.11) we infer that a K-submodule of $S^2 T_{\mathbb{C},x_0}^*$ can be written as a direct sum of submodules appearing in this table. It is easily seen that there are elements of the irreducible K-modules $E_{j,\mathbb{C},x_0}$, $(S^{2,0}T^*)_{x_0}^+$ and $(S^{0,2}T^*)_{x_0}^+$, with $0 \le j \le 3$, which do not belong to the K-module $E_{\mathbb{C},x_0}$. We also know that the K-modules $(S^{2,0}T^*)_{x_0}^-$ and $(S^{0,2}T^*)_{x_0}^-$ are submodules of $E_{\mathbb{C},x_0}$. From these remarks, we deduce the equality

(8.24) $$(S^2 T^*)^{--} = E.$$

Since the decomposition (8.7) is orthogonal, we see that the orthogonal complement of the sub-bundle $E_0 \oplus E_1 \oplus E_2$ in $S^2 T^*$ is equal to

$$F = (S^2 T^*)^- \oplus E_3.$$

From the decomposition (8.23) of $\mathfrak{g}$ into irreducible K-modules, the decompositions (8.2) and (8.11), and the preceding table, we obtain

(8.25) $\qquad \dim \mathrm{Hom}_K(\mathfrak{g}, \mathbb{C}) = 1, \qquad \dim \mathrm{Hom}_K(\mathfrak{g}, E_{j,\mathbb{C},x_0}) = 1,$

for $j = 0, 1, 2$, and

$$(8.26) \qquad \operatorname{Hom}_K(\mathfrak{g}, F_{\mathbb{C}, x_0}) = \{0\}.$$

Thus we have

$$(8.27) \qquad \dim \operatorname{Hom}_K(\mathfrak{g}, S_0^2 T^*_{\mathbb{C}, x_0}) = 2$$

(see Lemma 5.5 of [42]).

The highest weight of the irreducible G-module $\mathfrak{g}$ is equal to $\lambda_1 - \lambda_{m+n}$ and $E_{1,m+n}$ is a highest weight vector of $\mathfrak{g}$. If γ_1 is the element of the dual of the group G which is the equivalence class of this irreducible G-module, by (8.25)–(8.27) the Frobenius reciprocity theorem tells us that the G-modules $C^\infty_{\gamma_1}(X)$, $C^\infty_{\gamma_1}(E_{1,\mathbb{C}})$ and $C^\infty_{\gamma_1}(E_{2,\mathbb{C}})$ are irreducible and that the equalities

$$(8.28) \qquad \begin{aligned} C^\infty_{\gamma_1}(S^2 T^*_{\mathbb{C}}) &= C^\infty_{\gamma_1}((S^2 T^*)^+_{\mathbb{C}}), \\ C^\infty_{\gamma_1}(S_0^2 T^*_{\mathbb{C}}) &= C^\infty_{\gamma_1}(E_{1,\mathbb{C}}) \oplus C^\infty_{\gamma_1}(E_{2,\mathbb{C}}) \end{aligned}$$

hold.

Since the symmetric space X is irreducible and is not equal to a simple Lie group, from (2.25) and (8.28) we see that an element of $E(X)$ is a section of the sub-bundle $E_1 \oplus E_2$ of $S_0^2 T^*$; more precisely, we have

$$(8.29) \qquad E(X) = \{\, h \in C^\infty_{\gamma_1}(E_{1,\mathbb{C}}) \oplus C^\infty_{\gamma_1}(E_{2,\mathbb{C}}) \mid h = \bar{h},\ \operatorname{div} h = 0 \,\}.$$

From Proposition 2.40 and the equalities (8.27) and (8.29), we obtain the following result:

PROPOSITION 8.4. *Let X be the complex Grassmannian $G^{\mathbb{C}}_{m,n}$, with $m, n \geq 2$. Then the space $E(X)$ is an irreducible $SU(m+n)$-submodule of $C^\infty(E_1 \oplus E_2)$ isomorphic to the Lie algebra $\mathfrak{g}_0 = \mathfrak{su}(m+n)$.*

§4. Functions and forms on the complex Grassmannians

Let $m, n \geq 2$ be given integers. In this section, we describe explicit functions and symmetric 2-forms on the complex Grassmannian $X = G^{\mathbb{C}}_{m,n}$, which we view as a homogeneous space of the group $G = SU(m+n)$.

Let $S_{m,n}$ be the space of all complex $(m+n) \times m$ matrices A satisfying ${}^t\bar{A}A = I_m$. We view $S_{m,n}$ as the Stiefel manifold of all orthonormal m-frames in $\mathbb{C}^{m+n}$; the matrix A of $S_{m,n}$ determines the m-frame consisting of the m column vectors of A. The unitary group $U(m)$ acts on $S_{m,n}$ by right multiplication and we consider the quotient space $S_{m,n}/U(m)$. The mapping

$$\rho : S_{m,n} \to G^{\mathbb{C}}_{m,n},$$

sending the element A of $S_{m,n}$ into the m-plane spanned by the m column vectors of A, induces by passage to the quotient a diffeomorphism

$$\bar{\rho} : S_{m,n}/U(m) \to G^{\mathbb{C}}_{m,n}.$$

The group G acts on $S_{m,n}$ by left multiplication; clearly, the mappings ρ and $\bar{\rho}$ are G-equivariant. A function f on $S_{m,n}$ which is invariant under the right action of $U(m)$ determines a function $\tilde{f}$ on $G^{\mathbb{C}}_{m,n}$ satisfying $\rho^* \tilde{f} = f$.

For $1 \le j \le m+n$, we consider the $\mathbb{C}^m$-valued function Z_j on $S_{m,n}$, which sends a matrix of $S_{m,n}$ into its j-th row; if $1 \le l \le m$, we denote by Z_j^l the l-th component of this function. For $1 \le j, k \le m+n$, we consider the complex-valued function

$$f_{jk} = \langle Z_j, Z_k \rangle = \sum_{l=1}^{m} Z_j^l \bar{Z}_k^l$$

on $S_{m,n}$, which is invariant under the right action of $U(m)$. Let $\mathcal{H}$ be the space of all functions f on $S_{m,n}$ which are invariant under the right action of $U(m)$ and can be written in the form

(8.30)
$$f = \sum_{j,k=1}^{m+n} a_{jk} f_{jk},$$

where the coefficients $a_{jk} \in \mathbb{C}$ satisfy

(8.31)
$$\sum_{j=1}^{m+n} a_{jj} = 0.$$

Then $\mathcal{H}$ is a G-module isomorphic to $\mathfrak{g}$. In fact, the mapping

(8.32)
$$\mathcal{H} \to \mathfrak{g},$$

sending the function f given by (8.30) into the matrix (ia_{kj}) of $\mathfrak{g}$, is an isomorphism of G-modules; moreover, the image of the G-submodule

$$\mathcal{H}_0 = \{f \in \mathcal{H} \mid f = \bar{f}\}$$

of $\mathcal{H}$ under the isomorphism (8.32) is equal to the subalgebra $\mathfrak{g}_0 = \mathfrak{su}(m+n)$ of $\mathfrak{g}$. Thus $\mathcal{H}$ is an irreducible G-module and the G-submodule

$$\tilde{\mathcal{H}} = \{\tilde{f} \mid f \in \mathcal{H}\}$$

of $C^\infty(X)$ is isomorphic to $\mathcal{H}$ and to $\mathfrak{g}$. Therefore $\tilde{\mathcal{H}}$ is a G-submodule of the irreducible G-module $C^\infty_{\gamma_1}(X)$, and so we obtain the equality

$$(8.33) \qquad C^\infty_{\gamma_1}(X) = \tilde{\mathcal{H}}.$$

Clearly, the function $f_0 = f_{m+n,1} = \langle Z_{m+n}, Z_1 \rangle$ is a highest weight vector of $\mathcal{H}$, and so the function $\tilde{f}_0$ on X is a highest weight vector of the irreducible G-module $C^\infty_{\gamma_1}(X)$ (see [32]). By Lemma 2.39 and (8.6), we know that $\tilde{\mathcal{H}}$ is the eigenspace of the Laplacian Δ with eigenvalue $4(m+n)$.

Let $\zeta = (\zeta_1, \ldots, \zeta_{m+n})$ be the standard coordinate of $\mathbb{C}^{m+n}$. For $1 \le j, k \le m+n$, let $\tilde{Q}_{jk}$ be the sesquilinear form on $\mathbb{C}^{m+n}$ defined by

$$\tilde{Q}_{jk}(\zeta, \zeta') = \zeta_j \bar{\zeta}'_k,$$

for $\zeta, \zeta' \in \mathbb{C}^{m+n}$, and consider the $U(1)$-invariant complex polynomial Q_{jk} on $\mathbb{C}^{m+n}$ determined by

$$Q_{jk}(\zeta) = \tilde{Q}_{jk}(\zeta, \zeta) = \zeta_j \bar{\zeta}_k,$$

for all $\zeta \in \mathbb{C}^{m+n}$. If $\{v_1, \ldots, v_{m+n}\}$ is an orthonormal basis of $\mathbb{C}^{m+n}$, then we easily see that

$$(8.34) \qquad \sum_{l=1}^{m+n} Q_{jk}(v_l) = \delta_{jk}.$$

Now let $\{v_1, \ldots, v_m\}$ be an orthonormal system of vectors of $\mathbb{C}^{m+n}$ and s be the point of $S_{m,n}$ corresponding to this orthonormal m-frame of $\mathbb{C}^{m+n}$; then we have

$$(8.35) \qquad f_{jk}(s) = \sum_{l=1}^{m} Q_{jk}(v_l).$$

Let x be a point of X, and let $\{v_1, \ldots, v_m\}$ and $\{w_1, \ldots, w_n\}$ be orthonormal bases of the spaces V_x and W_x, respectively. Since

$$\{v_1, \ldots, v_m, w_1, \ldots, w_n\}$$

is an orthonormal basis of $\mathbb{C}^{m+n}$, according to (8.34) we have the equality

$$\sum_{l=1}^{m} Q_{jk}(v_l) + \sum_{r=1}^{n} Q_{jk}(w_r) = \delta_{jk}.$$

From the preceding equality and the relation (8.35), we infer that

$$(8.36) \qquad \tilde{f}_{jk}(x) = \sum_{l=1}^{m} Q_{jk}(v_l) = \delta_{jk} - \sum_{r=1}^{n} Q_{jk}(w_r).$$

We write $Q = Q_{m+n,1}$. In particular, the relation (8.36) tells us that

$$(8.37) \qquad \tilde{f}_0(x) = \sum_{l=1}^{m} Q(v_l) = - \sum_{r=1}^{n} Q(w_r).$$

We consider the G-module of all sesquilinear forms on $\mathbb{C}^{m+n}$ and its G-submodule $\mathcal{Q}$ consisting of all sesquilinear forms q on $\mathbb{C}^{m+n}$ which can be written in the form

$$(8.38) \qquad q = \sum_{j,k=1}^{m+n} a_{jk} \tilde{Q}_{jk},$$

where the coefficients $a_{jk} \in \mathbb{C}$ satisfy the relation (8.31). Then $\mathcal{Q}$ is a G-module isomorphic to $\mathfrak{g}$; in fact, the mapping

$$\mathcal{Q} \to \mathfrak{g},$$

sending the sesquilinear form q given by (8.38) into the matrix (ia_{kj}) of $\mathfrak{g}$, is an isomorphism of G-modules. Thus $\mathcal{Q}$ is an irreducible G-module and the sesquilinear form $\tilde{Q} = \tilde{Q}_{m+n,1}$ is a highest weight vector of $\mathcal{Q}$.

We consider the isomorphism of G-modules

$$\mathcal{H} \to \mathcal{Q}$$

sending the element f of $\mathcal{H}$ given by (8.30) into the element $q = Q(f)$ of $\mathcal{Q}$ given by (8.38). If f is a function of $\mathcal{H}$, we easily see that the form $Q(f)$ is Hermitian if and only if f belongs to $\mathcal{H}_0$. For $f \in \mathcal{H}$, we consider the sesquilinear forms $Q_1(f)$ on V^* and $Q_2(f)$ on W induced by $Q(f)$. Then we have $\tilde{Q} = Q(f_0)$, and we write $Q_1 = Q_1(f_0)$ and $Q_2 = Q_2(f_0)$. We see that

$$(8.39) \qquad Q_1(\alpha, \alpha) = Q(\alpha^\sharp), \qquad Q_2(w, w) = Q(w),$$

for $\alpha \in V^*$, $w \in W$.

For $f \in \mathcal{H}$, the symmetric 2-forms $g_1 \cdot Q_2(f)$ and $Q_1(f) \cdot g_2$ are sections of the vector bundles $\tilde{E}_{1,\mathbb{C}}$ and $\tilde{E}_{2,\mathbb{C}}$, respectively. By (8.36), we see that

$$(8.40) \qquad \text{Tr}\,(g_1 \cdot Q_2(f)) = -4m\tilde{f}, \qquad \text{Tr}\,(Q_1(f) \cdot g_2) = 4n\tilde{f},$$

for all $f \in \mathcal{H}$. Thus by (8.40), we have morphisms of G-modules

$$\kappa_1 : \mathcal{H} \to C^\infty(E_{1,\mathbb{C}}), \qquad \kappa_2 : \mathcal{H} \to C^\infty(E_{2,\mathbb{C}})$$

defined by

$$\kappa_1(f) = g_1 \cdot Q_2(f) + \frac{2}{n}\tilde{f} \cdot g, \quad \kappa_2(f) = Q_1(f) \cdot g_2 - \frac{2}{m}\tilde{f} \cdot g,$$

for $f \in \mathcal{H}$. Clearly the image of κ_j is contained in $C^\infty_{\gamma_1}(E_{j,\mathbb{C}})$, for $j = 1, 2$. If f is an element of $\mathcal{H}$, it is easily seen that $g_1 \cdot Q_2(f)$ (resp. $Q_1(f) \cdot g_2$) is a section of $\tilde{E}_1$ (resp. of $\tilde{E}_2$) if and only if $Q(f)$ is Hermitian, or equivalently if f belongs to $\mathcal{H}_0$. Thus if f belongs to $\mathcal{H}_0$, then $\kappa_j(f)$ is a section of E_j, for $j = 1, 2$.

Let x be a point of X. If the vectors e_1 and e_{m+n} belong to V_x and if q_2 is a sesquilinear form on W, we verify that

$$(Q_1 \cdot g_2)(\alpha_1 \otimes w, \alpha_2 \otimes w) = 1, \qquad (g_1 \cdot q_2)(\alpha_1 \otimes w, \alpha_2 \otimes w) = 0,$$

for all unit vectors $w \in W_x$, where α_1 and α_2 are the vectors of V_x determined by $\alpha_1^\sharp = e_1$ and $\alpha_2^\sharp = e_{m+n}$. If the vectors e_1 and e_{m+n} belong to W_x and if q_1 is a sesquilinear form on V, we verify that

$$(g_1 \cdot Q_2)(\alpha \otimes e_{m+n}, \alpha \otimes e_1) = 1, \qquad (q_1 \cdot g_2)(\alpha \otimes e_{m+n}, \alpha \otimes e_1) = 0,$$

for all unit vectors $\alpha \in V_x^*$. From these observations, if a, b are complex numbers which do not both vanish, we infer that the sections

$$a g_1 \cdot Q_2 + b Q_1 \cdot g_2, \qquad \kappa_1(f_0), \qquad \kappa_2(f_0),$$

are non-zero.

Since f_0 is a highest weight vector of $\mathcal{H}$, it follows that the sections $g_1 \cdot Q_2$ and $Q_1 \cdot g_2$ are highest weight vectors of the G-modules $C^\infty_{\gamma_1}(\tilde{E}_{1,\mathbb{C}})$ and $C^\infty_{\gamma_1}(\tilde{E}_{2,\mathbb{C}})$, respectively, and that the sections $\kappa_1(f_0)$ and $\kappa_2(f_0)$ are highest weight vectors of the G-modules $C^\infty_{\gamma_1}(E_{1,\mathbb{C}})$ and $C^\infty_{\gamma_1}(E_{2,\mathbb{C}})$, respectively. Since the G-modules $C^\infty_{\gamma_1}(E_{1,\mathbb{C}})$ and $C^\infty_{\gamma_2}(E_{2,\mathbb{C}})$ are irreducible, we obtain the equalities

$$(8.41) \qquad C^\infty_{\gamma_1}(E_{1,\mathbb{C}}) = \kappa_1(\mathcal{H}), \qquad C^\infty_{\gamma_2}(E_{2,\mathbb{C}}) = \kappa_2(\mathcal{H}).$$

Moreover, the sections $g_1 \cdot Q_2 + Q_1 \cdot g_2$ and $g_1 \cdot Q_2 - Q_1 \cdot g_2$ are highest weight vectors of the G-module $C^\infty_{\gamma_1}((S^2 T^*)^+_{\mathbb{C}})$; also since $\kappa_j(f_0)$ is a section of the vector bundle $E_{j,\mathbb{C}}$, the sections

$$\kappa_1(f_0) + \kappa_2(f_0), \qquad \kappa_1(f_0) - \kappa_2(f_0)$$

are highest weight vectors of the G-module $C^\infty_{\gamma_1}(S^2_0 T^*_{\mathbb{C}})$.

Since the differential operator $\text{Hess} : C^\infty(X) \to C^\infty(S^2 T^*_\mathbb{C})$ is homogeneous, it induces a morphism of G-modules $\text{Hess} : \mathcal{H} \to C^\infty_{\gamma_1}(S^2 T^*_\mathbb{C})$. Thus if f is an element of $\mathcal{H}$, from the equalities (8.28) and the decomposition (8.7) we infer that $\text{Hess}\,\tilde{f} = \pi_+ \text{Hess}\,\tilde{f}$ is a section of $(S^2 T^*)^+_\mathbb{C}$ and that

$$(8.42) \qquad \text{Hess}\,\tilde{f} = (\pi_1 + \pi_2)\text{Hess}\,\tilde{f} - \frac{2(m+n)}{mn}\,\tilde{f} \cdot g;$$

here we used the fact that $\tilde{f}$ is an eigenfunction of the Laplacian Δ with eigenvalue $4(m+n)$. By (1.35) and (8.6), we see that

$$(8.43) \qquad \text{div Hess}\,\tilde{f} = 2(m+n)d\tilde{f},$$

for $f \in \tilde{\mathcal{H}}$. From (8.42), (8.43) and (1.8), we obtain the identity

$$(8.44) \qquad \text{div}\,(\pi_1 + \pi_2)\text{Hess}\,\tilde{f} = 2(m+n)\left(1 - \frac{1}{mn}\right)d\tilde{f},$$

for $f \in \mathcal{H}$.

Let f be a non-zero element of $\mathcal{H}$. By (8.44), we know that the section $(\pi_1 + \pi_2)\text{Hess}\,\tilde{f}$ is non-zero. Since (8.7) is a direct sum decomposition, it follows that at least one of the two sections $\pi_1 \text{Hess}\,\tilde{f}$ and $\pi_2 \text{Hess}\,\tilde{f}$ does not vanish. Therefore the section $(\pi_1 - \pi_2)\text{Hess}\,\tilde{f}$ of $S^2_0 T^*_\mathbb{C}$ does not vanish. In §8, we shall prove the following result:

PROPOSITION 8.5. *We have*

$$\pi_1 \text{Hess}\,\tilde{f}_0 = \kappa_1(f_0), \qquad \pi_2 \text{Hess}\,\tilde{f}_0 = -\kappa_2(f_0).$$

This proposition implies that the two sections $\pi_1 \text{Hess}\,\tilde{f}_0$ and $\pi_2 \text{Hess}\,\tilde{f}_0$ do not vanish. Since Hess is a homogeneous differential operator, it follows that the morphism $\pi_j \text{Hess} : C^\infty(X) \to C^\infty(E_{j,\mathbb{C}})$ induces an isomorphism of G-modules

$$(8.45) \qquad \pi_j \text{Hess} : \mathcal{H} \to C^\infty_{\gamma_1}(E_{j,\mathbb{C}}),$$

for $j = 1, 2$. In fact, since κ_1 and κ_2 and the mappings (8.45) are isomorphisms of G-modules, from Proposition 8.5 we deduce that

$$(8.46) \qquad \pi_1 \text{Hess}\,\tilde{f} = \kappa_1(f), \qquad \pi_2 \text{Hess}\,\tilde{f} = -\kappa_2(f),$$

for all $f \in \mathcal{H}$. From (8.42) and (8.46), we now deduce that

$$(8.47) \qquad \text{Hess}\,\tilde{f} = g_1 \cdot Q_2(f) - Q_1(f) \cdot g_2.$$

For the remainder of this section, we suppose that $m = n$. Then we have

$$(\kappa_1 + \kappa_2)(f) = g_1 \cdot Q_2(f) + Q_1(f) \cdot g_2,$$

$$(\kappa_1 - \kappa_2)(f) = g_1 \cdot Q_2(f) - Q_1(f) \cdot g_2 + \frac{4}{n} \tilde{f} \cdot g,$$

for all $f \in \mathcal{H}$. According to (8.46), we have

$$(8.48) \qquad (\pi_1 - \pi_2)\text{Hess } \tilde{f} = g_1 \cdot Q_2(f) + Q_1(f) \cdot g_2,$$

for $f \in \mathcal{H}$.

We note that Proposition 8.5 is exploited here only in order to prove the equalities (8.47) and (8.48). Neither Proposition 8.5 nor these equalities shall be used in any of our subsequent proofs; they are presented here only for the sake of completeness.

Let x be a point of the Grassmannian $X = G_{n,n}^{\mathbb{C}}$. Let $\{v_1, \ldots, v_n\}$ and $\{w_1, \ldots, w_n\}$ be orthonormal bases of the vector spaces V_x and W_x, respectively. According to (8.35), for $1 \le j, k \le 2n$, we see that

$$\tilde{f}_{jk}(x) = \sum_{l=1}^{n} Q_{jk}(v_l), \qquad \tilde{f}_{jk}(\Psi(x)) = \sum_{r=1}^{n} Q_{jk}(w_r).$$

From these relations and (8.36), it follows that

$$\tilde{f}_{jk}(x) + \tilde{f}_{jk}(\Psi(x)) = \delta_{jk},$$

for $1 \le j, k \le 2n$. This equality implies that

$$\Psi^* \tilde{f} = -\tilde{f},$$

for all $f \in \mathcal{H}$, and thus the functions of $\tilde{\mathcal{H}}$ are odd. Since the G-module $C_{\gamma_1}^{\infty}(X)$ is irreducible, by (8.33) we see that

$$(8.49) \qquad C_{\gamma_1}^{\infty}(X) = C_{\gamma_1}^{\infty}(X)^{\text{odd}} = \tilde{\mathcal{H}}.$$

Moreover, since Hess is a homogeneous differential operator, if f is an element of $\mathcal{H}$, the symmetric 2-form Hess $\tilde{f}$ on X is odd.

We easily verify that

$$\Psi^*(g_1 \cdot Q_2(f)) = Q_1(f) \cdot g_2,$$

for all $f \in \mathcal{H}$. Therefore by (8.49), for $f \in \mathcal{H}$, we see that $(\kappa_1 + \kappa_2)(f)$ (resp. $(\kappa_1 - \kappa_2)(f)$) is an even (resp. an odd) section of $S_0^2 T_{\mathbb{C}}^*$. Thus the sections

$$(\kappa_1 + \kappa_2)(f_0) = g_1 \cdot Q_2 + Q_1 \cdot g_2,$$

$$(\kappa_1 - \kappa_2)(f_0) = g_1 \cdot Q_2 - Q_1 \cdot g_2 + \frac{4}{n} \tilde{f}_0 \cdot g$$

are highest weight vectors of the G-modules

$$C^\infty_{\gamma_1}(S^2_0 T^*_\mathbb{C})^{\mathrm{ev}}, \qquad C^\infty_{\gamma_1}(S^2_0 T^*_\mathbb{C})^{\mathrm{odd}},$$

respectively. Moreover, from (8.28) it follows that these G-modules are irreducible and the equalities

$$(8.50) \qquad C^\infty_{\gamma_1}(S^2_0 T^*_\mathbb{C})^{\mathrm{ev}} = (\kappa_1 + \kappa_2)\mathcal{H}, \qquad C^\infty_{\gamma_1}(S^2_0 T^*_\mathbb{C})^{\mathrm{odd}} = (\kappa_1 - \kappa_2)\mathcal{H}$$

hold; hence by (8.50), we have

$$(8.51) \qquad \begin{aligned} C^\infty_{\gamma_1}(S^2 T^*_\mathbb{C})^{\mathrm{ev}} &= C^\infty_{\gamma_1}(S^2_0 T^*_\mathbb{C})^{\mathrm{ev}} = (\kappa_1 + \kappa_2)\mathcal{H}, \\ C^\infty_{\gamma_1}(S^2 T^*_\mathbb{C})^{\mathrm{odd}} &= \tilde{\mathcal{H}} \cdot g \oplus (\kappa_1 - \kappa_2)\mathcal{H}. \end{aligned}$$

By (8.39), we see that

$$(8.52) \qquad (\kappa_1 + \kappa_2)(f_0)(\alpha \otimes w, \alpha \otimes w) = 2(Q(\alpha^\sharp) + Q(w)),$$

for all unit vectors $\alpha \in V$ and $w \in W$.

Since the involutive isometry Ψ of X is anti-holomorphic, we see that

$$(8.53) \qquad \Psi^* \partial = \bar{\partial} \Psi^*$$

on $\bigwedge^p T^*_\mathbb{C}$. From the relations (2.28), (8.53) and (8.49), we obtain the equalities

$$(8.54) \qquad \begin{aligned} C^\infty_{\gamma_1}(T_\mathbb{C})^{\mathrm{ev}} &= \mathcal{K}_\mathbb{C}, \\ C^\infty_{\gamma_1}(T^*_\mathbb{C})^{\mathrm{ev}} &= (\partial - \bar{\partial})\tilde{\mathcal{H}}, \qquad C^\infty_{\gamma_1}(T^*_\mathbb{C})^{\mathrm{odd}} = d\tilde{\mathcal{H}} \end{aligned}$$

of irreducible G-modules.

Let f be an element of $\mathcal{H}$. By (8.22) and (8.49), we see that the section $(\pi_1 + \pi_2)\mathrm{Hess}\, \tilde{f}$ of $S^2_0 T^*_\mathbb{C}$ is odd and that the section $(\pi_1 - \pi_2)\mathrm{Hess}\, \tilde{f}$ of $S^2_0 T^*_\mathbb{C}$ is even. If f is non-zero, we saw that these two sections do not vanish; thus the sections $(\pi_1 + \pi_2)\mathrm{Hess}\, \tilde{f}$ and $(\pi_1 - \pi_2)\mathrm{Hess}\, \tilde{f}$ are non-zero vectors of the G-modules $C^\infty_{\gamma_1}(S^2_0 T^*_\mathbb{C})^{\mathrm{odd}}$ and $C^\infty_{\gamma_1}(S^2_0 T^*_\mathbb{C})^{\mathrm{ev}}$, respectively. Since the G-modules $C^\infty_{\gamma_1}(S^2_0 T^*_\mathbb{C})^{\mathrm{odd}}$ and $C^\infty_{\gamma_1}(T_\mathbb{C})^{\mathrm{odd}}$ are irreducible, by (8.44) we therefore see that the homogeneous differential operator div induces an isomorphism of G-modules

$$(8.55) \qquad \mathrm{div} : C^\infty_{\gamma_1}(S^2_0 T^*_\mathbb{C})^{\mathrm{odd}} \to C^\infty_{\gamma_1}(T^*_\mathbb{C})^{\mathrm{odd}}.$$

We consider the symmetric space $Y = \bar{G}^\mathbb{C}_{n,n}$ and the natural projection $\varpi : X \to Y$. We consider the G-submodules

$$E(X)^{\mathrm{ev}} = E(X) \cap C^\infty(S^2 T^*)^{\mathrm{ev}}, \qquad E(X)^{\mathrm{odd}} = E(X) \cap C^\infty(S^2 T^*)^{\mathrm{odd}}$$

of $E(X)$. According to §4, Chapter II, the projection ϖ induces an isomorphism of G-modules $\varpi^* : E(Y) \to E(X)^{\mathrm{ev}}$ given by (2.7) and we have the decomposition (2.10) of $E(X)$.

Since the mapping (8.55) is an isomorphism, by (2.25) we see that $E(X)^{\mathrm{odd}} = \{0\}$; it follows that

$$E(X) = E(X)^{\mathrm{ev}} \subset C_{\gamma_1}^\infty(S_0^2 T_{\mathbb{C}}^*)^{\mathrm{ev}}.$$

Proposition 8.4 tells us that $E(X)$ is an irreducible G-module isomorphic to $\mathfrak{g}_0$, and so we have the equality

$$E(X)_{\mathbb{C}} = C_{\gamma_1}^\infty(S_0^2 T_{\mathbb{C}}^*)^{\mathrm{ev}}.$$

From the first equalities of (8.51), we obtain the relations (8.56) of the next proposition. Since $(\kappa_1 + \kappa_2)\mathcal{H}_0$ is a G-submodule of $C^\infty(S_0^2 T^*)^{\mathrm{ev}}$ isomorphic to $\mathfrak{g}_0$, it is therefore equal to $E(X)$. From the above discussion, we obtain the following:

PROPOSITION 8.6. *Let X be the complex Grassmannian $G_{n,n}^{\mathbb{C}}$, with $n \geq 2$, and Y be the symmetric space $\bar{G}_{n,n}^{\mathbb{C}}$. The spaces $E(X)$ and $E(Y)$ are irreducible $SU(2n)$-modules isomorphic to $\mathfrak{g}_0 = \mathfrak{su}(2n)$; moreover, $E(X)$ is equal to the $SU(2n)$-submodule $(\kappa_1 + \kappa_2)\mathcal{H}_0$ of $C^\infty(S_0^2 T^*)^{\mathrm{ev}}$, and we have*

$$(8.56) \qquad E(X)_{\mathbb{C}} = C_{\gamma_1}^\infty(S^2 T_{\mathbb{C}}^*)^{\mathrm{ev}} = C_{\gamma_1}^\infty(S_0^2 T_{\mathbb{C}}^*)^{\mathrm{ev}} = (\kappa_1 + \kappa_2)\mathcal{H},$$

$$(8.57) \qquad\qquad E(Y)_{\mathbb{C}} = C_{\gamma_1}^\infty(Y, S^2 T_{Y,\mathbb{C}}^*).$$

§5. The complex Grassmannians of rank two

In this section, we consider the complex Grassmannian $X = G_{2,n}^{\mathbb{C}}$, with $n \geq 2$, endowed with the metric g of §2. We view X as a homogeneous space of the group $G = SU(n+2)$. The standard Hermitian scalar product on $\mathbb{C}^{n+2}$ induces a Hermitian scalar product $\hat{g}$ on $\bigwedge^2 \mathbb{C}^{n+2}$, which in turn induces a Hermitian metric $\tilde{g}$ of constant holomorphic curvature 4 on the complex projective space $G_1^{\mathbb{C}}(\bigwedge^2 \mathbb{C}^{n+2})$ of all complex lines of $\bigwedge^2 \mathbb{C}^{n+2}$. If u is a non-zero vector of $\bigwedge^2 \mathbb{C}^{n+2}$, we denote by $\pi(u)$ the element of $G_1^{\mathbb{C}}(\bigwedge^2 \mathbb{C}^{n+2})$ corresponding to the complex line of $\bigwedge^2 \mathbb{C}^{n+2}$ generated by u. If u is a unit vector of $\bigwedge^2 \mathbb{C}^{n+2}$ and v is a vector of $\bigwedge^2 \mathbb{C}^{n+2}$ orthogonal to u, we consider the vector

$$\pi_*(u, v) = \frac{d}{dt} \pi(u + tv)_{|t=0}$$

tangent to the projective space $G_1^{\mathbb{C}}(\bigwedge^2 \mathbb{C}^{n+2})$ at $\pi(u)$. It is easily verified that the Plücker imbedding

$$\iota : G_{2,n}^{\mathbb{C}} \to G_1^{\mathbb{C}}(\bigwedge^2 \mathbb{C}^{n+2}),$$

which sends the complex 2-plane generated by the vectors v_1, v_2 of $\mathbb{C}^{n+2}$ into $\pi(v_1 \wedge v_2)$, is an isometric imbedding.

Throughout the remainder of this section, we shall suppose that $n = 2$. Let $\{e_1, e_2, e_3, e_4\}$ be the standard basis of $\mathbb{C}^4$. The complex quadratic form H on $\bigwedge^2 \mathbb{C}^4$ with values in the one-dimensional vector space $\bigwedge^4 \mathbb{C}^4$ defined by

$$H(\xi_1, \xi_2) = \xi_1 \wedge \xi_2,$$

for $\xi_1, \xi_2 \in \bigwedge^2 \mathbb{C}^4$, is non-degenerate. The vectors

$$\omega_1 = \frac{1}{\sqrt{2}}(e_1 \wedge e_2 + e_3 \wedge e_4), \quad \omega_2 = \frac{i}{\sqrt{2}}(e_1 \wedge e_2 - e_3 \wedge e_4),$$

$$\omega_3 = \frac{1}{\sqrt{2}}(e_2 \wedge e_3 + e_1 \wedge e_4), \quad \omega_4 = \frac{i}{\sqrt{2}}(e_2 \wedge e_3 - e_1 \wedge e_4),$$

$$\omega_5 = \frac{1}{\sqrt{2}}(e_1 \wedge e_3 - e_2 \wedge e_4), \quad \omega_6 = \frac{i}{\sqrt{2}}(e_1 \wedge e_3 + e_2 \wedge e_4)$$

form an orthonormal basis for $\bigwedge^2 \mathbb{C}^4$ (with respect to $\hat{g}$) which diagonalizes the quadratic form H; in fact, we have

$$H(\omega_j, \omega_k) = \delta_{jk}\, e_1 \wedge e_2 \wedge e_3 \wedge e_4,$$

for $1 \le j, k \le 6$. An element A of the group G acts on $\bigwedge^2 \mathbb{C}^4$ and preserves both the scalar product $\hat{g}$ and the quadratic form H. Then we may write $A\omega_j = \sum_{k=1}^6 c_j^k \omega_k$; from these properties of A, we easily deduce that the coefficients c_j^k are real.

We consider the complex hypersurface Z of $G_1^{\mathbb{C}}(\bigwedge^2 \mathbb{C}^4)$ defined by the homogeneous equation $H(\xi, \xi) = 0$, for $\xi \in \bigwedge^2 \mathbb{C}^4$. By Cartan's lemma, we easily see that the image of ι is equal to Z. The complex coordinate system $(\zeta_1, \ldots, \zeta_6)$ on $\bigwedge^2 \mathbb{C}^4$ determined by the orthonormal basis $\{\omega_1, \ldots, \omega_6\}$ allows us to identify $\bigwedge^2 \mathbb{C}^4$ with $\mathbb{C}^6$ and $G_1^{\mathbb{C}}(\bigwedge^2 \mathbb{C}^4)$ with $\mathbb{C}P^5$. Then Z is identified with the complex quadric Q_4 of the complex projective space $\mathbb{C}P^5$ defined by the homogeneous equation

$$\zeta_1^2 + \cdots + \zeta_6^2 = 0.$$

We consider the metric on Z induced by the metric $\tilde{g}$ of $G_1^{\mathbb{C}}(\bigwedge^2 \mathbb{C}^4)$; then $\iota : G_{2,2}^{\mathbb{C}} \to Z$ is an isometry, and the complex Grassmannian $G_{2,2}^{\mathbb{C}}$ is isometric via this mapping to the complex quadric Q_4 endowed with the Riemannian metric g of §2, Chapter V. Thus from Theorem 6.45, with $n = 4$, we deduce:

THEOREM 8.7. *The Grassmannian $G_{2,2}^{\mathbb{C}}$ is infinitesimally rigid.*

The complex conjugation of $\mathbb{C}^6$ induces an involution τ of $\bigwedge^2\mathbb{C}^4$: the point ζ of $\bigwedge^2\mathbb{C}^4$ with coordinates $(\zeta_1,\dots,\zeta_6)$ is sent into the point $\tau(\zeta)$ with coordinates $(\bar\zeta_1,\dots,\bar\zeta_6)$. In turn, this involution induces an involutive isometry τ of $G_1^{\mathbb{C}}(\bigwedge^2\mathbb{C}^4)$ which preserves the hypersurface Z. If Ψ is the isometry of $X = G_{2,2}^{\mathbb{C}}$ which sends a 2-plane in $\mathbb{C}^4$ into its orthogonal complement, we now verify that the diagram

$$
\begin{array}{ccc}
G_{2,2}^{\mathbb{C}} & \xrightarrow{\ \Psi\ } & G_{2,2}^{\mathbb{C}} \\
\Big\downarrow{\scriptstyle\iota} & & \Big\downarrow{\scriptstyle\iota} \\
Z & \xrightarrow{\ \tau\ } & Z
\end{array}
$$

(8.58)

is commutative. Indeed, if x_0 is the point of X corresponding to the 2-plane generated by $\{e_1, e_2\}$, then $\Psi(x_0)$ corresponds to its orthogonal complement, which is the 2-plane generated by $\{e_3, e_4\}$. Let ϕ be an element of G. Then the point $\phi(x_0)$ of X corresponds to the 2-plane generated by $\{\phi(e_1), \phi(e_2)\}$, while $\Psi\phi(x_0)$ corresponds to the 2-plane generated by $\{\phi(e_3), \phi(e_4)\}$. We verify that

$$
(8.59) \qquad e_1 \wedge e_2 = \frac{1}{\sqrt{2}}\,(\omega_1 - i\omega_2), \qquad e_3 \wedge e_4 = \frac{1}{\sqrt{2}}\,(\omega_1 + i\omega_2).
$$

We may write $\phi\omega_j = \sum_{k=1}^6 c_j^k \omega_k$, where the coefficients c_j^k are real. It follows that

$$
\phi(e_1 \wedge e_2) = \frac{1}{\sqrt{2}} \sum_{j=1}^6 (c_1^j - ic_2^j)\,\omega_j, \qquad \phi(e_3 \wedge e_4) = \frac{1}{\sqrt{2}} \sum_{j=1}^6 (c_1^j + ic_2^j)\,\omega_j.
$$

Thus the points $\iota\phi(x_0)$ and $\iota\Psi\phi(x_0)$ of $G_1^{\mathbb{C}}(\bigwedge^2\mathbb{C}^4)$ correspond to the points of $\mathbb{CP}^5$ with homogeneous coordinates $(\zeta_1,\dots,\zeta_6)$ and $(\zeta_1',\dots,\zeta_6')$, respectively, where

$$
\zeta_j = \frac{1}{\sqrt{2}}\,(c_1^j - ic_2^j), \qquad \zeta_j' = \frac{1}{\sqrt{2}}\,(c_1^j + ic_2^j).
$$

Since $\zeta_j' = \bar\zeta_j$, we see that $\iota\Psi\phi(x_0)$ is equal to $\tau\iota\phi(x_0)$. Since the group G acts transitively on the Grassmannian X, from this last observation we obtain the commutativity of the diagram (8.58).

According to (5.59), the quotient of the complex quadric Q_4 by the action of the group of isometries of Q_4 generated by τ is isometric to the real Grassmannian $G_{2,4}^{\mathbb{R}}$ endowed with the Riemannian metric $\frac{1}{2}g$, where

g is the metric on $G_{2,4}^{\mathbb{R}}$ considered in §1, Chapter IV. Hence the commutativity of diagram (8.58) implies that the space $\bar{G}_{2,2}^{\mathbb{C}}$ is isometric to the Grassmannian $G_{2,4}^{\mathbb{R}}$ endowed with this metric $\frac{1}{2}g$; moreover, a symmetric p-form u on the quadric Z is even (resp. odd) with respect to the involution τ if and only if the symmetric p-form $\iota^* u$ on $G_{2,2}^{\mathbb{C}}$ is even (resp. odd). From Proposition 4.14 and Theorems 6.47 and 6.71, with $n = 4$, we deduce the following three results, the first of which is also given by Theorem 2.24:

PROPOSITION 8.8. *The maximal flat Radon transform for functions on the symmetric space $\bar{G}_{2,2}^{\mathbb{C}}$ is injective.*

THEOREM 8.9. *The symmetric space $\bar{G}_{2,2}^{\mathbb{C}}$ is rigid in the sense of Guillemin.*

THEOREM 8.10. *A 1-form on the symmetric space $\bar{G}_{2,2}^{\mathbb{C}}$ satisfies the Guillemin condition if and only if it is exact.*

The mapping $\iota^* : C^\infty(S^2 T_Z^*) \to C^\infty(S^2 T^*)$ induces an isomorphism

$$\iota^* : E(Z) \to E(X).$$

According to the commutativity of diagram (8.58), from the equalities (8.56) and the relation (6.79) given by Proposition 6.63, it follows that

(8.60) $$\mathcal{N}_2 \cap E(X) = \mathcal{N}_2 \cap (\kappa_1 + \kappa_2)\mathcal{H} = \{0\}.$$

We remark that the first equality of (8.56) for $G_{2,2}^{\mathbb{C}}$ is also a consequence of relation (5.111) of Proposition 5.17.

We continue to identify Z with the quadric Q_4 as above. We also identify the vector bundle L of rank 2 over the complex quadric Q_4 defined in §3, Chapter V with a sub-bundle of $S^2 T_Z^*$. We now proceed to verify that, for all $x \in X$, the isomorphism $\iota^* : S^2 T_{Z,\iota(x)}^* \to S^2 T_{X,x}^*$ induces an isomorphism

(8.61) $$\iota^* : L_{\iota(x)} \to E_{X,x}.$$

Let x_1 be the point of X corresponding to the 2-plane generated by $\{e_3, e_4\}$. According to §2, Chapter V and the relations (8.59), the tangent vectors

$$\{\pi_*(e_3 \wedge e_4, \omega_j), \pi_*(e_3 \wedge e_4, i\omega_j)\},$$

with $3 \le j \le 6$, form an orthonormal basis for the tangent space of Z at the point $\tilde{x}_1 = \iota(x_1)$, and the unit tangent vector $\nu = \pi_*(e_3 \wedge e_4, -e_1 \wedge e_2)$ at $\tilde{x}_1$ is normal to Z. According to the relations (5.54), the action of the real structure K_ν of the quadric associated with the unit normal ν is given by

$$K_\nu \pi_*(e_3 \wedge e_4, \omega_j) = \pi_*(e_3 \wedge e_4, \omega_j), \quad K_\nu \pi_*(e_3 \wedge e_4, i\omega_j) = -\pi_*(e_3 \wedge e_4, i\omega_j),$$

for $3 \le j \le 6$. Since

$$e_1 \wedge e_4 = \frac{1}{\sqrt{2}}(\omega_3 + i\omega_4), \qquad e_2 \wedge e_3 = \frac{1}{\sqrt{2}}(\omega_3 - i\omega_4),$$

we see that

$$K_\nu \pi_*(e_3 \wedge e_4, e_1 \wedge e_4) = \pi_*(e_3 \wedge e_4, e_2 \wedge e_3).$$

The elements h_ν and $h_{J\nu}$ of $S^2 T^*_{Z,\tilde{x}_1}$ determined by

$$h_\nu(\xi, \eta) = \tilde{g}(K_\nu \xi, \eta), \qquad h_{J\nu}(\xi, \eta) = \tilde{g}(JK_\nu \xi, \eta),$$

for $\xi, \eta \in T_{Z,\tilde{x}_1}$, are generators of the fiber $L_{\tilde{x}_1}$. Thus we obtain

$$h_\nu(\pi_*(e_3 \wedge e_4, e_1 \wedge e_4), \pi_*(e_3 \wedge e_4, e_1 \wedge e_4))$$
$$= \tilde{g}(K_\nu \pi_*(e_3 \wedge e_4, e_1 \wedge e_4), \pi_*(e_3 \wedge e_4, e_1 \wedge e_4))$$
$$= \tilde{g}(\pi_*(e_3 \wedge e_4, e_2 \wedge e_3), \pi_*(e_3 \wedge e_4, e_1 \wedge e_4))$$
$$= 0;$$

on the other hand, we also have

$$h_{J\nu}(\pi_*(e_3 \wedge e_4, e_1 \wedge e_4), \pi_*(e_3 \wedge e_4, e_1 \wedge e_4))$$
$$= \tilde{g}(JK_\nu \pi_*(e_3 \wedge e_4, e_1 \wedge e_4), \pi_*(e_3 \wedge e_4, e_1 \wedge e_4))$$
$$= \tilde{g}(J\pi_*(e_3 \wedge e_4, e_2 \wedge e_3), \pi_*(e_3 \wedge e_4, e_1 \wedge e_4))$$
$$= 0.$$

It follows that

$$(8.62) \qquad h(\pi_*(e_3 \wedge e_4, e_1 \wedge e_4), \pi_*(e_3 \wedge e_4, e_1 \wedge e_4)) = 0,$$

for all $h \in L_{\tilde{x}_1}$.

Let x be an arbitrary point of X and let $\alpha \in V^*_x$ and $w \in W_x$ be unit vectors. We consider the tangent vector $\xi = \alpha \otimes w \in (V^* \otimes W)_x$. Then there are orthonormal bases $\{v_1, v_2\}$ of V_x and $\{w_1, w_2\}$ of W_x such that $\langle v_1, \alpha \rangle = 1$, $\langle v_2, \alpha \rangle = 0$ and $w_1 = w$. Then we easily see that

$$\iota_* \xi = \pi_*(v_1 \wedge v_2, w_1 \wedge v_2).$$

There exists an element ϕ of G satisfying

$$\phi(e_3) = v_1, \quad \phi(e_4) = v_2, \quad \phi(e_1) = w_1, \quad \phi(e_2) = w_2.$$

Then the isometries ϕ of X and $G_1^{\mathbb{C}}(\wedge^2\mathbb{C}^4)$ induced by this element of G satisfy $\iota \circ \phi = \phi \circ \iota$, $\phi(x_1) = x$, $\phi(\tilde{x}_1) = \iota(x)$; in turn, these isometries induce isomorphisms

$$\phi^* : L_{\iota(x)} \to L_{\tilde{x}_1}, \qquad \phi^* : E_{X,x} \to E_{X,x_1}.$$

Clearly we have

(8.63) $$\phi_* \pi_* (e_3 \wedge e_4, e_1 \wedge e_4) = \iota_* \xi.$$

From the relations (8.62) and (8.63), we see that

$$h(\iota_*\xi, \iota_*\xi) = 0,$$

for all $h \in L_{\iota(x)}$, and so we obtain the inclusion $\iota^*(L_{\iota(x)}) \subset E_{X,x}$. According to (8.21) and (8.24), we know that the bundle E_X is of rank 2. Therefore the equality $\iota^*(L_{\iota(x)}) = E_{X,x}$ holds, and we have verified that the mapping (8.61) is an isomorphism.

The commutativity of the diagram (8.58), the isomorphisms (8.61) and Theorem 6.52 give us the following:

THEOREM 8.11. *Let X be the complex Grassmannian $G_{2,2}^{\mathbb{C}}$. An even section of E_X over X, which satisfies the Guillemin condition, vanishes identically.*

§6. The Guillemin condition on the complex Grassmannians

Let $m, n \geq 1$ be given integers. In this section, we return to our study of the complex Grassmannian $X = G_{m,n}^{\mathbb{C}}$, endowed with the metric g, and continue to identify the tangent bundle T of X with the vector bundle $V^* \otimes_{\mathbb{C}} W$ as in §2.

Let F be a totally real subspace of $\mathbb{C}^{m+n}$ of dimension $m+n$. The standard Hermitian scalar product on $\mathbb{C}^{m+n}$ induces a positive definite scalar product on F; we consider the real Grassmannian $Y = G_m^{\mathbb{R}}(F)$ endowed with the Riemannian metric determined by this scalar product on F. We then have a totally geodesic imbedding

(8.64) $$\iota : G_m^{\mathbb{R}}(F) \to G_{m,n}^{\mathbb{C}},$$

which sends the real subspace F' of F of dimension m into the complex subspace of $\mathbb{C}^{m+n}$ of dimension m generated by F' over $\mathbb{C}$. If y is a point of Y, then $V_{\iota(y)}$ is the subspace of $\mathbb{C}^{m+n}$ generated by $V_{Y,y}$ over $\mathbb{C}$ and $W_{\iota(y)}$ is the subspace of $\mathbb{C}^{m+n}$ generated by $W_{Y,y}$ over $\mathbb{C}$. The scalar product on F allows us to identify the vector bundle V_Y with its dual bundle V_Y^* and therefore also $V_Y \otimes_{\mathbb{R}} W_Y$ with $\mathrm{Hom}_{\mathbb{R}}(V_Y, W_Y)$. Then we see that, for $y \in Y$, the mapping ι_{*y} induced by ι is equal to the injective mapping

$$(8.65) \qquad (V_Y \otimes_{\mathbb{R}} W_Y)_y \to (V^* \otimes_{\mathbb{C}} W)_{\iota(y)},$$

which sends $\theta \in (V_Y \otimes_{\mathbb{R}} W_Y)_y$, considered as an element of $\mathrm{Hom}_{\mathbb{R}}(V_Y, W_Y)_y$, into the unique element $\tilde{\theta}$ of $\mathrm{Hom}_{\mathbb{C}}(V, W)_{\iota(y)}$ whose restriction to $V_{Y,y}$ is equal to θ and which satisfies $\tilde{\theta} \circ J = J \circ \tilde{\theta}$, where J is the complex structure of $\mathbb{C}^{m+n}$. For $y \in Y$, we identify the space $(V_Y \otimes_{\mathbb{R}} W_Y)_y$ with its image in $(V^* \otimes_{\mathbb{C}} W)_{\iota(y)}$ under the mapping (8.65); then we have the equality

$$(8.66) \qquad \mathrm{Exp}_{\iota(y)}(V_Y \otimes_{\mathbb{R}} W_Y)_y = \iota(G_m^{\mathbb{R}}(F))$$

of closed totally geodesic submanifolds of $G_{m,n}^{\mathbb{C}}$. The tangent spaces of these submanifolds of $G_{m,n}^{\mathbb{C}}$ at $\iota(y)$ are equal. From the formula for the curvature of $G_{m,n}^{\mathbb{C}}$, we infer that $\mathrm{Exp}_{\iota(y)}(V_Y \otimes_{\mathbb{R}} W_Y)_y$ is a totally geodesic subman- ifold of $G_{m,n}^{\mathbb{C}}$ and a globally symmetric space. Clearly, the submanifold $\iota(G_m^{\mathbb{R}}(F))$ has these same properties. In fact, the subgroup of $SU(m+n)$ consisting of all elements of $SU(m+n)$ which preserve the subspace F of $\mathbb{C}^{m+n}$ acts transitively on these submanifolds by isometries. These var- ious observations yield the equality (8.66). From the above description of the mappings ι_{*y}, with $y \in Y$, we infer that, if h is a section of the vector bundle E of $S^2 T^*$ over X, then the symmetric 2-form $\iota^* h$ is a section of the vector bundle E_Y over Y. Since Y is a symmetric space of the same rank as X, if u is a symmetric p-form on X satisfying the Guillemin condition, then the symmetric p-form $\iota^* u$ on Y also satisfies the Guillemin condition. Since an arbitrary point of X belongs to the image of an imbedding of the form (8.64), from the preceding observation and Proposition 4.12 we deduce the following result, which is also given by Theorem 2.24:

PROPOSITION 8.12. *For $m, n \geq 2$, with $m \neq n$, the maximal flat totally geodesic Radon transform on the complex Grassmannian $G_{m,n}^{\mathbb{C}}$ is injective.*

Let x be a given point of $G_{m,n}^{\mathbb{C}}$. Let V' and W' be totally real non-zero subspaces of V_x^* and W_x of dimension p and q, respectively. Then there is a natural injective mapping

$$(8.67) \qquad V' \otimes_{\mathbb{R}} W' \to (V^* \otimes_{\mathbb{C}} W)_x;$$

we shall identify $V' \otimes_{\mathbb{R}} W'$ with its image under the mapping (8.67), which is a totally real subspace of $(V^* \otimes_{\mathbb{C}} W)_x$. We choose a totally real subspace

V_1 of V_x^* of dimension m containing V' and a totally real subspace W_1 of W_x of dimension n containing W'. We then consider the injective mapping

$$(8.68) \qquad V_1 \otimes_\mathbb{R} W_1 \to (V^* \otimes_\mathbb{C} W)_x$$

given by (8.67). Let V_1' be the totally real subspace of V_x consisting of the vectors v of V_x for which $\langle \alpha, v \rangle \in \mathbb{R}$, for all $\alpha \in V_1$, and let V'' be the subspace of V_1' equal to the image of V' under the isomorphism $V_1 \to V_1'$ induced by the scalar product on V_1'. We consider the totally real subspace $F = V_1' \oplus W_1$ of $\mathbb{C}^{m+n}$ and the corresponding totally geodesic isometric imbedding ι given by (8.64). If y is the point of $Y = G_m^\mathbb{R}(F)$ corresponding to the real subspace V_1' of F, then we have $\iota(y) = x$, and we see that $V_{Y,y} = V_1'$ and $W_{Y,y} = W_1$. Thus the image of the mapping (8.68) is equal to the image of the mapping (8.65) at y induced by this imbedding ι. We denote by Z the real Grassmannian $G_{p,q}^\mathbb{R}$ and let $j : Z \to Y$ be the isometric imbedding given by Lemma 4.6 whose image is equal to the submanifold $\mathrm{Exp}_y V'' \otimes_\mathbb{R} W'$ of Y. We easily see that the image of the restriction of the mapping ι_{*y} to the subspace $V'' \otimes_\mathbb{R} W'$ of $(V_Y \otimes_\mathbb{R} W_Y)_y$ is equal to the subspace $V' \otimes_\mathbb{R} W'$ of $(V^* \otimes_\mathbb{C} W)_x$. Therefore the image of the isometric imbedding $i = \iota \circ j : Z \to X$ is equal to the totally geodesic submanifold $X' = \mathrm{Exp}_x V' \otimes_\mathbb{R} W'$ of $G_{m,n}^\mathbb{C}$. If h is a section of the vector bundle E of $S^2 T^*$ over X, we have seen that the symmetric 2-form $\iota^* h$ is a section of the vector bundle E_Y over Y; by Lemma 4.6, we now infer that the symmetric 2-form $i^* h$ is a section of the vector bundle E_Z over Z. Thus we have proved the following lemma:

LEMMA 8.13. *Let x be a point of the complex Grassmannian $X = G_{m,n}^\mathbb{C}$, with $m, n \geq 2$. Let V' and W' be totally real non-zero subspaces of V_x^* and W_x of dimension p and q, respectively. Then $X' = \mathrm{Exp}_x V' \otimes_\mathbb{R} W'$ is a closed totally geodesic submanifold of X isometric to the real Grassmannian $Z = G_{p,q}^\mathbb{R}$. Moreover, there is an isometric imbedding $i : Z \to X$ whose image is equal to X' and which has the following property: if h is a section of the sub-bundle E of $S^2 T^*$ over X, then $i^* h$ is a section of the sub-bundle E_Z of $S^2 T_Z^*$.*

Assume that $p = m$ and that $q = n$. Then X' is a symmetric space of the same rank as X. Thus the restriction to X' of any symmetric form on X satisfying the Guillemin condition also satisfies the Guillemin condition. Moreover we may choose a maximal flat totally geodesic torus Z_0 of X contained in X' and containing the point x. If Z is an arbitrary maximal flat totally geodesic torus of X and z is a point of Z, there exists an element ϕ of $SU(m+n)$ such that $\phi(Z_0) = Z$ and $\phi(x) = z$; therefore Z is contained in the totally geodesic submanifold $\phi(X') = \mathrm{Exp}_z V_2 \otimes_\mathbb{R} W_2$ of X, which is isometric to the real Grassmannian $G_{m,n}^\mathbb{R}$, where V_2 is the totally real subspace $\phi^{-1*}(V')$ of V_z^* and W_2 is the totally real subspace $\phi(W')$ of W_z.

From Propositions 4.12 and 4.13 and Lemma 4.6, we then obtain the following:

PROPOSITION 8.14. *Let x be a point of the complex Grassmannian $X = G_{m,n}^{\mathbb{C}}$, with $2 \le m < n$. Let X' be a closed totally geodesic subman-ifold of X isometric to the real Grassmannian $G_{p,q}^{\mathbb{R}}$ which can be written in the form $\operatorname{Exp}_x V' \otimes_{\mathbb{R}} W'$, where V' and W' are totally real non-zero subspaces of V_x^* and W_x of dimension p and q, respectively. Assume either that $p = m$ or that $q = n$. If u is a symmetric form on X satisfying the Guillemin condition, then the restriction of u to X' satisfies the Guillemin condition.*

Let F_1 and F_2 be orthogonal complex subspaces of $\mathbb{C}^{m+n}$ and let p_1 and p_2 be given integers satisfying $0 \le p_j \le \dim F_j$, for $j = 1, 2$, and $p_1 + p_2 = m$. We suppose that F_2 is the orthogonal complement of F_1 in $\mathbb{C}^{m+n}$. For $j = 1, 2$, the space F_j is endowed with the Hermitian scalar induced by the standard Hermitian scalar product of $\mathbb{C}^{m+n}$; we consider the complex Grassmannians $Y = G_{p_1}^{\mathbb{C}}(F_1)$ and $Z = G_{p_2}^{\mathbb{C}}(F_2)$ endowed with the Hermitian metrics induced by these Hermitian scalar products. Then there is totally geodesic imbedding

$$(8.69) \qquad\qquad \iota : Z \times Y \to X,$$

sending the pair (z, y), where $z \in Z$ and $y \in Y$, into the point of X corresponding to the m-plane of $\mathbb{C}^{m+n}$ generated by the subspaces of F_2 and F_1 corresponding to the points z and y, respectively.

We now fix points $y \in Y$ and $z \in Z$; we write $x = \iota(z, y)$. By definition, we have

$$(8.70) \qquad V_x = V_{Y,y} \oplus V_{Z,z}, \qquad W_x = W_{Y,y} \oplus W_{Z,z}.$$

It is easily seen that the mapping $\iota_{*(z,y)}$ from the tangent space of $Z \times Y$ at (z, y) to the tangent space T_x induced by ι is identified with the mapping

$$(8.71) \qquad (V_Z^* \otimes W_Z)_z \oplus (V_Y^* \otimes W_Y)_y \to (V^* \otimes W)_x$$

sending $\theta_1 \oplus \theta_2$, where $\theta_j \in V_j^* \otimes W_j$ is considered as an element of $\operatorname{Hom}_{\mathbb{C}}(V_j, W_j)$, into the element θ of $(V^* \otimes W)_x$ determined by

$$\theta(v_1) = \theta_1(v_1), \qquad \theta(v_2) = \theta_2(v_2),$$

for all vectors $v_1 \in V_1$ and $v_2 \in V_2$. Let

$$\iota_z : V_{Y,y} \to V_x, \qquad \iota_z : W_{Y,y} \to W_x$$

be the inclusion mappings corresponding to the decompositions (8.70). The first decomposition of (8.70) also determines an injective mapping

$$\iota_z : V^*_{Y,y} \to V^*_x.$$

If α is a vector of $V^*_{Y,y}$, we consider the vector $\alpha^\sharp$ of $V_{Y,y}$ determined by α; then we have the relation

(8.72) $$\iota_z \alpha^\sharp = (\iota_z \alpha)^\sharp$$

among vectors of V_x.

We consider the totally geodesic imbedding

$$\varphi = \iota_z : Y \to X$$

defined by $\varphi(a) = \iota_z(a) = \iota(z, a)$, for all $a \in Y$; then we have $\varphi(y) = x$. Then we have the equality

(8.73) $$\varphi_*(\alpha \otimes w) = (\iota_z \alpha) \otimes (\iota_z w)$$

among vectors of $(V^* \otimes W)_x$, for all $\alpha \in V^*_{Y,y}$ and $w \in W_{Y,y}$. According to (8.73), if h is a section of the sub-bundle E of S^2T^* over X, we see that, for $z \in Z$, the symmetric 2-form φ^*h is a section of the sub-bundle E_Y of $S^2T^*_Y$.

If F_x denotes the image of $(V^*_Y \otimes W_Y)_y$ under the mapping (8.71), we have the equality

(8.74) $$\mathrm{Exp}_x F_x = \varphi(G^{\mathbb{C}}_{p_1}(F_1))$$

of closed totally geodesic submanifolds of $G^{\mathbb{C}}_{m,n}$. Indeed, using the above description of the mapping ι_* at (z, y) given by (8.71), we see that the tangent spaces of these two submanifolds of $G^{\mathbb{C}}_{m,n}$ at x are equal. From the formula for the curvature of $G^{\mathbb{C}}_{m,n}$, we infer that $\mathrm{Exp}_x F_x$ is a totally geodesic submanifold of $G^{\mathbb{C}}_{m,n}$ and a globally symmetric space. Clearly, the submanifold $\varphi(G^{\mathbb{C}}_{p_1}(F_1))$ has these same properties. In fact, the subgroup $SU(m + n, F_1)$ of $SU(m + n)$, consisting of all elements of $SU(m + n)$ which preserve the subspace F_1 and are the identity on the orthogonal complement of F_1, acts transitively on these submanifolds by isometries. These various observations yield the relation (8.74).

Now let V' be a complex subspace of V^*_x of dimension p and W' be a complex subspace of W_x of dimension q. Let V'_1 be the subspace of V_x equal to the image of V' under the mapping $V^*_x \to V_x$ sending $\alpha \in V^*_x$ into $\alpha^\sharp \in V_x$. In fact, V'_1 is the orthogonal complement of the subspace

$$V'' = \{ v \in V_x \mid \langle v, \alpha \rangle = 0, \text{ for all } \alpha \in V' \}$$

of V_x. Then the complex subspace $F_1 = V_1' \oplus W'$ of $\mathbb{C}^{m+n}$ of dimension $p + q$ is orthogonal to V''. Thus V'' is a subspace of dimension $m - p$ of the orthogonal complement F_2 of F_1 in $\mathbb{C}^{m+n}$. We consider the complex Grassmannians $Y = G_p^{\mathbb{C}}(F_1)$ and $Z = G_{p_2}^{\mathbb{C}}(F_2)$, with $p_2 = m - p$, and the totally geodesic imbedding $\iota : Z \times Y \to X$ given by (8.69), with $p_1 = p$.

Let y be the point of Y corresponding to the subspace V_1' of F_1 and let z be the point of Z corresponding to the subspace V'' of F_2. Then we have $V_{Y,y} = V_1'$ and $W_{Y,y} = W'$ and x is equal to $\iota(z, y)$. The totally geodesic imbedding $\varphi = \iota_z : Y \to X$ sends the point of Y corresponding to a p-plane F' of F_1 into the m-plane of $\mathbb{C}^{m+n}$ generated by the subspaces F' and V''. The image $(V_Y^* \otimes W_Y)_y$ under the mapping (8.71) is equal to the subspace $V' \otimes_{\mathbb{C}} W'$ of $(V^* \otimes_{\mathbb{C}} W)_x$. Therefore from (8.74), we obtain the equality

$$(8.75) \qquad \qquad \operatorname{Exp}_x V' \otimes_{\mathbb{C}} W' = \varphi(G_p^{\mathbb{C}}(F_1))$$

of closed totally geodesic submanifolds of $G_{m,n}^{\mathbb{C}}$.

From these remarks and the equality (8.75), we obtain:

LEMMA 8.15. *Let x be a point of the complex Grassmannian $X = G_{m,n}^{\mathbb{C}}$, with $m, n \geq 2$. Let V' and W' be complex non-zero subspaces of V_x^* and W_x of complex dimension p and q, respectively. Then $X' = \operatorname{Exp}_x V' \otimes_{\mathbb{C}} W'$ is a closed totally geodesic submanifold of X isometric to the complex Grassmannian $Y = G_{p,q}^{\mathbb{C}}$. Moreover, there is an isometric imbedding $i : Y \to X$ whose image is equal to X', a unique point y of Y satisfying $i(y) = x$, and isomorphisms $\varphi_1 : V_{Y,y}^* \to V'$, $\varphi_2 : W_{Y,y} \to W'$, which possess the following properties:*

(i) the mapping $i_* : (V_Y^* \otimes_{\mathbb{C}} W_Y)_y \to (V^* \otimes_{\mathbb{C}} W)_x$ induced by i is equal to $\varphi_1 \otimes \varphi_2$;

(ii) if h is a section of the sub-bundle E of $S^2 T^*$ over X, then $i^* h$ is a section of the sub-bundle E_Y of $S^2 T_Y^*$.

If we take $p = 1$ in the preceding lemma, then X' is isometric to the complex projective space $\mathbb{CP}^q$; moreover, the restriction to X' of an arbitrary section of the vector bundle E vanishes.

PROPOSITION 8.16. *Let x be a point of the complex Grassmannian $X = G_{m,n}^{\mathbb{C}}$, with $2 \leq m < n$. Let X' be a closed totally geodesic submanifold of X isometric to the complex Grassmannian $G_{p,q}^{\mathbb{C}}$ which can be written in the form $\operatorname{Exp}_x V' \otimes_{\mathbb{C}} W'$, where V' and W' are non-zero complex subspaces of V_x^* and W_x of dimension p and q, respectively. Assume either that $p = m$ or that $q = n$. If u is a symmetric form on X satisfying the Guillemin condition, then the restriction of u to X' satisfies the Guillemin condition.*

PROOF: Let u be a symmetric form on X satisfying the Guillemin condition. Let Z be a maximal flat totally geodesic torus of the submanifold X'. To prove the proposition, we need to show that the restriction of u

to Z satisfies the Guillemin condition. It suffices to consider the case when Z contains the point x. Indeed, according to the previous discussion, there is an element ϕ of $SU(m + n)$ preserving X' and such that $\phi(x)$ belongs to Z. We have seen above that there exist a totally real subspace V_1 of V' of dimension p and a totally real subspace W_1 of W_x of dimension q such that Z is contained in the totally geodesic submanifold $X_1 = \mathrm{Exp}_x V_1 \otimes_{\mathbb{R}} W_1$ of X', which by Lemma 8.13 is isometric to the real Grassmannian $G^{\mathbb{R}}_{p,q}$. By Proposition 8.14, the restriction of u to the submanifold X_1 satisfies the Guillemin condition. Since X_1 is a symmetric space of the same rank as X', it follows that Z is a maximal flat totally geodesic torus of the submanifold X_1. Hence the restriction of u to Z satisfies the Guillemin condition. This completes the proof of the proposition.

If we take either $p = 1$ and $q = n$, or $p = m$ and $q = 1$ in the preceding proposition, since the Grassmannians $G^{\mathbb{C}}_{m,n}$ and $G^{\mathbb{C}}_{m,n}$ are isometric, by (8.13) we obtain:

PROPOSITION 8.17. *Let x be a point of the complex Grassmannian* $X = G^{\mathbb{C}}_{m,n}$, *with $m, n \geq 2$ and $m \neq n$. Let X' be a closed totally geodesic submanifold of X isometric to a complex projective space of dimension ≥ 2, which can be written either in the form $\mathrm{Exp}_x \alpha \otimes W_x$, where α is a unit vector of V^*_x, or in the form $\mathrm{Exp}_x V^*_x \otimes w$, where w is a unit vector of W_x. If u is a symmetric form on X satisfying the Guillemin condition, then the restriction of u to X' satisfies the zero-energy condition.*

§7. Integrals of forms on the complex Grassmannians

Let $m, n \geq 2$ be given integers. We consider the complex Grassmannian $X = G^{\mathbb{C}}_{m,n}$, the group $G = SU(m+n)$ and the mapping $\rho : S_{m,n} \to X$ of §4.

This section and §9 are mainly devoted to results which lead to the following:

PROPOSITION 8.18. *Let X be the complex Grassmannian $G^{\mathbb{C}}_{m,n}$, with $m, n \geq 2$. Then the equality*

$$(8.76) \qquad\qquad \mathcal{N}_2 \cap E(X) = \{0\}$$

holds.

We remark that, when $m = n = 2$, the assertion of this proposition is already given by (8.60). According to Lemma 2.11, we see that Proposition 8.18 implies that the equality

$$(8.77) \qquad\qquad \mathcal{Z}_2 \cap E(X) = \{0\}$$

also holds.

We consider the sesquilinear form $\tilde{Q} = Q_{m+n,1}$ on $\mathbb{C}^{m+n}$ and the sesquilinear forms Q_1 on V^* and Q_2 on W induced by $\tilde{Q}$, which are defined in §4. We also consider the $U(1)$-invariant polynomial Q on $\mathbb{C}^{m+n}$ determined by $\tilde{Q}$ and the function $\tilde{f}_0$ on X, which are also defined in §4.

Let F be a complex subspace of $\mathbb{C}^{m+n}$ of dimension $q+1$, with $q \geq 1$. We consider the complex projective space $G_1^{\mathbb{C}}(F)$ of dimension q, the unit sphere $S(F)$ of F and the natural projection

$$\pi : S(F) \to G_1^{\mathbb{C}}(F),$$

which sends $u \in S(F)$ into the line generated by u. Clearly, a $U(1)$-invariant function on $S(F)$ induces by passage to the quotient a function on $G_1^{\mathbb{C}}(F)$. Let $\{u_1, \dots, u_p\}$ be an orthonormal basis for the orthogonal complement $F^\perp$ of F in $\mathbb{C}^{m+n}$; by an argument similar to the one which gives us the relation (8.37), we see that the expression $\sum_{l=1}^p Q(u_l)$ is independent of the choice of the basis for $F^\perp$. We consider the $U(1)$-invariant function f_1 on $S(F)$ defined by

$$f_1(u) = Q(u) + \sum_{l=1}^p Q(u_l),$$

for $u \in S(F)$. We consider the functions $\tilde{f}_1$ on $G_1^{\mathbb{C}}(F)$ induced by f_1. If F contains the vectors e_1 and e_{m+n}, we remark that $Q(v)$ vanishes for all $v \in F^\perp$, and so in this case we see that $f_1(u) = Q(u)$, for all $u \in S(F)$. If u, u' are vectors of $S(F)$, which are orthogonal with respect to the standard Hermitian scalar product on $\mathbb{C}^{m+n}$ and if $t \in \mathbb{R}$, we consider the unit vector

$$\sigma(t) = \cos t \cdot u + \sin t \cdot u'$$

of F; in §4, Chapter III, we remarked that the path $\gamma = \gamma_{u,u'}$ defined by $\gamma(t) = (\pi \circ \sigma)(t)$, with $0 \leq t \leq \pi$, is a closed geodesic of $G_1^{\mathbb{C}}(F)$.

Let F_1 be a complex subspace of $\mathbb{C}^{m+n}$ of dimension $n+1$; we consider the complex projective space $Y = G_1^{\mathbb{C}}(F_1)$. Let $F_1^\perp$ be the orthogonal complement of F_1 in $\mathbb{C}^{m+n}$ and let $\{u_1, \dots, u_{m-1}\}$ be an orthonormal basis of $F_1^\perp$. We consider the totally geodesic imbedding

$$\iota_1 : Y \to G_{m,n}^{\mathbb{C}},$$

which sends $\pi(u)$, with $u \in S(F_1)$, into the m-plane of $\mathbb{C}^{m+n}$ generated by u and the subspace $F_1^\perp$. We consider the functions $\tilde{f}_1$ on Y induced by the $U(1)$-invariant function f_1 on $S(F_1)$. According to the definition of ι_1 and the equality (8.37), we see that

(8.78) $\iota_1^* f_0 = f_1.$

Let u be a given element of $S(F_1)$; we consider the points $y = \pi(u)$ of Y and $x = \iota_1(y)$ of X. Then u belongs to V_x. Let α be the unit vector of V_x^* determined by the relations

$$\langle u, \alpha \rangle = 1, \qquad \langle v, \alpha \rangle = 0,$$

for all $v \in F_1^\perp$. Then we easily verify that α is the vector of V_x^* determined by the relation $\alpha^\sharp = u$. According to the relation (8.73), we see that

$$(8.79) \qquad \iota_{1*} T_{Y,y} = \alpha \otimes W_x.$$

Let u, u' be vectors of $S(F_1)$, which are orthogonal with respect to the standard Hermitian scalar product on $\mathbb{C}^{m+n}$. As above, we consider the unit vector

$$\sigma(t) = \cos t \cdot u + \sin t \cdot u'$$

of F_1, with $t \in \mathbb{R}$, and the closed geodesic $\gamma = \gamma_{u,u'}$ of Y defined by $\gamma(t) = (\pi \circ \sigma)(t)$, for $0 \le t \le \pi$. We write $x(t) = \iota_1 \gamma(t)$, for $0 \le t \le \pi$; the subspace $V_{x(t)}$ of $\mathbb{C}^{m+n}$ is generated by the vector $\sigma(t)$ and the space $F_1^\perp$. For $0 \le t \le \pi$, let $\alpha(t)$ be the unit vector of $V_{x(t)}^*$ determined by

$$\langle \sigma(t), \alpha(t) \rangle = 1, \qquad \langle v, \alpha(t) \rangle = 0,$$

for all $v \in F_1^\perp$; then we easily verify that $\alpha(t)^\sharp = \sigma(t)$. The unit vector $\dot\sigma(t) = -\sin t \cdot u + \cos t \cdot u'$ belongs to the space $W_{x(t)}$, and according to (8.73), we have

$$\iota_{1*} \dot\gamma(t) = \alpha(t) \otimes \dot\sigma(t) \in (V^* \otimes_{\mathbb{C}} W)_{x(t)}.$$

By (8.78), we see that

$$(8.80) \qquad \tilde f_0(x(t)) = \tilde f_1(\gamma(t)) = Q(\sigma(t)) + \sum_{l=1}^{m-1} Q(u_l).$$

Moreover, by (8.39) we obtain the relations

$$(8.81) \qquad \begin{aligned} (Q_1 \cdot g_2)(\alpha(t) \otimes \dot\sigma(t), \alpha(t) \otimes \dot\sigma(t)) &= 2Q(\sigma(t)), \\ (g_1 \cdot Q_2)(\alpha(t) \otimes \dot\sigma(t), \alpha(t) \otimes \dot\sigma(t)) &= 2Q(\dot\sigma(t)). \end{aligned}$$

If h is a section of $S^2 T_{\mathbb{C}}^*$ over X, we have

$$(8.82) \qquad \int_{\iota_1 \gamma} h = \int_0^\pi h(\alpha(t) \otimes \dot\sigma(t), \alpha(t) \otimes \dot\sigma(t)) \, dt.$$

We now suppose that the subspace F_1 of $\mathbb{C}^{m+n}$ contains the vectors e_1 and e_{m+n}. We also suppose that

$$u = \frac{1}{\sqrt{2}}(e_1 + e_{m+n})$$

and that u' is orthogonal to e_1 and e_{m+n}. We consider the closed geodesic $\delta_1 = \iota_1 \circ \gamma_{u,u'}$ of X. For $0 \le t \le \pi$, the unit vectors $\sigma(t)$ of $V_{x(t)}$ and $\dot\sigma(t)$ of $W_{x(t)}$ associated above to the geodesic $\gamma_{u,u'}$ of Y are given by

(8.83)
$$\sigma(t) = \frac{\cos t}{\sqrt{2}} \cdot (e_1 + e_{m+n}) + \sin t \cdot u',$$

$$\dot\sigma(t) = -\frac{\sin t}{\sqrt{2}} \cdot (e_1 + e_{m+n}) + \cos t \cdot u'.$$

From (8.83), we deduce that

$$Q(\sigma(t)) = \frac{1}{2}\cos^2 t, \qquad Q(\dot\sigma(t)) = \frac{1}{2}\sin^2 t.$$

According to (8.80), we see that

$$\tilde f_0(x(t)) = Q(\sigma(t)) = \frac{1}{2}\cos^2 t.$$

Hence by (8.81) and (8.82), we have

$$\int_{\delta_1} \tilde f_0 = \frac{\pi}{4}, \qquad \int_{\delta_1} g_1 \cdot Q_2 = \int_{\delta_1} Q_1 \cdot g_2 = \frac{\pi}{2}.$$

From these equalities, we obtain the relations

(8.84)
$$\int_{\delta_1} \kappa_1(f_0) = \frac{n+1}{2n}\,\pi, \qquad \int_{\delta_1} \kappa_2(f_0) = \frac{m-1}{2m}\,\pi.$$

Now let F_2 be a complex subspace of $\mathbb{C}^{m+n}$ of dimension $m+1$. We consider the Grassmannians $Z = G_m^{\mathbb{C}}(F_2)$ and $Z' = G_1^{\mathbb{C}}(F_2)$, the isometry $\Psi : Z' \to Z$ defined in §2 and the totally geodesic imbedding

$$\iota_2 : Z \to G_{m,n}^{\mathbb{C}},$$

sending an m-plane of F_2 into the m-plane of $\mathbb{C}^{m+n}$ which it determines. We consider the function $\tilde f_1$ on Z' induced by the $U(1)$-invariant function f_1 on $S(F_2)$. We now verify that the equality

(8.85)
$$\Psi^* \iota_2^* \tilde f_0 = -\tilde f_1$$

holds. Indeed, let $\{u_1, \ldots, u_{n-1}\}$ be an orthonormal basis for the orthogonal complement $F_2^{\perp}$ of F_2 in $\mathbb{C}^{m+n}$. Let u be an element of $S(F_2)$ and let $\{\varepsilon_1, \ldots, \varepsilon_m\}$ be an orthonormal basis for the orthogonal complement of the space $\mathbb{C}u$ in F_2. Then the element $\Psi(\pi(u))$ of Z is the m-plane of $\mathbb{C}^{m+n}$ generated by the vectors $\{\varepsilon_1, \ldots, \varepsilon_m\}$, and so by (8.37) we have

$$(8.86) \qquad (\iota_2^* \tilde{f}_0)(\Psi(\pi(u))) = \sum_{k=1}^{m} Q(\varepsilon_k).$$

Since

$$\{\varepsilon_1, \ldots, \varepsilon_m, u, u_1, \ldots, u_{n-1}\}$$

is an orthonormal basis of $\mathbb{C}^{m+n}$, by (8.37) we have

$$Q(u) + \sum_{k=1}^{m} Q(\varepsilon_k) + \sum_{j=1}^{n-1} Q(u_j) = 0.$$

From the preceding relation and (8.86), it follows that

$$\tilde{f}_1(\pi(u)) = -(\iota_2^* \tilde{f}_0)(\Psi(\pi(u))),$$

and so the equality (8.85) holds.

Let z be a point of Z and set $x = \iota_2(z)$. Let w be a non-zero vector of F_2 orthogonal to V_x. Then according to the relation (8.73), we see that

$$(8.87) \qquad \iota_{2*} T_{Z,z} = V_x^* \otimes w.$$

Let v, v' be vectors of $S(F_2)$, which are orthogonal with respect to the standard Hermitian scalar product on $\mathbb{C}^{m+n}$. As above, we consider the unit vector

$$\sigma(t) = \cos t \cdot v + \sin t \cdot v'$$

of F_2, with $t \in \mathbb{R}$, and the closed geodesic $\gamma = \gamma_{v,v'}$ of $G_1^{\mathbb{C}}(F_2)$ defined by $\gamma(t) = (\pi \circ \sigma)(t)$, for $0 \le t \le \pi$. We consider the closed geodesic $\gamma' = \iota_2 \circ \Psi \circ \gamma$ of X. Let V_1 be the subspace of F_2 orthogonal to the vectors v and v'. For $0 \le t \le \pi$, the space $V_{\gamma'(t)}$ is generated by V_1 and the unit vector $\dot{\sigma}(t) = -\sin t \cdot v + \cos t \cdot v'$ of F_2. The unit vector $\sigma(t)$ of F_2 belongs to $W_{\gamma'(t)}$. Let $\beta(t)$ be the unit vector of $V_{\gamma'(t)}^*$ determined by

$$\langle \dot{\sigma}(t), \beta(t) \rangle = -1, \qquad \langle v, \beta(t) \rangle = 0,$$

for all $v \in V_1$; we easily verify that $\beta(t)^{\sharp} = -\dot{\sigma}(t)$. Then by (8.13) and (8.73), we have

$$\dot{\gamma}'(t) = \iota_{2*} \Psi_* \dot{\gamma}(t) = \beta(t) \otimes \sigma(t) \in (V^* \otimes_{\mathbb{C}} W)_{\gamma'(t)}.$$

By (8.85), we see that

$$
(8.88) \qquad \tilde{f}_0(\gamma'(t)) = -\tilde{f}_1(\gamma(t)) = -Q(\sigma(t)) - \sum_{l=1}^{n-1} Q(u_l).
$$

Moreover, by (8.39) we obtain the relations

$$
(8.89) \qquad \begin{aligned}
(Q_1 \cdot g_2)(\beta(t) \otimes \sigma(t), \beta(t) \otimes \sigma(t)) &= 2Q(\dot{\sigma}(t)), \\
(g_1 \cdot Q_2)(\beta(t) \otimes \sigma(t), \beta(t) \otimes \sigma(t)) &= 2Q(\sigma(t)).
\end{aligned}
$$

If h is a section of $S^2 T_{\mathbb{C}}^*$ over X, we have

$$
(8.90) \qquad \int_{\gamma'} h = \int_0^\pi h(\beta(t) \otimes \sigma(t), \beta(t) \otimes \sigma(t)) \, dt.
$$

We now suppose that the subspace F_2 of C^{m+n} contains the vectors e_1 and e_{m+n}. We also suppose that

$$
v = \frac{1}{\sqrt{2}} (e_1 + e_{m+n})
$$

and that v' is orthogonal to e_1 and e_{m+n}. We consider the closed geodesic $\delta_2 = \iota_2 \circ \psi \circ \gamma_{v,v'}$ of X. For $0 \le t \le \pi$, the unit vectors $\sigma(t)$ of $W_{x(t)}$ and $\dot{\sigma}(t)$ of $V_{x(t)}$ associated above with the geodesic $\gamma_{v,v'}$ of Z' are given by

$$
(8.91) \qquad \begin{aligned}
\sigma(t) &= \frac{\cos t}{\sqrt{2}} \cdot (e_1 + e_{m+n}) + \sin t \cdot v', \\
\dot{\sigma}(t) &= -\frac{\sin t}{\sqrt{2}} \cdot (e_1 + e_{m+n}) + \cos t \cdot v'.
\end{aligned}
$$

From (8.91), we deduce that

$$
Q(\sigma(t)) = \frac{1}{2} \cos^2 t, \qquad Q(\dot{\sigma}(t)) = \frac{1}{2} \sin^2 t.
$$

According to (8.88), we see that

$$
\tilde{f}_0(\gamma'(t)) = -Q(\sigma(t)).
$$

Hence by (8.89) and (8.90), we have

$$
\int_{\delta_2} \tilde{f}_0 = -\frac{\pi}{4}, \qquad \int_{\delta_2} g_1 \cdot Q_2 = \int_{\delta_2} Q_1 \cdot g_2 = \frac{\pi}{2}.
$$

From these equalities, we obtain the relations

$$(8.92) \qquad \int_{\delta_2} \kappa_1(f_0) = \frac{n-1}{2n}\,\pi, \qquad \int_{\delta_2} \kappa_2(f_0) = \frac{m+1}{2m}\,\pi.$$

We suppose that the subspaces F_1 and F_2 of $\mathbb{C}^{m+n}$ both contain the vectors e_1 and e_{m+n}. We denote by $\tilde{Y}$ and $\tilde{Z}$ the totally geodesic submanifolds of X equal to the images of $\iota_1 : Y \to X$ and $\iota_2 : Z \to X$. We again suppose that

$$u = v = \frac{1}{\sqrt{2}}\,(e_1 + e_{m+n})$$

and that the vectors $u' \in S(F_1)$ and $v' \in S(F_2)$ are orthogonal to e_1 and e_{m+n}. We consider the closed geodesic $\delta_1 = \iota_1 \circ \gamma_{u,u'}$ of X contained in the submanifold $\tilde{Y}$ and the closed geodesic $\delta_2 = \iota_2 \circ \Psi \circ \gamma_{v,v'}$ of X contained in the submanifold $\tilde{Z}$.

Let $a, b \in \mathbb{C}$ and consider the section

$$h = (a\kappa_1 + b\kappa_2)(f_0)$$

of $S^2 T_{\mathbb{C}}^*$. According to (8.84) and (8.92), we have

$$(8.93) \qquad \int_{\delta_1} h = \frac{\pi}{2}\left(\frac{n+1}{n}\,a - \frac{m-1}{m}\,b\right),$$

$$(8.94) \qquad \int_{\delta_2} h = \frac{\pi}{2}\left(\frac{n-1}{n}\,a - \frac{m+1}{m}\,b\right).$$

Now suppose that the restrictions of h to the totally geodesic submanifolds $\tilde{Y}$ and $\tilde{Z}$ of X satisfy the zero-energy condition. According to Proposition 8.17, we know that this assumption on h holds either if h satisfies the zero-energy condition or if $m \neq n$ and h satisfies the Guillemin condition. Then from the equalities (8.93) and (8.94), we obtain the relations

$$\frac{n+1}{n}\,a - \frac{m-1}{m}\,b = 0, \qquad \frac{n-1}{n}\,a - \frac{m+1}{m}\,b = 0;$$

since the determinant of the matrix

$$\begin{pmatrix} n+1 & m-1 \\ n-1 & m+1 \end{pmatrix}$$

is equal to $2(m + n)$, it follows that h vanishes. We have thus proved the following result:

PROPOSITION 8.19. *Let X be the complex Grassmannian $G_{m,n}^{\mathbb{C}}$, with $m, n \geq 2$. Let*

$$h = (a\kappa_1 + b\kappa_2)(f_0)$$

be an element of $C^\infty(S_0^2 T_{\mathbb{C}}^*)$ corresponding to $a, b \in \mathbb{C}$.

(i) *If the symmetric 2-form h satisfies the zero-energy condition, then h vanishes.*

(ii) *If $m \neq n$ and if the symmetric 2-form h satisfies the Guillemin condition, then h vanishes.*

PROPOSITION 8.20. *Let X be the complex Grassmannian $G_{m,n}^{\mathbb{C}}$, with $m, n \geq 2$, and let h be an element of $C_{\gamma_1}^\infty(S_0^2 T_{\mathbb{C}}^*)$.*

(i) *If the symmetric 2-form h satisfies the zero-energy condition, then h vanishes.*

(ii) *If $m \neq n$ and if the symmetric 2-form h satisfies the Guillemin condition, then h vanishes.*

PROOF: Assume that the G-module $\mathcal{N}_{2,\mathbb{C}} \cap C_{\gamma_1}^\infty(S_0^2 T^*)$ is non-zero (resp. the G-module $\mathcal{Z}_{2,\mathbb{C}} \cap C_{\gamma_1}^\infty(S_0^2 T^*)$ is non-zero and that $m \neq n$). According to (8.28) and (8.41), a highest weight vector h of this non-zero G-module is a linear combination of the highest weight vectors $\kappa_1(f_0)$ and $\kappa_2(f_0)$ of $C_{\gamma_1}^\infty(S_0^2 T^*)$. By Proposition 8.19, we see that the section h must vanish, which is a contradiction.

From Proposition 8.20 and the relations (2.25), we deduce that the equality (8.77) holds whenever $m, n \geq 2$, and that the equality (8.76) of Proposition 8.18 is true when $m \neq n$.

§8. Relations among forms on the complex Grassmannians

The results of this section are used only to prove Proposition 8.5 and the equalities (8.47) and (8.48). They do not enter into any of our other proofs and are presented only for the sake of completeness.

Let $m, n \geq 2$ be given integers. We consider the complex Grassmannian $X = G_{m,n}^{\mathbb{C}}$. We consider the complex-valued function $\tilde{f}_0$ on X obtained from the function f_0 on $S_{m,n}$. In §4, we saw that the equalities (8.28) and the decomposition (8.7) imply that

$$h_0 = \operatorname{Hess} \tilde{f}_0 = \pi_+ \operatorname{Hess} \tilde{f}_0$$

is a section of $(S^2 T^*)_{\mathbb{C}}^+$.

This section is mainly devoted to the proof of the following result:

LEMMA 8.21. *Let x be a point of the Grassmannian $X = G_{m,n}^{\mathbb{C}}$, with $m, n \geq 2$. If $\alpha \in V_x^*$ and $w \in W_x$ are unit vectors, we have*

(8.95)
$$(\pi_1 \operatorname{Hess} \tilde{f}_0)(\alpha \otimes w, \alpha \otimes w) = \kappa_1(f_0)(\alpha \otimes w, \alpha \otimes w),$$

$$(\pi_2 \operatorname{Hess} \tilde{f}_0)(\alpha \otimes w, \alpha \otimes w) = -\kappa_2(f_0)(\alpha \otimes w, \alpha \otimes w).$$

From Lemma 8.21, we immediately obtain the result given by Proposition 8.5.

If x is a point of the Grassmannian X and if $\alpha \in V_x^*$ and $w \in W_x$ are unit vectors, by (8.39) we see that the formulas (8.95) are equivalent to

(8.96)
$$(\pi_1 \mathrm{Hess}\, \tilde{f}_0)(\alpha \otimes w, \alpha \otimes w) = \frac{2}{n}\, \tilde{f}_0(x) + Q(w),$$

$$(\pi_2 \mathrm{Hess}\, \tilde{f}_0)(\alpha \otimes w, \alpha \otimes w) = \frac{2}{m}\, \tilde{f}_0(x) - Q(\alpha^\sharp).$$

Let F_1 be a complex subspace of $\mathbb{C}^{m+n}$ of dimension $n+1$; we denote by $F_1^\perp$ the orthogonal complement of F_1 in $\mathbb{C}^{m+n}$. We consider the complex projective space $Y = G_1^{\mathbb{C}}(F_1)$ and the totally geodesic imbedding

$$\iota_1 : Y \to G_{m,n}^{\mathbb{C}}$$

of §7, which sends $\pi(u)$, with $u \in S(F_1)$, into the m-plane of $\mathbb{C}^{m+n}$ generated by u and the subspace $F_1^\perp$.

Let u be a given element of $S(F_1)$; we consider the points $y = \pi(u)$ of Y and $x = \iota_1(y)$ of X. Then we know that u belongs to V_x. Let α be the unit vector of V_x^* determined by the relation $\alpha^\sharp = u$. In §7, we saw that the equality (8.79) holds.

Let w be an arbitrary unit vector of W_x and let h be an element of $(S^2 T^*)_x^+$. According to the second equality of (8.9) and the equality (8.79), we have

(8.97)
$$(\pi_2 h)(\alpha \otimes w, \alpha \otimes w) = \frac{1}{2n}\left(\mathrm{Tr}_Y\, \iota_1^* h - \frac{1}{m}\, \mathrm{Tr}\, h \right).$$

If f is a real-valued function on X, then by (1.71) we have

$$\mathrm{Tr}_Y\, \iota_1^* \pi_+ \mathrm{Hess}\, f = \mathrm{Tr}_Y\, \iota_1^* \mathrm{Hess}\, f = -\Delta_Y \iota_1^* f;$$

thus if h is the section $\pi_+ \mathrm{Hess}\, f$ of $(S^2 T^*)^+$, by (1.71) and (8.97) we see that

(8.98)
$$(\pi_2 h)(\alpha \otimes w, \alpha \otimes w) = \frac{1}{2mn}\, (\Delta f)(x) - \frac{1}{2n}\, (\Delta_Y \iota_1^* f)(y).$$

Let F_2 be a complex subspace of $\mathbb{C}^{m+n}$ of dimension $m+1$. As in §7, we consider the Grassmannians $Z = G_m^{\mathbb{C}}(F_2)$ and $Z' = G_1^{\mathbb{C}}(F_2)$, the isometry $\Psi : Z' \to Z$ defined in §2 and the totally geodesic imbedding

$$\iota_2 : Z \to G_{m,n}^{\mathbb{C}},$$

sending an m-plane of F_2 into the m-plane of $\mathbb{C}^{m+n}$ which it determines.

Let z be a point of Z and set $x = \iota_2(z)$. Let w be a unit vector of F_2 orthogonal to V_x. In §7, we saw that the equality (8.87) holds. Let α be an arbitrary unit vector of V_x^* and let h be an element of $(S^2T^*)_x^+$. According to the first equality of (8.9) and the equality (8.87), we have

$$(8.99) \qquad (\pi_1 h)(\alpha \otimes w, \alpha \otimes w) = \frac{1}{2m}\left(\mathrm{Tr}_Z \iota_2^* h - \frac{1}{n}\mathrm{Tr}\, h\right).$$

If f is a real-valued function on X, then by (1.71) we have

$$\mathrm{Tr}_Z \iota_2^* \pi_+ \mathrm{Hess}\, f = \mathrm{Tr}_Z \iota_2^* \mathrm{Hess}\, f = -\Delta_Z \iota_2^* f;$$

thus if h is the section $\pi_+ \mathrm{Hess}\, f$ of $(S^2T^*)^+$, by (1.71) and (8.99) we see that

$$(8.100) \qquad (\pi_1 h)(\alpha \otimes w, \alpha \otimes w) = \frac{1}{2mn}(\Delta f)(x) - \frac{1}{2m}(\Delta_Z \iota_2^* f)(z).$$

Let x be a point of the Grassmannian X and let $\alpha \in V_x^*$ and $w \in W_x$ be given unit vectors. Let v be the vector of V_x equal to $\alpha^\sharp$. Let $\{v_1, \ldots, v_m\}$ and $\{w_1, \ldots, w_n\}$ be orthonormal bases of the spaces V_x and W_x, respectively. We proceed to verify the equalities (8.96).

We now suppose that F_1 is the subspace $W_x \oplus \mathbb{C}v$ of $\mathbb{C}^{m+n}$; the orthogonal complement $F_1^\perp$ of F_1 is equal to the orthogonal complement of $\mathbb{C}v$ in V_x. We easily see that $x = \iota_1(\pi(v))$. We consider the function $\tilde{f}_1$ on Y induced by the $U(1)$-invariant function f_1 on $S(F_1)$. Since $\tilde{f}_0$ is an eigenfunction of the Laplacian Δ with eigenvalue $4(m+n)$, according to formulas (8.78) and (8.98), with $u = v$, we have

$$(8.101) \qquad (\pi_2 h_0)(\alpha \otimes w, \alpha \otimes w) = \frac{2(m+n)}{mn}\tilde{f}_0(x) - \frac{1}{2n}(\Delta_Y \tilde{f}_1)(\pi(v)).$$

For $u \in S(F_1)$, we write

$$u = \zeta_0 v + \zeta_1 w_1 + \cdots + \zeta_n w_n,$$

where $(\zeta_0, \ldots, \zeta_n) \in \mathbb{C}^{n+1}$; we view $(\zeta_0, \zeta_1, \ldots, \zeta_n)$ as the homogeneous coordinates of the point $\pi(u)$ of Y. Since the homogeneous coordinates of the point $\pi(v)$ are equal to $(1, 0, \ldots, 0)$, by (3.24) we see that

$$(\Delta_Y \tilde{f}_1)(\pi(v)) = 4nQ(v) - 4\sum_{j=1}^{n} Q(w_j).$$

From the preceding equality and the relations (8.37) and (8.101), we obtain the second formula of (8.96).

We now suppose that F_2 is the subspace $V_x \oplus \mathbb{C}w$ of $\mathbb{C}^{m+n}$. If z is the point of Z corresponding to the m-plane V_x of F_2, we see that $\iota_2(z) = x$ and $\Psi(\pi(w)) = z$. We consider the function $\tilde{f}_1$ on Z' induced by the $U(1)$-invariant function f_1 on $S(F_2)$. By (8.85) we have

$$(\Delta_Z \iota_2^* \tilde{f}_0)(z) = -(\Delta_{Z'} \tilde{f}_1)(\pi(w)).$$

Since $\tilde{f}_0$ is an eigenfunction of the Laplacian Δ with eigenvalue $4(m+n)$, from (8.100) and the previous equality we obtain the relation

$$(8.102) \quad (\pi_1 h_0)(\alpha \otimes w, \alpha \otimes w) = \frac{2(m+n)}{mn} \tilde{f}_0(x) + \frac{1}{2m}(\Delta_{Z'}\tilde{f}_1)(\pi(w)).$$

For $u \in S(F_2)$, we write

$$u = \zeta_0 w + \zeta_1 v_1 + \cdots + \zeta_m v_m,$$

where $(\zeta_0, \zeta_1, \ldots, \zeta_m) \in \mathbb{C}^{m+1}$; we view $(\zeta_0, \zeta_1, \ldots, \zeta_m)$ as the homogeneous coordinates of the point $\pi(u)$ of Z'. Since the homogeneous coordinates of the point $\pi(w)$ of Z' are equal to $(1, 0, \ldots, 0)$, by (3.24) we see that

$$(\Delta_{Z'}\tilde{f}_2)(\pi(w)) = 4mQ(w) - 4\sum_{k=1}^m Q(v_k).$$

From the preceding equality and the relation (8.102), we obtain the first formula of (8.96). This completes the proof of Lemma 8.21.

§9. The complex Grassmannians $\bar{G}_{n,n}^{\mathbb{C}}$

Let $n \geq 2$ be a given integer. Let F be a complex vector space of dimension $2n$ endowed with a positive definite Hermitian scalar product. We consider the complex Grassmannian $X = G_n^{\mathbb{C}}(F)$ endowed with the Hermitian metric induced by the Hermitian scalar product of F. Let U be a complex subspace of F of dimension $2n-2$. The orthogonal complement $U^\perp$ of U in F is two-dimensional. We consider the complex Grassmannians $Y = G_{n-1}^{\mathbb{C}}(U)$ and $Z = G_1^{\mathbb{C}}(U^\perp)$. Since $\dim U^\perp = 2$, the manifold Z is isometric to $\mathbb{CP}^1$. There is a totally geodesic imbedding

$$\iota : Z \times Y \to X,$$

given by (8.69), sending the pair (z, y), where $z \in Z$ and $y \in Y$, into the point of X corresponding to the n-plane of F generated by the subspaces of $U^\perp$ and U corresponding to the points z and y, respectively. For $z \in Z$, let $\iota_z : Y \to X$ be the mapping sending $y \in Y$ into $\iota(z, y)$.

We consider the involutive isometries Ψ_X, Ψ_Y and Ψ_Z of X, Y and Z, respectively. Clearly, the diagram

$$
\begin{array}{ccc}
G_1^{\mathbb{C}}(U^\perp) \times G_{n-1}^{\mathbb{C}}(U) & \xrightarrow{\ \iota\ } & G_n^{\mathbb{C}}(F) \\
\Big\downarrow{\scriptstyle \Psi_Z \times \Psi_Y} & & \Big\downarrow{\scriptstyle \Psi_X} \\
G_1^{\mathbb{C}}(U^\perp) \times G_{n-1}^{\mathbb{C}}(U) & \xrightarrow{\ \iota\ } & G_n^{\mathbb{C}}(F)
\end{array}
$$

is commutative. For $z \in Z$, the commutativity of this diagram gives us the relation

(8.103) $$\Psi_X \circ \iota_z = \iota_{\Psi_Z(z)} \circ \Psi_Y,$$

as mappings from Y to X.

If γ is a closed geodesic of $Z = G_1^{\mathbb{C}}(U^\perp)$, a symmetric p-form θ on X determines a symmetric p-form $\theta_{U,\gamma}$ on Y as follows. If $\xi_1, \ldots, \xi_p \in T_Y$, we consider the real-valued function f on Z defined by

$$f(z) = (\iota_z^* \theta)(\xi_1, \ldots, \xi_p),$$

for $z \in Z$, and we define $\theta_{U,\gamma}$ by setting

$$\theta_{U,\gamma}(\xi_1, \ldots, \xi_p) = \frac{1}{2\pi} \int_\gamma (f + \Psi_Z^* f) = \frac{1}{2\pi}\left(\int_\gamma f + \int_{\gamma'} f \right),$$

where γ' is the closed geodesic $\Psi_Z \circ \gamma$ of Z.

If h is a section of the sub-bundle E of $S^2 T^*$ over X, in §6 we saw that, for $z \in Z$, the symmetric 2-form $\iota_z^* h$ is a section of the sub-bundle E_Y of $S^2 T_Y^*$; therefore if γ is a closed geodesic of Z, the symmetric 2-form $h_{U,\gamma}$ on Y is also a section of E_Y.

Let θ be a symmetric p-form on X. Suppose that there exists a symmetric p-form $\tilde{\theta}$ on Y such that

$$\iota_z^* \theta = \tilde{\theta},$$

for all $z \in Z$; since the length of an arbitrary closed geodesic of Z is equal to π, we see that

$$\theta_{U,\gamma} = \tilde{\theta},$$

for all closed geodesics γ of Z.

We consider the symmetric p-form $\theta' = \Psi_X^* \theta$ on X; if $\xi_1, \ldots, \xi_p$ are vectors of T_Y, we also consider the real-valued functions f_1 and f_2 on Z defined by

$$f_1(z) = (\iota_z^* \theta')(\xi_1, \ldots, \xi_p), \qquad f_2(z) = (\iota_z^* \theta)(\Psi_{Y*}\xi_1, \ldots, \Psi_{Y*}\xi_p),$$

for $z \in Z$. From (8.103), it follows that

$$f_1 = \Psi^*_Z f_2.$$

Since Ψ_Z is an involution, this relation implies that

$$\theta'_{U,\gamma}(\xi_1, \ldots, \xi_p) = \frac{1}{2\pi} \int_\gamma (f_1 + \Psi^*_Z f_1) = \frac{1}{2\pi} \int_\gamma (f_2 + \Psi^*_Z f_2)$$

$$= \theta_{U,\gamma}(\Psi_{Y*}\xi_1, \ldots, \Psi_{Y*}\xi_p),$$

for all closed geodesics γ of Z. Thus we have shown that

(8.104) $$(\Psi^*_X \theta)_{U,\gamma} = \Psi^*_Y (\theta_{U,\gamma}),$$

for all closed geodesics γ of Z. The following lemma is a direct consequence of the preceding equality.

LEMMA 8.22. *Let $n \geq 2$ be a given integer. Let F be a complex vector space of dimension $2n$ endowed with a positive definite Hermitian scalar product. Let U be a complex subspace of F of dimension $2n - 2$ and let $U^\perp$ be the orthogonal complement of U in F; let γ be a closed geodesic of $G^{\mathbb{C}}_1(U^\perp)$. If θ is an even symmetric p-form on $G^{\mathbb{C}}_n(F)$, then the symmetric p-form $\theta_{U,\gamma}$ on $G^{\mathbb{C}}_{n-1}(U)$ is even.*

Let Y' be a maximal flat totally geodesic torus of Y and let γ be a closed geodesic of Z. We consider the images Z' of the closed geodesic γ and Z'' of the closed geodesic $\Psi_Z \circ \gamma$. Then we see that $X' = \iota(Z' \times Y')$ and $X'' = \iota(Z'' \times Y')$ are maximal flat tori of X. Let $\xi_1, \ldots, \xi_p$ be arbitrary parallel vector fields on Y'; they induce parallel vector fields $\xi'_1, \ldots, \xi'_p$ on X' and $\xi''_1, \ldots, \xi''_p$ on X'' determined by

$$\xi'_j(\iota(z', y)) = \iota_{z'*}\xi_j(y), \qquad \xi''_j(\iota(z'', y)) = \iota_{z''*}\xi_j(y),$$

for $z' \in Z'$, $z'' \in Z''$, $y \in Y'$ and $1 \leq j \leq p$. Then from the definition of $\theta_{U,\gamma}$ and Fubini's theorem, we infer that

$$\int_{Y'} \theta_{U,\gamma}(\xi_1, \ldots, \xi_p) \, dY'$$

$$= \frac{1}{2\pi} \left(\int_{X'} \theta(\xi'_1, \ldots, \xi'_p) \, dX' + \int_{X''} \theta(\xi''_1, \ldots, \xi''_p) \, dX'' \right).$$

From this equality, we deduce the following result:

LEMMA 8.23. *Let $n \geq 2$ be a given integer. Let F be a complex vector space of dimension $2n$ endowed with a positive definite Hermitian scalar product. Let U be a complex subspace of F of dimension $2n - 2$ and let $U^{\perp}$ be the orthogonal complement of U in F; let γ be a closed geodesic of $G_1^{\mathbb{C}}(U^{\perp})$. If θ is a symmetric p-form on $G_n^{\mathbb{C}}(F)$ satisfying the Guillemin condition, then the symmetric p-form $\theta_{U,\gamma}$ on $G_{n-1}^{\mathbb{C}}(U)$ satisfies the Guillemin condition.*

We now suppose that F is equal to the complex space $\mathbb{C}^{2n}$ endowed with the standard Hermitian scalar product. Then X is equal to the Grassmannian $G_{n,n}^{\mathbb{C}}$. We view X as a homogeneous space of the group $G = SU(2n)$. We consider the even symmetric 2-form

$$h_1^{(n)} = (\kappa_1 + \kappa_2)(f_0)$$

on $X = G_{n,n}^{\mathbb{C}}$.

We also suppose that U is the complex subspace of $\mathbb{C}^{2n-2}$ generated by the vectors $\{e_1, e_4, \ldots, e_{2n}\}$. The orthogonal complement $U^{\perp}$ of U in $\mathbb{C}^{2n+2}$ is the subspace generated by e_2 and e_3. We consider the isomorphism $\varphi : \mathbb{C}^{2n-2} \to U$ which sends the vector $(\zeta_1, \ldots, \zeta_{2n-2})$ of $\mathbb{C}^{2n-2}$ into the vector

$$\zeta_1 e_1 + \sum_{j=2}^{2n-2} \zeta_j e_{j+2}$$

of U. If Q' is the polynomial on $\mathbb{C}^{2n-2}$ determined by

$$Q'(\zeta) = \zeta_{2n-2}\bar{\zeta}_1,$$

for $\zeta = (\zeta_1, \ldots, \zeta_{2n-2}) \in \mathbb{C}^{2n-2}$, then we have

(8.105) $$\varphi^* Q = Q'.$$

The isomorphism φ allows us to identify the Grassmannians $Y = G_{n-1}^{\mathbb{C}}(F)$ and $G_{n-1,n-1}^{\mathbb{C}}$ and to view the symmetric 2-form $h_1^{(n-1)}$ as a symmetric 2-form on Y. We now proceed to prove that

(8.106) $$\iota_z^* h_1^{(n)} = h_1^{(n-1)},$$

for all $z \in Z$. Indeed, let y be a point of Y and z be a point of Z; we write $x = \iota(z, y)$; if $\alpha \in V_{Y,y}$ and $w \in W_{Y,y}$ are unit vectors, by (8.52), (8.72), (8.73) and (8.105) we have

$$\left(\iota_z^* h_1^{(n)}\right)(\alpha \otimes w, \alpha \otimes w) = 2(Q((\iota_z \alpha)^{\sharp}) + Q(\iota_z w))$$
$$= 2(Q(\alpha^{\sharp}) + Q(w))$$
$$= h_1^{(n-1)}(\alpha \otimes w, \alpha \otimes w).$$

The relation (8.106) is an immediate consequence of these equalities.

If γ is a closed geodesic of Z and h is the symmetric 2-form $h_1^{(n)}$ on X, according to (8.106) and the remark preceding Lemma 8.22 we see that

$$(8.107) \qquad\qquad h_{U,\gamma} = h_1^{(n-1)}.$$

LEMMA 8.24. *Let $n \geq 2$ be a given integer. The symmetric 2-form $h_1^{(n)}$ on $G_{n,n}^{\mathbb{C}}$ does not satisfy the Guillemin condition.*

PROOF: Suppose that the symmetric 2-form $h_1^{(n)}$ on $G_{n,n}^{\mathbb{C}}$ satisfies the Guillemin condition. If $n \geq 3$, according to the relation (8.107) and Lemma 8.23 the symmetric 2-form $h_1^{(n-1)}$ on $G_{n-1,n-1}^{\mathbb{C}}$ also satisfies the Guillemin condition. Our assumption therefore implies that the symmetric 2-form $h_1^{(2)}$ on $G_{2,2}^{\mathbb{C}}$ satisfies the Guillemin condition. The equalities (8.60) then tell us that $h_1^{(2)}$ vanishes. Thus our assumption leads to a contradiction.

PROPOSITION 8.25. *Let X be the complex Grassmannian $G_{n,n}^{\mathbb{C}}$, with $n \geq 2$, and let h be an element of $C_{\gamma_1}^{\infty}(S^2 T_{\mathbb{C}}^*)^{\mathrm{ev}}$. If h satisfies the Guillemin condition, then h vanishes.*

PROOF: Assume that the G-module $M = \mathcal{N}_{2,\mathbb{C}} \cap C_{\gamma_1}^{\infty}(S^2 T^*)^{\mathrm{ev}}$ is non-zero. Since the 2-form $h_1^{(n)}$ is a highest weight vector of the irreducible G-module $C_{\gamma_1}^{\infty}(S^2 T^*)^{\mathrm{ev}}$, a highest weight vector of the G-module M is a non-zero of the multiple of $h_1^{(n)}$. This implies that $h_1^{(n)}$ satisfies the Guillemin condition. Now Lemma 8.24 leads to a contradiction.

From Propositions 8.6 and 8.25, we now deduce that the equality (8.76) of Proposition 8.18 holds when $m = n \geq 2$. This last fact, together with the remarks appearing at the end of §7, which explain the consequences of Proposition 8.20, completes the proof of Proposition 8.18. According to Lemma 2.11, this result also gives us a proof of the equality (8.77) when $m = n$, which we had already derived from Proposition 8.20.

CHAPTER IX

THE RIGIDITY OF THE COMPLEX GRASSMANNIANS

§1. The rigidity of the complex Grassmannians

Let $m, n \geq 2$ be given integers. Let X be the complex Grassmannian $G^{\mathbb{C}}_{m,n}$ endowed with its Kähler metric g. As in §2, Chapter VIII, we view this Grassmannian as an irreducible symmetric space and as a homogeneous space of the group $G = SU(m + n)$, and we identify the tangent bundle T of X with the complex vector bundle $V^* \otimes_{\mathbb{C}} W$.

Let x be a point of X. Let $\mathcal{F}^1_x$ be the family of all closed connected totally geodesic surfaces of X passing through x of the form $\mathrm{Exp}_x F$, where F is generated (over $\mathbb{R}$) by the vectors $\{\alpha_1 \otimes w_1, \alpha_2 \otimes w_2\}$, where $\{\alpha_1, \alpha_2\}$ is an orthonormal set of elements belonging to a totally real subspace of V^*_x and $\{w_1, w_2\}$ is an orthonormal set of elements belonging to a totally real subspace of W_x.

Let $\mathcal{F}^2_x$ be the family of all closed connected totally geodesic surfaces of X passing through x of the form $\mathrm{Exp}_x F$, where F is a totally real subspace of the tangent space T_x satisfying one of the following two conditions:

(i) $F = V_1 \otimes w$, where V_1 is a totally real two-dimensional subspace of V^*_x and w is a unit vector of W_x;

(ii) $F = \alpha \otimes W_1$, where α is a unit vector of V^*_x and W_1 is a totally real two-dimensional subspace of W_x.

Let $\mathcal{F}^3_x$ be the family of all closed connected totally geodesic surfaces of X passing through x of the form $\mathrm{Exp}_x F$, where F is the complex subspace of T_x determined by a unit vector α of V^*_x and a unit vector w of W_x and generated by the vectors $\{\alpha \otimes w, J(\alpha \otimes w)\}$.

According to the expression for the curvature of the complex Grassmannian $G^{\mathbb{C}}_{m,n}$ given in §2, Chapter VIII, we know that a surface of $\mathcal{F}^1_x$ is a totally geodesic flat 2-torus; on the other hand, by Lemmas 8.13 and 8.15 a surface of $\mathcal{F}^2_x$ is isometric to a 2-sphere of constant curvature 1, while a surface of $\mathcal{F}^3_x$ is isometric to a 2-sphere of constant curvature 4.

Let $\mathcal{F}^4_x$ be the family of all closed connected totally geodesic submanifolds of X passing through x of the form $\mathrm{Exp}_x F$, where F is a totally real subspace of T_x equal to $V_1 \otimes_{\mathbb{R}} W_1$, where V_1 and W_1 are totally real subspaces of V^*_x and W_x of dimension 2 and n, respectively. Let $\mathcal{F}^5_x$ be the family of all closed connected totally geodesic submanifolds of X passing through x of the form $\mathrm{Exp}_x F$, where F is the complex subspace of T_x determined by a unit vector α of V^*_x and equal to $\alpha \otimes W_x$.

Clearly, each surface of $\mathcal{F}^1_x$ or of $\mathcal{F}^2_x$ is contained in a totally geodesic submanifold of X belonging to the family $\mathcal{F}^4_x$, while a surface of $\mathcal{F}^3_x$ is contained in a totally geodesic submanifold of X belonging to the family $\mathcal{F}^5_x$.

By Lemmas 8.13 and 8.15, we know that a submanifold of X belonging to the family $\mathcal{F}_x^4$ is isometric to the real Grassmannian $G_{2,n}^{\mathbb{R}}$ and that a submanifold of X belonging to the family $\mathcal{F}_x^5$ is isometric to the complex projective space $\mathbb{CP}^n$ of dimension n endowed with its Fubini-Study metric of constant holomorphic curvature 4.

For $1 \leq j \leq 4$, we consider the G-invariant family

$$\mathcal{F}^j = \bigcup_{x \in X} \mathcal{F}_x^j$$

of closed connected totally geodesic submanifolds of X; we write $\mathcal{F}'' = \mathcal{F}^4$ and also consider the G-invariant families

$$\mathcal{F} = \mathcal{F}^1 \cup \mathcal{F}^2 \cup \mathcal{F}^3, \qquad \tilde{\mathcal{F}} = \mathcal{F}^2 \cup \mathcal{F}^3, \qquad \mathcal{F}' = \mathcal{F}^4 \cup \mathcal{F}^5$$

of closed connected totally geodesic submanifolds of X. Clearly, every surface of $\mathcal{F}$ is contained in a totally geodesic submanifold of X belonging to the family $\mathcal{F}'$; moreover, we have $\tilde{\mathcal{F}} \subset \mathcal{F}$ and $\mathcal{F}'' \subset \mathcal{F}'$.

If Z is a submanifold of X belonging to the family $\mathcal{F}^j$, with $1 \leq j \leq 4$, there is a subgroup of G which acts transitively on Z. Thus for $1 \leq j \leq 4$, we see that an element u of $\bigotimes^q T_x^*$, with $x \in X$, vanishes when restricted to an arbitrary submanifold belonging to the family $\mathcal{F}^j$ if and only if it vanishes when restricted to an arbitrary submanifold belonging to the family $\mathcal{F}_x^j$.

LEMMA 9.1. *Let $m, n \geq 2$ be given integers and let X be the complex Grassmannian $G_{m,n}^{\mathbb{C}}$. Let u be an element of $\bigotimes^2 T_x^*$, with $x \in X$. If the restriction of u to an arbitrary submanifold of the family $\mathcal{F}$ vanishes, then u vanishes.*

PROOF: Let $\{\alpha_1, \ldots, \alpha_m\}$ (resp. $\{w_1, \ldots, w_n\}$) be an orthonormal basis for a totally real subspace of V_x^* (resp. of W_x) of dimension m (resp. of dimension n). Let $1 \leq i, j \leq m$ and $1 \leq k, l \leq n$ be given integers. The vectors $\{\alpha_i \otimes w_k, \alpha_j \otimes w_l\}$ (resp. the vectors $\{J(\alpha_i \otimes w_k), J(\alpha_j \otimes w_l)\}$) are tangent to a surface of X belonging to the family $\mathcal{F}^1 \cup \mathcal{F}^2$. The vectors $\{\alpha_i \otimes w_k, J(\alpha_j \otimes w_l)\}$ are also tangent to a surface of X belonging to the family $\mathcal{F}^1 \cup \mathcal{F}^2$ unless $i = j$ and $k = l$. Suppose that the restriction of u to an arbitrary submanifold of the family $\mathcal{F}$ vanishes; then we see that

$$u(\alpha_i \otimes w_k, \alpha_j \otimes w_l) = u(J(\alpha_i \otimes w_k), J(\alpha_j \otimes w_l)) = 0;$$

moreover if $i \neq j$ or $k \neq l$, we also have

$$u(\alpha_i \otimes w_k, J(\alpha_j \otimes w_l)) = 0.$$

On the other hand, the vectors $\{\alpha_i \otimes w_k, \alpha_i \otimes J w_k\}$ are tangent to a submanifold belonging to the family $\mathcal{F}^3$, and so we see that

$$u(\alpha_i \otimes w_k, J(\alpha_i \otimes w_k)) = 0.$$

It follows that u vanishes.

LEMMA 9.2. *Let $m, n \geq 2$ be given integers and let X be the complex Grassmannian $G^{\mathbb{C}}_{m,n}$. Let u be an element of $\bigwedge^2 T_x^*$, with $x \in X$. If the restriction of u to an arbitrary submanifold of the family $\tilde{\mathcal{F}}$ vanishes, then u vanishes.*

PROOF: Assume that the restriction of u to an arbitrary submanifold of the family $\tilde{\mathcal{F}}$ vanishes. In view of Lemma 9.1, it suffices to prove that the restriction of u to an arbitrary submanifold of the family $\mathcal{F}^1$ vanishes. In fact, let $\{\alpha_1, \alpha_2\}$ be an orthonormal set of elements belonging to a totally real subspace of V_x^* and $\{w_1, w_2\}$ be an orthonormal set of elements belonging to a totally real subspace of W_x. Our assumption on u tells us that

$$u(\alpha \otimes w_1, \alpha \otimes w_2) = 0, \qquad u(\alpha_1 \otimes w, \alpha_2 \otimes w) = 0,$$

with $\alpha = \alpha_1$, α_2 or $\alpha_1 + \alpha_2$ and $w = w_1$, w_2 or $w_1 + w_2$. The first set of relations implies that

$$u(\alpha_1 \otimes w_1, \alpha_2 \otimes w_2) + u(\alpha_2 \otimes w_1, \alpha_1 \otimes w_2) = 0,$$

while the second set tells us that

$$u(\alpha_1 \otimes w_1, \alpha_2 \otimes w_2) + u(\alpha_1 \otimes w_2, \alpha_2 \otimes w_1) = 0.$$

Since u belongs to $\bigwedge^2 T_x^*$, these two equalities imply that

$$u(\alpha_1 \otimes w_1, \alpha_2 \otimes w_2) = 0.$$

If $\hat{\mathcal{F}}$ is a G-invariant family of closed connected totally geodesic surfaces of X, we consider the sub-bundle $N_{\hat{\mathcal{F}}}$ of B consisting of those elements of B, which vanish when restricted to the closed totally geodesic submanifolds of $\hat{\mathcal{F}}$, which was introduced in §8, Chapter II. We write $N = N_{\mathcal{F}}$.

LEMMA 9.3. *Let $m, n \geq 2$ be given integers and let X be the complex Grassmannian $G^{\mathbb{C}}_{m,n}$. Then we have*

$$\operatorname{Tr} N \subset E.$$

PROOF: Let x be a point of X. Let α be an arbitrary unit vector of V_x^* and w be an arbitrary unit vector of W_x. Let $\{\alpha_1, \ldots, \alpha_m\}$

(resp. $\{w_1, \ldots, w_n\}$) be an orthonormal basis for a totally real subspace of V_x^* of dimension m (resp. of dimension n) containing α (resp. w), with $\alpha_1 = \alpha$ (resp. $w_1 = w$). As we have seen in the proof of Lemma 9.1, for $1 \le j \le m$ and $1 \le k \le n$, the two vectors $\alpha \otimes w$ and $\alpha_j \otimes w_k$ (resp. $\alpha \otimes w$ and $J(\alpha_j \otimes w_k)$) are tangent to a surface belonging to the family $\mathcal{F}$. Therefore if u is an element of N_x, we see that

$$(\operatorname{Tr} u)(\alpha \otimes w, \alpha \otimes w)$$
$$= \sum_{\substack{1 \le j \le m \\ 1 \le k \le n}} \{u(\alpha \otimes w, \alpha_j \otimes w_k, \alpha \otimes w, \alpha_j \otimes w_k)$$
$$+ u(\alpha \otimes w, J(\alpha_j \otimes w_k), \alpha \otimes w, J(\alpha_j \otimes w_k))\}$$
$$= 0.$$

Hence $\operatorname{Tr} N_x$ is a subspace of E_x.

PROPOSITION 9.4. *Let h be a section of E over the complex Grassmannian $X = G_{m,n}^{\mathbb{C}}$, with $m \ge 2$ and $n \ge 3$. If the restriction of h to an arbitrary submanifold Z of X belonging to the family $\mathcal{F}''$ is a Lie derivative of the metric of Z, then h vanishes.*

PROOF: Assume that h belongs to $\mathcal{L}(\mathcal{F}'')$. We consider the real Grassmannian $Z = G_{2,n}^{\mathbb{R}}$. Let X' be a submanifold of X belonging to the family $\mathcal{F}^4$. According to Lemma 8.13, there is a totally geodesic isometric imbedding $i : Z \to X$ whose image is equal to X' such that i^*h is a section of the sub-bundle E_Z of $S^2 T_Z^*$. Our hypotheses imply that the symmetric 2-form i^*h on Z satisfies the Guillemin condition. By Theorem 6.48, we infer that $i^*h = 0$ and hence that the restriction of h to X' vanishes. According to Lemma 8.15, the restriction of h to a submanifold of X belonging to the family $\mathcal{F}^5$ vanishes. Thus the restriction of h to an arbitrary submanifold of X belonging to the family $\mathcal{F}'$ or to the family $\mathcal{F}$ vanishes. The desired result is now a consequence of Lemma 9.1.

We now suppose that $n \ge 3$. The Guillemin rigidity of the real Grassmannian $G_{2,n}^{\mathbb{R}}$, given by Theorem 6.47, and the infinitesimal rigidity of the complex projective space of dimension n, given by Theorem 3.39, tell us that the family $\mathcal{F}'$ possesses properties (II) and (III) of §8, Chapter II. When $m < n$, according to Propositions 8.14 and 8.17, the family $\mathcal{F}'$ also possesses property (I) of §8, Chapter II. By (8.12), we know that E is a G-invariant sub-bundle of $S_0^2 T^*$. According to Lemma 9.3 and Proposition 9.4, the families $\mathcal{F}$ and $\mathcal{F}'$ and the vector bundle E satisfy the relations (2.33) and (2.37) of Theorem 2.48. Proposition 8.18 tells us that the equalities (2.35) and (2.36) hold. Thus the families $\mathcal{F}$ and $\mathcal{F}'$ and the vector bundle E satisfy the hypotheses of Theorem 2.49,(ii) and, when $m < n$, they also satisfy the hypotheses of Theorem 2.49,(i). Hence from Theorem 2.49, we deduce the infinitesimal rigidity of $G_{m,n}^{\mathbb{C}}$ and, when $m < n$, the Guillemin rigidity of $G_{m,n}^{\mathbb{C}}$. On the other hand, according to Theorem 8.7 the space $G_{2,2}^{\mathbb{C}}$ is infinitesimally rigid. Since the Grassmannian $G_{p,q}^{\mathbb{C}}$ is isometric to $G_{q,p}^{\mathbb{C}}$, we have therefore proved the following two theorems:

THEOREM 9.5. *The complex Grassmannian $G_{m,n}^{\mathbb{C}}$, with $m, n \geq 2$ and $m \neq n$, is rigid in the sense of Guillemin.*

THEOREM 9.6. *The complex Grassmannian $G_{m,n}^{\mathbb{C}}$, with $m, n \geq 2$, is infinitesimally rigid.*

Let $\check{\mathcal{F}}$ be the family of all closed totally geodesic submanifolds of X which can be written in the form $\mathrm{Exp}_x V' \otimes W'$, where $x \in X$ and where V' is a totally real subspace of V_x^* of dimension m and W' is a totally real subspace of W_x of dimension n. According to Lemma 8.13, a member of the family $\check{\mathcal{F}}$ is isometric to the real Grassmannian $X = G_{m,n}^{\mathbb{R}}$.

PROPOSITION 9.7. *Let $m, n \geq 2$ be given integers. If X is the complex Grassmannian $G_{m,n}^{\mathbb{C}}$, the sequence (1.24) is exact.*

PROOF: First, suppose that $m + n \geq 5$. Let γ be a closed geodesic of X; then there is a maximal flat totally geodesic torus Z of X containing γ. According to the remarks following Lemma 8.13, Z is contained in a totally geodesic submanifold Y of X belonging to the family $\check{\mathcal{F}}$; let $i : Y \to X$ be the natural inclusion. According to Proposition 7.10 and remarks made in §1, Chapter VII, we know that the sequence (1.24) for Y is exact. By Theorem 9.6, the family $\check{\mathcal{F}}$ satisfies all the hypotheses of Proposition 2.46; thus the desired conclusion is a consequence of this proposition when $m + n \geq 5$. On the other hand, the complex Grassmannian $G_{2,2}^{\mathbb{C}}$ is isometric to the complex quadric Q_4 (see §5, Chapter VIII), and so we know that the sequence (1.24) is exact when $m = n = 2$ (see §6, Chapter V).

According to Lemma 9.1, the sub-bundle $C_{\mathcal{F}'}$ of $\bigwedge^2 T^*$, consisting of those elements of $\bigwedge^2 T^*$ which vanish when restricted to the closed totally geodesic submanifolds of $\mathcal{F}'$, vanishes.

We now suppose that $n \geq 3$. By Theorems 3.40 and 7.12, we know that $\mathcal{F}'$ possesses property (VI) of §8, Chapter II. When $m < n$, Propositions 8.14 and 8.16, with $q = n$, tell us that the family $\mathcal{F}'$ possesses property (IV) of §8, Chapter II; moreover by Theorems 3.40 and 7.11, we know that $\mathcal{F}'$ also possesses property (V) of §8, Chapter II. Since the Grassmannian $G_{2,2}^{\mathbb{C}}$ is isometric to the complex quadric Q_4 (see §5, Chapter VIII), from these observations, Theorem 2.51, and Theorem 6.72, with $n = 4$, we obtain the following two theorems:

THEOREM 9.8. *Let $m, n \geq 2$ be given integers, with $m \neq n$. A form of degree 1 on the complex Grassmannian $G^{\mathbb{C}}_{m,n}$ satisfies the Guillemin condition if and only if it is exact.*

THEOREM 9.9. *Let $m, n \geq 2$ be given integers, with $m, n \geq 2$. A form of degree 1 on the complex Grassmannian $G^{\mathbb{C}}_{m,n}$ satisfies the zero-energy condition if and only if it is exact.*

Theorem 9.9 is given by Theorem 3 of [20].

§2. On the rigidity of the complex Grassmannians $\bar{G}^{\mathbb{C}}_{n,n}$

We consider the complex Grassmannian $X = G^{\mathbb{C}}_{n+1,n+1}$, with $n \geq 1$. We view X as an irreducible symmetric space and as a homogeneous space of the group $G = SU(2n+2)$ and we consider the involutive isometry Ψ_X of X. We consider the G-invariant families $\mathcal{F}$, $\mathcal{F}^j$ and $\tilde{\mathcal{F}}$ of closed connected totally geodesic surfaces of X. We also consider the sub-bundles $N = N_{\mathcal{F}}$ and $N_{\tilde{\mathcal{F}}}$ of B corresponding to the families $\mathcal{F}$ and $\tilde{\mathcal{F}}$, which were introduced in §1, and the associated differential operators

$$D_{1,\mathcal{F}} : S^2 T^* \to B/\mathcal{N}, \qquad D_{1,\tilde{\mathcal{F}}} : S^2 T^* \to B/N_{\tilde{\mathcal{F}}}$$

of §8, Chapter II.

Let U be a complex subspace of $\mathbb{C}^{2n+2}$ of dimension $2n$. The orthogonal complement $U^{\perp}$ of U in $\mathbb{C}^{2n+2}$ is two-dimensional. We consider the complex Grassmannians $Y = G^{\mathbb{C}}_n(U)$ and $Z = G^{\mathbb{C}}_1(U^{\perp})$, the unit sphere $S(U^{\perp})$ of $U^{\perp}$ and the natural projection $\pi : S(U^{\perp}) \to G^{\mathbb{C}}_1(U^{\perp})$, which sends $u \in S(U^{\perp})$ into the line generated by u. Since $\dim U^{\perp} = 2$, the manifold Z is isometric to $\mathbb{CP}^1$. We also consider the involutive isometry Ψ_Z of Z. There is totally geodesic imbedding

$$(9.1) \qquad\qquad \iota : Z \times Y \to X,$$

given by (8.69), sending the pair (z, y), where $z \in Z$ and $y \in Y$, into the point of X corresponding to the n-plane of F generated by the subspaces of $U^{\perp}$ and U corresponding to the points z and y, respectively. For $z \in Z$, let $\iota_z : Y \to X$ be the mapping sending $y \in Y$ into $\iota(z, y)$. If γ is a closed geodesic of Z and θ is a symmetric p-form on X, we consider the symmetric p-form $\theta_{U,\gamma}$ on Y defined in §9, Chapter VIII.

LEMMA 9.10. *Let f be a real-valued function, θ be a 1-form and h be a symmetric 2-form on the complex Grassmannian $X = G^{\mathbb{C}}_{n+1,n+1}$, with $n \geq 2$. Let U be a complex subspace of $\mathbb{C}^{2n+2}$ of dimension $2n$ and let $U^{\perp}$ be the orthogonal complement of U in $\mathbb{C}^{2n+2}$. Let $\iota : Z \times Y \to X$ be the mapping given by (9.1), where $Z = G^{\mathbb{C}}_1(U^{\perp})$ and $Y = G^{\mathbb{C}}_n(U)$, and let $\Psi' = \Psi_Z$ be the involutive isometry of Z.*

(i) *Suppose that, for any closed geodesic γ of Z, the function $f_{U,\gamma}$ on Y vanishes. Then we have*

$$\iota_z^* f + \iota_{\Psi'(z)}^* f = 0,$$

for all $z \in Z$.

(ii) *Suppose that, for any closed geodesic γ of Z, the symmetric 2-form $h_{U,\gamma}$ on Y vanishes. Then we have*

$$\iota_z^* h + \iota_{\Psi'(z)}^* h = 0,$$

for all $z \in Z$.

(iii) *Suppose that, for any closed geodesic γ of Z, the 1-form $\theta_{U,\gamma}$ on Y is closed. Then we have*

$$\iota_z^* d\theta + \iota_{\Psi'(z)}^* d\theta = 0.$$

(iv) *Suppose that, for any closed geodesic γ of Z, the symmetric 2-form $h_{U,\gamma}$ on Y is a Lie derivative of the metric of Y. Let y be a point of Y and let ξ_1, ξ_2 be vectors of $T_{Y,y}$ which are tangent to a totally geodesic surface of Y. Then we have*

$$\left(\iota_z^* D_g h + \iota_{\Psi'(z)}^* D_g h\right)(\xi_1, \xi_2, \xi_2, \xi_2) = 0,$$

for all $z \in Z$.

PROOF: We denote by g_Y the Riemannian metric of the Grassmannian Y. Let y be a point of Y and let ξ_1, ξ_2 be vectors of $T_{Y,y}$. We consider the real-valued functions f_1 and f_2 on Z defined by

$$f_1(z) = (\iota_z^* h)(\xi_1, \xi_2), \quad f_2(z) = (\iota_z^* d\theta)(\xi_1, \xi_2),$$
$$f_3(z) = (\iota_z^* D_g h)(\xi_1, \xi_2, \xi_1, \xi_2),$$

for $z \in Z$. We write $f_0 = f$ and $\tilde{f}_j = f_j + \Psi'^* f_j$, for $0 \le j \le 3$. Then by the relation (1.57), we have

$$f_3(z) = (D_{g_Y} \iota_z^* h)(\xi_1, \xi_2, \xi_1, \xi_2),$$

for $z \in Z$. Let γ be a closed geodesic of Z. We easily verify that the equalities

$$\frac{1}{2\pi} \int_\gamma \tilde{f}_1 = h_{U,\gamma}(\xi_1, \xi_2), \qquad \frac{1}{2\pi} \int_\gamma \tilde{f}_2 = (d\theta_{U,\gamma})(\xi_1, \xi_2),$$

$$\frac{1}{2\pi} \int_\gamma \tilde{f}_3 = (D_{g_Y} h_{U,\gamma})(\xi_1, \xi_2, \xi_1, \xi_2)$$

hold. If the hypothesis of the j-th assertion of the lemma holds, with $j = 1, 2$ or 3, then the even function $\tilde{f}_{j-1}$ on Z satisfies the zero-energy condition. If the symmetric 2-form $h_{U,\gamma}$ is a Lie derivative of the metric g_Y and if ξ_1, ξ_2 are tangent to a totally geodesic surface of Y, which is necessarily of constant curvature, by Lemma 1.1 and the relations (1.49) and (1.57) the expression $(D_{g_Y} h_{U,\gamma})(\xi_1, \xi_2, \xi_1, \xi_2)$ vanishes. Thus if the hypothesis of assertion (iv) holds and if ξ_1, ξ_2 are tangent to a totally geodesic surface of Y, the even function $\tilde{f}_3$ on Z satisfies the zero-energy condition. Proposition 3.29,(i) tells us that, if the function $\tilde{f}_j$, with $0 \le j \le 3$, on Z satisfies the zero-energy condition, then it vanishes identically. This gives us the desired results.

We henceforth suppose that $n \ge 2$. Let x be a point of X and let $\{v_1, \ldots, v_{n+1}\}$ and $\{w_1, \ldots, w_{n+1}\}$ be orthonormal bases of the complex vector spaces V_x and W_x, respectively. For $1 \le j \le n+1$, let V_j be the complex subspace of V_x of dimension n generated by the vectors $\{v_1, \ldots, v_{j-1}, v_{j+1}, \ldots, v_{n+1}\}$, and let W' be the complex subspace of W_x of codimension one generated by the vectors $\{w_1, \ldots, w_n\}$. For $1 \le j \le n+1$, we consider the complex subspaces

$$U_j = V_j \oplus W', \qquad \tilde{U}_j = V_j \oplus \mathbb{C}w_{n+1}$$

of $\mathbb{C}^{2n+2}$; the orthogonal complement $U_j^{\perp}$ of U_j in $\mathbb{C}^{2n+2}$ is generated by the vectors $\{v_j, w_{n+1}\}$. Let U' be the complex subspace of $\mathbb{C}^{2n+2}$ of dimension $2n$ whose orthogonal complement $U'^{\perp}$ in $\mathbb{C}^{2n+2}$ is generated by the vectors $\{v_2, v_{n+1}\}$.

For $1 \le j \le n+1$, we consider the mappings

$$\iota_j : G_1^{\mathbb{C}}(U_j^{\perp}) \times G_n^{\mathbb{C}}(U_j) \to X, \qquad \iota' : G_1^{\mathbb{C}}(U'^{\perp}) \times G_n^{\mathbb{C}}(U') \to X$$

given by (9.1). Let x_j be the point of X corresponding to the $(n+1)$-plane $\tilde{U}_j$, and let y_j be the point of $G_n^{\mathbb{C}}(U_j)$ corresponding to the n-plane V_j and z_j be the point $\pi(v_j)$ of $Z_j = G_1^{\mathbb{C}}(U_j^{\perp})$. The image z_j' of z_j under the involutive isometry Ψ_{Z_j} of Z_j is equal to $\pi(w_{n+1})$. Then we have

$$(9.2) \qquad \iota_j(z_j, y_j) = x, \qquad \iota_j(z_j', y_j) = x_j.$$

Let y' be the point of $G_n^{\mathbb{C}}(U')$ corresponding to the n-plane generated by the vectors $\{v_1, w_{n+1}\}$ when $n = 2$ and by the vectors $\{v_1, v_3, \ldots, v_n, w_{n+1}\}$ when $n \ge 3$, and let z' be the point $\pi(v_2)$ of $Z' = G_1^{\mathbb{C}}(U'^{\perp})$. The image z'' of z' under the involutive isometry $\Psi_{Z'}$ of Z' is equal to $\pi(v_{n+1})$. Then we have

$$(9.3) \qquad \iota'(z', y') = x_{n+1}, \qquad \iota'(z'', y') = x_2.$$

Let f be a real-valued function on X. Suppose that, for any complex subspace U of $\mathbb{C}^{2n+2}$ of dimension $2n$ and any closed geodesic γ of $G_1^{\mathbb{C}}(U^{\perp})$, the function $f_{U,\gamma}$ on $G_n^{\mathbb{C}}(U)$ vanishes. Then by Lemma 9.10,(i) and the relations (9.2) and (9.3), we see that

$$f(x) = -f(x_j), \qquad f(x_{n+1}) = -f(x_2),$$

for $1 \le j \le n+1$. These equalities imply that $f(x) = 0$. Thus we have proved the following:

PROPOSITION 9.11. *Let f be a real-valued function on the complex Grassmannian $X = G_{n+1,n+1}^{\mathbb{C}}$, with $n \ge 2$. Suppose that, for any complex subspace U of $\mathbb{C}^{2n+2}$ of dimension $2n$ and any closed geodesic γ of $G_1^{\mathbb{C}}(U^{\perp})$, where $U^{\perp}$ is the orthogonal complement of U in $\mathbb{C}^{2n+2}$, the function $f_{U,\gamma}$ on $G_n^{\mathbb{C}}(U)$ vanishes. Then the function f vanishes.*

We now return to the discussion which precedes Proposition 9.11. Let w_1' and w_2' be given vectors of W' and consider the vectors α_1 and α_2 of V_x^* determined by

$$\langle v_k, \alpha_1 \rangle = \delta_{k1}, \qquad \langle v_k, \alpha_2 \rangle = \delta_{k2},$$

for $1 \le k \le n+1$. The tangent vectors $\xi = \alpha_1 \otimes w_1'$ and $\eta = \alpha_1 \otimes w_2'$ belonging to T_x satisfy

$$\langle v_k, \xi \rangle = \delta_{k1} w_1', \qquad \langle v_k, \eta \rangle = \delta_{k1} w_2',$$

for $1 \le k \le n+1$. For $1 \le j \le n+1$, since $V_{x_j} = \tilde{U}_j$ and W' is a subspace of W_{x_j}, we may consider the tangent vectors ξ_j and η_j of T_{x_j} determined by

$$\langle v_k, \xi_j \rangle = \delta_{k1} w_1', \quad \langle v_k, \eta_j \rangle = \delta_{k1} w_2', \quad \langle w_{n+1}, \xi_j \rangle = \langle w_{n+1}, \eta_j \rangle = 0,$$

for $1 \le k \le n+1$, with $k \ne j$. For $1 \le j \le n+1$, we consider the point y_j of the Grassmannian $Y_j = G_n^{\mathbb{C}}(U_j)$; since $V_{Y_j,y_j} = V_j$ and $W_{Y_j,y_j} = W'$, we may consider the vectors $\tilde{\xi}_j$ and $\tilde{\eta}_j$ tangent to Y_j at y_j which are determined by

$$\langle v_k, \tilde{\xi}_j \rangle = \delta_{k1} w_1', \qquad \langle v_k, \tilde{\eta}_j \rangle = \delta_{k1} w_2',$$

for $1 \le k \le n+1$, with $k \ne j$. We consider the point y' of the Grassmannian $Y' = G_n^{\mathbb{C}}(U')$; since W' is a subspace of $W_{y'}$, we may consider the vectors ξ' and η' tangent to Y' at the point y' which are determined by

$$\langle v_k, \xi' \rangle = \delta_{k1} w_1', \quad \langle v_k, \eta' \rangle = \delta_{k1} w_2', \quad \langle w_{n+1}, \xi' \rangle = \langle w_{n+1}, \eta' \rangle = 0,$$

for $k = 1$ and $3 \le k \le n$.

If the vectors w_1' and w_2' belong to a totally real subspace of W_x (resp. if the vector w_2' is equal to Jw_1'), then the vectors ξ and η are tangent to a totally geodesic surface of X belonging to the family $\mathcal{F}^2$ (resp. the family $\mathcal{F}^3$); moreover, for $1 \le j \le n+1$, the vectors ξ_j and η_j are tangent to a totally geodesic surface of Y_j, while the vectors ξ' and η' are tangent to a totally geodesic surface of Y'.

For $1 \le j \le n+1$, we consider the mappings

$$\phi_j = \iota_{j,z_j} : G_n^{\mathbb{C}}(U_j) \to X, \qquad \phi_j' = \iota_{j,z_j'} : G_n^{\mathbb{C}}(U_j) \to X,$$

$$\phi' = \iota_{z'}' : G_n^{\mathbb{C}}(U') \to X, \qquad \phi'' = \iota_{z''}' : G_n^{\mathbb{C}}(U') \to X;$$

then we easily see that the relations

$$(9.4) \quad \begin{array}{cccc} \phi_{j*}\tilde{\xi}_j = \xi, & \phi_{j*}\tilde{\eta}_j = \eta, & \phi_{j*}'\tilde{\xi}_j = \xi_j, & \phi_{j*}'\tilde{\eta}_j = \eta_j, \\ \phi_*'\xi' = \xi_{n+1}, & \phi_*'\eta' = \eta_{n+1} & \phi_*''\xi' = \xi_2, & \phi_*''\eta' = \eta_2 \end{array}$$

hold for $2 \le j \le n+1$.

We now suppose that $n \ge 3$. We consider the complex subspace V' of V_x of dimension n generated by the vectors $\{v_1, v_2, v_4, \ldots, v_{n+1}\}$. For $1 \le j \le n+1$, we consider the complex subspace $\tilde{U}_j$ of $\mathbb{C}^{2n+2}$ of dimension $2n$ generated by V' and the vectors $\{w_1, \ldots, w_{j-1}, w_{j+1}, \ldots, w_{n+1}\}$, whose orthogonal complement $\tilde{U}_j^\perp$ in $\mathbb{C}^{2n+2}$ is generated by the vectors $\{v_3, w_j\}$; we also consider the complex subspace $\tilde{U}$ of $\mathbb{C}^{2n+2}$ of dimension $2n$ generated by V_x and the vectors $\{w_1, \ldots, w_{n-1}\}$, whose orthogonal complement $\tilde{U}^\perp$ in $\mathbb{C}^{2n+2}$ is generated by the vectors $\{w_n, w_{n+1}\}$. For $1 \le j \le n+1$, we consider the mappings

$$\tilde{\iota}_j : G_1^{\mathbb{C}}(\tilde{U}_j^\perp) \times G_n^{\mathbb{C}}(\tilde{U}_j) \to X, \qquad \tilde{\iota} : G_1^{\mathbb{C}}(\tilde{U}^\perp) \times G_n^{\mathbb{C}}(\tilde{U}) \to X$$

given by (9.1). Let $\tilde{x}_j$ be the point of X corresponding to the $(n+1)$-plane $V' \oplus \mathbb{C}w_j$; let $\tilde{y}_j$ be the point of $G_n^{\mathbb{C}}(\tilde{U}_j)$ corresponding to the n-plane V' and let $\tilde{z}_j$ be the point $\pi(v_3)$ of $\tilde{Z}_j = G_1^{\mathbb{C}}(\tilde{U}_j^\perp)$. The image $\tilde{z}_j'$ of $\tilde{z}_j$ under the involutive isometry $\Psi_{\tilde{z}_j}$ of $\tilde{Z}_j$ is equal to $\pi(w_j)$. Then we have

$$\tilde{\iota}_j(\tilde{z}_j, \tilde{y}_j) = x, \qquad \tilde{\iota}_j(\tilde{z}_j', \tilde{y}_j) = \tilde{x}_j.$$

Let $\tilde{y}$ be the point of $G_n^{\mathbb{C}}(\tilde{U})$ corresponding to the n-plane V' and let $\tilde{z}$ be the point $\pi(w_n)$ of $\tilde{Z} = G_1^{\mathbb{C}}(\tilde{U}^\perp)$. The image $\tilde{z}'$ of $\tilde{z}$ under the involutive isometry $\Psi_{\tilde{z}}$ of $\tilde{Z}$ is equal to $\pi(w_{n+1})$. Then we have

$$\tilde{\iota}(\tilde{z}, \tilde{y}) = \tilde{x}_n, \qquad \tilde{\iota}(\tilde{z}', \tilde{y}) = \tilde{x}_{n+1}.$$

The tangent vectors $\zeta = \alpha_1 \otimes w_1$ and $\zeta' = \alpha_2 \otimes w_2$ belonging to T_x satisfy

$$\langle v_k, \zeta \rangle = \delta_{k1} w_1, \qquad \langle v_k, \zeta' \rangle = \delta_{k2} w_2,$$

for $1 \leq k \leq n+1$. For $3 \leq j \leq n+1$, since $V_{\tilde{x}_j} = V' \oplus \mathbb{C} w_j$ and since the vectors w_1 and w_2 belong to $W_{\tilde{x}_j}$, we may consider the tangent vectors ζ_j and ζ'_j of $T_{\tilde{x}_j}$ determined by

$$\langle v_k, \zeta_j \rangle = \delta_{k1} w_1, \quad \langle v_k, \zeta'_j \rangle = \delta_{k2} w_2, \quad \langle w_j, \zeta_j \rangle = \langle w_j, \zeta'_j \rangle = 0,$$

for $1 \leq k \leq n+1$, with $k \neq 3$. For $3 \leq j \leq n+1$, we consider the point $\tilde{y}_j$ of the Grassmannian $\tilde{Y}_j = G_n^{\mathbb{C}}(\tilde{U}_j)$; since $V_{\tilde{Y}_j, \tilde{y}_j} = V'$ and since the vectors w_1 and w_2 belong to $W_{\tilde{Y}_j, \tilde{y}_j}$, we may consider the vectors $\tilde{\zeta}_j$ and $\tilde{\zeta}'_j$ tangent to $\tilde{Y}_j$ at $\tilde{y}_j$ which are determined by

$$\langle v_k, \tilde{\zeta}_j \rangle = \delta_{k1} w_1, \qquad \langle v_k, \tilde{\zeta}'_j \rangle = \delta_{k2} w_2,$$

for $1 \leq k \leq n+1$, with $k \neq 3$. We consider the point $\tilde{y}$ of the Grassmannian $\tilde{Y} = G_n^{\mathbb{C}}(\tilde{U})$; since $V_{\tilde{Y}, \tilde{y}} = V'$ and since the vectors w_1 and w_2 belong to $W_{\tilde{Y}, \tilde{y}}$, we may consider the vectors $\tilde{\zeta}$ and $\tilde{\zeta}'$ tangent to $\tilde{Y}$ at the point $\tilde{y}$ which are determined by

$$\langle v_k, \tilde{\zeta} \rangle = \delta_{k1} w_1, \qquad \langle v_k, \tilde{\zeta}' \rangle = \delta_{k1} w_2,$$

for $1 \leq k \leq n+1$, with $k \neq 3$.

Since $\{\alpha_1, \alpha_2\}$ is an orthonormal set of vectors belonging to a totally real subspace of V_x^* and $\{w_1, w_2\}$ is an orthonormal set of vectors belonging to a totally real subspace of W_x, the vectors ζ and ζ' are tangent to a totally geodesic surface of X belonging to the family $\mathcal{F}^1$; moreover, for $3 \leq j \leq n+1$, the vectors $\tilde{\zeta}_j$ and $\tilde{\zeta}'_j$ are tangent to a totally geodesic surface of $\tilde{Y}_j$, while the vectors $\tilde{\zeta}$ and $\tilde{\zeta}'$ are tangent to a totally geodesic surface of $\tilde{Y}$.

For $3 \leq j \leq n+1$, we consider the mappings

$$\tilde{\phi}_j = \tilde{\iota}_{j, \tilde{z}_j} : G_n^{\mathbb{C}}(\tilde{U}_j) \to X, \qquad \tilde{\phi}'_j = \tilde{\iota}_{j, \tilde{z}'_j} : G_n^{\mathbb{C}}(\tilde{U}_j) \to X,$$

$$\tilde{\phi} = \tilde{\iota}_{\tilde{z}} : G_n^{\mathbb{C}}(\tilde{U}) \to X, \qquad \tilde{\phi}' = \tilde{\iota}_{\tilde{z}'} : G_n^{\mathbb{C}}(\tilde{U}) \to X;$$

then we easily see that the relations

$$
\begin{array}{llll}
\tilde{\phi}_{j*} \tilde{\zeta}_j = \zeta, & \tilde{\phi}_{j*} \tilde{\zeta}'_j = \zeta', & \tilde{\phi}'_{j*} \tilde{\zeta}_j = \zeta_j, & \tilde{\phi}'_{j*} \tilde{\zeta}'_j = \zeta'_j, \\
\tilde{\phi}_* \tilde{\zeta} = \zeta_n, & \tilde{\phi}_* \tilde{\zeta}' = \zeta'_n & \tilde{\phi}'_* \tilde{\zeta} = \zeta_{n+1}, & \tilde{\phi}'_* \tilde{\zeta}' = \zeta'_{n+1}
\end{array}
$$

(9.5)

hold.

We again suppose that $n \geq 2$. Let θ be a 1-form and h be a symmetric 2-form on X. We first assume that, for any complex subspace U of $\mathbb{C}^{2n+2}$ of dimension $2n$ and any closed geodesic γ of $G^{\mathbb{C}}_1(U^\perp)$, the symmetric 2-form $h_{U,\gamma}$ on $G^{\mathbb{C}}_n(U)$ vanishes and the 1-form $\theta_{U,\gamma}$ on $G^{\mathbb{C}}_n(U)$ is closed. Let u be a form on X which is equal either to h or to $\beta = d\theta$. Since $\Psi_{Z_j}(z_j) = z'_j$ and $\Psi_{Z'}(z') = z''$, by Lemma 9.10,(ii) or (iii), we see that

$$(9.6) \qquad (\phi^*_j u + \phi'^*_j u)(\tilde{\xi}_j, \tilde{\eta}_j) = 0, \qquad (\phi'^* u + \phi''^* u)(\xi', \eta') = 0,$$

for $1 \leq j \leq n+1$; moreover when $n \geq 3$, since $\Psi_{\tilde{Z}_j}(\tilde{z}_j) = \tilde{z}'_j$ and $\Psi_{\tilde{Z}}(\tilde{z}) = \tilde{z}'$, we have

$$(9.7) \qquad (\tilde{\phi}^*_j h + \tilde{\phi}'^*_j h)(\tilde{\zeta}_j, \tilde{\zeta}'_j) = 0, \qquad (\tilde{\phi}^* h + \tilde{\phi}'^* h)(\tilde{\zeta}, \tilde{\zeta}') = 0,$$

for $3 \leq j \leq n+1$. Then from the equalities (9.4) and (9.6), we obtain the relations

$$u(\xi, \eta) = -u(\xi_j, \eta_j), \qquad u(\xi_{n+1}, \eta_{n+1}) = -u(\xi_2, \eta_2),$$

for $2 \leq j \leq n+1$. On the other hand, when $n \geq 3$, from the equalities (9.5) and (9.7) we obtain the relations

$$h(\zeta, \zeta') = -h(\zeta_j, \zeta'_j), \qquad h(\zeta_n, \zeta'_n) = -h(\zeta_{n+1}, \zeta'_{n+1}),$$

for $3 \leq j \leq n+1$. Thus our assumptions on h and θ imply that

$$(9.8) \qquad\qquad h(\xi, \eta) = 0, \qquad \beta(\xi, \eta) = 0,$$

and, when $n \geq 3$, that

$$(9.9) \qquad\qquad h(\zeta, \zeta') = 0.$$

From the equalities (9.8) we infer that

$$(9.10) \qquad h(\alpha \otimes W_x, \alpha \otimes W_x) = 0, \qquad \beta(\alpha \otimes W_x, \alpha \otimes W_x) = 0,$$

for all vectors $\alpha \in V^*_x$. According to the relation (8.104), our assumptions on h and θ imply that the two forms $\Psi^*_X h$ and $\Psi^*_X \theta$ also satisfy these assumptions; therefore since

$$(9.11) \qquad \Psi_{X*}(\alpha \otimes w) = -w^\flat \otimes \alpha^\sharp, \qquad \Psi_{X*}(\alpha \otimes W_x) = V^*_{\Psi(x)} \otimes \alpha^\sharp,$$

for $\alpha \in V^*_x$ and $w \in W_x$, the relations (9.10) lead us to

$$(9.12) \qquad h(V^*_x \otimes w, V^*_x \otimes w) = 0, \qquad \beta(V^*_x \otimes w, V^*_x \otimes w) = 0,$$

for all vectors $w \in W_x$. According to (9.10) and (9.12), the restrictions of h and $d\theta$ to an arbitrary submanifold of X belonging to the family $\tilde{\mathcal{F}}$ vanishes. When $n \geq 3$, according to (9.9) we see that the restriction of h to an arbitrary submanifold of X belonging to the family $\mathcal{F}^1$ vanishes; by Lemma 9.1, it follows that h vanishes.

Next, we suppose that the form h satisfies the following weaker condition: for any complex subspace U of $\mathbb{C}^{2n+2}$ of dimension $2n$ and any closed geodesic γ of $G_1^{\mathbb{C}}(U^{\perp})$, the symmetric 2-form $h_{U,\gamma}$ on $G_n^{\mathbb{C}}(U)$ is a Lie derivative of the metric of $G_n^{\mathbb{C}}(U)$. We also suppose either that the vectors w_1' and w_2' belong to a totally real subspace of W_x or that $w_2' = Jw_1'$; as we remarked above, for $1 \leq j \leq n+1$, the vectors $\tilde{\xi}_j$ and $\tilde{\eta}_j$ are tangent to a totally geodesic surface of Y_j, while the vectors ξ' and η' are tangent to a totally geodesic surface of Y'. Since $\Psi_{Z_j}(z_j) = z_j'$ and $\Psi_{Z'}(z') = z''$, by Lemma 9.10,(iv), we see that

$$
(9.13) \quad
\begin{aligned}
(\phi_j^* D_g h + \phi_j'^* D_g h)(\tilde{\xi}_j, \tilde{\eta}_j, \tilde{\xi}_j, \tilde{\eta}_j) &= 0, \\
(\phi'^* D_g h + \phi''^* D_g h)(\xi', \eta', \xi', \eta') &= 0,
\end{aligned}
$$

for $1 \leq j \leq n+1$; moreover when $n \geq 3$, since the vectors $\tilde{\zeta}_j$ and $\tilde{\zeta}_j'$ are tangent to a totally geodesic surface of $\tilde{Y}_j$ and the vectors $\tilde{\zeta}$ and $\tilde{\zeta}'$ are tangent to a totally geodesic surface of $\tilde{Y}$, and since $\Psi_{\tilde{Z}_j}(\tilde{z}_j) = \tilde{z}_j'$ and $\Psi_{\tilde{Z}}(\tilde{z}) = \tilde{z}'$, we have

$$
(9.14) \quad
\begin{aligned}
(\tilde{\phi}_j^* D_g h + \tilde{\phi}_j'^* D_g h)(\tilde{\zeta}_j, \tilde{\zeta}_j', \tilde{\zeta}_j, \tilde{\zeta}_j') &= 0, \\
(\tilde{\phi}^* D_g h + \tilde{\phi}'^* D_g h)(\tilde{\zeta}, \tilde{\zeta}', \tilde{\zeta}, \tilde{\zeta}') &= 0,
\end{aligned}
$$

for $3 \leq j \leq n+1$. Then from the equalities (9.4) and (9.13), we obtain the relations

$$
(D_g h)(\xi, \eta, \xi, \eta) = -(D_g h)(\xi_j, \eta_j, \xi_j, \eta_j),
$$
$$
(D_g h)(\xi_{n+1}, \eta_{n+1}, \xi_{n+1}, \eta_{n+1}) = -(D_g h)(\xi_2, \eta_2, \xi_2, \eta_2),
$$

for $2 \leq j \leq n+1$. On the other hand, when $n \geq 3$, from the equalities (9.5) and (9.14) we obtain the relations

$$
(D_g h)(\zeta, \zeta', \zeta, \zeta') = -(D_g h)(\zeta_j, \zeta_j', \zeta_j, \zeta_j'),
$$
$$
(D_g h)(\zeta_n, \zeta_n', \zeta_n, \zeta_n') = -(D_g h)(\zeta_{n+1}, \zeta_{n+1}', \zeta_{n+1}, \zeta_{n+1}'),
$$

for $3 \leq j \leq n+1$. Thus our assumption on h implies that

$$
(9.15) \quad (D_g h)(\xi, \eta, \xi, \eta) = 0,
$$

and, when $n \geq 3$, that

$$(9.16) \qquad\qquad (D_g h)(\zeta, \zeta', \zeta, \zeta') = 0.$$

The equality (9.15) implies that the restriction of $D_g h$ to a totally geodesic surface of X belonging to the family $\mathcal{F}^3$, or to a subspace of T_x of the form $\alpha \otimes W_1$, where α is a unit vector of V_x^* and W_1 is a totally real two-dimensional subspace of W_x, vanishes. According to the relation (8.104), our assumption on h implies that the form $\Psi_X^* h$ also satisfies this assumption; therefore by (9.11), the restriction of $D_g h$ to a subspace of T_x of the form $V_1 \otimes w$, where V_1 is a totally real 2-dimensional subspace of V_x^* and w is a unit vector of W_x, vanishes. Thus the restriction of $D_g h$ to a totally geodesic surface of X belonging to the family $\mathcal{F}^2$ vanishes; therefore so does the restriction of $D_g h$ to a surface belonging to the family $\tilde{\mathcal{F}}$. According to Proposition 1.14,(ii), we have shown that $D_{1,\tilde{\mathcal{F}}} h = 0$. When $n \geq 3$, according to (9.16) we see that the restriction of $D_g h$ to a totally geodesic surface of X belonging to the family $\mathcal{F}^1$ vanishes; therefore according to Proposition 1.14,(ii), we know that $D_{1,\mathcal{F}} h = 0$.

In the course of the previous discussion, we have proved the following proposition:

PROPOSITION 9.12. *Let h be a symmetric 2-form and θ be a 1-form on the complex Grassmannian $X = G^{\mathbb{C}}_{n+1,n+1}$, with $n \geq 2$.*

(i) Suppose that, for any complex subspace U of $\mathbb{C}^{2n+2}$ of dimension $2n$ and any closed geodesic γ of $G^{\mathbb{C}}_1(U^\perp)$, where $U^\perp$ is the orthogonal complement of U in $\mathbb{C}^{2n+2}$, the symmetric 2-form $h_{U,\gamma}$ on $G^{\mathbb{C}}_n(U)$ vanishes. Then the restriction of the symmetric form h to an arbitrary submanifold of X belonging to the family $\tilde{\mathcal{F}}$ vanishes; moreover, when $n \geq 3$, the symmetric form h vanishes.

(ii) Suppose that, for any complex subspace U of $\mathbb{C}^{2n+2}$ of dimension $2n$ and any closed geodesic γ of $G^{\mathbb{C}}_1(U^\perp)$, where $U^\perp$ is the orthogonal complement of U in $\mathbb{C}^{2n+2}$, the symmetric 2-form $h_{U,\gamma}$ on $G^{\mathbb{C}}_n(U)$ is a Lie derivative of the metric of $G^{\mathbb{C}}_n(U)$. Then the relation

$$D_{1,\tilde{\mathcal{F}}} h = 0$$

holds; moreover, when $n \geq 3$, the relation

$$D_{1,\mathcal{F}} h = 0$$

holds.

(iii) Suppose that, for any complex subspace U of $\mathbb{C}^{2n+2}$ of dimension $2n$ and any closed geodesic γ of $G^{\mathbb{C}}_1(U^\perp)$, where $U^\perp$ is the orthogonal complement of U in $\mathbb{C}^{2n+2}$, the 1-form $\theta_{U,\gamma}$ on $G^{\mathbb{C}}_n(U)$ is closed. Then the

restriction of the 2-form $d\theta$ to an arbitrary submanifold of X belonging to the family $\tilde{\mathcal{F}}$ vanishes.

Let n be an integer ≥ 2, and let U be a complex vector space of dimension $2n$ endowed with a positive definite scalar product. According to an observation made in §2, Chapter VIII, if all even functions on $G_{n,n}^{\mathbb{C}}$ satisfying the Guillemin condition vanish, then the analogous result is also true for the Grassmannian $G_n^{\mathbb{C}}(U)$; moreover if all even symmetric 2-forms (resp. 1-forms) on $G_{n,n}^{\mathbb{C}}$ satisfying the Guillemin condition are Lie derivatives of the metric (resp. are exact), then the analogous result is also true for the Grassmannian $G_n^{\mathbb{C}}(U)$. We shall use these remarks in the course of the proofs of the next three propositions.

PROPOSITION 9.13. For $n \geq 2$, the maximal flat Radon transform for functions on the symmetric space $\bar{G}_{n,n}^{\mathbb{C}}$ is injective.

PROOF: We proceed by induction on $n \geq 2$. Proposition 8.8 tells us that the desired result is true for $n = 2$. Next, let $n \geq 2$ be a given integer and suppose that the maximal flat Radon transform for functions on the symmetric space $\bar{G}_{n,n}^{\mathbb{C}}$ is injective. Let f be an even real-valued function on $X = \bar{G}_{n+1,n+1}^{\mathbb{C}}$ satisfying the Guillemin condition. Let U be an arbitrary subspace of $\mathbb{C}^{2n+2}$ of dimension $2n$; we denote by $U^{\perp}$ its orthogonal complement in $\mathbb{C}^{2n+2}$. Let γ be a closed geodesic of $G_1^{\mathbb{C}}(U^{\perp})$. According to Lemmas 8.22 and 8.23, the function $f_{U,\gamma}$ on $G_n^{\mathbb{C}}(U)$ is even and satisfies the Guillemin condition. From Lemma 8.1 and our induction hypothesis, we infer that the function $f_{U,\gamma}$ vanishes. Then by Proposition 9.11, we know that f vanishes. According to Lemma 8.1, this argument gives us the desired result for the space X.

The preceding proposition is also given by Theorem 2.24.

PROPOSITION 9.14. Let n be an integer ≥ 3 and suppose that all even symmetric 2-forms on $G_{n,n}^{\mathbb{C}}$ satisfying the Guillemin condition are Lie derivatives of the metric. Then an even symmetric 2-form on $X = G_{n+1,n+1}^{\mathbb{C}}$ satisfying the Guillemin condition is a Lie derivative of the metric.

PROOF: Let k be an even symmetric 2-form on X satisfying the Guillemin condition. According to the decomposition (1.11), we may decompose k as

$$k = h + D_0\xi,$$

where h is an even symmetric 2-form on X satisfying $\operatorname{div} h = 0$, which is uniquely determined by k, and where ξ is a vector field on X. Then by Lemma 2.10, h also satisfies the Guillemin condition. Let U be an arbitrary subspace of $\mathbb{C}^{2n+2}$ of dimension $2n$; we denote by $U^{\perp}$ its orthogonal complement in $\mathbb{C}^{2n+2}$. We consider the Grassmannian $Y = G_n^{\mathbb{C}}(U)$. Let γ be a closed geodesic of $G_1^{\mathbb{C}}(U^{\perp})$. According to Lemmas 8.22 and 8.23 and our hypothesis, we see that the symmetric 2-form $h_{U,\gamma}$ on Y is even and is a Lie derivative of the metric of Y. Therefore by Proposition 9.12,(ii), we know that $D_{1,\mathcal{F}}h = 0$. According to the relation (8.12) and Lemma 9.3, the vector bundle E and the symmetric 2-form h satisfy the hypotheses of Theorem 2.48,(i), with $\mathcal{F}' = \mathcal{F}$. Then Theorem 2.48,(i) tells us that we may write $h = h_1 + h_2$, where h_1 is an element of $E(X)$ which satisfies the Guillemin condition and h_2 is a section of E. According to Proposition 8.18, we see that $h_1 = 0$. Therefore h is a section of E, and so $h_{U,\gamma}$ is an even section of E_Y over Y. Proposition 9.4 gives us the vanishing of the form $h_{U,\gamma}$. Then Proposition 9.12,(i) tells us that h vanishes. Therefore the symmetric 2-form k is a Lie derivative of the metric of X.

PROPOSITION 9.15. *Let n be an integer ≥ 2 and suppose that all even 1-forms on $G_{n,n}^{\mathbb{C}}$ satisfying the Guillemin condition are exact. Then an even 1-form on $X = G_{n+1,n+1}^{\mathbb{C}}$ satisfying the Guillemin condition is exact.*

PROOF: Let θ be an even 1-form on X satisfying the Guillemin condition. Let U be an arbitrary subspace of $\mathbb{C}^{2n+2}$ of dimension $2n$; we denote by $U^\perp$ its orthogonal complement in $\mathbb{C}^{2n+2}$. We consider the Grassmannian $Y = G_n^{\mathbb{C}}(U)$. Let γ be a closed geodesic of $G_1^{\mathbb{C}}(U^\perp)$. According to Lemmas 8.22 and 8.23 and our hypothesis, we see that the 1-form $\theta_{U,\gamma}$ on Y is closed. Therefore by Proposition 9.12,(iii), we know that the restriction of $d\theta$ to an arbitrary submanifold of X belonging to the family $\tilde{\mathcal{F}}$ vanishes. Then Lemma 9.2 tells us that $d\theta = 0$. Since the cohomology group $H^1(X,\mathbb{R})$ vanishes, the form θ is exact.

The following theorem is a direct consequence of Propositions 8.2 and 9.14.

THEOREM 9.16. *If the symmetric space $\bar{G}_{n,n}^{\mathbb{C}}$, with $n \geq 3$, is rigid in the sense of Guillemin, then so is the symmetric space $\bar{G}_{n+1,n+1}^{\mathbb{C}}$.*

The following theorem is a direct consequence of Theorem 8.10 and Propositions 2.20 and 9.15.

THEOREM 9.17. *Let n be an integer ≥ 2. Then a 1-form on the symmetric space $X = \bar{G}_{n,n}^{\mathbb{C}}$ satisfies the Guillemin condition if and only if it is exact.*

§3. The rigidity of the quaternionic Grassmannians

Let $m, n \geq 1$ be given integers. We consider the quaternions $\mathbb{H}$ and we denote by $\bar{x}$ the conjugate of a quaternion $x \in \mathbb{H}$. We consider the space $\mathbb{H}^{m+n}$ as a right $\mathbb{H}$-module and we endow $\mathbb{H}^{m+n}$ with the Hermitian inner product defined as follows: the inner product $\langle x, y \rangle$ of the vectors

$x = (x_1, \ldots, x_{m+n})$ and $y = (y_1, \ldots, y_{m+n})$ of $\mathbb{H}^{m+n}$ is given by

$$\langle x, y \rangle = \sum_{j=1}^{m+n} \bar{x}_j \cdot y_j.$$

The quaternionic Grassmannian $X = G^{\mathbb{H}}_{m,n}$ consists of all right $\mathbb{H}$-submodules of $\mathbb{H}^{m+n}$ of dimension m (over $\mathbb{H}$). We denote by V_x the right submodule of $\mathbb{H}^{m+n}$ corresponding to the point x of X; then its orthogonal complement W_x is also a right $\mathbb{H}$-submodule of $\mathbb{H}^{m+n}$ and the dual $\mathbb{H}$-module V_x^* of V_x is a left $\mathbb{H}$-module. We shall identify the tangent space T_x at $x \in X$ with the real vector space

$$\operatorname{Hom}_{\mathbb{H}}(V_x, W_x) = W_x \otimes_{\mathbb{H}} V_x^*.$$

By restriction, the Hermitian inner product on $\mathbb{H}^{m+n}$ induces Hermitian inner products on V_x and W_x and therefore also on V_x^*, which we denote by $\langle\,,\,\rangle$. If θ is an element of $\operatorname{Hom}_{\mathbb{H}}(V_x, W_x)$, we consider its adjoint ${}^t\theta$ defined in terms of these Hermitian inner products which is an element of $\operatorname{Hom}_{\mathbb{H}}(W_x, V_x)$. If ϕ is an endomorphism of V_x over $\mathbb{H}$, we denote by $\operatorname{Tr} \phi$ the trace of ϕ viewed as an endomorphism of the real vector space V_x. We consider the Riemannian metric g on X determined by

$$g(\theta, \theta') = \frac{1}{4} \operatorname{Tr} {}^t\theta' \circ \theta,$$

for $\theta, \theta' \in \operatorname{Hom}_{\mathbb{H}}(V_x, W_x)$. We say that a real subspace V_1 of V_x^* (resp. W_1 of W_x) is totally real if $V_1 \cap (q \cdot V_1) = \{0\}$ (resp. if $W_1 \cap (W_1 \cdot q) = \{0\}$), for all $q \in \mathbb{H}$ satisfying $\operatorname{Re} q = 0$.

The left action of the symplectic group $G = Sp(m + n)$ on $\mathbb{H}^{m+n}$ induces a left action on X which is transitive; the metric g is easily seen to be G-invariant. The space X is isometric to the irreducible symmetric space $Sp(m + n)/Sp(m) \times Sp(n)$. We consider the G-invariant sub-bundle $E = E_X$ of $S^2 T^*$ consisting of all elements h of $S^2 T^*$ which satisfy

$$h(\xi, \xi) = 0,$$

for all elements ξ of $W \otimes_{\mathbb{H}} V^*$ of rank one. Clearly, we have

(9.17) $$\operatorname{Tr} E = \{0\}.$$

We also consider the Grassmannian $G^{\mathbb{H}}_{n,m}$. There is a natural mapping

$$\Psi : G^{\mathbb{H}}_{m,n} \to G^{\mathbb{H}}_{n,m},$$

sending $x \in G_{m,n}^{\mathbb{H}}$ into the right $\mathbb{H}$-module W_x of dimension n, which is easily seen to be an isometry. The Grassmannian $G_{n,m}^{\mathbb{H}}$ is also a homogeneous space of the group G and it is easily verified that the isometry Ψ satisfies

$$\Psi \circ \phi = \phi \circ \Psi,$$

for all $\phi \in G$. Thus $G_{m,n}^{\mathbb{H}}$ and $G_{n,m}^{\mathbb{H}}$ are isometric as symmetric spaces.

When $m = n$, the isometry Ψ of $X = G_{n,n}^{\mathbb{H}}$ is an involution. The group Λ of isometries of X generated by Ψ, which is of order 2, acts freely on X and we may consider the Riemannian manifold $\bar{X} = \bar{G}_{n,n}^{\mathbb{H}}$ equal to the quotient X/Λ endowed with the Riemannian metric $g_{\bar{X}}$ induced by g. The natural projection $\varpi : X \to \bar{X}$ is a two-fold covering. The action of the group $Sp(2n)$ on X passes to the quotient $\bar{X}$ and acts transitively on $\bar{X}$. The manifold $\bar{X}$ is a symmetric space of compact type of rank n, which is irreducible and equal to the adjoint space of X. We say that a section u of S^pT^* over X is even (resp. odd) if $\Psi^*u = \varepsilon u$, where $\varepsilon = 1$ (resp. $\varepsilon = -1$). Such a tensor u is even if and only if we can write $u = \varpi^*u'$, where u' is a symmetric p-form on $\bar{X}$. This notion of even or odd form on X (with respect to the involutive isometry Ψ) coincides with the one considered in §3, Chapter II.

We now suppose that m, n are arbitrary integers ≥ 2. Let x be a point of X. Let $\mathcal{F}_x$ be the family of all closed connected totally geodesic surfaces of X passing through x of the form $\mathrm{Exp}_x F$, where F is a real subspace of the tangent space T_x satisfying one of the following three conditions:

(i) F is generated (over $\mathbb{R}$) by the vectors $\{w_1 \otimes \alpha_1, w_2 \otimes \alpha_2\}$, where $\{\alpha_1, \alpha_2\}$ are unit vectors of V_x^* satisfying $\langle \alpha_1, \alpha_2 \rangle = 0$ and $\{w_1, w_2\}$ are unit vectors of W_x satisfying $\langle w_1, w_2 \rangle = 0$;

(ii) F is generated (over $\mathbb{R}$) by the vectors $\{w \otimes \alpha_1, w \otimes \alpha_2\}$, where $\{\alpha_1, \alpha_2\}$ are unit vectors of V_x^* satisfying $\langle \alpha_1, \alpha_2 \rangle = 0$ and w is a unit vector of W_x;

(iii) F is generated (over $\mathbb{R}$) by the vectors $\{w_1 \otimes \alpha, w_2 \otimes \alpha\}$, where α is a unit vector of V_x^* and $\{w_1, w_2\}$ are unit vectors of W_x satisfying $\langle w_1, w_2 \rangle = 0$;

(iv) F is generated (over $\mathbb{R}$) by the vectors $\{w \otimes \alpha, wq \otimes \alpha\}$, where q is a non-zero element of $\mathbb{H}$ satisfying $\mathrm{Re}\, q = 0$ and where α is a unit vector of V_x^* and w is a unit vector of W_x.

A surface of $\mathcal{F}_x$ corresponding to a subspace F of T_x of type (i) is a totally geodesic flat 2-torus; on the other hand, a surface of $\mathcal{F}_x$ corresponding to a subspace F of T_x of type (ii) or of type (iii) is isometric to a 2-sphere of constant curvature 1, while a surface of $\mathcal{F}_x$ corresponding to a subspace F of T_x of type (iv) is isometric to a 2-sphere of constant curvature 4.

Let $\mathcal{F}_x'$ be the family of all closed connected totally geodesic submanifolds of X passing through x of the form $\mathrm{Exp}_x F$, where F is a totally real subspace of T_x generated by the subspaces $W_1 \otimes_{\mathbb{R}} V_1$ and $W_1 q \otimes_{\mathbb{R}} V_1$, where V_1 and W_1 are totally real subspaces of V_x^* and W_x of dimension m and n, respectively, and where q is a non-zero element of $\mathbb{H}$ satisfying $\operatorname{Re} q = 0$. Clearly, each surface of $\mathcal{F}_x$ is contained in a totally geodesic submanifold of X belonging to the family $\mathcal{F}_x'$. A submanifold of X belonging to the family $\mathcal{F}_x'$ is isometric to the complex Grassmannian $Z = G_{m,n}^{\mathbb{C}}$. In fact, if Z' is a submanifold of X belonging to the family $\mathcal{F}_x'$, then there exists a totally geodesic isometric imbedding $\iota : Z \to X$ whose image is equal to Z' such that $\iota^* E_X = E_Z$. Moreover Z' has the same rank as X; therefore, if u is a symmetric p-form on X satisfying the Guillemin condition, the restriction of u to Z' also satisfies the Guillemin condition. When $m = n$, we may assume that the imbedding ι has the following additional property: if u is an even section of $S^p T^*$ over X, then $\iota^* u$ is an even p-form on Z.

We consider the G-invariant families

$$\mathcal{F} = \bigcup_{x \in X} \mathcal{F}_x, \qquad \mathcal{F}' = \bigcup_{x \in X} \mathcal{F}_x'$$

of closed connected totally geodesic submanifolds of X. Clearly, every surface of $\mathcal{F}$ is contained in a totally geodesic submanifold of X belonging to the family $\mathcal{F}'$. By the above remarks concerning a surface of the family $\mathcal{F}_x'$, with $x \in X$, we see that the family $\mathcal{F}'$ possesses property (I) of §8, Chapter II; hence by Proposition 8.12, we obtain the following result, which is also given by Theorem 2.24:

PROPOSITION 9.18. *For $m, n \geq 2$, with $m \neq n$, the maximal flat totally geodesic Radon transform on the quaternionic Grassmannian $G_{m,n}^{\mathbb{H}}$ is injective.*

We consider the sub-bundle $N = N_{\mathcal{F}}$ of B consisting of those elements of B, which vanish when restricted to the closed totally geodesic submanifolds of $\mathcal{F}$, which was introduced in §8, Chapter II. The proofs of the following two lemmas are similar to those of Lemmas 9.1 and 9.3 and shall be omitted.

LEMMA 9.19. *Let X be the quaternionic Grassmannian $G_{m,n}^{\mathbb{H}}$, with $m, n \geq 2$. Let u be an element of $\bigotimes^2 T_x^*$, with $x \in X$. If the restriction of u to an arbitrary submanifold of the family $\mathcal{F}$ vanishes, then u vanishes.*

LEMMA 9.20. *Let $m, n \geq 2$ be given integers and let X be the quaternionic Grassmannian $G_{m,n}^{\mathbb{H}}$. Then we have*

$$\operatorname{Tr} N \subset E.$$

PROPOSITION 9.21. *Let h be a section of E over the quaternionic Grassmannian $X = G_{m,n}^{\mathbb{H}}$, with $m, n \geq 2$. Suppose that the restriction of*

h to an arbitrary submanifold Z of X belonging to the family $\mathcal{F}'$ is a Lie derivative of the metric of Z. If either $n \geq 3$, or if $m = n = 2$ and h is an even section of E, then h vanishes.

PROOF: Let X' be a submanifold of X belonging to the family $\mathcal{F}'$. We consider the complex Grassmannian $Z = G_{m,n}^{\mathbb{C}}$. Then there is a totally geodesic isometric imbedding $i : Z \to X$ whose image is equal to X' such that i^*h is a section of the sub-bundle E_Z of $S^2 T_Z^*$; moreover, when $m = n$ and h is even, we may suppose that the 2-form i^*h on Z is even. Our hypotheses imply that the symmetric 2-form i^*h on Z is a Lie derivative of the metric of Z. When $n \geq 3$, according to Proposition 9.4 we see that i^*h vanishes. If $m = n = 2$ and h is even, the vanishing of the even section i^*h over Z is given by Theorem 8.11. Thus the restriction of h to X' vanishes. The desired result follows from Lemma 9.19.

By Theorems 9.5 and 9.6, we know that the family $\mathcal{F}'$ possesses property (III) of §8, Chapter II and that, when $m \neq n$, it also possesses property (II) of §8, Chapter II. Also Theorem 1.22 gives us the vanishing of the space $E(X)$.

When $n \geq 3$, according to Proposition 9.21, we know that

$$(9.18) \qquad \mathcal{L}(\mathcal{F}') \cap C^\infty(E) = \{0\}.$$

Since the Grassmannian $G_{n,2}^{\mathbb{H}}$ is isometric to $G_{2,n}^{\mathbb{H}}$, the following two theorems are direct consequences of Theorem 2.49, the relation (9.17) and Lemma 9.20.

THEOREM 9.22. *The quaternionic Grassmannian $G_{m,n}^{\mathbb{H}}$, with $m, n \geq 2$ and $m \neq n$, is rigid in the sense of Guillemin.*

THEOREM 9.23. *Let $m, n \geq 2$ be given integers, with $m + n \geq 5$. The quaternionic Grassmannian $G_{m,n}^{\mathbb{H}}$ is infinitesimally rigid.*

When $m = n = 2$, according to Proposition 9.21 we know that

$$(9.19) \qquad \mathcal{L}(\mathcal{F}') \cap C^\infty(E)^{\mathrm{ev}} = \{0\}.$$

By Proposition 2.18 and Theorem 8.9, if either $n = 2$ or if $n \geq 3$ and the space $\bar{G}_{n,n}^{\mathbb{C}}$ is rigid in the sense of Guillemin, an even symmetric 2-form on the space $X = G_{n,n}^{\mathbb{H}}$ satisfying the Guillemin condition belongs to $\mathcal{L}(\mathcal{F}')$. The following two results are a consequence of Proposition 2.18, Theorem 2.48,(ii), the relations (9.17)–(9.19), Lemma 9.20 and Theorem 9.16.

THEOREM 9.24. *The symmetric space $\bar{G}_{2,2}^{\mathbb{H}}$ is rigid in the sense of Guillemin.*

THEOREM 9.25. *Let n_0 be an integer ≥ 3. If the space $\bar{G}_{n_0,n_0}^{\mathbb{C}}$ is rigid in the sense of Guillemin, then the symmetric spaces $\bar{G}_{n,n}^{\mathbb{H}}$ are rigid in the sense of Guillemin, for all $n \geq n_0$.*

The proof of the following result is similar to that of Proposition 9.7 and shall therefore be omitted.

PROPOSITION 9.26. *Let $m, n \geq 2$ be given integers, with $m + n \geq 5$. If X is the quaternionic Grassmannian $G_{m,n}^{\mathbb{H}}$, the sequence (1.24) is exact.*

According to Lemma 9.19, the sub-bundle $C_{\mathcal{F}'}$ of $\bigwedge^2 T^*$, consisting of those elements of $\bigwedge^2 T^*$ which vanish when restricted to the closed totally geodesic submanifolds of $\mathcal{F}'$, vanishes. By Theorem 9.9, we know that $\mathcal{F}'$ possesses property (VI) of §8, Chapter II. Since the rank of a submanifold of X belonging to the family $\mathcal{F}'$ is equal to the rank of X, the family $\mathcal{F}'$ possesses property (IV) of §8, Chapter II. When $m \neq n$, by Theorem 9.8, we know that $\mathcal{F}'$ also possesses property (V) of §8, Chapter II. From these observations and Theorem 2.51, we obtain the following two theorems:

THEOREM 9.27. *Let $m, n \geq 2$ be given integers, with $m \neq n$. A form of degree 1 on the quaternionic Grassmannian $G_{m,n}^{\mathbb{H}}$ satisfies the Guillemin condition if and only if it is exact.*

THEOREM 9.28. *Let $m, n \geq 2$ be given integers. A form of degree 1 on the quaternionic Grassmannian $G_{m,n}^{\mathbb{H}}$ satisfies the zero-energy condition if and only if it is exact.*

The next theorem is a consequence of Theorem 9.17, and its proof is similar to that of Theorem 9.27 and shall be omitted.

THEOREM 9.29. *Let n be an integer ≥ 2. A form of degree 1 on the symmetric space $\bar{G}_{n,n}^{\mathbb{H}}$ satisfies the Guillemin condition if and only if it is exact.*

CHAPTER X

PRODUCTS OF SYMMETRIC SPACES

§1. Guillemin rigidity and products of symmetric spaces

Let Y and Z be two manifolds; we consider the product $X = Y \times Z$ and the natural projections pr_Y and pr_Z of X onto Y and Z, respectively. If θ is a section of $\bigotimes^p T_Y^*$ over Y (resp. of $\bigotimes^p T_Z^*$ over Z), we shall also denote by θ the section $\mathrm{pr}_Y^* \theta$ (resp. the section $\mathrm{pr}_Z^* \theta$) of $\bigotimes^p T^*$ over X; a vector field ξ on Y (resp. on Z) induces a vector field on the product X, which we shall also denote by ξ. If θ_1 is a symmetric p-form on Y and θ_2 is a symmetric q-form on Z, we shall consider the symmetric $(p+q)$-form $\theta_1 \cdot \theta_2$ on X. We identify the bundles $\mathrm{pr}_Y^{-1} T_Y$ and $\mathrm{pr}_Z^{-1} T_Z$ with sub-bundles of T, which we also denote by T_Y and T_Z, respectively; similarly, we identify the bundles $\mathrm{pr}_Y^{-1} T_Y^*$ and $\mathrm{pr}_Z^{-1} T_Z^*$ with sub-bundles of T^*, which we also denote by T_Y^* and T_Z^*, respectively. We then have the direct sum decompositions

$$(10.1) \qquad T = T_Y \oplus T_Z, \qquad T^* = T_Y^* \oplus T_Z^*.$$

We denote by $\pi_Y : T \to T_Y$ and by $\pi_Z : T \to T_Z$ the natural projections of T onto T_Y and T_Z, respectively.

The fiber of the vector bundle $T_Y^* \otimes T_Z^*$ at the point $x = (y, z)$ of X, with $y \in Y$ and $z \in Z$, is equal to $T_{Y,y}^* \otimes T_{Z,z}^*$. We identify $T_Y^* \otimes T_Z^*$ with the sub-bundle of $S^2 T^*$, which is equal to the image of the injective morphism of vector bundles over X

$$\phi : T_Y^* \otimes T_Z^* \to S^2 T^*,$$

defined by

$$(\phi v)(\xi, \eta) = v(\pi_Y \xi, \pi_Z \eta) + v(\pi_Y \eta, \pi_Z \xi),$$

for all $v \in T_Y^* \otimes T_Z^*$ and $\xi, \eta \in T$. Then if θ_1 is a 1-form on Y and θ_2 is a 1-form on Z, the symmetric 2-form $\theta_1 \cdot \theta_2$ on X is equal to $\phi(\theta_1 \otimes \theta_2)$. Also we have the decomposition

$$(10.2) \qquad S^2 T^* = S^2 T_Y^* \oplus (T_Y^* \otimes T_Z^*) \oplus S^2 T_Z^*.$$

Let (Y, g_Y) and (Z, g_Z) be two Riemannian manifolds; we consider the product manifold $X = Y \times Z$ endowed with the product metric $g = g_Y + g_Z$. The Riemann curvature tensor R of (X, g) is given by

$$R = R_Y + R_Z,$$

where R_Y and R_Z are the Riemann curvature tensors of the manifolds (Y, g_Y) and (Z, g_Z), respectively.

We now assume that $X_1 = Y$ and $X_2 = Z$ are symmetric spaces of compact type. For $j = 1, 2$, there is a Riemannian symmetric pair (G_j, K_j) of compact type, where G_j is a compact, connected semi-simple Lie group and K_j is a closed subgroup of G_j such that the space X_j is isometric to the homogeneous space G_j/K_j endowed with a G_j-invariant metric. Let Γ_j be the dual of the group G_j. For all $\gamma \in \Gamma_j$, we recall that the multiplicity of the G_j-module $C_\gamma^\infty(X_j)$ is ≤ 1.

For the remainder of this section, we suppose that (X, g) is the symmetric space of compact type equal to the product $X_1 \times X_2$ endowed with the product metric $g = g_1 + g_2$. We view X as the homogeneous space G/K, where G is the group $G_1 \times G_2$ and K is the group $K_1 \times K_2$. We may identify Γ with $\Gamma_1 \times \Gamma_2$. For all $\gamma_1 \in \Gamma_1$ and $\gamma_2 \in \Gamma_2$, we easily see that the equality

$$(10.3) \qquad C_{(\gamma_1,\gamma_2)}^\infty(T_{\mathbb{C}}) = C_{\gamma_1}^\infty(X_1) \cdot C_{\gamma_2}^\infty(T_{X_2,\mathbb{C}}) + C_{\gamma_2}^\infty(X_2) \cdot C_{\gamma_1}^\infty(T_{X_1,\mathbb{C}})$$

of G-modules holds.

We denote by pr_1 the natural projection of X onto X_1. Let u' be a symmetric p-form on X_1 and consider the symmetric p-form $u = \mathrm{pr}_1^* u'$ on X. A maximal flat totally geodesic torus Z of X is equal to the product $Z_1 \times Z_2$, where Z_j is a maximal flat totally geodesic torus of X_j. A parallel vector field ξ on Z can be written as the sum $\xi = \xi_1 + \xi_2$, where ξ_j is a parallel vector field on Z_j, for $j = 1, 2$. Then we have

$$\int_Z u(\xi, \xi, \ldots, \xi)\, dZ = \mathrm{Vol}\,(Z_2, g_2) \cdot \int_{Z_1} u'(\xi_1, \xi_1, \ldots, \xi_1)\, dZ_1.$$

Let γ be a closed geodesic of X. Then the path $\mathrm{pr}_1 \circ \gamma$ in X_1 determines a closed geodesic γ' of X_1, and it is easily verified that the integral of the symmetric p-form u over the closed geodesic γ is a constant multiple of the integral of u' over the closed geodesic γ'. From these remarks, it follows that, if the symmetric p-form u' on X_1 satisfies the Guillemin (resp. the zero-energy) condition, then so does the symmetric p-form u on X.

PROPOSITION 10.1. *Let X_1 and X_2 be symmetric spaces of compact type. Suppose that the symmetric space X_1 is not rigid in the sense of Guillemin (resp. is not infinitesimally rigid). Then the symmetric space $X = X_1 \times X_2$ is not rigid in the sense of Guillemin (resp. is not infinitesimally rigid).*

PROOF: Let h' be a symmetric 2-form on X_1, which satisfies the Guillemin (resp. the zero-energy) condition and is not a Lie derivative of the metric g_1. According to the previous discussion, the symmetric 2-form $h = \mathrm{pr}_1^* h'$ on X satisfies the Guillemin (resp. the zero-energy) condition. Let z be a given point of X_2 and let $\iota : X_1 \to X$ be the totally geodesic imbedding sending $y \in X_1$ into the point (y, z) of X; then we have $\iota^* h = h'$. If there exists a vector field on X such that $\mathcal{L}_\xi g = h$, then we consider the vector field η on X_1 determined by

$$\iota_* \eta(y) = \mathrm{pr}_{1*} \xi(y, x_2),$$

for all $y \in X_1$; by Lemma 1.1, we obtain the equality $\mathcal{L}_\eta g_1 = h'$, which leads us to a contradiction.

PROPOSITION 10.2. *Let X_1 and X_2 be symmetric spaces of compact type. Suppose that the maximal flat Radon transform for functions on the space X_j is injective, for $j = 1, 2$. Then the maximal flat Radon transform for functions on the symmetric space $X = X_1 \times X_2$ is injective.*

PROOF: Let f be a real-valued function on X, whose maximal flat Radon transform vanishes. For $z \in X_2$, we consider the real-valued function f_z on X_1 defined by

$$f_z(y) = f(y, z),$$

for all $y \in X_1$. Let Z_1 be a maximal flat totally geodesic torus of X_1; we define a real-valued function f_{Z_1} on X_2 by

$$(10.4) \qquad\qquad f_{Z_1}(z) = \int_{Z_1} f_z \, dZ_1,$$

for all $z \in X_2$. If Z_2 is a maximal flat totally geodesic torus of X_2, then $Z_1 \times Z_2$ is a maximal flat totally geodesic torus of $X_1 \times X_2$, and we have

$$\int_{Z_2} f_{Z_1} \, dZ_2 = \int_Z f \, dZ = 0.$$

Therefore the function f_{Z_1} on X_2 satisfies the Guillemin condition, and so, according to our hypothesis on X_2, it vanishes. From the equality (10.4), for $z \in X_2$, we see that the function f_z on X_1 satisfies the Guillemin condition, and so, according to our hypothesis on X_1, it vanishes. Thus we have shown that the function f vanishes.

PROPOSITION 10.3. *Let X_1 and X_2 be symmetric spaces of compact type. Let p be an integer ≥ 0. If f is a complex-valued function on X_1 and θ is a complex symmetric p-form on X_2, then the complex symmetric $(p+1)$-form $u = df \cdot \theta$ on the product $X = X_1 \times X_2$ satisfies the Guillemin condition.*

PROOF: Let Z be a maximal flat totally geodesic torus of X and ξ be a parallel vector field on Z. The torus Z is equal to the product $Z_1 \times Z_2$, where Z_j is a maximal flat totally geodesic torus of X_j, and the vector field ξ can be written as the sum $\xi = \xi_1 + \xi_2$, where ξ_j is a parallel vector field on Z_j, for $j = 1, 2$. Since the 1-form df on X_1 satisfies the Guillemin condition, we have

$$\int_Z u(\xi, \xi, \ldots, \xi)\, dZ = \int_{Z_1} (\xi_1 \cdot f)\, dZ_1 \cdot \int_{Z_2} \theta(\xi_2, \xi_2, \ldots, \xi_2)\, dZ_2 = 0.$$

We fix a maximal torus T_j of the group G_j and a system of positive roots Δ_j for the complexification of the Lie algebra of G_j. We then consider the maximal torus $T_1 \times T_2$ of G and the system $\Delta_1 \cup \Delta_2$ of positive roots for the complexification of the Lie algebra of G.

We now suppose that X_1 and X_2 are irreducible symmetric spaces which are not equal to simple Lie groups. Then the complexifications $\mathfrak{g}_1$ and $\mathfrak{g}_2$ of the Lie algebras of the groups G_1 and G_2 are simple. For $j = 1, 2$, let γ_j^1 be the element of Γ_j which is the equivalence class of the irreducible G_j-module $\mathfrak{g}_j$.

PROPOSITION 10.4. *Let X_1 and X_2 be irreducible symmetric spaces of compact type which are not equal to simple Lie groups. Let γ_j be an element of Γ_j, for $j = 1, 2$; let θ_j be a highest weight vector of the G_j-module $C^\infty_{\gamma_j}(T^*_{X_j, \mathbb{C}})$. Suppose that the following two conditions hold:*

(i) *either the G_1-module $C^\infty_{\gamma_1}(X_1)$ vanishes or $\gamma_2 \neq \gamma_2^1$;*

(ii) *either the G_2-module $C^\infty_{\gamma_2}(X_2)$ vanishes or $\gamma_1 \neq \gamma_1^1$.*

Then the symmetric 2-form $\theta_1 \cdot \theta_2$ on $X = X_1 \times X_2$ does not belong to the space $D_0 C^\infty(T_{\mathbb{C}})$.

PROOF: Suppose that the symmetric 2-form $h = \theta_1 \cdot \theta_2$ on X is equal to $D_0 \xi$, where ξ is a section of $T_{\mathbb{C}}$ over X. Since h is a highest weight vector of $C^\infty_{(\gamma_1, \gamma_2)}(S^2 T^*_{\mathbb{C}})$ and the differential operator D_0 is homogeneous, we may suppose that ξ is a highest weight vector of $C^\infty_{(\gamma_1, \gamma_2)}(T_{\mathbb{C}})$. Since the multiplicities of the modules $C^\infty_{\gamma_1}(X_1)$ and $C^\infty_{\gamma_2}(X_2)$ are ≤ 1, according to the equality (10.3) we may write

$$\xi = f_1 \xi_2 + f_2 \xi_1,$$

where $f_j \in C^\infty_{\gamma_j}(X_j)$ and $\xi_j \in C^\infty_{\gamma_j}(T_{X_j, \mathbb{C}})$ are either highest weight vectors of these modules or equal to 0. Then we have

$$D_0 \xi = f_1 D_{0, X_2} \xi_2 + f_2 D_{0, X_1} \xi_1 + df_1 \cdot g_2^\flat(\xi_2) + df_2 \cdot g_1^\flat(\xi_1).$$

Since $h(\eta_j, \eta_j) = 0$, for all $\eta_j \in T_{X_j}$ and $j = 1, 2$, we obtain the relations

$$f_1 D_{0, X_2} \xi_2 = f_2 D_{0, X_1} \xi_1 = 0.$$

Thus if f_1 is non-zero, we see that $D_{0,X_2}\xi_2$ vanishes; hence ξ_2 is a Killing vector field on X_2 and so belongs to the G_2-module $C_\gamma^\infty(T_{X_2,\mathbb{C}})$, where $\gamma = \gamma_2^1$. According to our hypothesis (i), we know that $\gamma_2 \neq \gamma_2^1$ and therefore ξ_2 vanishes. Similarly, from the hypothesis (ii) we deduce that either f_2 or ξ_1 vanishes. Thus we have shown that the vector field ξ and the symmetric form h on X vanish, which leads to a contradiction.

THEOREM 10.5. *Let X_1 and X_2 be irreducible symmetric spaces of compact type which are not equal to simple Lie groups. The symmetric space $X = X_1 \times X_2$ is not rigid in the sense of Guillemin.*

PROOF: We choose elements $\gamma_1 \in \Gamma_1$ and $\gamma_2 \in \Gamma_2$ satisfying $\gamma_1 \neq \gamma_1^1$ and $\gamma_2 \neq \gamma_2^1$. We also suppose that γ_1 and γ_2 do not correspond to the trivial representations of G_1 and G_2, and that the modules $C_{\gamma_1}^\infty(X_1)$ and $C_{\gamma_2}^\infty(X_2)$ do not vanish. Let f_1 and f_2 be highest weight vectors of $C_{\gamma_1}^\infty(X_1)$ and $C_{\gamma_2}^\infty(X_2)$, respectively. Then we know that df_1 and df_2 are highest weight vectors of $C_{\gamma_1}^\infty(T_{X_1,\mathbb{C}}^*)$ and $C_{\gamma_2}^\infty(T_{X_2,\mathbb{C}}^*)$. According to Proposition 10.3, with $p = 1$, the real and imaginary parts of the complex 2-form $h = df_1 \cdot df_2$ on X satisfy the Guillemin condition. By Proposition 10.4, the form h does not belong to the space $D_0 C^\infty(T_\mathbb{C})$. Therefore either the real or the imaginary part of h is not equal to a Lie derivative of the metric of X, and so X is not rigid in the sense of Guillemin.

If the irreducible spaces X_1 and X_2 of Theorem 10.5 are equal to their adjoint spaces, then so is their product $X = X_1 \times X_2$; under this hypothesis, by Theorem 2.24 we know that the maximal flat Radon transform for functions on X is injective. For example, according to Theorem 10.5 the product $\mathbb{RP}^n \times \mathbb{RP}^n$, with $n \geq 2$, is not rigid in the sense of Guillemin, while the maximal flat Radon transform for functions on this space is injective. On the other hand, we shall see that this space is infinitesimally rigid (see Theorem 10.19).

From Theorem 10.5 and Proposition 10.1, we infer that the product

$$X_1 \times X_2 \times \cdots \times X_p,$$

with $p \geq 2$ and where each factor X_j, with $1 \leq j \leq p$, is a symmetric space of compact type, and where X_1 and X_2 are irreducible spaces which are not equal to simple Lie groups, is not rigid in the sense of Guillemin.

THEOREM 10.6. *Let X_1 and X_2 be symmetric spaces of compact type. There exists a 1-form on the symmetric space $X = X_1 \times X_2$ which satisfies the Guillemin condition and is not exact.*

PROOF: For $j = 1, 2$, let f_j be a non-constant real-valued function on X_j. Then the 1-form $u = f_2 df_1$ clearly satisfies

$$du = -df_1 \wedge df_2 \neq 0.$$

According to Proposition 10.3, with $p = 0$, the 1-form u on X satisfies the Guillemin condition.

§2. Conformally flat symmetric spaces

Let (X, g) be a Riemannian manifold of dimension n. We consider the automorphism of the vector bundle $T^* \otimes T$ which sends $u \in T^* \otimes T$ into $u^\natural$ and is determined by the following: if $u = \beta \otimes \xi$, where $\beta \in T^*$ and $\xi \in T$, then $u^\natural$ is equal to $g^\flat(\xi) \otimes g^\sharp(\beta)$. Let B_1 be the sub-bundle of B consisting of those elements v of B for which the relation $v(\xi_1, \xi_2, \xi_3, \xi_4) = 0$ holds, with $\xi_1, \xi_2, \xi_3, \xi_4 \in T$, whenever all the vectors ξ_j are tangent to the same factor or whenever two of the ξ_j are tangent to Y and the other two are tangent to Z.

We recall that, if X is a simply-connected symmetric space, then X is isometric to the product $X_0 \times X^+ \times X^-$, where X_0 is a Euclidean space, and X^+ and X^- are symmetric spaces of compact and non-compact type, respectively (see Proposition 3.4 in Chapter V of [36]); we call X_0 the Euclidean factor of X. If X is a locally symmetric space, we say that X does not admit a Euclidean factor at a point $x \in X$ if there exists a neighborhood of x isometric to a subset of a product $X^+ \times X^-$, where X^+ and X^- are symmetric spaces of compact and non-compact type, respectively.

Let (Y, g_Y) and (Z, g_Z) be two Riemannian manifolds. We suppose that X is the product manifold $Y \times Z$ endowed with the product metric $g = g_Y + g_Z$. Here we use the conventions concerning tensors on a product established in §1. As above, we identify $T_Y^* \otimes T_Z^*$ with the sub-bundle of $S^2 T^*$; then it is easily verified that

$$\hat{\tau}_B(T_Y^* \otimes T_Z^*) \subset B_1.$$

We consider the sub-bundles

$$g_1^Y = \{\, u \in T_Y^* \otimes T_Y \mid \rho(u)g_Y = 0 \,\}, \quad g_1^Z = \{\, v \in T_Z^* \otimes T_Z \mid \rho(v)g_Z = 0 \,\}.$$

By means of the decompositions (10.1), we identify the bundles $\mathrm{pr}_Y^{-1} g_1^Y$ and $\mathrm{pr}_Z^{-1} g_1^Z$ with sub-bundles of $T^* \otimes T$, which we also denote by g_1^Y and g_1^Z, respectively. The sub-bundle

$$g_1^{Y,Z} = \{\, u - u^\natural \mid u \in T_Y^* \otimes T_Z \,\}$$

of $T^* \otimes T$ is isomorphic to $T_Y^* \otimes T_Z$; it is clear that $g_1^{Y,Z} \subset g_1$. Moreover, we have:

LEMMA 10.7. Let (Y, g_Y) and (Z, g_Z) be two Riemannian manifolds. Let (X, g) be the Riemannian product $(Y \times Z, g_Y + g_Z)$. Then we have the equality

$$g_1 = g_1^Y \oplus g_1^Z \oplus g_1^{Y,Z}.$$

We now suppose that (Y, g_Y) and (Z, g_Z) are connected locally symmetric spaces. We set

$$\tilde{B}_{Y,Z} = \rho(g_1^{Y,Z})R;$$

then we have the surjective morphism of vector bundles

(10.5) $$T_Y^* \otimes T_Z \to \tilde{B}_{Y,Z},$$

sending u into $\rho(u - u^\natural)R$. It is easily verified that

$$\tilde{B}_{Y,Z} \subset B_1.$$

In fact, $\tilde{B}_{Y,Z}$ is the sub-bundle of B_1 consisting of all elements v of B_1 for which there exists an element $u \in T_Y^* \otimes T_Z$ such that the equalities

$$v(\xi_1, \eta_1, \eta_2, \eta_3) = R_Z(u(\xi_1), \eta_1, \eta_2, \eta_3),$$
$$v(\eta_1, \xi_1, \xi_2, \xi_3) = -R_Y(u^\natural(\eta_1), \xi_1, \xi_2, \xi_3)$$

hold for all $\xi_1, \xi_2, \xi_3 \in T_Y$ and $\eta_1, \eta_2, \eta_3 \in T_Z$.

The following result is given by Lemma 1.2 of [19].

LEMMA 10.8. *Let (Y, g_Y) and (Z, g_Z) be two connected locally symmetric spaces. Let (X, g) be the Riemannian product $(Y \times Z, g_Y + g_Z)$. Then we have*

(10.6) $$\tilde{B} = \tilde{B}_Y \oplus \tilde{B}_Z \oplus \tilde{B}_{Y,Z}.$$

Let $x = (y, z)$ be a point of X, with $y \in Y$ and $z \in Z$; if Y (or Z) does not admit a Euclidean factor at y (or z), then the mapping (10.6) is an isomorphism at the point x.

When Y and Z have constant curvature equal to K_Y and K_Z, respectively, we know that $\tilde{B}_Y = \{0\}$ and $\tilde{B}_Z = \{0\}$ (see §3, Chapter I); from the above discussion and the equality (10.6), for $u \in T_Y^* \otimes T_Z$, we deduce that

$$\rho(u - u^\natural)R = -2\tau_B(u' \otimes (K_Y g_Y - K_Z g_Z)),$$

where $u' = (\mathrm{id} \otimes g_Z^\flat)u$, and hence that

(10.7) $$\tilde{B} = \tau_B((T_Y^* \otimes T_Z^*) \otimes (K_Y g_Y - K_Z g_Z)).$$

The first assertion of the following proposition is a direct consequence of the equality (10.7) and its second assertion is proven in [16, §2].

PROPOSITION 10.9. *Let* (Y, g_Y) *and* (Z, g_Z) *be two Riemannian manifolds of constant curvature* K_Y *and* K_Z, *respectively. Assume that the dimension of* Y *is* ≥ 1 *and that* $K_Z \neq 0$. *Let* (X, g) *be the Riemannian product* $(Y \times Z, g_Y + g_Z)$. *If either* $\dim Y = 1$ *or if* $K_Z = -K_Y$, *then we have*

$$(10.8) \qquad\qquad \tilde{B} = \hat{\tau}_B(T_Y^* \otimes T_Z);$$

moreover, when the dimension of X *is* ≥ 4, *the equality* (1.48) *holds.*

In fact, under the hypotheses of the preceding proposition, when the dimension of X is equal to 3, the rank of the vector bundle $H \cap (T^* \otimes \tilde{B})$ is equal to 3.

We no longer suppose that X is a product manifold; we now assume that the dimension n of X is ≥ 3. We consider the orthogonal decomposition

$$B = \hat{\tau}_B(S^2 T^*) \oplus B^0$$

given in §1, Chapter I.

The following result is given by Lemma 3.4 of [13].

LEMMA 10.10. *Let* x *be a point of* X *and let* v *be an element of* B_x. *Then the relation* $\rho(u)v = 0$ *holds for all* $u \in g_{1,x}$ *if and only if there exists* $a \in \mathbb{R}$ *such that* $v = a\hat{\tau}_B(g)(x)$.

The Weyl tensor W of (X, g) is the section of the sub-bundle B^0 which is the orthogonal projection of the Riemann curvature tensor R onto B^0. Then we have

$$(10.9) \qquad\qquad R = W + \hat{\tau}_B(h),$$

where h is the section of $S^2 T^*$ given by

$$h = \frac{1}{n-2}\left(\frac{r(g)}{n-1} g - 2\operatorname{Ric}\right).$$

If (X, g) is an Einstein manifold, from formula (10.9) we infer that the Weyl tensor W of X vanishes if and only if X has constant curvature. When $n \geq 4$, a classic result due to Weyl asserts that the metric g is conformally flat if and only if the Weyl tensor W vanishes (see [15, §3]).

LEMMA 10.11. *The Weyl tensor* W *vanishes if and only if we have the inclusion*

$$(10.10) \qquad\qquad \tilde{B} \subset \hat{\tau}_B(S^2 T^*).$$

PROOF: Let x be a point of X and let ξ be an element of T_x satisfying $(\mathcal{L}_\xi g)(x) = 0$. By (10.9), we obtain the equality

$$(\mathcal{L}_\xi R)(x) = (\mathcal{L}_\xi W)(x) + \hat{\tau}_B((\mathcal{L}_\xi h)(x)),$$

where $(\mathcal{L}_\xi W)(x)$ belongs to B^0 and $\hat{\tau}_B((\mathcal{L}_\xi h)(x))$ is an element of $S^2 T^*$. Hence $(\mathcal{L}_\xi R)(x)$ belongs to $\hat{\tau}_B(S^2 T^*)$ if and only if $(\mathcal{L}_\xi W)(x) = 0$. We therefore see that the inclusion $\tilde{B} \subset \hat{\tau}_B(S^2 T^*)$ is valid if and only if the relation $(\mathcal{L}_\xi W)(x) = 0$ holds for all $x \in X$ and $\xi \in T_x$ satisfying $(\mathcal{L}_\xi g)(x) = 0$. Now according to formula (1.1), this last condition is equivalent to the fact that the relation $\rho(u)W = 0$ holds for all $u \in \mathfrak{g}_1$. By Lemma 10.10, we know that it is equivalent to $W = 0$.

The following theorem is due to S. T. Yau (see [57, §5]).

THEOREM 10.12. *Let (Y, g_Y) and (Z, g_Z) be two connected Riemannian manifolds. Let (X, g) be the Riemannian product $(Y \times Z, g_Y + g_Z)$. If the dimension of X is ≥ 3 and if neither Y nor Z is reduced to a point, then the following assertions are equivalent:*

(i) *The manifold (X, g) is conformally flat.*

(ii) *The manifolds (Y, g_Y) and (Z, g_Z) have constant curvature, and if both factors Y and Z are of dimension ≥ 2, the curvature of (Y, g_Y) is the negative of the curvature of (Z, g_Z).*

The following theorem is a direct consequence of Proposition 10.9 and Theorem 10.12.

THEOREM 10.13. *Let (Y, g_Y) and (Z, g_Z) be two connected Riemannian manifolds of dimension ≥ 1. Let (X, g) be the Riemannian product $(Y \times Z, g_Y + g_Z)$. Assume that (X, g) is a conformally flat Riemannian manifold of dimension $n \geq 3$. Then the inclusion (10.10) holds; moreover, if $n \geq 4$, then the equality (1.48) holds.*

In fact, under the hypotheses of Theorem 10.13, if (X, g) is not flat, then the equality (10.8) holds.

THEOREM 10.14. *Let (X, g) be a Riemannian manifold of dimension ≥ 4. The following assertions are equivalent:*

(i) *The manifold (X, g) is locally symmetric and conformally flat.*

(ii) *The inclusion (10.10) and the equality (1.48) hold.*

PROOF: First, Lemmas 1.4 and 10.11 give us the implication (ii) $\Rightarrow$ (i). Conversely, we begin by remarking that a locally homogeneous irreducible Riemannian manifold is Einstein; moreover, if the Weyl tensor of such a manifold vanishes, we saw above that it has constant curvature and so $\tilde{B} = \{0\}$ (see §3, Chapter I). In general, if (i) holds, then X is locally isometric to a product of symmetric spaces which are either irreducible or flat; if this product is non-trivial, according to Theorem 10.12 the manifold X is locally isometric to a product of two Riemannian manifolds (Y, g_Y) and (Z, g_Z) satisfying condition (ii) of Theorem 10.12. In this case, the desired result follows from Theorem 10.13.

§3. Infinitesimal rigidity of products of symmetric spaces

Let (X, g) be a compact symmetric space. For $x \in X$, the set $C_{X,x}$ of vectors ξ of $T_x - \{0\}$, for which $\mathrm{Exp}_x \mathbb{R}\xi$ is a closed geodesic of X, is a dense subset of T_x.

Let (Y, g_Y) and (Z, g_Z) be two compact symmetric spaces. We suppose that (X, g) is equal to the Riemannian product $(Y \times Z, g_Y + g_Z)$.

Let h be a symmetric 2-form on X. For $y \in Y$ and $\xi \in C_{Y,y}$, we consider the 1-form h_ξ on Z defined by

$$h_\xi(\eta) = \frac{1}{L} \int_0^L h(\dot\gamma(t), \eta) \, dt,$$

for $\eta \in T_Z$, where $\gamma(t) = \mathrm{Exp}_x t\xi$ and $\dot\gamma(t)$ is the tangent vector to the closed geodesic γ of length L. We have $h_{\lambda\xi} = \lambda h_\xi$, for all $\xi \in C_{Y,y}$ and $\lambda \in \mathbb{R}$, with $\lambda \neq 0$.

We say that the product $Y \times Z$ satisfies condition (A) if, for all symmetric 2-forms $h \in C^\infty(T_Y^* \otimes T_Z^*)$ on X satisfying the zero-energy condition, there exists a section h_1 of $T_Y^* \otimes T_Z^*$ over X such that

(10.11) $h_1(\xi, \eta) = h_\xi(\eta),$

for all $\xi \in C_{Y,y}$, with $y \in Y$, and $\eta \in T_Z$. This condition on the product $Y \times Z$ means that averaging the sections of $T_Y^* \otimes T_Z^*$ satisfying the zero-energy condition over the closed geodesics of Y is a C^∞-process.

The following general result is proved in [19]:

THEOREM 10.15. *Let Y and Z be two compact symmetric spaces. Suppose that the following conditions hold:*

(i) *either the universal covering space of Y or the universal covering space of Z does not admit a Euclidean factor;*

(ii) *the spaces Y and Z are infinitesimally rigid;*

(iii) *the 1-forms on Y and on Z satisfying the zero-energy condition are exact;*

(iv) *the product $Y \times Z$ satisfies condition (A).*
Then the product $X = Y \times Z$ is infinitesimally rigid.

If Y is a flat torus, or a projective space different from a sphere, or the complex quadric Q_n of dimension $n \geq 3$, in [19] we showed that the product $Y \times Z$ satisfies condition (A). In fact if Y is a projective space different from a sphere, the geodesic flow φ_t on the unit sphere bundle S_Y of T_Y of Y is periodic of period π; in this case, if h is a section of $C^\infty(T_Y^* \otimes T_Z^*)$ over X satisfying the zero-energy condition, we define a function h_1 on $T_Y \times T_Z$ by

$$h_1(\xi, \eta) = \frac{1}{\pi} \int_0^\pi h(\varphi_t\xi, \eta)\, dt,$$

for $\xi \in S_Y$ and $\eta \in T_Z$, and by setting $h_1(\lambda\xi, \eta) = \lambda h_1(\xi, \eta)$ for $\xi \in S_Y$, $\eta \in T_Z$ and $\lambda \in \mathbb{R}$. In [19], we verified that h_1 arises from a C^∞-section of $T_Y^* \otimes T_Z^*$ which we also denote by h_1. Since $C_{Y,y} = T_{Y,y} - \{0\}$, for $y \in Y$, we see that the relation (10.11) holds, and therefore so does condition (A). Since projective spaces, different from spheres, flat tori and the complex quadrics of dimension ≥ 3 are infinitesimally rigid and the 1-forms on these spaces satisfying the zero-energy condition are exact (see Chapters III and VI), from the preceding theorem we deduce:

THEOREM 10.16. *Let Y be a compact symmetric space which is either a projective space different from a sphere, or a flat torus, or a complex quadric of dimension ≥ 3. Let Z be a symmetric space of compact type which is infinitesimally rigid; assume that the 1-forms on Z satisfying the zero-energy condition are exact. If Y is a flat torus, assume moreover that the universal covering space of Z does not admit a Euclidean factor. Then the product $X = Y \times Z$ is infinitesimally rigid.*

From Theorems 7.4, 7.12, 9.6, 9.9, 9.23 and 9.28 and the preceding theorem, we deduce the following:

THEOREM 10.17. *Let m, n be given integers. Let Y be a compact symmetric space which is either a projective space different from a sphere, or a flat torus, or a complex quadric of dimension ≥ 3. Let Z be a symmetric space of compact type which is either the Grassmannian $\widetilde{G}_{m,n}^{\mathbb{R}}$, with $m, n \geq 5$, or the Grassmannian $G_{m,n}^{\mathbb{C}}$, with $m, n \geq 2$, or the Grassmannian $G_{m,n}^{\mathbb{H}}$, with $m, n \geq 5$. Then the product $X = Y \times Z$ is infinitesimally rigid.*

The following theorem is also proved in [19].

THEOREM 10.18. *Let Y and Z be two compact symmetric spaces which are infinitesimally rigid. Assume that the 1-forms on Y and Z satisfying the zero-energy condition are exact. Then a 1-form on the product manifold $X = Y \times Z$ satisfies the zero-energy condition if and only if it is exact.*

Since the 1-forms on a projective space, different from a sphere, on a flat torus, or on the complex quadric Q_n of dimension $n \geq 3$, which satisfy the zero-energy condition, are exact, the following theorem is a direct consequence of Theorems 10.16 and 10.18.

THEOREM 10.19. *A product of Riemannian manifolds*

$$X = X_1 \times X_2 \times \cdots \times X_r,$$

where each factor X_j is either a projective space different from a sphere, or a flat torus, or a complex quadric of dimension ≥ 3, is infinitesimally rigid.

§4. The infinitesimal rigidity of $G_{2,2}^{\mathbb{R}}$

We consider the product manifold $X = Y \times Z$, where the manifolds Y and Z are equal to the sphere S^2. We endow Y and Z with the metric g_0 of S^2 of constant curvature 1 and consider the product metric on X. We denote by τ the anti-podal involution of S^2; let σ be the involution $\tau \times \tau$ of X. We consider the Riemannian metrics g' on the real Grassmannians $\widetilde{G}_{2,2}^{\mathbb{R}}$ and $G_{2,2}^{\mathbb{R}}$ defined in §1, Chapter IV and denoted there by g. By Proposition 4.3, we know that the Grassmannian $\widetilde{G}_{2,2}^{\mathbb{R}}$ endowed with the metric $2g'$ is isometric to the manifold $X = S^2 \times S^2$ endowed with the product metric. According to Proposition 2.22, the sphere S^2 is not infinitesimally rigid; then Proposition 10.1 tells us that the Grassmannian $\widetilde{G}_{2,2}^{\mathbb{R}}$ is *not* infinitesimally rigid. If Λ is the group of isometries of X generated by σ, in §9, Chapter V we saw that the Riemannian manifold X/Λ endowed with the metric induced by the metric of X is isometric to the real Grassmannian $G_{2,2}^{\mathbb{R}}$ endowed with the metric $2g'$.

This section is devoted to the proof of the following two theorems:

THEOREM 10.20. *The real Grassmannian $G_{2,2}^{\mathbb{R}}$ is infinitesimally rigid.*

THEOREM 10.21. *A differential form of degree 1 on the real Grassmannian $G_{2,2}^{\mathbb{R}}$ satisfies the zero-energy condition if and only if it is exact.*

By Proposition 4.3, we know that the space $\mathbb{RP}^2 \times \mathbb{RP}^2$ endowed with the product metric is isometric to the quotient of the Grassmannian $G_{2,2}^{\mathbb{R}}$, endowed with the metric $2g'$, by its group of isometries generated by the involutive isometry Ψ considered in §1, Chapter IV. Thus from Theorem 10.20 and Proposition 2.19, we deduce that the space $\mathbb{RP}^2 \times \mathbb{RP}^2$ is infinitesimally rigid; this last result is also given by Theorem 10.17. From Theorem 10.21, we infer that a 1-form on $\mathbb{RP}^2 \times \mathbb{RP}^2$ satisfies the zero-energy condition if and only if it is exact; this last result can also be obtained from Theorems 3.26 and 10.18.

According to Propositions 2.19 and 2.20, we know that Theorems 10.20 and 10.21 are respectively equivalent to the following two results:

THEOREM 10.22. *A symmetric 2-form on the product $S^2 \times S^2$, which is even with respect to the involution σ and satisfies the zero-energy condition, is a Lie derivative of the metric.*

THEOREM 10.23. *A differential 1-form on the product $S^2 \times S^2$, which is even with respect to the involution σ and satisfies the zero-energy condition, is exact.*

We consider the Fubini-Study metric $\tilde{g}$ on the complex projective space $\mathbb{CP}^1$ of constant curvature 4, and various objects and notions associated to this space in §4, Chapter III. In particular, we consider the isometry φ from (S^2, g_0) to $(\mathbb{CP}^1, 4\tilde{g})$ and the involutive isometry Ψ of $\mathbb{CP}^1$. We say that a symmetric p-form on $\mathbb{CP}^1$ is even (resp. odd) if $\Psi^* u = \varepsilon u$, where $\varepsilon = 1$ (resp. $\varepsilon = -1$). We identify $\mathbb{CP}^1$ with the Hermitian symmetric space $SU(2)/K'$, where K' is the subgroup $S(U(1) \times U(1))$ of $SU(2)$. Let Γ' denote the dual of the group $SU(2)$. If k is an integer ≥ 0, let γ_k be the element of Γ' corresponding to the irreducible $SU(2)$-module $\mathcal{H}_k$; we also consider the complex-valued function $\tilde{f}_k$ on $\mathbb{CP}^1$, which belongs to $\mathcal{H}_k$, and we recall that $\tilde{f}_k = (\tilde{f}_1)^k$.

For the remainder of this section, we let X be the product manifold $Y \times Z$, where the manifolds Y and Z are equal to the complex projective line $\mathbb{CP}^1$. We endow Y and Z with the Fubini-Study metrics g_Y and g_Z of constant curvature 4, respectively, and we consider the product metric g on X. We denote by Ψ' the involutive isometry $\Psi \times \Psi$ of $\mathbb{CP}^1 \times \mathbb{CP}^1$. We say that a symmetric p-form u on X is even (resp. odd) if $\Psi'^* u = \varepsilon u$, where $\varepsilon = 1$ (resp. $\varepsilon = -1$). We consider the subspaces $\mathcal{H}_{Y,k} = C^\infty_{\gamma_k}(Y)$ of $C^\infty(Y)$ and $\mathcal{H}_{Z,k} = C^\infty_{\gamma_k}(Z)$ of $C^\infty(Z)$, which correspond to the space $\mathcal{H}_k$ of functions on $\mathbb{CP}^1$; we also denote by f_Y (resp. by f_Z) the function on Y belonging to $\mathcal{H}_{Y,1}$ (resp. on Z belonging to $\mathcal{H}_{Z,1}$) corresponding to the function $\tilde{f}_1$ on $\mathbb{CP}^1$.

We view X as the symmetric space G/K of compact type, where G is the group $SU(2) \times SU(2)$ and $K = K' \times K'$. We identify the dual Γ of the group G with $\Gamma' \times \Gamma'$. The complexification $\mathfrak{g}$ of the Lie algebra of G is equal to the Lie algebra $\mathfrak{sl}(2) \oplus \mathfrak{sl}(2)$. The group of all diagonal matrices of G is a maximal torus of G, and the complexification $\mathfrak{t}$ of the Lie algebra of this torus is a Cartan subalgebra of the semi-simple Lie algebra $\mathfrak{g}$. For $1 \leq j \leq 4$, the linear form $\lambda_j : \mathfrak{t} \to \mathbb{C}$, sending the diagonal matrix, whose diagonal entries are a_1, a_2, a_3, a_4, into a_j, is purely imaginary on $\mathfrak{t}_0$. Then $\{\lambda_1 - \lambda_2, \lambda_3 - \lambda_4\}$ is a positive system for the roots of $\mathfrak{g}$ with respect to $\mathfrak{t}$.

According to the commutativity of diagram (3.25), in order to prove Theorem 10.21 (resp. Theorem 10.22) it suffices to show that an even symmetric 2-form (resp. 1-form) on the product $X = \mathbb{CP}^1 \times \mathbb{CP}^1$, which satisfies the zero-energy condition, is a Lie derivative of the metric (resp. is exact). The remainder of this section is devoted to the proof of these two assertions.

We use the notation introduced in §1. We view the metrics g_Y and g_Z as sections over X of the sub-bundles $S^2 T^*_Y$ and $S^2 T^*_Z$ of $S^2 T^*$; then we have $g = g_Y + g_Z$. Let F_0 be the sub-bundle of rank 2 of $S^2 T^*$ generated by the sections g_Y and g_Z; clearly, $F = F_0 \oplus (T_Y^* \otimes T_Z^*)$ is a homogeneous sub-bundle of $S^2 T^*$ invariant under the isometry Ψ'.

Let h be a symmetric 2-form on X. According to the decomposition (10.2), we may write $h = k_1 + k_2 + k_3$, where k_1, k_2, k_3 are sections of $S^2 T_Y^*$, $S^2 T_Z^*$ and $T_Y^* \otimes T_Z$, respectively. For $y \in Y$ and $z \in Z$, by means of the isometry φ, we apply the equality (1.64) to the restriction of k_1 to $Y \times \{z\}$ and to the restriction of k_2 to $\{y\} \times Z$. We then see that there exist a vector field ξ and a section h_1 of F over X such that

$$h = \mathcal{L}_\xi g + h_1.$$

Since the differential operator D_0 is homogeneous, by (2.1) it follows that

(10.12)
$$C_\gamma^\infty(S^2 T_{\mathbb{C}}^*) = D_0 C_\gamma^\infty(T_{\mathbb{C}}) + C_\gamma^\infty(F_{\mathbb{C}}),$$
$$C_\gamma^\infty(S^2 T_{\mathbb{C}}^*)^{\mathrm{ev}} = D_0 C_\gamma^\infty(T_{\mathbb{C}})^{\mathrm{ev}} + C_\gamma^\infty(F_{\mathbb{C}})^{\mathrm{ev}},$$

for all $\gamma \in \Gamma$.

Let f_Y' and f_Z' be given elements of $\mathcal{H}_{1,Y}$ and $\mathcal{H}_{1,Z}$, respectively; according to the relation (3.31), we know that there exist complex Killing vector fields ξ_Y on Y and ξ_Z on Z such that

$$g_Y^\flat(\xi_Y) = (\partial - \bar{\partial}) f_Y', \qquad g_Z^\flat(\xi_Z) = (\partial - \bar{\partial}) f_Z'.$$

Thus if f_Y'' and f_Z'' are arbitrary complex-valued functions on Y and Z, respectively, we easily see that

(10.13) $$D_0(f_Y'' \xi_Z) = df_Y'' \cdot (\partial - \bar{\partial}) f_Z', \qquad D_0(f_Z'' \xi_Y) = (\partial - \bar{\partial}) f_Y \cdot df_Z'';$$

moreover, according to formulas (1.5) and (3.33), and the derivation of the relation (3.34), we see that

(10.14)
$$D_0(f_Z''(df_Y')^\sharp) = -8 f_Y' f_Z'' g_Y + df_Y' \cdot df_Z'',$$
$$D_0(f_Y''(df_Z')^\sharp) = -8 f_Y'' f_Z' g_Z + df_Y'' \cdot df_Z'.$$

Let γ be an element of Γ. Unless γ is of the form (γ_k, γ_l), where k, l are integers ≥ 0, we know that $C_\gamma^\infty(X) = \{0\}$ and, by Proposition 3.28, we have $C_\gamma^\infty(F_{\mathbb{C}}) = \{0\}$; moreover, according to Proposition 3.28 and the equality (10.3), we see that $C_\gamma^\infty(T_{\mathbb{C}}^*) = \{0\}$, unless γ is of the form (γ_k, γ_l), where k, l are integers ≥ 0 satisfying $k + l > 0$.

Now let $k, l \geq 0$ be given integers and suppose that γ is the element (γ_k, γ_l) of Γ. From the relations (3.29), we obtain the following equalities among irreducible G-modules

(10.15) $$C_\gamma^\infty(X) = C_{\gamma_k}^\infty(Y) C_{\gamma_l}^\infty(Z) = \begin{cases} C_\gamma^\infty(X)^{\mathrm{ev}} & \text{if } k + l \text{ is even,} \\ C_\gamma^\infty(X)^{\mathrm{odd}} & \text{if } k + l \text{ is odd;} \end{cases}$$

moreover, we see that the function $f_Y^k f_Z^l$ on X is a highest weight vector of this G-module. From the relations (10.15), it follows that

$$C_\gamma^\infty(F_\mathbb{C}) = C_{\gamma_k}^\infty(Y)C_{\gamma_l}^\infty(Z)g_Y + C_{\gamma_k}^\infty(Y)C_{\gamma_l}^\infty(Z)g_Z + C_{\gamma_k}^\infty(T_{Y,\mathbb{C}}^*) \cdot C_{\gamma_l}^\infty(T_{Z,\mathbb{C}}^*);$$

moreover, we infer that the G-module $C_\gamma^\infty(F_\mathbb{C})^{\mathrm{ev}}$ is given by

$$
\begin{aligned}
C_\gamma^\infty(F_\mathbb{C})^{\mathrm{ev}} = {} & C_{\gamma_k}^\infty(Y)C_{\gamma_l}^\infty(Z)g_Y + C_{\gamma_k}^\infty(Y)C_{\gamma_l}^\infty(Z)g_Z \\
& + C_{\gamma_k}^\infty(T_{Y,\mathbb{C}}^*)^{\mathrm{ev}} \cdot C_{\gamma_l}^\infty(T_{Z,\mathbb{C}}^*)^{\mathrm{ev}} \\
& + C_{\gamma_k}^\infty(T_{Y,\mathbb{C}}^*)^{\mathrm{odd}} \cdot C_{\gamma_l}^\infty(T_{Z,\mathbb{C}}^*)^{\mathrm{odd}}
\end{aligned}
$$

when $k + l$ is an even integer, and by

$$C_\gamma^\infty(F_\mathbb{C})^{\mathrm{ev}} = C_{\gamma_k}^\infty(T_{Y,\mathbb{C}}^*)^{\mathrm{ev}} \cdot C_{\gamma_l}^\infty(T_{Z,\mathbb{C}}^*)^{\mathrm{odd}} + C_{\gamma_k}^\infty(T_{Y,\mathbb{C}}^*)^{\mathrm{odd}} \cdot C_{\gamma_l}^\infty(T_{Z,\mathbb{C}}^*)^{\mathrm{ev}}$$

when $k + l$ is an odd integer. By Proposition 3.28, it follows that the G-module $C_\gamma^\infty(F_\mathbb{C})^{\mathrm{ev}}$ vanishes whenever one of the integers k and l is equal to 0 and the other one is odd. According to (3.30), we also see that

$$C_\gamma^\infty(F_\mathbb{C})^{\mathrm{ev}} = (\partial - \bar\partial)\mathcal{H}_{1,Y} \cdot d\mathcal{H}_{l,Z}$$

when $k = 1$ and the integer l is even, and that

$$C_\gamma^\infty(F_\mathbb{C})^{\mathrm{ev}} = d\mathcal{H}_{k,Y} \cdot (\partial - \bar\partial)\mathcal{H}_{1,Z}$$

when $l = 1$ and the integer k is even. From the previous relations and the equalities (10.3) and (10.13), we then obtain the inclusion

$$(10.16) \qquad C_\gamma^\infty(F_\mathbb{C})^{\mathrm{ev}} \subset D_0 C_\gamma^\infty(T_\mathbb{C})$$

whenever one of the integers k and l is equal to 1 and the other one is even.

Let h be a highest weight vector of the G-module $C_\gamma^\infty(F_\mathbb{C})^{\mathrm{ev}}$. According to the equalities (3.30), the observations concerning highest weight vectors made in §4, Chapter III and the descriptions of this G-module given above, if the integers k and l are positive, we see that the section h of $F_\mathbb{C}$ can be written in the form

$$h = f_Y^{k-1} f_Z^{l-1}\big(f_Y f_Z(a_1 g_Y + a_2 g_Z) + b_1 df_Y \cdot df_Z + b_2(\partial - \bar\partial)f_Y \cdot (\partial - \bar\partial)f_Z\big),$$

with $a_1, a_2, b_1, b_2 \in \mathbb{C}$, when $k + l$ is even, and in the form

$$(10.17) \qquad h = f_Y^{k-1} f_Z^{l-1}\big(b_1 df_Y \cdot (\partial - \bar\partial)f_Z + b_2(\partial - \bar\partial)f_Y \cdot df_Z\big),$$

with $b_1, b_2 \in \mathbb{C}$, when $k + l$ is odd. Moreover, if one of the integers k and l is equal to 0 and the other one is even, we may write

$$(10.18) \qquad\qquad h = f_Y^k f_Z^l (a_1 g_Y + a_2 g_Z),$$

with $a_1, a_2 \in \mathbb{C}$.

According to formulas (10.14), when l is an odd integer and $k = 1$, we know that the section

$$D_0(f_Z^l (df_Y)^\sharp) = -8 f_Y f_Z^l g_Y + df_Y \cdot df_Z^l$$

of $F_{\mathbb{C}}$ belongs to $D_0 C_\gamma^\infty(T_{\mathbb{C}})^{\mathrm{ev}}$ and is a highest weight vector of $C_\gamma^\infty(F_{\mathbb{C}})^{\mathrm{ev}}$; on the other hand, when k is an odd integer and $l = 1$, the section

$$D_0(f_Y^k (df_Z)^\sharp) = -8 f_Y^k f_Z g_Z + df_Y^k \cdot df_Z$$

of $F_{\mathbb{C}}$ belongs to $D_0 C_\gamma^\infty(T_{\mathbb{C}})^{\mathrm{ev}}$ and is a highest weight vector of $C_\gamma^\infty(F_{\mathbb{C}})^{\mathrm{ev}}$.

According to (10.3), (3.29) and (3.30), the G-module $C_\gamma^\infty(T_{\mathbb{C}}^*)^{\mathrm{ev}}$ is given by

$$C_\gamma^\infty(T_{\mathbb{C}}^*)^{\mathrm{ev}} = \mathcal{H}_{k,Y} \cdot d\mathcal{H}_{l,Z} + \mathcal{H}_{l,Z} \cdot d\mathcal{H}_{k,Y}$$

when $k + l$ is an even positive integer, and by

$$C_\gamma^\infty(T_{\mathbb{C}}^*)^{\mathrm{ev}} = \mathcal{H}_{k,Y} \cdot (\partial - \bar{\partial})\mathcal{H}_{l,Z} + \mathcal{H}_{l,Z} \cdot (\partial - \bar{\partial})\mathcal{H}_{k,Y}$$

when $k + l$ is an odd integer. If $k + l \geq 1$, according to observations concerning highest weight vectors made in §4, Chapter III, a highest weight vector β of the G-module $C_\gamma^\infty(T_{\mathbb{C}}^*)^{\mathrm{ev}}$ can be written in the form

$$\beta = a_1 f_Y^k \, df_Z^l + a_2 f_Z^l \, df_Y^k,$$

with $a_1, a_2 \in \mathbb{C}$, when $k + l$ is even, and in the form

$$\beta = a_1 f_Y^k \, (\partial - \bar{\partial})f_Z^l + a_2 f_Z^l \, (\partial - \bar{\partial})f_Y^k,$$

with $a_1, a_2 \in \mathbb{C}$, when $k + l$ is odd. Moreover when $k + l \geq 1$, we see that the section $d(f_Y^k \, df_Z^l)$ of $T_{\mathbb{C}}^*$ is a highest weight vector of the G-submodule $dC_\gamma^\infty(X)$ of $C_\gamma^\infty(T_{\mathbb{C}}^*)$.

Let $k, l \geq 0$ be given integers. We consider the real-valued function $F_{k,l}$ on $\mathbb{R}$ defined by

$$F_{k,l}(t) = \sin^k t \cdot \cos^l t,$$

for $t \in \mathbb{R}$, and we set

$$A_{k,l} = \int_0^\pi F_{k,l}(t) \, dt, \qquad B_{k,l} = \int_0^\pi F_{k,l}(2t) \, dt.$$

We know that $A_{k,l} = B_{k,l} > 0$ when k and l are both even, and that $A_{k,l}$ and $B_{k,l}$ vanish when k and l are both odd. From the relation

$$F_{k,l} = F_{k+2,l} + F_{k,l+2},$$

we obtain the formula

(10.19)
$$A_{k,l} = A_{k+2,l} + A_{k,l+2}.$$

Clearly, we have

$$F_{k,l}(\pi/2 - t) = F_{l,k}(t),$$

for $t \in \mathbb{R}$; when $k + l$ is even, the function $F_{k,l}$ is π-periodic and so, from the preceding relation, we deduce the equality

(10.20)
$$A_{k,l} = A_{l,k}.$$

We consider the function $F_k = F_{k,k}$ and we write $A_k = A_{k,k}$; when $k \geq 1$, we also consider the real-valued function $\tilde{F}_k$ on $\mathbb{R}$ defined by

$$\tilde{F}_k(t) = (k - 1) \cos^2 2t - \sin^2 2t.$$

Then for $t \in \mathbb{R}$, we have

$$F_k(t) = \frac{1}{2^k} F_{k,0}(2t), \qquad A_k = \frac{1}{2^{k+1}} \int_0^{2\pi} \sin^k t \, dt;$$

hence we easily verify that

(10.21)
$$F_1''(t) = -2 \sin 2t, \qquad F_k'' = k F_{k-2} \cdot \tilde{F}_k,$$

for all $t \in \mathbb{R}$ and $k \geq 2$, and, using integration by parts, we obtain the formula

(10.22)
$$A_k = 4 \frac{k + 2}{k + 1} A_{k+2}.$$

The relations (10.19) and (10.20) imply that

(10.23)
$$A_k = 2 A_{k+2,k}.$$

According to (10.19) and (10.23), we have

$$A_{k+4,k} = A_{k+2,k} - A_{k+2};$$

hence by (10.22), we see that

$$(10.24) \qquad\qquad A_{k+4,k} = \frac{k+3}{k+1} A_{k+2}.$$

We easily verify that

$$F_k(t) \cos^2 2t = F_{k+4,k}(t) + F_{k,k+4}(t) - 2F_{k+1}(t),$$

for $t \in \mathbb{R}$; then by (10.20) and (10.24), we obtain the formula

$$(10.25) \qquad\qquad \int_0^\pi F_k(t) \cos^2 2t \, dt = \frac{4}{k+1} A_{k+2}.$$

The closed geodesics of $\mathbb{CP}^1$ are described in §4, Chapter III. Let δ_1 (resp. δ_2) be the closed geodesic $\gamma_{u,u'}$ of $\mathbb{CP}^1$ corresponding to the pair of unit vectors $u = (1,0)$ and $u' = (0,1)$ (resp. $u = (0,1)$ and $u' = (1,0)$). For $\theta \in \mathbb{R}$, let δ^θ be the closed geodesic $\gamma_{u,u'}$ of $\mathbb{CP}^1$ corresponding to the pair of unit vectors

$$u = \frac{1}{\sqrt{2}} (1,1), \qquad u' = \frac{e^{i\theta}}{\sqrt{2}} (-1,1);$$

when $\theta = \pi$, we write $\delta_3 = \delta^\theta$. For all $0 \le t \le \pi$, by (3.20) and (3.21) we easily verify that

$$(10.26) \qquad \begin{aligned} \tilde{f}_1(\delta_1(t)) &= \tilde{f}_1(\delta_2(t)) = \sin t \cdot \cos t, \\ \langle d\tilde{f}_1, \dot\delta_1(t) \rangle &= \langle d\tilde{f}_1, \dot\delta_2(t) \rangle = \cos 2t, \\ \langle (\partial - \bar\partial)\tilde{f}_1, \dot\delta_1(t) \rangle &= -\langle (\partial - \bar\partial)\tilde{f}_1, \dot\delta_2(t) \rangle = 1, \\ \tilde{f}_1(\delta^\theta(t)) &= \tfrac{1}{2} (\cos 2t + i \sin\theta \cdot \sin 2t), \\ \langle d\tilde{f}_1, \dot\delta^\theta(t) \rangle &= -\sin 2t + i \sin\theta \cdot \cos 2t, \\ \langle (\partial - \bar\partial)\tilde{f}_1, \dot\delta^\theta(t) \rangle &= \cos\theta. \end{aligned}$$

In particular, for $0 \le t \le \pi$, we have

$$\tilde{f}_1(\delta_3(t)) = \tfrac{1}{2} \cos 2t, \quad \langle d\tilde{f}_1, \dot\delta_3(t) \rangle = -\sin 2t, \quad \langle (\partial - \bar\partial)\tilde{f}_1, \dot\delta_3(t) \rangle = -1.$$

For $\theta \in \mathbb{R}$, we consider the closed geodesics $\eta_1, \eta_2, \eta_3, \xi_1^\theta$ and ξ_2^θ of X defined by

$$\begin{aligned} \eta_1(t) &= (\delta_1(t), \delta_1(t)), \quad \eta_2(t) = (\delta_1(t), \delta_2(t)), \quad \eta_3(t) = (\delta_1(t), \delta_3(t)), \\ \xi_1^\theta(t) &= (\delta_1(t), \delta^\theta(t)), \quad \xi_2^\theta(t) = (\delta^\theta(t), \delta_1(t)), \end{aligned}$$

for all $0 \leq t \leq \pi$. If η is one of these closed geodesics of X and u is a section of $S^p T^*$ over X, then we have

$$(10.27) \qquad \int_\eta u = \frac{1}{2^{p/2}} \int_0^\pi u(\dot\eta(t), \dot\eta(t), \ldots, \dot\eta(t)) \, dt.$$

Let $k, l \geq 0$ be given integers. We define a function $\psi_{k,l}$ on $\mathbb{R}^2$ by

$$\psi_{k,l}(t, \theta) = \frac{1}{2^{k+l-1}} \cos\theta \cdot F_{k,1}(2t) \cdot (\cos 2t + i \sin\theta \cdot \sin 2t)^l,$$

for $t, \theta \in \mathbb{R}$; we see that

$$(10.28) \qquad \frac{\partial \psi_{k,l}}{\partial\theta}(t, 0) = \frac{il}{2^{k+l-1}} F_{k+1,l}(2t),$$

for $t \in \mathbb{R}$.

We now consider the sections

$$h_1 = f_Y^k f_Z^l \, df_Y \cdot (\partial - \bar\partial) f_Z, \qquad h_2 = f_Y^k f_Z^l \, (\partial - \bar\partial) f_Y \cdot df_Z$$

of $F_{\mathbb{C}}$. According to the formulas (10.26), we see that

$$h_1(\dot\eta_3(t), \dot\eta_3(t)) = -\frac{1}{2^{k+l-1}} F_{k,l+1}(2t), \qquad h_1(\dot\xi_1^\theta(t), \dot\xi_1^\theta(t)) = \psi_{k,l}(t, \theta),$$

$$h_2(\dot\eta_3(t), \dot\eta_3(t)) = -\frac{1}{2^{k+l-1}} F_{k+1,l}(2t), \qquad h_2(\dot\xi_2^\theta(t), \dot\xi_2^\theta(t)) = \psi_{l,k}(t, \theta),$$

for all $0 \leq t \leq \pi$ and $\theta \in \mathbb{R}$. For $j = 1, 2$, we consider the real-valued function P_j on $\mathbb{R}$ defined by

$$P_j(\theta) = \int_{\xi_j^\theta} h_j.$$

According to the relations (10.27) and (10.28), we see that

$$(10.29) \qquad P_1'(0) = \frac{il}{2^{k+l}} B_{k+1,l}, \qquad P_2'(0) = \frac{ik}{2^{k+l}} B_{l+1,k}.$$

LEMMA 10.24. *Let $k, l \geq 1$ be given integers. Suppose that $k + l$ is odd.*

(i) *If l is even, then the section*

$$f_Y^k f_Z^l \, df_Y \cdot (\partial - \bar\partial) f_Z$$

of $F_{\mathbb{C}}$ does not satisfy the zero-energy condition.

(ii) *If k is even, then the section*

$$f_Y^k f_Z^l (\partial - \bar{\partial}) f_Y \cdot df_Z$$

of $F_{\mathbb{C}}$ does not satisfy the zero-energy condition.

PROOF: If the integer l (resp. k) is even and positive, according to the equalities (10.29) and the properties of the coefficients $B_{r,s}$ given above, we see that $P_1'(0)$ (resp. $P_2'(0)$) is non-zero; therefore there exists an element $\theta_1 \in \mathbb{R}$ (resp. $\theta_2 \in \mathbb{R}$) such that the integral $P_1(\theta_1)$ (resp. $P_2(\theta_2)$) does not vanish.

LEMMA 10.25. *Let k, l be given integers and b_1, b_2 be given complex numbers. Suppose that $k + l$ is odd and that the section*

$$h = b_1 f_Y^k f_Z^l \, df_Y \cdot (\partial - \bar{\partial}) f_Z + b_2 f_Y^k f_Z^l (\partial - \bar{\partial}) f_Y \cdot df_Z$$

of $F_{\mathbb{C}}$ satisfies the zero-energy condition. Then the coefficient b_1 vanishes when k is even, and the coefficient b_2 vanishes when k is odd.

PROOF: According to the formulas (10.26), we see that

$$h(\dot{\eta}_3(t), \dot{\eta}_3(t)) = -\frac{1}{2^{k+l-1}} \left(b_1 F_{k,l+1} + b_2 F_{k+1,l} \right)(2t),$$

for all $0 \le t \le \pi$. Thus by (10.27), we have

$$\int_{\eta_3} h = -\frac{1}{2^{k+l}} \left(b_1 B_{k,l+1} + b_2 B_{k+1,l} \right).$$

The desired conclusion is a direct consequence of this equality and the properties of the coefficients $B_{r,s}$ given above.

From Lemmas 10.24 and 10.25, we immediately deduce the following result:

LEMMA 10.26. *Let $k, l \ge 2$ be given integers and b_1, b_2 be given complex numbers. Suppose that $k + l$ is odd and that the section*

$$h = b_1 f_Y^k f_Z^l \, df_Y \cdot (\partial - \bar{\partial}) f_Z + b_2 f_Y^k f_Z^l (\partial - \bar{\partial}) f_Y \cdot df_Z$$

of $F_{\mathbb{C}}$ satisfies the zero-energy condition. Then we have $b_1 = b_2 = 0$ and the section h vanishes.

We consider the sections

$$h_3 = f_Y^k f_Z^l \, df_Y \cdot df_Z, \qquad h_4 = f_Y^k f_Z^l (\partial - \bar{\partial}) f_Y \cdot (\partial - \bar{\partial}) f_Z$$

of $F_{\mathbb{C}}$. According to the formulas (10.26), we see that

$$(10.30) \qquad \begin{aligned} h_3(\dot{\eta}_1(t), \dot{\eta}_1(t)) &= h_3(\dot{\eta}_2(t), \dot{\eta}_2(t)) = 2F_{k+l}(t) \cdot \cos^2 2t, \\ h_4(\dot{\eta}_1(t), \dot{\eta}_1(t)) &= -h_4(\dot{\eta}_2(t), \dot{\eta}_2(t)) = 2F_{k+l}(t), \end{aligned}$$

for all $0 \le t \le \pi$.

LEMMA 10.27. *Let $k, l \ge 1$ be given integers and a_1, a_2, b_1, b_2 be given complex numbers. Suppose that $k + l$ is even and that the section*

$$h = f_Y^{k-1} f_Z^{l-1} \big(f_Y f_Z (a_1 g_Y + a_2 g_Z) + b_1 df_Y \cdot df_Z + b_2 (\partial - \bar{\partial}) f_Y \cdot (\partial - \bar{\partial}) f_Z \big)$$

of $F_{\mathbb{C}}$ satisfies the zero-energy condition. Then the coefficient b_2 vanishes, and the relation

$$(10.31) \qquad (k + l - 1)(a_1 + a_2) + 8b_1 = 0$$

holds.

PROOF: According to the formulas (10.26) and (10.30), we see that

$$\begin{aligned} h(\dot{\eta}_1(t), \dot{\eta}_1(t)) &= (a_1 + a_2)F_{k+l}(t) + 2(b_1 \cos^2 2t + b_2)F_{k+l-2}(t), \\ h(\dot{\eta}_2(t), \dot{\eta}_2(t)) &= (a_1 + a_2)F_{k+l}(t) + 2(b_1 \cos^2 2t - b_2)F_{k+l-2}(t), \end{aligned}$$

for all $0 \le t \le \pi$. Hence by means of the equalities (10.22), (10.25) and (10.27), we have

$$\int_{\eta_1} h = \frac{1}{2}\left((a_1 + a_2) + \frac{8}{k + l - 1}(b_1 + (k + l)b_2)\right) \cdot A_{k+l},$$

$$\int_{\eta_2} h = \frac{1}{2}\left((a_1 + a_2) + \frac{8}{k + l - 1}(b_1 - (k + l)b_2)\right) \cdot A_{k+l}.$$

Since $k + l$ is even, we know that $A_{k+l} > 0$; as h satisfies the zero-energy condition, we therefore obtain the relation

$$(k + l - 1)(a_1 + a_2) + 8(b_1 + \varepsilon(k + l)b_2) = 0,$$

for $\varepsilon = 1, -1$. This implies the desired assertion.

LEMMA 10.28. *Let $k, l \geq 1$ be given integers and a_1, a_2 be given complex numbers. Suppose that $k + l$ is odd and that the section*

(10.32)
$$\beta = a_1 f_Y^k (\partial - \bar{\partial}) f_Z^l + a_2 f_Z^l (\partial - \bar{\partial}) f_Y^k$$

of $T_{\mathbb{C}}^*$ satisfies the zero-energy condition. Then the section β vanishes.

PROOF: According to the formulas (10.26), we see that

$$\beta(\dot{\eta}_1(t)) = (la_1 + ka_2) F_{k+l-1}(t), \qquad \beta(\dot{\eta}_2(t)) = (ka_2 - la_1) F_{k+l-1}(t),$$

for all $0 \leq t \leq \pi$. Thus by (10.27), we have

$$\int_{\eta_1} \beta = \frac{1}{\sqrt{2}} (la_1 + ka_2) A_{k+l-1}, \qquad \int_{\eta_2} \beta = \frac{1}{\sqrt{2}} (ka_2 - la_1) A_{k+l-1}.$$

Since $k + l$ is odd, we know that $A_{k+l-1} > 0$; the desired conclusion is a consequence of the preceding equalities.

We consider the group of translations Λ of $\mathbb{R}^2$ generated by the vectors $(\pi, 0)$ and $(0, \pi)$ and the torus $W = \mathbb{R}^2/\Lambda$ of dimension 2, which is the quotient of $\mathbb{R}^2$ by the group Λ. We endow W with the flat metric g' induced by the standard Euclidean metric of $\mathbb{R}^2$. Let $\varpi : \mathbb{R}^2 \to W$ be the natural projection. Let (θ, φ) be the standard coordinate system of $\mathbb{R}^2$. Clearly, we have

$$\varpi^* g' = d\theta \otimes d\theta + d\varphi \otimes d\varphi.$$

If h' is a symmetric 2-form on W, then according to formula (3.1), we know that

(10.33)
$$\varpi^* (D_{g'} h') \left(\frac{\partial}{\partial \theta}, \frac{\partial}{\partial \varphi}, \frac{\partial}{\partial \theta}, \frac{\partial}{\partial \varphi} \right)$$
$$= \frac{1}{2} \left\{ \frac{\partial^2}{\partial \theta^2} (\varpi^* h') \left(\frac{\partial}{\partial \varphi}, \frac{\partial}{\partial \varphi} \right) + \frac{\partial^2}{\partial \varphi^2} (\varpi^* h') \left(\frac{\partial}{\partial \theta}, \frac{\partial}{\partial \theta} \right) \right.$$
$$\left. - 2 \frac{\partial^2}{\partial \theta \partial \varphi} (\varpi^* h') \left(\frac{\partial}{\partial \theta}, \frac{\partial}{\partial \varphi} \right) \right\}.$$

Let
$$\delta^1 : [0, \pi] \to \mathbb{CP}^1, \qquad \delta^2 : [0, \pi] \to \mathbb{CP}^1$$

be closed geodesics of $\mathbb{CP}^1$ parametrized by their arc-length. Let α_1 and α_2 be the π-periodic real-valued functions on $\mathbb{R}$ determined by

$$\alpha_j(t) = \tilde{f}_1(\delta^j(t)),$$

for $j = 1, 2$. The imbedding

$$\iota : W \to X$$

sending the point $\varpi(\theta, \varphi)$ of W into $(\delta^1(\theta), \delta^2(\varphi))$, with $0 \leq \theta, \varphi \leq \pi$, is totally geodesic.

Let $k, l \geq 1$ be given integers and a_1, a_2, b be given complex numbers. We consider the section

$$(10.34) \qquad h = f_Y^{k-1} f_Z^{l-1} \big(f_Y f_Z (a_1 g_Y + a_2 g_Z) + b\, df_Y \cdot df_Z \big)$$

of $F_{\mathbb{C}}$ and the section

$$(10.35) \qquad \beta = a_1 f_Y^k\, df_Z^l + a_2 f_Z^l\, df_Y^k$$

of $T_{\mathbb{C}}^*$. Then we have

$$\varpi^* \iota^* h = \alpha_1^k \alpha_2^l (a_1\, d\theta \otimes d\theta + a_2\, d\varphi \otimes d\varphi) + b\alpha_1^{k-1} \alpha_2^{l-1} \frac{d\alpha_1}{d\theta} \frac{d\alpha_2}{d\varphi}\, d\theta \cdot d\varphi.$$

According to formula (10.33), we see that the equality $D_{g'} \iota^* h = 0$ is equivalent to the relation

$$(10.36) \qquad a_1 \alpha_1^k \frac{d^2 \alpha_2^l}{d\varphi^2} + a_2 \frac{d^2 \alpha_1^k}{d\theta^2} \alpha_2^l = \frac{2b}{kl} \frac{d^2 \alpha_1^k}{d\theta^2} \cdot \frac{d^2 \alpha_2^l}{d\varphi^2}.$$

Also we have

$$(10.37) \qquad \varpi^* d\iota^* \beta = (a_1 - a_2) \frac{d\alpha_1^k}{d\theta} \frac{d\alpha_2^l}{d\varphi}\, d\theta \wedge d\varphi.$$

We consider the functions $\Phi_{k,l}$ on $\mathbb{R}^2$ and Ψ_l on $\mathbb{R}$ defined by

$$\Phi_{k,l}(\theta, \varphi) = la_1 \sin^2 2\theta \cdot \tilde{F}_l(\varphi) + ka_2 \sin^2 2\varphi \cdot \tilde{F}_k(\theta) - 8b\tilde{F}_k(\theta) \cdot \tilde{F}_l(\varphi),$$

$$\Psi_l(\varphi) = (la_1 + 8b) \tilde{F}_l(\varphi) - a_2 \sin^2 2\varphi,$$

for $\theta, \varphi \in \mathbb{R}$. Then we have

$$(10.38) \qquad \begin{aligned} \Phi_{k,l}(0,0) &= -8(k-1)(l-1)b, \\ \Phi_{k,l}(\pi/4, 0) &= (l-1)(la_1 + 8b) = \Psi_l(0), \\ \Phi_{k,l}(0, \pi/4) &= (k-1)(ka_2 + 8b), \\ \Psi_l(\pi/4) &= -(la_1 + a_2 + 8b). \end{aligned}$$

We now suppose that the geodesics δ^1 and δ^2 are equal to the geodesic δ_1 defined above; then we have

$$\alpha_1(t) = \alpha_2(t) = F_1(t),$$

for all $t \in \mathbb{R}$. In this case, by (10.21) the equality $D_{g'}\iota^* h = 0$ or the relation (10.36) is equivalent to the equality

$$F_{k-2}(\theta) \cdot F_{l-2}(\varphi) \cdot \Phi_{k,l}(\theta, \varphi) = 0,$$

for all $\theta, \varphi \in \mathbb{R}$, when $k, l \geq 2$, and to the equality

$$\sin 2\theta \cdot F_{l-2}(\varphi) \cdot \Psi_l(\varphi) = 0,$$

for all $\theta, \varphi \in \mathbb{R}$, when $k = 1$ and $l \geq 2$. Thus when $k, l \geq 2$, the equality $D_{g'}\iota^* h = 0$ is equivalent to the vanishing of the function $\Phi_{k,l}$; furthermore when $k = 1$ and $l \geq 2$, the equality $D_{g'}\iota^* h = 0$ is equivalent to the vanishing of the function Ψ_l.

Now suppose that the symmetric 2-form h on X given by (10.34) satisfies the zero-energy condition. According to Proposition 3.6, we know that $D_{g'}\iota^* h = 0$. We first suppose that $k, l \geq 2$; then the function $\Phi_{k,l}$ vanishes. From the relations (10.38), we deduce the vanishing of the coefficients a_1, a_2 and b. Therefore the section h vanishes in this case. Next, we suppose that $k = 1$ and $l \geq 2$; then the function Ψ_l vanishes. The relations (10.38) give us the equalities $la_1 + 8b = 0$ and $a_2 = 0$. By a similar reasoning, when $l = 1$ and $k \geq 2$, we obtain the equalities $ka_2 + 8b = 0$ and $a_1 = 0$.

On the other hand, when $k = l = 1$, by (10.21) the relation (10.36) is easily seen to be equivalent to the equality

$$(a_1 + a_2 + 8b) \sin 2\theta \cdot \sin 2\varphi = 0,$$

for all $\theta, \varphi \in \mathbb{R}$. Our hypothesis on h therefore implies that this last identity holds; thus we obtain the relation $a_1 + a_2 + 8b = 0$, which is also given by Lemma 10.27.

The previous discussion, together with the formulas (10.14), gives us the following result:

LEMMA 10.29. *Let $k, l \geq 1$ be given integers and a_1, a_2, b be given complex numbers. Suppose that the section h of $F_{\mathbb{C}}$ given by (10.34) satisfies the zero-energy condition.*

(i) *If $k, l \geq 2$, then the section h vanishes.*

(ii) *If $k = 1$ and $l \geq 2$, then we have $a_2 = 0$ and $la_1 = -8b$, and the section h is given by*

(10.39) $$h = -\frac{a_1}{8} D_0(f_Z^l(df_Y)^\sharp).$$

(iii) If $l = 1$ and $k \geq 2$, then we have $a_1 = 0$ and $ka_2 = -8b$, and the section h is given by

$$(10.40) \qquad h = -\frac{a_2}{8} D_0(f^k_Y (df_Z)^\sharp).$$

(iv) If $k = l = 1$, then we have $a_1 + a_2 + 8b = 0$ and the section h is given by

$$(10.41) \qquad h = -\frac{1}{8} \left(a_1 D_0(f_Z(df_Y)^\sharp) + a_2 D_0(f_Y(df_Z)^\sharp) \right).$$

LEMMA 10.30. *Let $k, l \geq 1$ be given integers and a_1, a_2 be given complex numbers. Suppose that the section β of $T^*_{\mathbb{C}}$ given by (10.35) satisfies the zero-energy condition. Then we have $a_1 = a_2$ and the section β is given by $\beta = a_1 d(f^k_Y f^l_Z)$.*

PROOF: We consider the torus W defined above and the imbedding $\iota : W \to X$ corresponding to the geodesics $\delta^1 = \delta_1$ and $\delta^2 = \delta_1$. According to Proposition 3.6, we know that $d\iota^*\beta = 0$. The equality (10.37) tells us that

$$kl(a_1 - a_2)F_{k-1}(\theta) \cdot F_{l-1}(\varphi) \cdot \cos 2\theta \cdot \cos 2\varphi = 0,$$

for all $\theta, \varphi \in \mathbb{R}$. This implies the desired relation $a_1 = a_2$.

LEMMA 10.31. *Let $k \geq 0$ be a given even integer and a_1, a_2 be given complex numbers. Suppose that the section*

$$h = f^k_Y(a_1 g_Y + a_2 g_Z)$$

of $F_{\mathbb{C}}$ satisfies the zero-energy condition. Then we have $a_1 = a_2 = 0$ and the section h vanishes.

PROOF: Let z be a given point of Z and let y be a point of Y such that $f_Y(y) \neq 0$. We consider the closed geodesics η' and η'' of X defined by

$$\eta'(t) = (\delta_1(t), z), \qquad \eta''(t) = (y, \delta_1(t)),$$

for $0 \leq t \leq \pi$. Then we see that

$$h(\dot\eta'(t), \dot\eta'(t)) = a_1 F_k(t), \qquad h(\dot\eta''(t), \dot\eta''(t)) = a_2 f^k_Y(y),$$

for $0 \leq t \leq \pi$; it follows that

$$\int_{\eta'} h = a_1 A_k, \qquad \int_{\eta''} h = \pi a_2 f^k_Y(y).$$

Since A_k is positive and $f_Y(y)$ is non-zero, the vanishing of these integrals implies that $a_1 = a_2 = 0$.

Let $k \geq 0$ be an even integer; the proof of the preceding lemma also shows that, if the section

$$f_Z^k(a_1 g_Y + a_2 g_Z)$$

of $F_{\mathbb{C}}$ satisfies the zero-energy condition, then it vanishes.

PROPOSITION 10.32. *Let $k, l \geq 1$ be given odd integers and let γ be the element (γ_k, γ_l) of Γ. If one of the integers k, l is equal to 1, then we have the inclusion*

$$\mathcal{Z}_{2,\mathbb{C}} \cap C_{\gamma}^{\infty}(F_{\mathbb{C}})^{\mathrm{ev}} \subset D_0 C_{\gamma}^{\infty}(T_{\mathbb{C}})^{\mathrm{ev}}.$$

PROOF: Above we provided an explicit highest weight vector of the G-submodule $D_0 C_{\gamma}^{\infty}(T_{\mathbb{C}})^{\mathrm{ev}} \cap C_{\gamma}^{\infty}(F_{\mathbb{C}})^{\mathrm{ev}}$ of $\mathcal{Z}_{2,\mathbb{C}} \cap C_{\gamma}^{\infty}(F_{\mathbb{C}})^{\mathrm{ev}}$. Now let h be an arbitrary highest weight vector of the G-module $\mathcal{Z}_{2,\mathbb{C}} \cap C_{\gamma}^{\infty}(F_{\mathbb{C}})^{\mathrm{ev}}$. According to the description of the highest weight vectors of the G-module $C_{\gamma}^{\infty}(F_{\mathbb{C}})^{\mathrm{ev}}$ given above, by Lemma 10.27 there are coefficients $a_1, a_2, b \in \mathbb{C}$ satisfying (10.31), with $b_1 = b$, such that the equality (10.34) holds. When $k = 1$ and $l \geq 3$ (resp. $l = 1$ and $k \geq 3$), Lemma 10.29,(ii) (resp. Lemma 10.29,(iii)) tells us that the relation (10.39) (resp. the relation (10.40)) holds, and so h belongs to $D_0 C_{\gamma}^{\infty}(T_{\mathbb{C}})^{\mathrm{ev}}$. When $k = l = 1$, the relation (10.31) says that $a_1 + a_2 + 8b = 0$; Lemma 10.29,(iv) tells us that the relation (10.41) holds, and so h belongs to $D_0 C_{\gamma}^{\infty}(T_{\mathbb{C}})^{\mathrm{ev}}$. These observations imply the desired inclusion.

PROPOSITION 10.33. *Let $k, l \geq 0$ be given integers and let γ be the element (γ_k, γ_l) of Γ. Suppose that $k, l \geq 2$, or that one of the integers k, l is equal to 0 and that the other one is even. Then the G-module $\mathcal{Z}_{2,\mathbb{C}} \cap C_{\gamma}^{\infty}(F_{\mathbb{C}})^{\mathrm{ev}}$ vanishes.*

PROOF: Suppose that the G-module $\mathcal{Z}_{2,\mathbb{C}} \cap C_{\gamma}^{\infty}(F_{\mathbb{C}})^{\mathrm{ev}}$ does not vanish and let h be a highest weight vector of this module. We now exploit the description of the highest weight vectors of the G-module $C_{\gamma}^{\infty}(F_{\mathbb{C}})^{\mathrm{ev}}$ given above. We first suppose that $k, l \geq 2$. When $k + l$ is even, by Lemma 10.27 we know that h may be written in the form (10.34), with $a_1, a_2, b \in \mathbb{C}$; then Lemma 10.29 gives us the vanishing of h, which leads us to a contradiction. When $k + l$ is odd, the symmetric 2-form h may be written in the form (10.17), with $b_1, b_2 \in \mathbb{C}$; then Lemma 10.26 gives us the vanishing of h, which leads us to a contradiction. Finally, assume that one of the integers k, l is equal to 0 and that the other one is even. The symmetric 2-form h may be written in the form (10.18), with $a_1, a_2 \in \mathbb{C}$;

then Lemma 10.31, together with the remark following it, implies that h vanishes; this leads us once again to a contradiction.

Propositions 10.32 and 10.33 and the inclusions (10.16), together with the results stated above concerning the vanishing of certain G-modules of the form $C_\gamma^\infty(F_{\mathbb{C}})^{\mathrm{ev}}$, with $\gamma \in \Gamma$, imply that the inclusion

$$\mathcal{Z}_{2,\mathbb{C}} \cap C_\gamma^\infty(F_{\mathbb{C}})^{\mathrm{ev}} \subset D_0 C_\gamma^\infty(T_{\mathbb{C}})^{\mathrm{ev}}$$

holds for all $\gamma \in \Gamma$. The second equality of (10.12) and Lemma 2.6 then tell us that

$$\mathcal{Z}_{2,\mathbb{C}} \cap C_\gamma^\infty(S^2 T_{\mathbb{C}}^*)^{\mathrm{ev}} = D_0 C_\gamma^\infty(T_{\mathbb{C}})^{\mathrm{ev}},$$

for all $\gamma \in \Gamma$. According to Proposition 2.30,(ii), with $\Sigma = \{\Psi'\}$ and $\varepsilon = +1$, we have therefore shown that the equality

$$\mathcal{Z}_{2,\mathbb{C}} \cap C^\infty(S^2 T_{\mathbb{C}}^*)^{\mathrm{ev}} = D_0 C^\infty(T_{\mathbb{C}})^{\mathrm{ev}}$$

holds. Thus we have proved that an even symmetric 2-form on the product $X = \mathbb{CP}^1 \times \mathbb{CP}^1$, which satisfies the zero-energy condition, is a Lie derivative of the metric. As we have seen above, this result implies both Theorems 10.22 and 10.20.

PROPOSITION 10.34. *Let $k, l \geq 0$ be given integers, with $k + l \geq 1$, and let γ be the element (γ_k, γ_l) of Γ.*

(i) *When $k + l$ is odd, the G-module $\mathcal{Z}_{1,\mathbb{C}} \cap C_\gamma^\infty(T_{\mathbb{C}}^*)^{\mathrm{ev}}$ vanishes.*

(ii) *When $k + l$ is even, we have the inclusion*

$$\mathcal{Z}_{1,\mathbb{C}} \cap C_\gamma^\infty(T_{\mathbb{C}}^*)^{\mathrm{ev}} \subset dC_\gamma^\infty(X).$$

PROOF: First, assume that $k + l$ is odd. Suppose that the G-module $\mathcal{Z}_{1,\mathbb{C}} \cap C_\gamma^\infty(T_{\mathbb{C}}^*)^{\mathrm{ev}}$ does not vanish and let β be a highest weight vector of this module. According to the description of the highest weight vectors of the G-module $C_\gamma^\infty(T_{\mathbb{C}}^*)^{\mathrm{ev}}$ given above, there are coefficients $a_1, a_2 \in \mathbb{C}$ such that the equality (10.32) holds. Then Lemma 10.28 gives us the vanishing of β, which leads us to a contradiction. Thus we have proved assertion (i). Next, assume that $k + l$ is even. Then we know that the 1-form $d(f_Y^k f_Z^l)$ is a highest weight vector of the G-submodule $dC_\gamma^\infty(X)^{\mathrm{ev}}$ of $\mathcal{Z}_{1,\mathbb{C}} \cap C_\gamma^\infty(T_{\mathbb{C}}^*)^{\mathrm{ev}}$. Let β be an arbitrary highest weight vector of the G-module $\mathcal{Z}_{1,\mathbb{C}} \cap C_\gamma^\infty(T_{\mathbb{C}}^*)^{\mathrm{ev}}$. According to the description of the highest weight vectors of the G-module $C_\gamma^\infty(T_{\mathbb{C}}^*)^{\mathrm{ev}}$ given above, there are coefficients $a_1, a_2 \in \mathbb{C}$ such that the equality (10.35) holds. Then Lemma 10.30 tells us that β is a multiple of the section $d(f_Y^k f_Z^l)$ of $T_{\mathbb{C}}^*$, and so β belongs to $dC_\gamma^\infty(X)^{\mathrm{ev}}$. These observations imply assertion (ii).

Proposition 10.34 and Lemma 2.6, together with the results stated above concerning the vanishing of certain G-modules of the form $C_\gamma^\infty(T_{\mathbb{C}}^*)^{\text{ev}}$, with $\gamma \in \Gamma$, imply that the equality

$$\mathcal{Z}_{1,\mathbb{C}} \cap C_\gamma^\infty(T_{\mathbb{C}}^*)^{\text{ev}} = dC_\gamma^\infty(X)^{\text{ev}}$$

holds for all $\gamma \in \Gamma$. According to Proposition 2.32,(ii), with $\Sigma = \{\Psi'\}$ and $\varepsilon = +1$, we have therefore shown that the equality

$$\mathcal{Z}_{1,\mathbb{C}} \cap C^\infty(T_{\mathbb{C}}^*)^{\text{ev}} = dC^\infty(X)^{\text{ev}}$$

holds. Thus we have proved that an even 1-form on the product $\mathbb{CP}^1 \times \mathbb{CP}^1$, which satisfies the zero-energy condition, is exact. As we have seen above, this result implies both Theorems 10.23 and 10.21.

REFERENCES

[1] T. BAILEY and M. EASTWOOD, Zero-energy fields on real projective space, *Geom. Dedicata*, 67 (1997), 245–258.

[2] L. BÉRARD-BERGERY, J. P. BOURGUIGNON and J. LAFONTAINE, Déformations localement trivialement triviales des variétés riemanniennes, *Proc. Sympos. Pure Math.*, Vol. 27, Amer. Math. Soc., Providence, RI, 1975, 3–32.

[3] M. BERGER and D. EBIN, Some decompositions of the space of symmetric tensors on a Riemannian manifold, *J. Differential Geom.*, 3 (1969), 379–392.

[4] M. BERGER, P. GAUDUCHON and E. MAZET, *Le spectre d'une variété riemannienne*, Lect. Notes in Math., Vol. 194, Springer-Verlag, Berlin, Heidelberg, New York, 1971.

[5] A. BESSE, *Manifolds all of whose geodesics are closed*, Ergeb. Math. Grenzgeb., Bd. 93, Springer-Verlag, Berlin, Heidelberg, New York, 1978.

[6] A. BESSE, *Einstein manifolds*, Ergeb. Math. Grenzgeb., 3 Folge, Bd. 10, Springer-Verlag, Berlin, Heidelberg, New York, 1987.

[7] R. BRYANT, S. S. CHERN, R. GARDNER, H. GOLDSCHMIDT and P. GRIFFITHS, *Exterior differential systems*, Math. Sci. Res. Inst. Publ., Vol. 18, Springer-Verlag, New York, Berlin, Heidelberg, 1991.

[8] E. CALABI, On compact, Riemannian manifolds with constant curvature. I, *Proc. Sympos. Pure Math.*, Vol. 3, Amer Math. Soc., Providence, RI, 1961, 155-180.

[9] E. CALABI and E. VESENTINI, On compact, locally symmetric Kähler manifolds, *Ann. of Math.*, 71 (1960), 472–507.

[10] Y. DIENG, Quelques résultats de rigidité infinitésimale pour les quadriques complexes, *C. R. Acad. Sci. Paris Sér. I Math.*, 304 (1987), 393–396.

[11] H. DUISTERMAAT and V. GUILLEMIN, The spectrum of positive elliptic operators and periodic bicharacteristics, *Invent. Math.*, 29 (1975), 39–79.

[12] P. ESTEZET, *Tenseurs symétriques à énergie nulle sur les variétés à courbure constante*, Thèse de doctorat de troisième cycle, Université de Grenoble I, 1988.

[13] J. GASQUI and H. GOLDSCHMIDT, Déformations infinitésimales des espaces riemanniens localement symétriques. I, *Adv. in Math.*, 48 (1983), 205–285.

[14] J. GASQUI and H. GOLDSCHMIDT, Déformations infinitésimales des espaces riemanniens localement symétriques. II. La conjecture infinitésimale de Blaschke pour les espaces projectifs complexes, *Ann. Inst. Fourier (Grenoble)*, 34, 2 (1984), 191–226.

[15] J. GASQUI and H. GOLDSCHMIDT, *Déformations infinitésimales des structures conformes plates*, Progress in Math., Vol. 52, Birkhäuser, Boston, Basel, Stuttgart, 1984.

[16] J. GASQUI and H. GOLDSCHMIDT, Infinitesimal rigidity of $S^1 \times \mathbb{RP}^n$, *Duke Math. J.*, 51 (1984), 675–690.

[17] J. GASQUI and H. GOLDSCHMIDT, Une caractérisation des formes exactes de degré 1 sur les espaces projectifs, *Comment. Math. Helv.*, 60 (1985), 46–53.

[18] J. GASQUI and H. GOLDSCHMIDT, Rigidité infinitésimale des espaces projectifs et des quadriques complexes, *J. Reine Angew. Math.*, 396 (1989), 87–121.

[19] J. GASQUI and H. GOLDSCHMIDT, Infinitesimal rigidity of products of symmetric spaces, *Illinois J. Math.*, 33 (1989), 310–332.

[20] J. GASQUI and H. GOLDSCHMIDT, Critère d'exactitude pour les formes de degré 1 sur les quadriques complexes, *Bull. Soc. Math. France*, 117 (1989), 103–119.

[21] J. GASQUI and H. GOLDSCHMIDT, On the geometry of the complex quadric, *Hokkaido Math. J.*, 20 (1991), 279–312.

[22] J. GASQUI and H. GOLDSCHMIDT, The infinitesimal rigidity of the complex quadric of dimension four, *Amer. Math. J.*, 116 (1994), 501–539.

[23] J. GASQUI and H. GOLDSCHMIDT, Radon transforms and spectral rigidity on the complex quadrics and the real Grassmannians of rank two, *J. Reine Angew. Math.*, 480 (1996), 1–69.

[24] J. GASQUI and H. GOLDSCHMIDT, The infinitesimal spectral rigidity of the real Grassmannians of rank two, in *CR-geometry and overdetermined systems*, edited by T. Akahori et al., Advanced Studies in Pure Mathematics, Vol. 25, Math. Soc. of Japan, Kinokuniya Co., Tokyo, 1997, 122–140.

[25] J. Gasqui and H. Goldschmidt, The Radon transform and spectral rigidity of the Grassmannians, *Contemp. Math.*, 251 (2000), 205–221.

[26] J. Gasqui and H. Goldschmidt, Injectivité de la transformation de Radon sur les grassmanniennes, *Séminaire de théorie spectrale*, Vol. 18, Université de Grenoble I, 2000, 27–41.

[27] J. Gasqui and H. Goldschmidt, Some remarks on the infinitesimal rigidity of the complex quadric, in *Lie groups, geometric structures and differential equations – One hundred years after Sophus Lie –*, edited by T. Morimoto et al., Advanced Studies in Pure Mathematics, Vol. 37, Math. Soc. of Japan, Tokyo, 2002, 79–97.

[28] H. Goldschmidt, Existence theorems for analytic linear partial differential equations, *Ann. of Math.*, 86 (1967), 246–270.

[29] H. Goldschmidt, The Radon transform for symmetric forms on real projective spaces, *Contemp. Math.*, 113 (1990), 81–96.

[30] H. Goldschmidt, On the infinitesimal rigidity of the complex quadrics, *Contemp. Math.*, 140 (1992), 49–63.

[31] E. Grinberg, Spherical harmonics and integral geometry on projective spaces, *Trans. Amer. Math. Soc.*, 279 (1983), 187–203.

[32] E. Grinberg, On images of Radon transforms, *Duke. Math. J.*, 52 (1985), 939–972.

[33] E. Grinberg, Aspects of flat Radon transforms, *Contemp. Math.*, 140 (1992), 73–85.

[34] E. Grinberg, Flat Radon transforms on compact symmetric spaces with application to isospectral deformations (preprint).

[35] V. Guillemin, On micro-local aspects of analysis on compact symmetric spaces, in *Seminar on micro-local analysis*, by V. Guillemin, M. Kashiwara and T. Kawai, Ann. of Math. Studies, No. 93, Princeton University Press, University of Tokyo Press, Princeton, NJ, 1979, 79–111.

[36] S. Helgason, *Differential geometry, Lie groups, and symmetric spaces*, Academic Press, Orlando, FL, 1978.

[37] S. Helgason, *Groups and geometric analysis*, Academic Press, Orlando, FL, 1984.

[38] K. Kiyohara, Riemannian metrics with periodic geodesic flows on projective spaces, *Japan J. Math. (N.S.)*, 13, 2 (1987), 209–234.

[39] A. W. KNAPP, *Lie groups beyond an introduction*, Progress in Math., Vol. 140, Birkhäuser, Boston, Basel, Berlin, 1996.

[40] S. KOBAYASHI and K. NOMIZU, *Foundations of differential geometry*, Vol. II, Interscience Publishers, New York, 1969.

[41] N. KOISO, Rigidity and stability of Einstein metrics – The case of compact symmetric spaces, *Osaka J. Math.*, 17 (1980), 51–73.

[42] N. KOISO, Rigidity and infinitesimal deformability of Einstein metrics, *Osaka J. Math.*, 19 (1982), 643–668.

[43] A. LICHNEROWICZ, *Géométrie des groupes de transformations*, Dunod, Paris, 1958.

[44] A. LICHNEROWICZ, Propagateurs et commutateurs en relativité générale, *Inst. Hautes Études Sci. Publ. Math.*, 10 (1961), 1–52.

[45] R. MICHEL, Problèmes d'analyse géométrique liés à la conjecture de Blaschke, *Bull. Soc. Math. France*, 101 (1973), 17–69.

[46] R. MICHEL, (a) Un problème d'exactitude concernant les tenseurs symétriques et les géodésiques, *C. R. Acad. Sci. Sér. A*, 284 (1977), 183–186; (b) Tenseurs symétriques et géodésiques, *C. R. Acad. Sci. Sér. A*, 284 (1977), 1065–1068.

[47] R. MICHEL, Sur quelques problèmes de géométrie globale des géodésiques, *Bol. Soc. Bras. Mat.*, 9 (1978), 19–38.

[48] M. OBATA, Certain conditions for a Riemannian manifold to be isometric with a sphere, *J. Math. Soc. Japan*, 14 (1962), 333–340.

[49] B. SMYTH, Differential geometry of complex hypersurfaces, *Ann. of Math.*, 85 (1967), 246–266.

[50] R. STRICHARTZ, The explicit Fourier decomposition of $L^2(SO(n)/SO(n-m))$, *Canad. J. Math.*, 27 (1975), 294–310.

[51] S. TANNO, Eigenvalues of the Laplacian of Riemannian manifolds, *Tôhoku Math. J.*, 25 (1973), 391–403.

[52] L. TELA NLENVO, *Formes impaires et un problème de rigidité infinitésimale pour la quadrique complexe de dimension 3*, Thèse de Doctorat, Université de Grenoble I, 1997.

[53] C. TSUKAMOTO, Infinitesimal Blaschke conjectures on projective spaces, *Ann. Sci. École Norm. Sup. (4)*, 14 (1981), 339–356.

[54] C. TSUKAMOTO, Spectra of Laplace-Beltrami operators on $SO(n+2)/SO(2) \times SO(n)$ and $Sp(n+1)/Sp(1) \times Sp(n)$, *Osaka J. Math.*, 18 (1981), 407–426.

[55]   V. S. VARADARAJAN, *Lie groups, Lie algebras, and their represen-
       tations*, Graduate Texts in Math., Vol. 102, Springer-Verlag, New
       York, Berlin, Heidelberg, 1984.

[56]   N. WALLACH, *Harmonic analysis on homogeneous spaces*, Marcel
       Dekker, New York, 1973.

[57]   S. T. YAU, Remarks on conformal transformations, *J. Differential
       Geom.*, 8 (1973), 369–381.

INDEX